Plant Viruses

Diversity, Interaction and Management

Plant Viruses

Diversity, Interaction and Management

Edited by
Rajarshi Kumar Gaur
Khurana SMP
Yuri Dorokhov

CRC Press is an imprint of the
Taylor & Francis Group, an **informa** business

CRC Press
Taylor & Francis Group
6000 Broken Sound Parkway NW, Suite 300
Boca Raton, FL 33487-2742

Printed on acid-free paper

International Standard Book Number-13: 978-1-138-06151-4 (Hardback)

Visit the Taylor & Francis Web site at
http://www.taylorandfrancis.com

and the CRC Press Web site at
http://www.crcpress.com

Contents

SECTION II Host Interaction

SECTION III Management

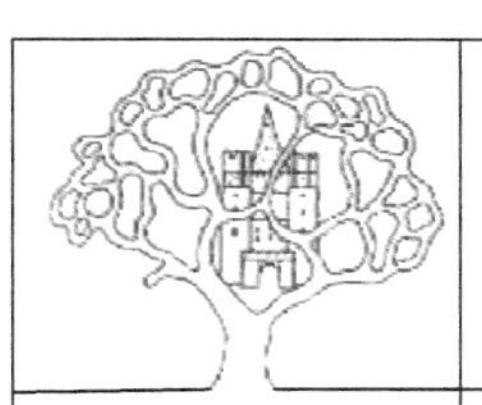

BOTANISCHES INSTITUT
DER UNIVERSITÄT BASEL
Schoenbeinstr 6, 4056 Basel
SWITZERLAND

Prof. Thomas Hohn
Phone +41 (61) 701 75 02
e-mail hohn@fmi.ch

Foreword

Plant Viruses: Diversity, Interaction and Management is an excellent compilation of authoritative accounts by an impressive team of experts on plant viruses. Prof. Gaur, Prof. Khurana, and Prof. Dorokhov have directed and coordinated the project with meticulous editing. It is a substantial volume of high standard in this field. This multi-authored book is a collection of 22 reviews on selected areas of molecular plant pathology that have received much attention, and where significant advances have been made. The book opens with three chapters dedicated to diversity of plant viruses, molecular aspects of interaction or pathogenicity, and management. A series of plant viruses of economic importance affecting major crop plants are described. By far, the greatest economic damage is caused by members of the genus *Potyviruses*, transmitted by aphids (see Chapters 2, 3, 5, 9, 11, 17, and 19).

This book is for anyone who has ever wondered about the viruses they find on their plants—what they are doing there and how best to minimize the damage, if that is really necessary and economically sensible, in the most environmentally responsible way. In Section II, Host Interaction, the chapters describe how different viruses infecting the same host can interact with each other. If related, they may cause exclusion of one of them, for instance through silencing; if unrelated, they may mutually enhance their replication and spreading, for instance through silencing suppression.

Thousands of viruses of crops and plants have currently been detected, but only a few of the economically most important ones are represented in this book. The second most economically important group of viruses—Geminiviruses—are discussed, with special focus on those found in the Indian subcontinent (Chapters 6, 12, and 18). It is discussed how fast these viruses evolve and invade new crops through recombination, resorting to mutation. The last few chapters are concise but effective, giving a critical overview on how novel technology can help to improve our understanding of plant–virus interactions. I feel these are the most useful chapters in the book, as this information will help advanced research students as well as more seasoned scientists to understand the management strategies likely to be used in the next decade or so in molecular plant pathology.

There is a short discussion on economic impact, with particular emphasis on newly emerging viruses. These viruses have had an impact because of changing agricultural practices, particularly the widespread planting of new, susceptible varieties. This book also deals with strict sanitary control practices to avoid viral epidemics. It describes host resistance, which can be achieved through conventional breeding and transgenic practices, the latter unfortunately not accepted everywhere.

This book will inevitably be updated and modified as more information becomes available and as later editions come out, but it will undoubtedly remain the benchmark and foundational text in this field for decades to come.

Prof. Thomas Hohn

Thomas Hohn *is Professor Emeritus at the Botanical Institute, University Basel. He has studied at the Max-Planck Institute for Developmental Biology in Tubingen, Germany and performed post-doctoral studies in Stanford University, California. He was junior group leader at the Bicocenter of the University Basel and group leader at the Friedrich Miescher Institute, Basel. His interests are in virology, originally of bacteriophages, where he studied self-assembly of small bacteriophages as well as morphogenesis of bacteriophage. The latter work, performed together with his wife (Prof. Barbara Hohn), led to DNA packaging. He later shifted to plant viruses, where he recognized the first plant pararetrovirus (CaMV) and detected special viral translation strategies. He recently became interested in the topic of RNA-interference and transgenesis. He has published more than 200 papers during his career. For several years he has been involved in the Indo-Swiss Collaboration in Biotechnology project, working together with Indian scientists to apply biotechnology for the improvement of pulses (leguminosae) and Cassava for use by subsistence farmers.*

Preface

All crops get infected with economically important viruses and serious virus diseases responsible for heavy yield losses. Plant viruses in the ecosystem are recognized as important elements for the emergence of new pathogens. However, occurrence of new viruses and strains of existing viruses, along with changing contexts due to agricultural escalation and microclimate variations, have created new challenges and demand much effort to overcome hurdles to increase agricultural productivity, food availability, and economic development.

This book includes chapters on the evolution of plant viruses, genomic structure, diversity, plant–virus and vector–virus interactions, subcellular movement, and so forth. The subject matter widely covers virus identification, characterization, and detection, and management. Virus profiles briefly describe the major properties of the viruses including their taxonomic position, biology, virions, and genomes. References are provided to enable researchers and students to acquire additional information on the targeted viruses.

The 22 chapters in this volume are divided into three sections, forming a practical sequence for gaining insight into the subject. Section I outlines some of the vital features of plant–virus interactions. The virus capsids play a pivotal role in the protection of the viral genome, assembly and disassembly events, and also serve to interact with vector component wherever vector-mediated transmission is involved (Chapter 1). Cowpea aphid-borne mosaic potyviruses are important pests of passion fruit and cowpea in Brazil (Chapter 2). Several isolates were found infecting either cowpea or passion fruit, having Vigna sp. as common host. Potato virus Y infects a variety of solanaceaeous plants worldwide, causing enormous crop damage (Chapter 3). Chapter 4 reviews the present knowledge of tobacco mosaic virus (TMV) and its related tobamoviruses affecting members of several plant families. The strategies of gene expression, including the mechanisms of internal translation and the production of subgenomic RNAs, are discussed. Chapter 5 describes the means of diagnostic and antiviral measures, and cultural practices concerning the Sharka or plum pox potyviruses. In the Indian sub-continent, we focused on new viruses that invade new crops though recombination, resorting and mutation (Chapter 6). The book describes series of plant virus of economic importance affecting major crops such as Badnavirus (Chapter 7), Tritimovirus (Chapter 8) and Potyvirus (Chapter 9). The strategies of gene expression, including the mechanisms of internal translation and the production of subgenomic RNAs, are discussed. They are also important for systemic spread and in some cases for local movement within the plant. The virus life cycle can only be fully understood in the context of its interaction with the host, while the different developmental stages of a plant can provide further insight into the roles of the different viral components.

Section II describes the symptoms and host ranges of plant viruses and various other agents that induce virus-like diseases, together with some of the practicalities of working on biological aspects of these pathogens. Plant-virus interactions are highly dynamic, and the selective pressure exerted by plant resistance frequently favors the emergence of adapted virus populations (Chapters 10 through 12). The new emerging viruses cause enormous economic losses in many important crops, including banana, grapes, paddy, sugarcane, papaya, and others. Economic interests predetermined the discovery of tobacco mosaic as the first virus and its continued use as an object of virology and molecular and cellular biology. Proteomic approaches, described in Chapter 13, help in identification of the target viruses and disease biomarkers, which may prove useful for managing viral diseases and developing diagnostic tools. Current knowledge of the host factors, cellular structures, and virus-induced protein complexes, essential in protecting genomes from degradation and helping to achieve translation, are summarized in Chapter 14. The presence of overlapping genes is explained not only by the need for genomic compression but also by interactions with the host plant (Chapters 15 and 16). The evolutionary processes driving the emergence of these viral genomic components include genetic recombination (intra-species and inter-species) followed by mutation and nucleotide substitution. This section also answers important questions on interactions between the virus and infected host plants.

Section III contains various approaches for management of plant viruses. Adopting a strategic integrated virus management system by cultural practices can be helpful, such as using healthy (virus-free) seedlings of a tolerant cultivar, selecting the season for transplanting when the vector population is low, planting border crops, selective roguing of infected plants, and controlling the population of aphid vectors (Chapters 17 and 19). The integrated practices are described such that infection can be avoided in the early stages of plants.

The authors have accumulated knowledge and current methods of diagnostics that allow specialists to not only detect the pathogen but also identify its strain characteristics and geographical distribution of the dominant strains (Chapter 22). An efficient way to fight against a virus disease is to increase crop resistance. Generally, two forms of resistance exist; one is to use transgenic plants and the other is to use resistance genes in traditional crossing methods. RNAi is a promising technology for control of viral infection, and plant "immunization" is currently being developed. Virus infections frequently affect plant growth and plant development, for instance by depletion of resources, by side effects of silencing and silencing suppression, and by induction of micro-RNAs (Chapter 17). A large number of plant viruses have been modified to develop new VIGS vectors in the last decade, which has led to better understanding of the functions of numerous plant genes. The causes of genetic diversity such as reassortment, recombination, and mutations have been discussed for developing effective future virus management strategies. Natural machineries which allow the emergence of virus populations adapted to plants have also been reviewed.

This edited book provides in-depth knowledge of plant virus genes' interactions with host, and their localization and expression. The most recent information regarding advances in plant virus evolution, crop responses to viruses and crop improvement are provided. The book also focuses on different strategies for plant virus management necessary for food security. This book will surely be beneficial for molecular biologists and plant virologists because it combines characterization of plant viruses with their disease management. When these topics are discussed together, it is easy to compare all aspects of diversity, interaction, and management strategies.

We are deeply grateful to a large number of colleagues for their contributions, helpful discussion on various topics, and for providing their unpublished material.

The editors are also deeply grateful to the authorities of their respective universities and institutes for encouragement, freedom, and full support for working on the book and to successfully complete the venture in time.

R.K. Gaur
Khurana SMP
Yuri Leonidovich Dorokhov

Editors

Prof. (Dr.) Rajarshi Kumar Gaur is presently working as Professor and Head, Department of Bioscience, Mody University of Science and Technology, Lakshmangath, Sikar, Rajasthan, India. He earned his PhD on molecular characterization of sugarcane viruses of India, in which he partially characterized three sugarcane viruses: sugarcane mosaic virus, sugarcane streak mosaic virus, and sugarcane yellow luteovirus. He received a MASHAV Fellowship in 2004 from the government of Israel for his postdoctoral studies, and joined The Volcani Center in Israel, and then BenGurion University (Negev, Israel). In 2007, he received the Visiting Scientist fellowship from the Swedish Institute, for one year to work in Umeå University (Umeå, Sweden). He is also a recipient of ICGEB, Italy Post-Doctoral fellowship (2008). He worked on the development of a marker-free transgenic plant against cucumber viruses and has made significant contributions on sugarcane viruses. He has published 130 national/international papers and presented approximately 50 papers in national and international conferences. He was awarded Fellow of the Linnean Society (London, UK). He has visited Thailand, New Zealand, London, Canada, the United States, and Italy to attend various conferences and workshops. He is currently working on various projects funded by the government of India on plant viruses and disease management.

Prof. (Dr.) Khurana SMP, earned his PhD in botany-virology (1969) on papaya viruses under Prof. K.S. Bhargava, followed by two years (1970–1972) of postdoctoral research in advanced plant virology at Kyushu University (Fukuoka, Japan) with Prof. Zyun Hidaka, followed by one year as a visiting scientist at the University of Minnesota (St. Paul; 1987–1988) with Prof. E.E. Banttari, where he specialized in immunodiagnostic technology. He served from scientist to Director at the Central Potato Research Institute (Shimla, India; 1973–2004); consultant at International Potato Center, Lima, Peru/ Food and Agriculture Organization of the United Nations, Rome (1992, 1996, 1997); vice-chancellor, Rani Durgavati University, (Jabalpur, India; 2004–2009), Director, Amity Institute of Biotechnology (Amity University Uttar Pradesh, Noida, India; 2009–2010), director, Amity Institute of Biotechnology (2010–2015), Dean, Faculty of Science, Engineering, and Technology, Amity University Haryana (2013–2016). He is currently Professor of Biotechnology and Head, University Science Instrumentation Centre, Manesar, Gurgaon, Amity University Haryana, Gurgaon (2015–present). He has published more than 200 research papers, 95 reviews/chapters and has authored or edited 15 books and 12 technical bulletins. He was also an editor/chief editor for the Indian Virology Society from 1989 to 2000.

Prof. Yuri Leonidovich Dorokhov is an expert in the molecular biology of plant viruses. He graduated from the Sechenov Moscow Medical Institute with honors (1969) and earned a Doctor of Biological Sciences (1986). His candidate thesis (Russian analog of PhD) was "Propagation of bacteriophages in bacteria deficient in the dark repair system" (1973), and the subject of his doctoral dissertation was "Transport of infection and resistance of plants to viruses" (1985). He is presently a professor in the specialty of virology (1999) and is head of the Laboratory of Molecular Biology of Viruses, A.N. Belozersky Institute of Physico-Chemical Biology, Moscow State University, Russia. He has supervised more than 40 graduate students and 13 PhD theses and has published more than 80 scientific papers in the fields of virology and molecular biology. He is a Laureate of the State Prize of the Russian Federation (1994) for studies on plant virus cell-to-cell movement. His scientific interests include the expression of the viral genome and the intercellular transport of proteins and RNA. He determined the complete nucleotide sequence of a new cruciferous-infecting tobamovirus and discovered a new mechanism for the internal translation initiation of tobamovirus. In the field of applied research, he created effective plant viral-based vectors for producing vaccines and anti-cancer monoclonal antibodies in plants. He developed an effective system for producing trastuzumab and pertuzumab as plant biosimilars for the treatment of HER2-positive breast cancer. He has authored more than 30 patents on the production of pharmaceutical proteins in plants.

Contributors

Akhtar Ali
Department of Biological Sciences
University of Tulsa
Tulsa, Oklahoma

Asztéria Almási
Plant Protection Institute, Centre for Agricultural Research
Hungarian Academy of Sciences
Budapest, Hungary

C. Anuradha
Molecular Virology Lab
Indian Council of Agriculture Research (ICAR)-National Research Centre for Banana
Tiruchirapalli, Tamil Nadu, India

V. Balasubramanian
Molecular Virology Lab
Indian Council of Agriculture Research (ICAR)-National Research Centre for Banana
Tiruchirapalli, Tamil Nadu, India

Virendra Kumar Baranwal
Advanced Centre for Plant Virology, Division of Plant Pathology
ICAR-Indian Agricultural Research Institute
New Delhi, India

S. Chakraborty
Molecular Virology Laboratory, School of Life Sciences
Jawaharlal Nehru University
New Delhi, India

V. Celia Chalam
Division of Plant Quarantine
Indian Council of Agriculture Research (ICAR)-National Bureau of Plant Genetic Resources (NBPGR), Pusa Campus
New Delhi, India

Indranil Dasgupta
Department of Plant Molecular Biology
University of Delhi South Campus
New Delhi, India

Swarnalok De
Department of Food and Environmental Sciences
University of Helsinki
Helsinki, Finland

Francisco de Assis Câmara Rabelo Filho
Departamento de Fitotecnia, Centro de Ciência Agrárias
Universidade Federal do Ceara, Campus do Pici
Fortaleza, Brazil

Ravi Kant
Department of Plant Molecular Biology
University of Delhi South Campus
New Delhi, India

R.K. Khetarpal
CABI – South Asia
NASC Complex
Pusa New Delhi, India

Tatiana V. Komarova
Vavilov Institute of General Genetics RAS
Moscow State University
Moscow, Russia

P. Vignesh Kumar
Advanced Centre for Plant Virology, Division of Plant Pathology
Indian Council of Agriculture Research (ICAR)-Indian Agricultural Research Institute
New Delhi, India

R. Vinoth Kumar
Molecular Virology Laboratory, School of Life Sciences
Jawaharlal Nehru University
New Delhi, India

Sandeep Kumar
AICRP on Medicinal and Aromatic Plants and Betelvine
Orissa University of Agriculture and Technology
Bhubaneswar, India

Jiban Kumar Kundu
Plant Virus and Vector Interactions Group, Division of Crop Protection and Plant Health
Crop Research Institute
Prague, Czech Republic

José Albérsio de Araújo Lima, PhD
Bolsista Pesquisador 1A do CNPq, Laboratório de Virologia Vegetal
Universidade Federal do Ceará
Fortaleza-CE, Brazil

Andres Lõhmus
Department of Food and Environmental Sciences
University of Helsinki
Helsinki, Finland

Laianny Morais Maia, MS
Laboratório de Virologia Vegetal Centro de Ciências Agrárias
Universidade Federal do Ceará
Fortaleza-CE, Brazil

Kristiina Mäkinen
Department of Food and Environmental Sciences
University of Helsinki
Helsinki, Finland

V.G. Malathi
Department of Plant Pathology
Tamil Nadu Agricultural University
Coimbatore, Tamil Nadu, India

Avinash Marwal
Department of Biosciences, College of Arts, Science and Humanities
Mody University of Science and Technology
Lakshmangath, Rajasthan, India

Ulrich Melcher
Department of Biochemistry and Molecular Biology
Oklahoma State University
Stillwater, Oklahoma

Mathur R.N. Murthy
Molecular Biophysics Unit
Indian Institute of Science
Bengaluru, India

Aline Kelly Queiroz do Nascimento, Dr.
Laboratório de Virologia Vegetal
Centro de Ciências Agrárias
Universidade Federal do Ceará
Fortaleza-CE, Brazil

Katalin Nemes
Plant Protection Institute, Centre for Agricultural Research
Hungarian Academy of Sciences
Budapest, Hungary

László Palkovics
Faculty of Horticultural Science, Department of Plant Pathology
Szent István University
Budapest, Hungary

Nikolay M. Petrov
Department of Natural Sciences
New Bulgarian University
Sofia, Bulgaria

Maija Pollari
Department of Food and Environmental Sciences
University of Helsinki
Helsinki, Finland

Fernando Ponz
Centro de Biotecnología y Genómica de Plantas (CBGP, UPM-INIA)
Campus de Montegancedo
Madrid, Spain

S. Rageshwari
Department of Plant Pathology
Tamil Nadu Agricultural University
Coimbatore, Tamil Nadu, India

Richa Rai
Advanced Centre for Plant Virology, Division of Plant Pathology
Indian Council of Agriculture Research (ICAR)-Indian Agricultural Research Institute
New Delhi, India

Govind Pratap Rao
Advanced Centre for Plant Virology, Division of Plant Pathology
ICAR-Indian Agricultural Research Institute
New Delhi, India

P. Renukadevi
Department of Plant Pathology
Tamil Nadu Agricultural University
Coimbatore, Tamil Nadu, India

Shreya Saha
Department of Food and Environmental Sciences
University of Helsinki
Helsinki, Finland

Katalin Salánki
Plant Protection Institute, Centre for Agricultural Research
Hungarian Academy of Sciences
Budapest, Hungary

Flora Sánchez
Centro de Biotecnología y Genómica de Plantas (CBGP, UPM-INIA)
Campus de Montegancedo
Madrid, Spain

Handanahal S. Savithri
Department of Biochemistry
Indian Institute of Science
Bengaluru, India

R. Selvarajan
Molecular Virology Lab
Indian Council of Agriculture Research (ICAR)-National Research Centre for Banana
Tiruchirapalli, Tamil Nadu, India

Sunil Kumar Sharma
Indian Council of Agriculture Research (ICAR)-Indian Agricultural Research Institute
Regional Station
Aundh, Pune, India

Susheel Kumar Sharma
Indian Council of Agricultural Research (ICAR)-Research Complex for NEH Region
Manipur Centre
Imphal, Manipur, India

Ekaterina V. Sheshukova
Vavilov Institute of General Genetics RAS
Moscow, Russia

Khushwant Singh
Plant Virus and Vector Interactions Group, Division of Crop Protection and Plant Health
Crop Research Institute
Prague, Czech Republic

Antoniy Stoev
Institute of Soil Science "Nikola Pushkarov," Agrotechnologies and Plant Protection
Sofia, Bulgaria

Mariya I. Stoyanova
Institute of Soil Science "Nikola Pushkarov," Agrotechnologies and Plant Protection
Sofia, Bulgaria

István Tóbiás
Plant Protection Institute, Centre for Agricultural Research
Hungarian Academy of Sciences
Budapest, Hungary

Savarni Tripathi
ICAR-Indian Agricultural Research Institute, Regional Station
Aundh, Pune, India

Sangita Venkataraman
Amity Institute of Integrative Sciences and Health
Amity University Haryana
Gurgaon, India

Rakesh Kumar Verma
Department of Biosciences, College of Arts, Science and Humanities
Mody University of Science and Technology
Lakshmangarh, Rajasthan, India

Zhimin Yin
Plant Breeding and Acclimatization Institute–National Research Institute
Młochów Research Center
Młochów, Poland

Section I

Diversity

1 Structural Aspects of Plant Viruses

Sangita Venkataraman, Handanahal S. Savithri, and Mathur R.N. Murthy

CONTENTS

1.1 INTRODUCTION

Viruses are nucleoprotein complexes that rely on specific hosts for their propagation. The protein component of the virus serves the primary purpose of protecting the genome during transmission from one susceptible host to another. Specific epitopes present on viral capsids serve to recognize host receptors during the process of cell entry. Apart from these roles, the capsids function to encapsidate the correct nucleic acid genome during the process of assembly. They disassemble to release the nucleic acid genome upon entry into a new host. In many arboviruses, the capsid serves to interact with the insect vectors during the course of transmission. In 1956, Crick and Watson realized that viruses are capable of coding for only a few species of protein molecules of limited size due to the relatively small size of their genomes (Crick and Watson 1956). Therefore, a protein shell large enough to encapsidate the genome has to be built from several copies of identical, smaller protein molecules. They also suggested that the use of identical subunits would lead to identical packing environment of subunits using symmetry principles which in turn would facilitate error-free assembly and disassembly. The basic principles underlying the design of viral capsids is discussed in the following sections.

1.1.1 PRINCIPLES OF CAPSID DESIGN

Viral capsids are made up of multiple copies of one or more protein subunits arranged with symmetry that allows their error-free assembly and stability. Based on the nature of the capsid, viruses may broadly be classified as helical, polyhedral, or complex.

Helical viruses are designed by employing helical symmetry, which is a useful way of arranging a large number of identical protein subunits in invariant environments (Figure 1.1). Helices are formed by stacking repeated components (protein subunits) around the helix axis with a constant spatial relationship between the neighboring subunits (angle of rotation and rise per subunit along the helix

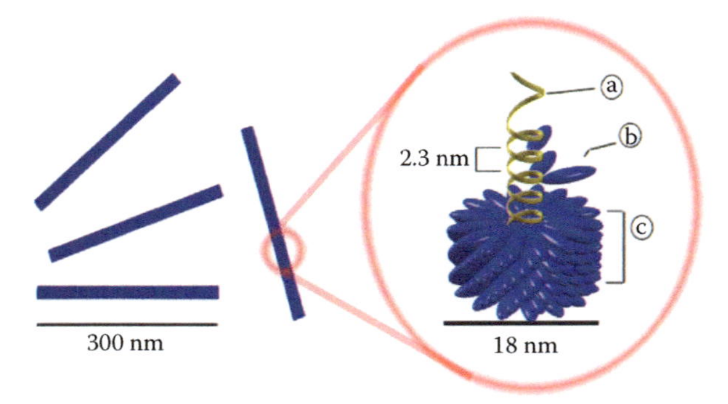

FIGURE 1.1 Schematic model of TMV showing helical symmetry. (a) Nucleic acid. (b) Coat protein subunit. (c) Capsid. (https://commons.wikimedia.org/wiki/File:Tobacco_mosaic_virus_structure.png.)

axis). The length of a helical virus is determined by the size of the genome. The symmetry of a helical assembly of subunits is defined by u, the number of subunits per turn, and p, the displacement along the helix axis between one subunit and the next. P, the pitch of the helix, is the distance along the helix axis that corresponds to exactly one turn of the helix; $P = u\,Xp$. Tobacco mosaic virus (TMV) is the classical example of a helical virus. In TMV, u is 16.33, p is 0.14 nm, and hence, P is 2.3 nm. TMV particles are rigid, rod-like structures (Franklin and Holmes 1956; Namba and Stubbs 1986). However, longer helical virus particles are often seen to be curved or bent and they display considerable flexibility.

Polyhedral viruses are non-enveloped (lacking a membrane bilayer) viruses whose capsids form geometric shapes with faces that appear flat when viewed under an electron microscope and possess icosahedral symmetry. Use of icosahedral symmetry is the best way of producing a shell of equivalently bonded identical structures with the largest interior volume for packaging the nucleic acid. Therefore, such capsids are likely to correspond to low free energy structures. Isometric viruses are built on icosahedral symmetry that requires a definite number (60) of structural units to complete the shell. Crick and Watson (1956) pointed out that viruses are likely to be built on 532 symmetry with 60 protein subunits (Figure 1.2).

In an icosahedral organization of 60 molecules, all subunits are related identically to their neighbors. Use of multiple copies of one or a few unique species of subunits for the formation of the shell is

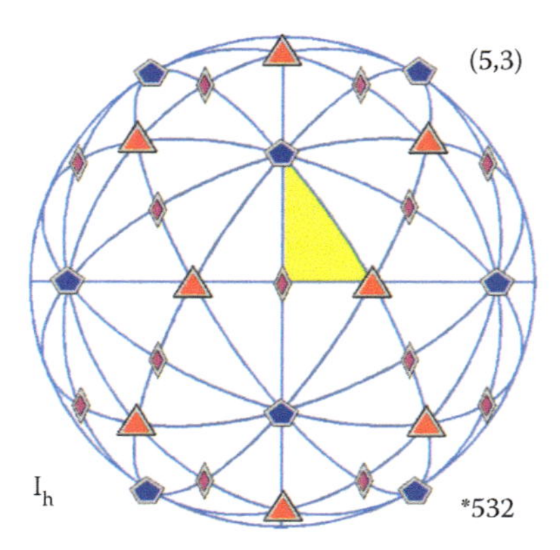

FIGURE 1.2 The orientation of symmetry axes in an icosahedron. Blue stars represent 5-fold axes, red triangles represent 3-fold axes, and red ovals represent 2-fold axes of the icosahedral capsid. The yellow triangle represents the asymmetric unit of the icosahedral particle. (https://commons.wikimedia.org/wiki/File:Sphere_symmetry_group_ih.png.)

an efficient strategy to minimize the genetic content required to code for the capsid. Correct structure can then be formed with minimum wastage, because if a subunit is synthesized or folded incorrectly, only that unit needs to be discarded to avoid incorrect assembly.

Major technical advances were made in the 1950s in electron microscopic examination of biological samples. The first high-resolution micrographs of negatively stained icosahedral viruses (Horne et al. 1959; Huxley and Zubay 1960) presented a structural paradox. The number of morphological units observed on the surface of known icosahedral viruses at that time was seldom 60, and was some number not anticipated on the basis of icosahedral symmetry, contradictory to what Watson and Crick had proposed. In addition, the capsomers (prominent structural units visible in the micrographs) themselves appeared to be symmetrical and were located on symmetry axes. These paradoxes led to the proposal of the "quasi-equivalence" hypothesis as the underlying principle in the construction of symmetrical viral capsids by Caspar and Klug (1962). According to this hypothesis, the protein subunits on the capsid need not be in identical environments. Only approximate equivalence, or quasi-equivalence, was assumed to be required for the formation and stability of these structures. The deformations from exact equivalence were thought to manifest in the form of small changes in intersubunit packing, requiring a low energy cost. According to the theory of quasi-equivalence, the icosahedral virus capsid could have only $20T$ structural units, where T is the triangulation number given by the rule: $T = Pf^2$, where $P = h^2 + hk + k^2$, for all pairs of integers h and k with no common factors, and f is any integer. The structural units could be clustered as $12T$ pentamers, $20T$ trimers, $30T$ dimers, or separated as $60T$ monomers. Different clustering patterns of the subunits result in different but characteristic appearance of capsids and accounted for all observed patterns in electron micrographs. However, certain deltahedra with T equal to or greater than 7 are skewed and exhibit right- and left-handed forms.

1.1.2 Capsid Morphologies Prevalent in Viruses

Most viruses from the plant kingdom, especially the group IV viruses, have helical capsids. Among these, Potyviruses, Flexiviruses, and Closteroviruses (Kendall et al. 2008) exhibit flexible filamentous morphology. Viruses from *Virgaviridae* possess capsids that are rigid rods (Chapman 2013). Among bacteriophages, filamentous capsids are observed with M13, fd, and f1 phages (Acheson 2011). Not many animal viruses have capsids with filamentous morphology. However, members of orthomyxoviruses (Dadonaite et al. 2016) and filoviruses do possess filamentous capsids. Indeed, filamentous morphology has dominated the design principles of plant viruses, while most viruses from the animal kingdom have evolved to possess isometric design for the construction of their capsids.

Polyhedral viruses (Figure 1.3) are common among animal viruses and include the Adenoviruses, Polyomaviruses, Parvoviruses, Picornaviruses, and so forth.

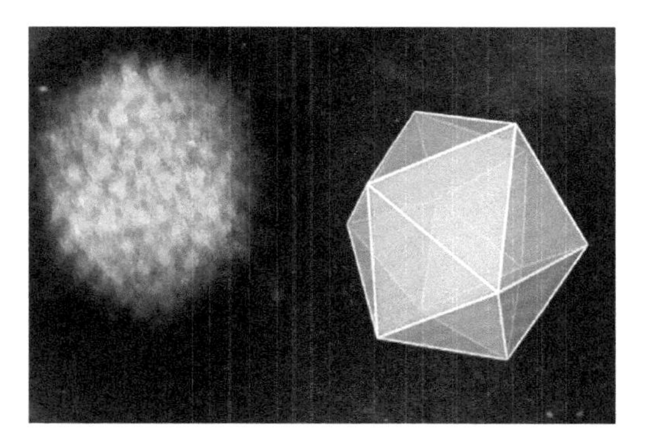

FIGURE 1.3 An electron micrograph of avian adenovirus showing isometric capsids. The figure on the right hand side illustrates an idealized model of the capsid. (https://commons.wikimedia.org/wiki/File:Icosahedral_Adenoviruses.jpg.)

Bacteriophages such as PhiX174, and plant viruses such as Sobemoviruses, Tombusviruses, and Tymoviruses are some examples of polyhedral viruses (Mahy and Van Regenmortel 2010). Depending upon the complexity of the capsid, the assembly is either self-driven or aided by scaffolding proteins that generally are not part of the final structure (Fujisawa and Hayashi 1977). Complex viruses such as the bacteriophages T4, T2, T7, and so forth have capsids that possess both helical and icosahedral features (Figure 1.4). The head regions of such bacteriophages harbor the nucleic acid genome and are isometric in nature, while the tail components are helical. These viruses are assembled from subassemblies generated during the process of capsid maturation. The head and tail are built independently and later assembled to form an elegant capsid. Animal viruses such as the adenoviruses have capsids that are polyhedral, but the fibers embedded on the capsids that aid in attachment of the virus to its receptor are helical (Reddy et al. 2010; Pache et al. 2008a,b).

Viral capsids belonging to dsDNA phages such as HK97 conform to T=7 icosahedral arrangement (Wikoff et al. 2000). Polyomaviruses, including SV40, have 360 subunits, which is disallowed on the basis of the Caspar and Klug quasi-equivalent hypothesis. At positions where hexameric clusters of protein subunits are expected in a perfect T=7 lattice, only pentameric clusters are observed in these capsids (Stehle et al. 1996). Other viruses that tend to violate quasi-equivalence include ds RNA viruses belonging to *Reoviridae* group. The structure of Blue tongue virus (BTV) sub-core (Grimes et al. 1998) and Reovirus core particles (Reinisch et al. 2000) have 120 protein subunits with pseudo T=2 symmetry. Similar architecture has also been found in the plant virus, Rice dwarf virus (RDV; Nakagawa et al. 2003) where the protein subunits are arranged as parallel dimers with highly non-equivalent intersubunit interactions. Although the wild-type particles of Brome mosaic virus (BMV) and Cowpea chlorotic mottle virus (CCMV) are normal T=3 capsids, capsids containing 120 non-equivalent protein subunits are assembled from their recombinant or dissociated capsid proteins (CPs; Krol et al. 1999; Zlotnick et al. 2000). Sulfolobus turreted icosahedral virus 1 has a unique triangulation number, T=31 (Veesler et al. 2013; Figure 1.5).

Several distinct groups of viruses, including many animal viruses, have capsids with a lipid envelope. These enveloped viruses have a layer of lipid surrounding their protein shell. The lipid layer is acquired from the host cell during morphogenesis. While such viruses are common in the animal kingdom, there are just a few representative genera of these in the plant kingdom, such as the Tospovirus, Nucleorhabdovirus, and Cytorhabdovirus (Moyer et al. 1999). An example of enveloped viruses with helical symmetry is the Influenza virus. Understandably, the lipid envelopes of these viruses aid during host entry and exit. Viruses such as those from *Poxviridae* are brick-shaped and acquire multiple lipid membranes during their replication and assembly in the cytoplasm (Joklik 1966). These viruses look like lipid droplets and are unique in morphology (Figure 1.6).

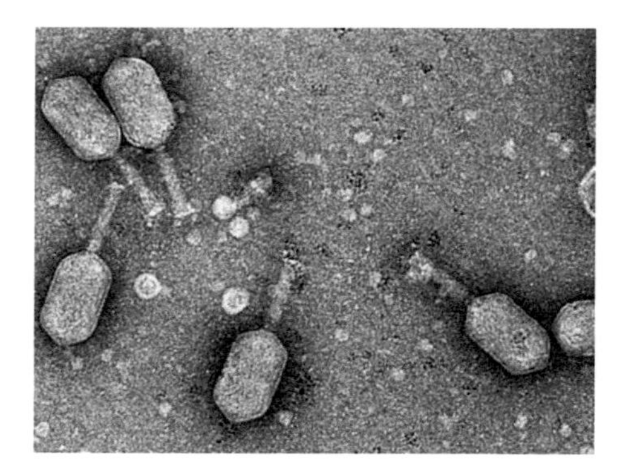

FIGURE 1.4 Electron micrograph of the complex Vibrio phage φpp2. (https://commons.wikimedia.org/wiki/File:Phage.phiPP2.jpeg.)

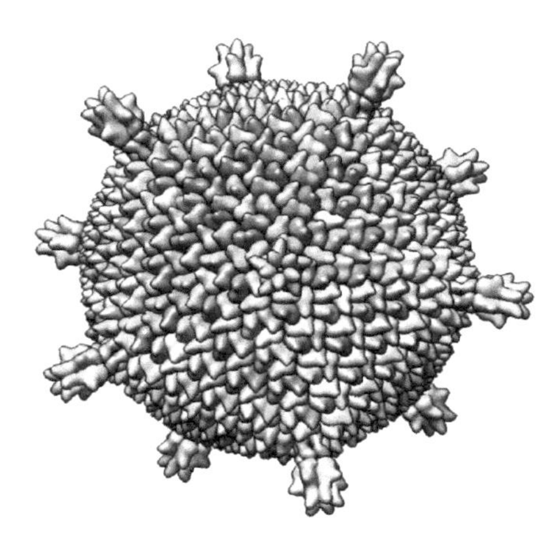

FIGURE 1.5 4.5 Å structure of Sulfolobus Turreted Icosahedral Virus determined by cryoEM. (https://commons.wikimedia.org/wiki/File:Emd-5584.jpg.)

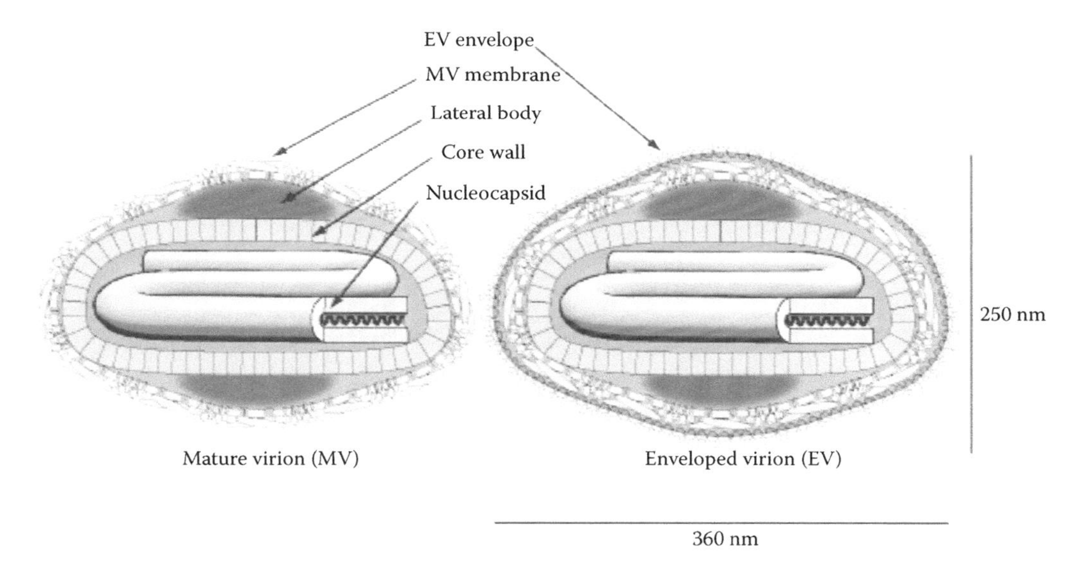

FIGURE 1.6 Schematic representation of the structure of poxviruses. (Adapted with permission from Hulo C, de Castro E, Masson P, Bougueleret L, Bairoch A, Xenarios I, LeMercier P (2011). ViralZone: A knowledge resource to understand virus diversity. *Nucl Acids Res* 39(Database issue): D576–582. http://viralzone.expasy.org.)

1.2 COAT PROTEIN STRUCTURE

The capsids discussed in previous sections are formed by the assembly of individual CP subunits whose three-dimensional structure reveals fascinating details. CPs of most of the isometric viruses from different domains of life possess a jellyroll motif comprising eight anti-parallel β-strands arranged in a sandwich-like fold. In spite of the significant variations in the biology of these viruses, the CP fold has been maintained, although the reasons for the observed structural similarity are far from certain. It is plausible that such a fold offers the necessary compactness and suitability for being accommodated in a triangular facet and formation of an isometric capsid. The CPs of many viruses may be divided into different domains including RNA-binding (R), shell (S), and projecting (P) domains. Different viruses contain one or more of these domains. The S domains of the CPs of several plant and animal icosahedral viruses adopt a β-barrel (jellyroll) motif (Figure 1.7a). This canonical β-barrel consists of two β-sheets, each consisting of four strands. The strands are designated as BIDG and CHEF, following the nomenclature used for Tomato bushy stunt virus (TBSV; Harrison et al. 1978). A few viruses have one or two extra β-strands. The structure of the CP of bacteriophage *MS2* (Golmohammadi et al. 1993) is unlike the canonical jellyroll fold found in other icosahedral viruses.

Unlike the CPs of isometric viruses, the helical viruses that have been studied possess CPs that are all α-helical (Figure 1.7b; Solovyev and Makarov 2016). The stark contrast in the conformation of the CPs of helical and isometric viruses is indicative of the architectural features that the two kinds of capsids demand (Figure 1.7). Apart from the nature of secondary structures, the isometric viral CPs are characterized by metal mediated interactions that are not as promiscuous in helical viruses.

1.2.1 Techniques to Study Viral Morphology

The very first visualization of viruses and the studies on their various morphologies began with the invention of electron microscopes (EMs). To a large extent, discovery of newer viruses should also

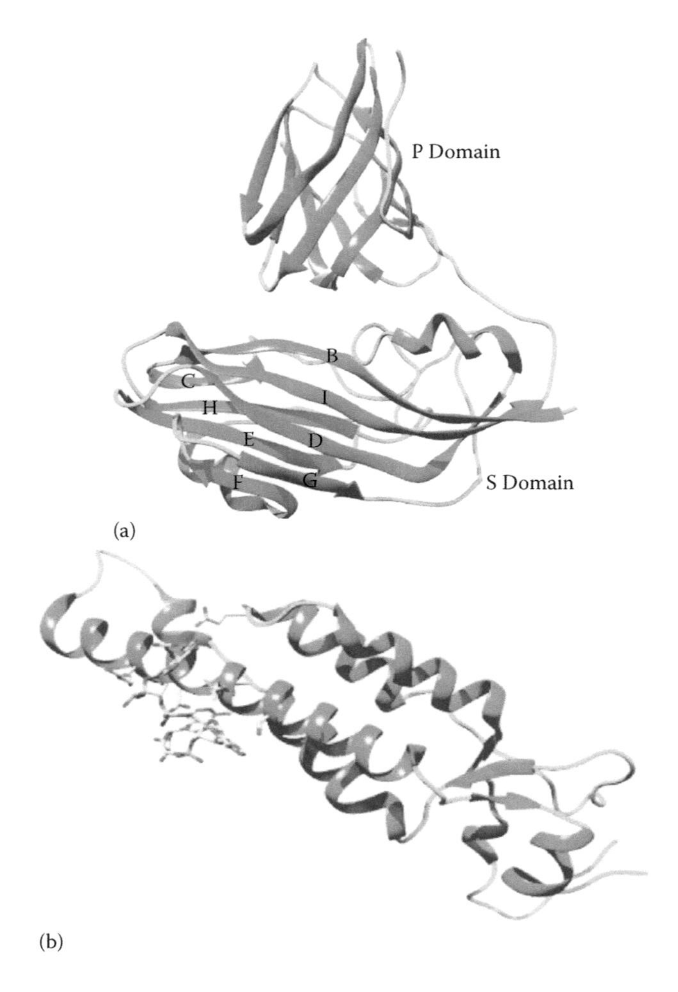

FIGURE 1.7 (a) Subunit structure in TBSV (PDB ID: 2tbv) showing the S and P domain and β-sandwich motif of S domain. (b) CP conformation of TMV (PDB ID: 4udv) showing helical conformation. Figures were generated using Chimera. (From Petterson EF et al. *J Comput Chem* 25(13): 1605–1612, 2004.)

be credited to EMs. TMV was the first virus to be observed under the EM by Ruska (1986) (Figure 1.8; Goldsmith and Miller 2009). The general outline and surface features of viruses are directly and easily visualized using EM. In the negative staining technique, virions can be visualized by the exclusion of electron-absorbing molecules such as uranyl acetate. Alternatively, positive staining could be used to study proteins, nucleic acids, and lipids in plasticized thin sections of infected cells (Acheson 2011). However, in either case, staining results in distorting or blurring the structural details of virions and there is considerable damage to the virions.

More recently, the technique of CryoEM has enabled the determination of virus structures without introducing distortion (in solution conditions) of many large isometric and helical viruses. The technique involves embedding of viral specimens in vitreous ice and subjecting it to imaging using electron microscopy. Apart from the use of EM, many advances in techniques and instrumentation have facilitated the study of viral capsids. X-ray crystallography and X-ray fiber diffraction have played an important role in gaining insights into the atomic details of simple viruses such as TMV, to most elaborate and exquisite structures such as those of bacteriophage PRD1 (Benson et al. 2001). The technique of X-ray crystallography involves crystallization of the desired viral capsid, obtaining the X-ray diffraction data from the crystal, and then solving the crystal structure through the techniques of molecular replacement, multiple isomorphous replacement or multiple anomalous dispersion (Brunger et al. 1998). X-ray fiber diffraction is used to study the molecular structure of long assemblies of identical subunits such as the helical viruses. The technique of X-ray diffraction indeed gave the most fundamental breakthrough in the understanding of capsid assembly and architecture at high resolution. Due to the large sizes of viral capsids, none of the capsid structures has so far been attempted for study using nuclear magnetic resonance (NMR) technology.

1.2.1.1 Viral Capsids as Rigid Rods

One of the preferred morphologies for the capsids of plant viruses is helical. Viruses in about eight genera show the morphology of helical rigid rods. Apart from the similarity in morphology, viruses in these genera share the common feature of possessing single-stranded (ss) RNA genomes, with most of them being positive-sense (+) with the exception of the genus Varicosa virus that has a negative-sense (–) RNA genome (Chapman 2013). The *Virgaviridae* (from the Latin virga = rod) family epitomizes the family with rod-shaped particles with almost half of all the definitive species in the Tobamovirus genus. The other genera are Furovirus, Hordeivirus, Pecluvirus, Pomovirus, and Tobravirus. The chief characteristics of members of the family include the alpha-like replication proteins, (+) ssRNA with a 3′ t-RNA-like structure and no polyA tail, rod-shaped virions that are 20–25 nm in diameter, and CPs of 19–24 kDa size (Adams et al. 2009). The members of the genus Tobamovirus are known to be

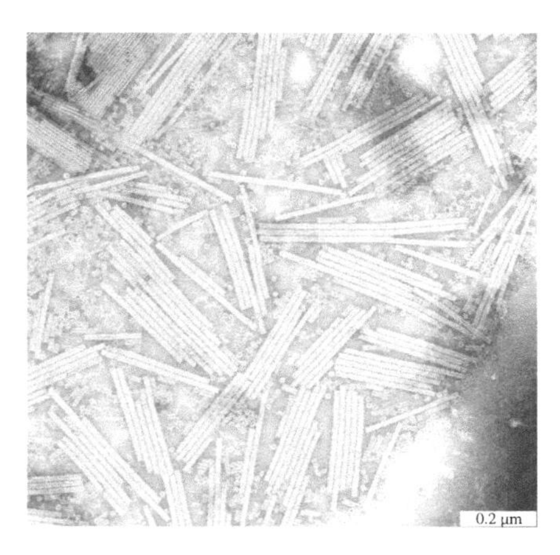

FIGURE 1.8 EM picture of TMV showing capsids that are rigid rods. (https://commons.wikimedia.org/wiki/File:TMV.jpg.)

transmitted by mechanical means, while the members of other genera are transmitted by soil-borne vectors or through pollen and seed. The genus *Benyvirus* (type member Beet necrotic yellow vein virus), which includes viruses with rod-shaped morphology, is more distantly related to other genera with rod-shaped particles (Figure 1.1).

The helical order of TMV was examined to a resolution of 2.3 Å in cubic ice, prepared by controlled devitrification (Cyrklaff and Kühlbrandt 1994). TMV was studied to a resolution better than 5 Å by the development of single-particle based electron microscopic helical reconstruction using 200,000 asymmetric units (Sachse et al. 2007) and improved to a resolution of 3.3 Å by using a total of 1,900,000 asymmetric units recorded on film (Ge and Zhou 2011). The overall structure of the capsid includes 2130 molecules of CPs and one molecule of genomic ssRNA, 6400 bases long. The CP folds into a rod-like helical structure around the RNA with 16.3 subunits per helix turn. The model described by Namba et al. (1989) identified the positions of all of the non-hydrogen atoms of the CP and the RNA as well as 71 water molecules and two calcium ions. The CP monomer consists of 158 amino acids that are assembled into four main α-helices, two minor helices, and one tiny β sheet joined by loops, one of which runs radially and binds RNA (Figure 1.7b; Ge and Zhou 2011). Similar to TMV, Ribgrass mosaic virus (RMV; Wang et al. 1997) and Cucumber green mottle mosaic virus (CGMMV; Wang and Stubbs 1994) also show CP conformation of four helix bundles. Each CP monomer binds to three RNA nucleotides (Klug 1999). Interactions of CP with RNA in a repetitive fashion leads to the formation of disks that in turn assemble as right-handed helical rigid rods 3000 Å in length and 180 Å in diameter. A distinct inner channel of ~4 nm diameter is evident from the EM pictures of TMV that mark the central cavity or channel.

The assembly of TMV has been shown to begin with a two-turn helical CP complex, called the 20S aggregate, or the "disk" (Caspar and Namba 1999) with the recognition of an initiation motif containing AAG repeats from the 3′ end of the genome (called the nucleation site) (Figure 1.9). The assembly is dependent on pH and ionic strength, as evident from conformational changes associated with changes in pH and the concentration of calcium ions (Ge and Zhou 2011).

The continuous interactions of the CP with the genomic RNA leads to dynamic rearrangements around the helix axis bringing about the transformation from the nucleating disk into a proto-helix. The growth of the helix in both the directions of the nucleated rod using disk subassembly determines the final rigid morphology of TMV rods, thus establishing the pivotal role of the CPs in genome recognition and capsid assembly.

The structure of Barley stripe mosaic virus (BSMV), the type member of the genus Hordeivirus, was determined by fiber diffraction and cryo-electron microscopy (Kendall et al. 2013). The helical symmetry of BSMV was determined to be 23.2 subunits per turn of the viral helix, and is very different from those of the well-characterized tobamoviruses (16.3 subunits per turn), though the CP structure was found to be similar.

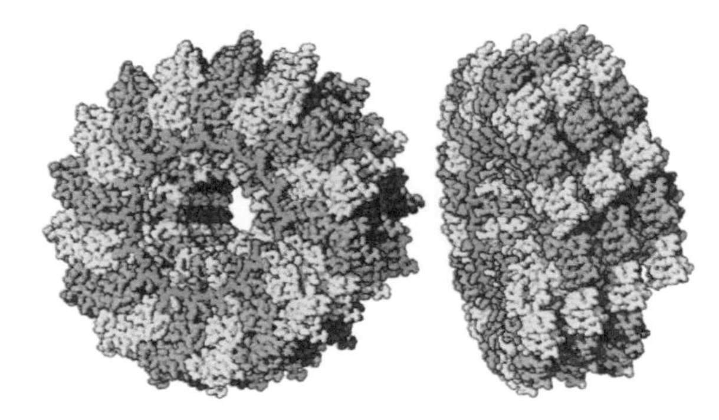

FIGURE 1.9 Discs formed during assembly in TMV. (https://commons.wikimedia.org/wiki/File:Tobacco_mosaic_virus_tmv2.png.)

1.2.1.2 Flexuous Filamentous Viruses

The flexible filamentous plant viruses are widely found and are responsible for more than half of the viral crop damage in the world. Flexible filamentous plant viruses include 19 recognized genera (Fauquet et al. 2005), distributed in the families of (+)ssRNA viruses, the *Potyviridae*, the *Alfaflexiviridae*, the *Betaflexiviridae*, and the *Closteroviridae*. Members of the family *Potyviridae* account for almost one-third of the total known plant virus species (Fauquet et al. 2005) infecting most of the economically important crops (Hull 2001). The flexuous and filamentous capsids of these viruses contain several copies of helically arrayed CP subunits protecting the genome (Figure 1.10). The structural studies on these viruses are mostly from EM data, as they are difficult to study using other methods due to their flexuous nature. The viruses cannot be crystallized and are too flexible for high-resolution X-ray fiber diffraction or cryo-EM investigations. The earliest published structural studies of these viruses dates back to 1941 using Potato virus X (PVX; Bernal and Fankunchen 1941).

Low resolution X-ray fiber diffraction studies and various forms of EM examinations suggested that all potexviruses may share a common architecture, with slightly less than nine protein subunits per helical turn (Richardson et al. 1981). A number of low-resolution models of flexible plant viruses have been generated with the assumption that the virions have a right-handed helical pitch, as observed in TMV (Kendall et al. 2008, 2013). Though there is very little sequence similarity between the CP of flexuous viruses in different families, there have been suggestions of structural and evolutionary relationships (Dolja et al. 1991; Koonin et al. 2015).

Flexible plant viruses appear to possess common architectural features on the basis of structural studies of virions using X-ray fiber diffraction and cryoEM. The CPs of PVX, and Narcissus mosaic virus (NMV) were found to be helical, as suggested by vibrational circular dichroism studies (Shanmugam et al. 2005). The cryo-EM studies on PVX and Soybean mosaic Potyvirus (SMV; Kendall et al. 2008) suggest that they are helical structures with a symmetry of nine subunits per turn a diameter of 130 ~ 140 Å and a central channel of ~ 30 Å. The PVX virion was shown to have a deeply grooved surface (Parker et al. 2002).The N- and C-termini of potyviruses are exposed and are easily degraded by trypsin to yield a trypsin-resistant core. It was shown in the case of pepper vein banding Potyvirus that these surface exposed regions are crucial for initiation of assembly. It was proposed that the head to tail electrostatic interaction between these regions of CP monomers result in dimers, which in turn form the 16S ring intermediate (Anindya and Savithri 2003). AFM studies have shown the presence of VPg, HC-Pro, and CI at the tip structures in potyvirus virions (Torrance et al. 2006; Gabrenaite-Verkhovskaya et al. 2008). These proteins assist the virions in translation initiation, viral movement, and vector transmission. The structures of two of the members of family *Alphaflexiviridae* and genus Potexvirus that include the Bamboo mosaic virus (BaMV; DiMaio et al. 2016) and Papaya mosaic virus (PapMV; Yang et al. 2012) were determined to near atomic resolution using cryo-EM and X-ray crystallography. The structure of PapMV suggests a particle diameter of 135 Å with a helical symmetry of ~10 subunits per turn. The PapMV CP showed an all-helix fold with a seven-α-helical bundle (Figure 1.11) that is different from the four-helix bundle fold of TMV (Figure 1.7b). Studies on BaMV suggest that it has a highly intertwined structure where each subunit makes direct contact with eight other subunits. BaMV has a pitch of 35 Å with ~8.8 subunits per turn.

FIGURE 1.10 Schematic representation of Potato virus Y. (Adapted with permission from Hulo C, de Castro E, Masson P, Bougueleret L, Bairoch A, Xenarios I, LeMercier P (2011). ViralZone: A knowledge resource to understand virus diversity. *Nucl Acids Res* 39(Database issue): D576–582. http://viralzone.expasy.org.)

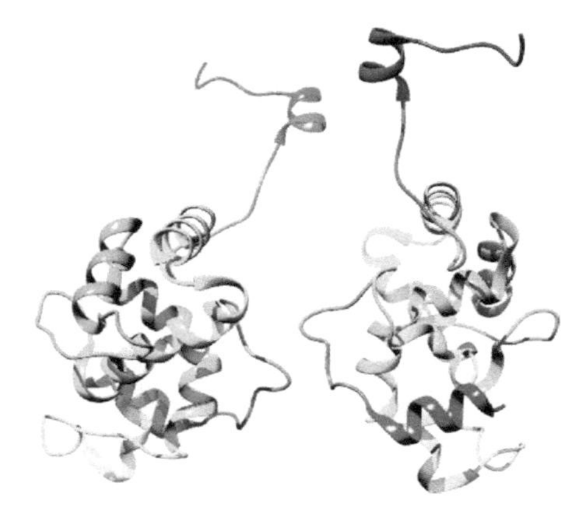

FIGURE 1.11 Structure of PapMV (PDB ID: 3dox) rendered in ribbon using Chimera. (From Pettersen EF et al. *J Comput Chem* 25(13): 1605–1612, 2004.)

A fiber diffraction study of the tritimovirus, Wheat streak mosaic virus (WSMV), suggested that WSMV has 6.9 subunits per turn of the viral helix (Parker et al. 2005). Recently, the cryo EM structure of Pepino mosaic virus (PepMV) virions was determined at 3.9 Å resolution (Agirrezabala et al. 2015). The CP fold of PepMV was found to resemble that of nucleoproteins (NPs) from the genus Phlebovirus (family *Bunyaviridae*), suggesting plausible gene transfer between eukaryotic (+) and (–) ssRNA viruses.

1.2.1.3 Bacilliform Viruses

Bacilliform, or bullet-shaped, viruses are less prevalent among plant viruses. However, the family *Rhabdoviridae* (Figure 1.12) consists of mostly enveloped, bacilliform virus particles with a (–), ssRNA genome that infects vertebrates, invertebrates, or plants. The plant rhabdoviruses are grouped under two genera, Nucleorhabdovirus and Cytorhabdovirus, that are distinguished on the basis of their intracellular site of morphogenesis in plant cells.

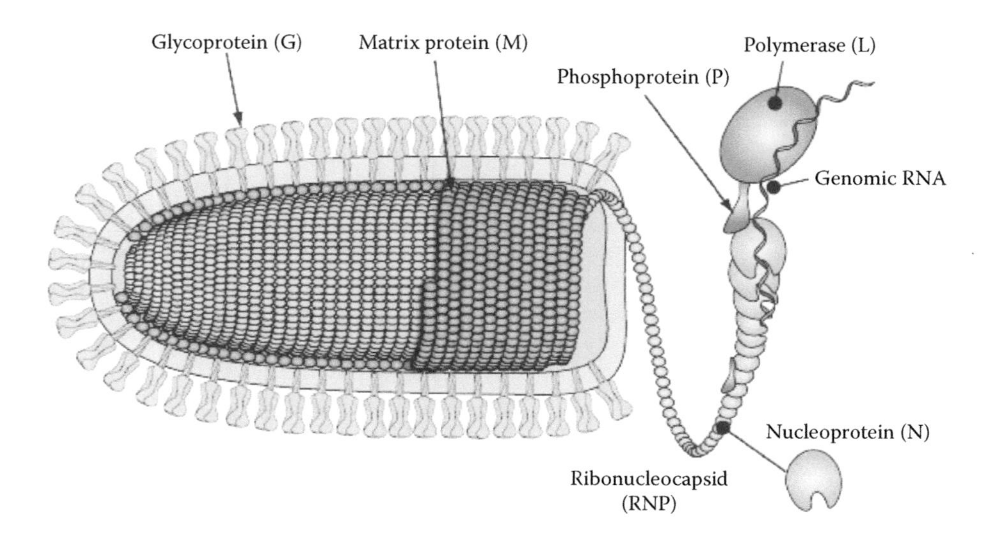

FIGURE 1.12 Bacilliform morphology of a nucleorhabdovirus. (Adapted with permission from Hulo C, de Castro E, Masson P, Bougueleret L, Bairoch A, Xenarios I, LeMercier P (2011). ViralZone: A knowledge resource to understand virus diversity. *Nucl Acids Res* 39(Database issue): D576–582. http://viralzone.expasy.org.)

These viruses replicate in both their insect and plant hosts. All plant rhabdoviruses carry analogs of the five core genes: the nucleocapsid (N), phosphoprotein (P), matrix (M), glycoprotein (G), and large-domain or polymerase (L). The particles are 180 nm long and 75 nm wide. Certain plant rhabdoviruses are bacilliform in shape, and their lengths are almost twice their breadths (Jackson et al. 2005). Some of the important bacilliviruses belonging to rhabdoviruses are Lettuce necrotic yellow virus (LNYV), Northern cereal mosaic virus (NCMV), and Potato yellow dwarf virus (PYDV). Plant rhabdoviruses are transmitted mainly via aphid, leafhopper, or plant-hopper vectors.

Badnavirus and Tungrovirus, within the family *Caulimoviridae*, are viruses with bacilliform morphology (Figure 1.13). They are of length 60–900 nm and width 35–50 nm. Their genomes are circular double-stranded DNA of approximately 7.5 kbp with one or more single-stranded discontinuities (Borah et al. 2013). Detailed structural information on bacilliform viruses is yet to be obtained. The available information has been obtained mainly from EM studies.

The Alfalfa mosaic virus (AMV), a member of the family *Bromoviridae*, forms four types of particles composed of one of the four genomic RNAs and a large number of single species of CP subunits. AMV is a multipartite virus and is composed of four morphologically distinct particles (three bacilliform and one spheroidal; Bisby et al. 2008).The particles are 180 Å in diameter and range in length from 300Å to 570Å. After mild treatment with trypsin and the removal of RNA, the capsid subunit readily forms T=1 particles. The X-ray crystal structure of these particles has been determined (Kumar et al. 1997). The subunit is a simple β-barrel in the axial orientation with extended termini at the base resembling the CP structure of CCMV. Unlike CCMV and BMV, Tobacco steak virus (TSV), type member of the ilarvirus genus is labile and pleomorphic and therefore was difficult to crystallize. The TSV CP expressed in *Escherichia coli* formed virus-like particles (VLPs). Mutational analysis of the N-terminal disordered domain showed that 26 amino acid residues from the amino-terminus could be crucial for capsid heterogeneity, the zinc-binding domain was essential for assembly, and the N-terminal arginine-rich motif (N-ARM) was required for binding to nucleic acid (Mathur et al. 2014). It was possible to crystallize the N-terminal 26 amino acids deleted TSV CP dimers in two distinct forms (space groups $P2_1$ and C2) and determine their X-ray crystal structures at 2.4 Å and 2.1 Å resolutions, respectively. The TSV CP was found to be structurally similar to that of AMV, accounting for a similar mechanism of genome activation in alfamo and ilarviruses. The C-terminal arms (C-arm)

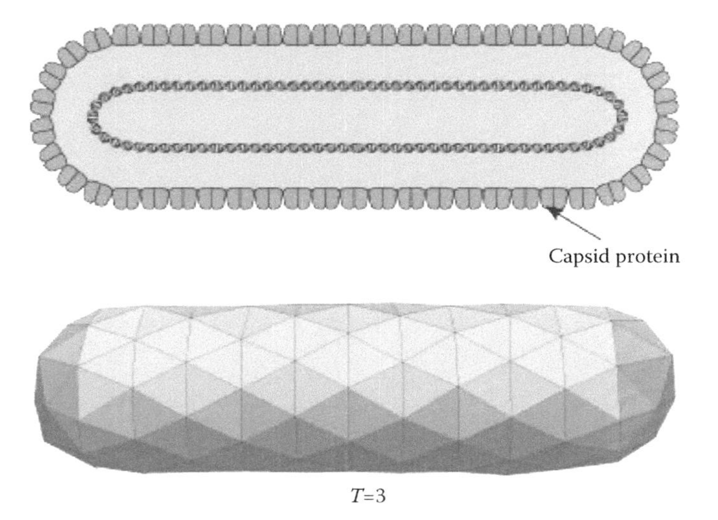

FIGURE 1.13 Bacilliform morphology of Badnavirus belonging to Caulimoviridae. (Adapted with permission from Hulo C, de Castro E, Masson P, Bougueleret L, Bairoch A, Xenarios I, LeMercier P (2011). ViralZone: A knowledge resource to understand virus diversity. *Nucl Acids Res* 39(Database issue): D576–582. http://viralzone.expasy.org.)

were observed to be swapped in both the crystal forms. Mutation of a residue in the hinge region in the C-arm which makes it flexible showed that it could be responsible for the polymorphic and pleomorphic nature of TSV capsids (Gulati et al. 2015). Another plant virus genus, Tospovirus, is also pleomorphic, with 80–120 nm diameter. However, in this case the virus particles are enveloped, displaying the glycoproteins Gn and Gc on the surface of the membrane derived from the endoplasmic reticulum of host cells (Ribeiro et al. 2009). The three segments of the viral genome are encapsidated by the nucleocapsid protein. The pseudo-circular viral nucleocapsids of the *Bunyaviridae* family to which Tospoviruses belong are 2–2.5 nm in diameter, 200–3000 nm length, and they usually display helical symmetry. The L protein, the RNA-dependent RNA polymerase, is also present within the virions, as these are (–) RNA viruses.

1.2.1.4 Isometric Viruses

Though the idea that the structure of a virus could be solved by X-ray diffraction originated with J.D. Bernal and his colleagues in the 1930s, the goal was realized only in 1978 when Steve Harrison and his group at Harvard University determined the structure of TBSV to a resolution of 2.9 Å (Harrison et al. 1978). Following this, in 1980, Michael Rossmann's group at Purdue University reported the structure of another spherical plant virus, Southern bean mosaic virus (SBMV; Abad-Zapatero et al. 1980). These were the first reports on the structures of spherical viruses. These two plant virus structures revealed unexpected similarity in the folds of the S domains of their CPs.

In T=1 viruses, symmetrical arrangement of 60 subunits leads to identical bonding interactions. However, this limits the volume inside the capsids available for packing the genome. Many satellite viruses (those that need another helper virus for multiplication) are built on a T=1 construct. The simplest capsids are those of T=1 plant satellite viruses. Satellites are subviral agents composed of ssRNA of ~ 1000 nt as their genetic material. Detailed structures are available for three plant satellite viruses, Satellite tobacco necrosis virus (STNV; Jones and Liljas 1984), STMV (Larson et al. 1998; Figure 1.14) and Satellite panicum mosaic virus (SPMV; Ban and McPherson 1995) to atomic resolutions. The CP monomer has the typical jellyroll motif in all the satellites and at least five basic residues at the N-termini, which are implicated in interaction with RNA. The nucleic acid–protein interactions are more pronounced in STMV, while they are less so in the case of SPMV and STNV (Larson et al. 1998). Apart from the satellite viruses, AMV, belonging to *Bromoviridae*, also possesses a T=1 capsid architecture (Kumar et al. 1997). The family *Nanoviridae* includes viruses with a multipartite ssDNA genome that encapsidates individual genomic components in separate T=1 particles. Detailed structural information is unavailable for this group of viruses. Arabis mosaic virus

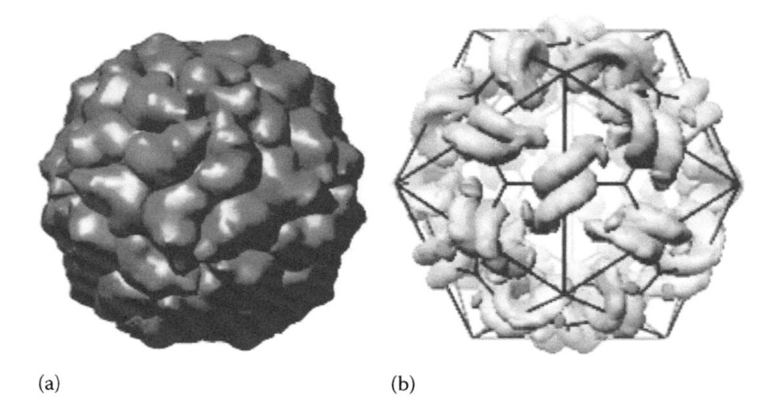

FIGURE 1.14 (a) Capsid and (b) nucleic acid organization in STMV. (Adapted with permission from Carrillo-Tripp M, Shepherd CM, Borelli IA, Venkataraman S, Lander G, Natarajan P, Johnson JE, Brooks III CL, Reddy VS (2009). VIPERdb2: An enhanced and web API enabled relational database for structural virology. *Nucl Acid Res* 37: D436–D442. http://mmtsb.scripps.edu/viper.)

(ArMV) and Grapevine fan leaf virus (GFLV) are two picorna-like viruses from the genus *Nepovirus*, possessing a bipartite RNA genome encapsidated into a 30-nm icosahedral viral particle formed by 60 copies of a single CP. Recently, a 6.5 Å resolution cryo-electron microscopy structures of ArMV was determined, and homology modeling and flexible fitting approaches were employed to build its pseudo-atomic structure (Lai-Kee-Him et al. 2013). Previously determined structure of GFLV revealed important components of the CP that might play a pivotal role in nematode transmission (Schellenberger et al. 2011).

Most of the self-sufficient viruses (viruses that do not need another helper virus for multiplication) are made of 180 CP subunits organized on a T=3 shell. In order to attain optimal bonding interactions, the subunits are arranged in quasi-equivalent environments involving slight conformational differences associated with the subunits and intersubunit packing adjustments. In a T=3 particle, the identical CP subunits display three different conformations, designated as A, B, and C. Structures are now available for a large number of T=3 viruses from different families infecting plants, animals, insects, and bacteria. Among the plant viral families, bromoviridae form isometric virions encapsidating four segments of linear (+) ssRNA. The multipartite genome is divided among more than one type of particle. The structures of BMV (Lucas et al. 2002a; Figure 1.15), CCMV (Speir et al. 1995), Cucumber mosaic virus (CMV; Smith et al. 2000) and Tomato aspermy virus (TAV; Lucas et al. 2002b) have been determined to near atomic resolutions. The capsid architectures are more or less similar between CCMV and BMV and between TAV and CMV. TAV and CMV could be regarded as "swollen states" of BMV and CCMV, respectively. These viruses have an abundance of positively charged residues at the interior surface of the capsids that may be involved in RNA binding. The ordered N-terminal residues of B and C subunits of BMV and CCMV form a distinctive β-hexamer about the threefold axes at the center of each hexameric capsomere, which are replaced by a unique bundle of six amphipathic helices in TAV and CMV.

The capsids of tombusviruses enclose one or two segments of linear, (+) ssRNA. Structures of TBSV (Olson et al. 1983; Harrison et al. 1978), Tomato crinkle virus (TCV; Hogle et al. 1986), Carnation mottle virus (CMtV; Morgunova et al. 1994), Tobacco necrosis virus (TNV; Oda et al. 2000), and Cowpea mottle virus (CPMoV; Ke et al. 2004) have been determined using X-ray crystallography. The R, S, and P domains of the CP are prominent features of these viruses. TNV lacks the usual protrusion of the capsids made by the pentameric and hexameric capsomeres observed in the other members of this family. The orientation of the P domains of TCV and CPMoV differ from that of TBSV and give the protruding dimers more of a "hammerhead"-like shape, whereas the dimers in TBSV form a smooth knob-like shape (Figure 1.16a).

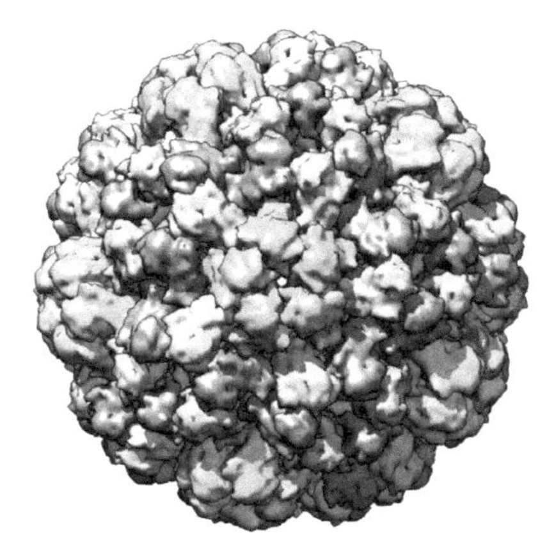

FIGURE 1.15 Capsid architecture of BMV (PDB ID: 3J7L). (Adapted from Berman HM et al. *The Protein Data Bank Nucleic Acids Research* 28: 235–242, 2000. www.rcsb.org/pdb. With permission.)

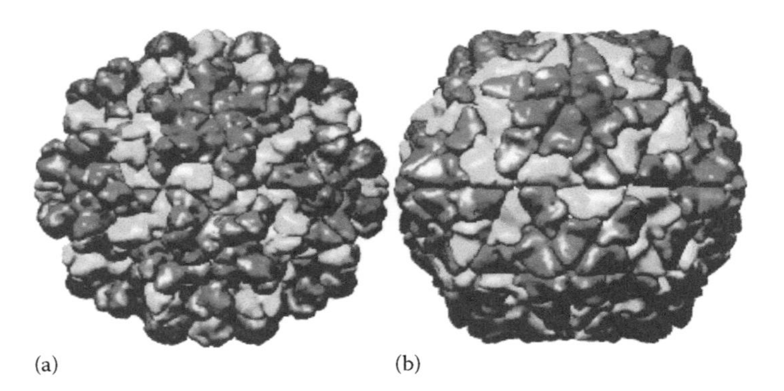

FIGURE 1.16 Capsid architecture in (a) TBSV and (b) TNV. (Adapted with permission from Carrillo-Tripp M, Shepherd CM, Borelli IA, Venkataraman S, Lander G, Natarajan P, Johnson JE, Brooks III CL, Reddy VS (2009). VIPERdb2: An enhanced and web API enabled relational database for structural virology. *Nucl Acid Res* 37: D436–D442. http://mmtsb.scripps.edu/viper.)

A β-annulus structure is observed at the hexameric axes of TBSV, TCV, and TNV formed by the N-terminal arms of three C-type subunits. This is absent in CMtV and CPMoV due to the lack of order in the N-terminal arms of the C subunits. TNV resembles the members of *Sobemoviridae* family more closely in its tertiary structure and capsid features (Figure 1.16b). The structure of Melon necrotic spot virus (MNSV) belonging to genus Carmovirus that has been determined at 2.8 Å resolution shows a higher degree of similarity to TBSV than to other members of the genus Carmovirus (Wada et al. 2008). Thus, the classification of the family *Tombusviridae* at the genus level conflicts with the patterns of similarity among CP structures. A recently discovered isometric virus isolated from a cultivated Adonis plant (*A. ramosa*), 28 nm in diameter and composed of a single coat of protein and a single RNA genome of 3991 nucleotides is proposed to belong to Carmovirus group (Yasaki et al. 2017).

Sobemoviruses infect a narrow range of monocotyledonous and dicotyledonous plants (~15–20 plant families) and are transmitted by beetles or mechanical inoculation. The virions possess a single molecule of (+) ssRNA of length 4100–5700 nt as their genome. Apart from the genomic RNA, many sobemoviruses have been shown to encapsidate a viroid-like satellite RNA of size 220–390 nt that needs carrier virus for its replication (Tamm and Truve 2000).

The structures of Southern cowpea mosaic virus (SCPMV; Abad-Zapatero et al. 1980; Silva and Rossmann 1987), earlier named as SBMV, Sesbania mosaic virus (SeMV; Bhuvaneshwari et al. 1995), Rice yellow mosaic virus (RYMV, Qu et al. 2000) and Cocksfoot mottle virus (CftMV; Tars et al. 2003) have been determined to near atomic resolutions. The CP units of sobemoviruses have R- and S- domains, of which the latter adopts the jellyroll fold. A polypeptide segment in the R-domain consists of basic residues such as arginines and lysines, and is disordered. The chemically identical CP subunits in these viruses are present in three quasi-equivalent conformations designated as A, B, and C (Figure 1.17). The A subunits interact at the icosahedral fivefold axes and form 12 pentamers. The B and C subunits interact at the icosahedral threefold axes to form 20 hexamers. The virus is stabilized by RNA protein, protein-protein, and metal ion-mediated protein–protein interactions. While in viruses stabilized predominantly by protein–protein interactions, empty shells are formed in vivo in SeMV; however, the VLPs are formed when the CP is expressed in *E. coli* by encapsidation of the 23S *E. coli* ribosomal RNA. The amino terminal segment residues 44-71 (numbering according to SeMV), which is ordered only in the C subunit, interact with one another at the quasi sixfold axes to make a β-annulus-like structure (Figure 1.18).

Extensive studies have been carried out to understand the mechanism of assembly of SeMV, using a large number of deletion and substitution mutants of the CP. While recombinant SeMV and the N-terminal deletion mutant CPNΔ22 were found to assemble in *E. coli* as VLPs similar to native capsids (Sangita et al. 2005a), deletion mutants, CPNΔ36, and CPNΔ65 mostly formed smaller *T*=1

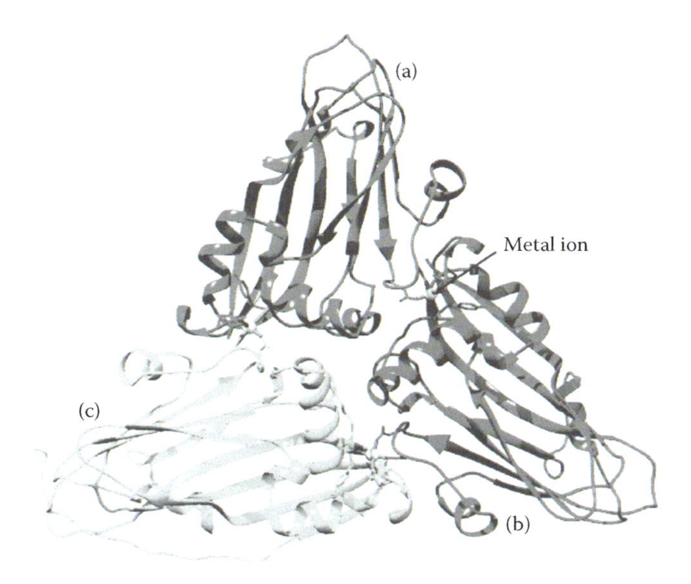

FIGURE 1.17 Asymmetric unit of SeMV (PDB ID: 1SMV) generated using Chimera. (From Petterson EF et al. *J Comput Chem* 25(13): 1605–1612, 2004.) The metal ions present at the interface of subunits coordinate with the side chains of aspartate residues.

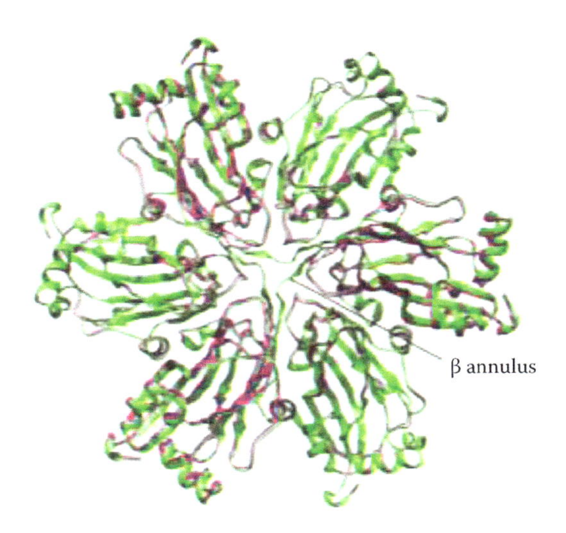

FIGURE 1.18 β-annulus structure of SeMV (shown in green color, PDB ID: 1SMV) at the hexameric axis overlaid with the structure of SeMV CP delta 48-59 mutant (PDB ID: 2VQO) shown in magenta, in which the residues corresponding to the β-annulus were deleted. The figure was generated using Chimera. (From Petterson EF et al. *J Comput Chem* 25(13): 1605–1612, 2004.)

particles consisting of 60 protein subunits (Sangita et al. 2005b). It is interesting to note that even though the sequences corresponding to arginine-rich motif and β-annulus (earlier implicated in proper assembly of T=3 particles) were still present in CPNΔ36, they form only smaller T=1 capsids, suggesting that the length of R domain is critical for T=3 assembly. The polypeptide folds of the subunits of all VLPs were found to closely resemble that of the S-domain of the native virus.

The recombinant particles bind calcium ions in a manner indistinguishable from that of the native capsids. The structures revealed major differences in the quaternary organization responsible for the formation of T=1 against T=3 particles (Sangita et al. 2004). This is substantiated by the recent work of Gulati et al. (2016), where replacement of R-domain with other sequences of similar length led to assembly of T=3 particles that are heterogeneous. The stability of the particle was significantly affected upon the substitution of calcium-binding aspartates (D146 and D149) by asparagines, although the mutated CP still formed VLPs (Satheshkumar et al. 2004). Deletion of residues forming the characteristic β-annulus did not affect the assembly of VLPs. The structure of these VLPs when overlaid on SeMV VLPs revealed that except for the absence of β-annulus, all other regions were very similar (Figure 1.18). These studies imply that the β-annulus is only a consequence of T=3 assembly and is not a prerequisite. Mutation of a single tryptophan that is present near the icosahedral fivefold axis resulted in the disruption of the capsid, leading to soluble dimers. The structures of these dimers were like that of the AB dimers (Pappachan et al. 2009). Thus it is possible that the AB dimers are the initial units of assembly which form a 10-mer via the interactions of the tryptophan at the five fold. The 10-mer could then associate with the RNA via the arginine-rich motif. This imposes order in the amino terminal segments of dimers added to the 10-mer complex, leading to the formation of the β-annulus and CC dimers. Subsequent addition of CP dimers leads to the formation of swollen T=3 particles. Calcium binding to assembling or assembled particles results in compact T=3 particles (Savithri and Murthy 2010). Contrary to other sobemoviruses, the βA arm of the C subunits in RYMV is swapped with the twofold-related βA arm to a similar, noncovalent bonding environment that leads to the so-called 3D domain swapping and produces long-range interactions throughout the capsid surface.

Tymoviruses form a distinct group of isometric plant viruses possessing monopartite ssRNA genome of size ~6.3 kb. Protein–protein interactions are very strong in tymoviral capsids, which accounts for the occurrence of empty capsids in natural preparations as well as under varied conditions (Sastri et al. 1997). The structures of three members of this group of viruses have been determined. These include Turnip yellow mosaic virus (TYMV; Canady et al. 1983), Desmodium yellow mottle virus (DYMV; Larson et al. 2000), and Physalis mottle virus (PhMV; Krishna et al. 1999; Figure 1.19) determined to resolutions of 3.2 Å, 2.7 Å, and 3.4 Å, respectively. The capsids of tymoviruses have prominent protrusions of pentamers and hexamers on their surface and deep valleys at the quasi threefold axes. Tymoviruses lack the basic N-terminal arm, which is characteristic of viruses stabilized by RNA–protein interactions. The β-hexamer observed at threefold axes is formed by the N-terminal arms of B and C subunits in TYMV and PhMV and of A and B subunits in DYMV.

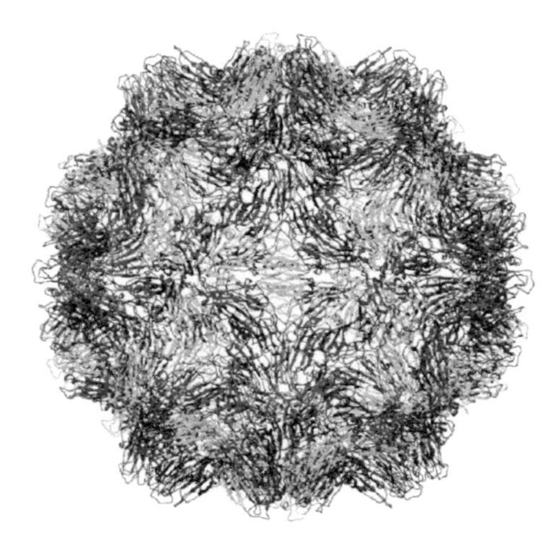

FIGURE 1.19 Capsid architecture in PhMV (PDB ID: 2XPJ) generated using Chimera. (From Petterson EF et al. *J Comput Chem* 25(13): 1605–1612, 2004.)

Detailed analysis of the deletion and site-specific mutants of PhMV CP has shown that CP folding and assembly could be concerted events in viruses stabilized predominantly by protein–protein interactions (Sastri et al. 1997; Umashankar et al. 2006).

The family *Reoviridae* includes viruses with dsRNA genome from both plant and animal hosts. The Phytoreoviruses (RDV), Fijiviruses (Fiji disease virus, FDV), and Oryzavirus (Rice ragged stunt virus, RRSV) are the reoviruses infecting plants. These isometric viruses are double layered with an outer capsid of a T=13 icosahedral symmetry and an inner capsid of T=2 symmetry. There are 260 trimers of the major outer capsid P8 protein and 60 dimers of the inner capsid P3 protein comprising each shell of the capsid. The capsid is assembled on subviral particles. The atomic structure of RDV was determined at 3.5 Å resolution by X-ray crystallography (Nakagawa et al. 2003). The atomic structure shows structural and electrostatic complementarities between both homologous (P3-P3 and P8-P8) and heterologous (P3-P8) interactions. There are overall conformational changes found in P3-P3 dimer caused by the insertion of amino-terminal loop regions of one of the P3 proteins into the other. The varied interactions suggest how the 900 (120 P3 and 780 P8) protein components are built into a higher-ordered virus core structure (Figure 1.20).

In several isometric viruses, three distinct protein subunits occur in the icosahedral asymmetric unit. The structures of these distinct units are similar and their disposition closely resembles those of T=3 arrangement. Therefore these true T=1 capsids are considered pseudo T=3 and denoted as P=3. Plant Comoviruses and animal Picornaviruses have P=3 symmetry (Venkataraman et al. 2008). This group of viruses, in the Picornavirus superfamily, comprises icosahedral particles encapsidating bipartite, (+) ssRNA genomes. In the shell, the domains of two protein subunits, S and L, occupy positions analogous to quasi-equivalent A, B, and C subunits of T=3 viruses. The structure of three members that have been elucidated include Cowpea mosaic virus (CPMV; Lin et al. 1999), Bean pod mottle virus (BPMV; Chen et al. 1989), and Rice clover mottle virus (RCMV; Lin et al. 2000) to resolutions of 2.8 Å, 3.0 Å, and 2.4 Å, respectively.

A notable surface feature of BPMV and other comoviruses is the protrusion centered on the viral fivefold axes formed by the pentameric S subunits (Figure 1.21). Toward the interior opening of the fivefold axis, three N-terminal residues of S subunits form a pentameric annulus structure. This feature is not seen in BPMV due to shorter N-termini of the S subunit. A putative metal ion site was also identified in both CPMV and RCMV at the center of the pentameric annuli.

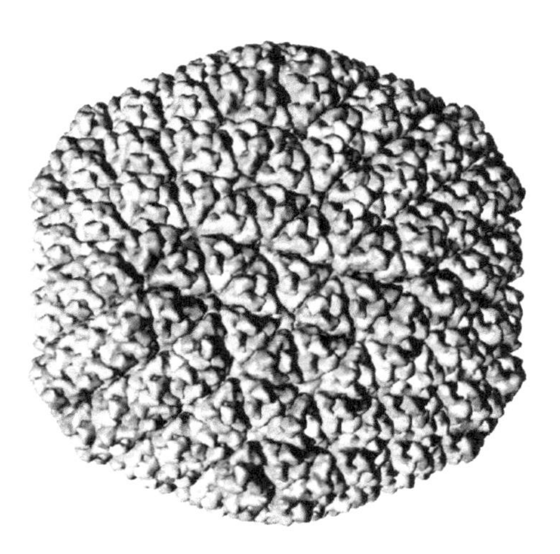

FIGURE 1.20 Capsid structure of RDV (PDBID: 1uf2). (From Berman HM et al. *The Protein Data Bank Nucleic Acids Research* 28: 235–242, 2000. www.rcsb.org/pdb. With permission.)

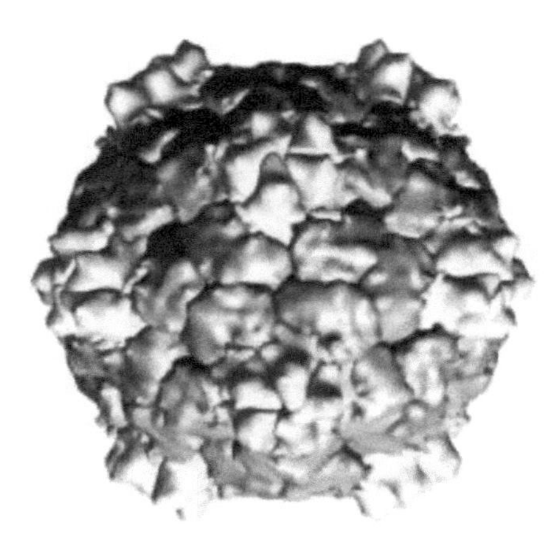

FIGURE 1.21 Capsid structure of BPMV (PDB ID: 1BMV). (From Berman HM et al. *The Protein Data Bank Nucleic Acids Research* 28: 235–242, 2000. www.rcsb.org/pdb. With permission.)

The other viruses with (+) ssRNA genomes and having isometric capsids are found in the family *Luteoviridae* (e.g., Barley yellow dwarf virus, Potato leaf roll virus, Pea enation mosaic virus). They comprise spherical virions of about 25–30 nm in diameter with $T=3$ icosahedral symmetry composed of 180 CP proteins. The family *Caulimoviridae* (e.g., Cauliflower mosaic virus, CaMV) comprises viruses with dsDNA genome co-encapsidated with reverse transcriptase. The capsids are not enveloped but are either bacilliform (30 × 60–900 nm) or isometric (45–50 nm in diameter; Figure 1.22). CaMV is transmitted through insect vectors and in this process the virion-associated proteins P2 and P3 are implicated. The crystal structure of P3 determined to a resolution of 2.3Å showed tetrameric parallel coiled coil arrangement with a unique organization showing two successive four-stranded sub-domains with opposite supercoiling handedness stabilized by a ring of interchain disulfide bridges (Figure 1.22). Homology modeling studies of virus-liganded P3 showed anti-parallel coiled-coil topology coating the virus surface (Hoh et al. 2010).

Geminiviridae is a family of plant viruses that derive their name from the geminate (gemini = twin moons) morphology of their capsids. There are currently 325 species in this family, divided among seven genera. It is the largest known family of ssDNA viruses. Symptoms associated with diseases caused by viruses of this family include yellow mosaic, yellow mottle, leaf curling, stunting, streaks, and reduced yields. They have circular ssDNA genomes, encoding genes that diverge in both directions from a virion strand origin of replication (Hulo et al. 2011). Virions in this family have paired incomplete ($T=1$) icosahedral capsids, composed of subunits of a single CP (29 kd) (Figure 1.23). Some members of the group can also form trimers and tetramers, but geminate structures are

FIGURE 1.22 Crystal structure of P3 protein of CaMV (PDB ID: 3K4T). (From Berman HM et al. *The Protein Data Bank Nucleic Acids Research* 28: 235–242, 2000. www.rcsb.org/pdb. With permission.)

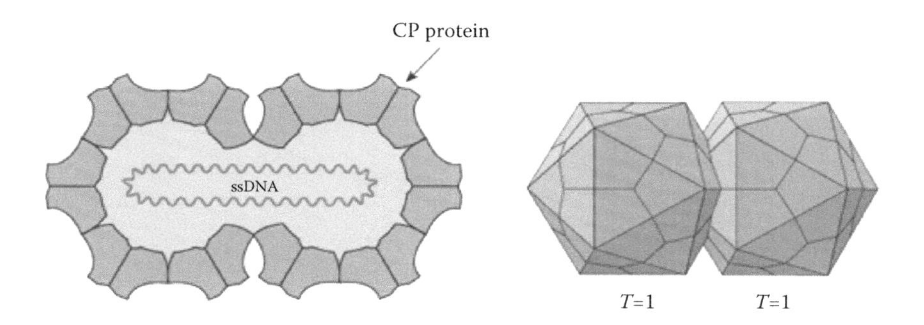

FIGURE 1.23 Representation of geminate particles of geminiviruses. (Adapted with permission from Hulo C, de Castro E, Masson P, Bougueleret L, Bairoch A, Xenarios I, LeMercier P (2011). ViralZone: A knowledge resource to understand virus diversity. *Nucl Acids Res* 39(Database issue): D576–582. http://viralzone.expasy.org.)

most common and are typical of the group. The structure of Maize streak virus (MSV), determined to a resolution of 25 Å, reveals two incomplete *T*=1 icosahedra with 532 point-group symmetry. The geminate MSV particles are of 220 × 380 Å in dimension consisting of two joined, incomplete *T*=1 icosahedral "heads" and a total of 110 CP subunits, organized as 22 pentameric capsomers (Figure 1.23). The CP has an eight-stranded, antiparallel β-barrel motif with an N-terminal α-helix (Zhang et al. 2001).

One molecule of the DNA genome is encapsidated within each geminate particle. The nuclear localization signal (NLS) sequence present in the N-terminal region of CP has been shown to be involved in the transport of viral DNA into the nucleus (Kunik et al. 1998; Palanichelvam et al. 1998). Deletions of the nuclear targeting sequences present at the N-terminus of the African cassava mosaic virus (ACMV) CP prevented twinned particle assembly (Unseld et al. 2004). Tomato leaf curl Bangalore virus and Cotton leaf curl Kokhran virus-Dabawali recombinant CPs were shown to bind to ssDNA. A histidine residue (H85) present within the conserved C2H2 type zinc finger motif was shown to be most crucial for binding of DNA (Kirthi and Savithri 2003; Priyadarshini and Savithri 2009; Godara et al. 2015, 2017). It has not been possible to crystallize these viruses and therefore high resolution structures of none of the members of the *Geminviridae* family have been reported thus far.

1.3 CONCLUSIONS

Viruses encapsidate their nucleic acid genomes in protein shells. Plant viruses are constructed from one or a few types of identical subunits, giving rise to highly symmetrical capsid morphology. A small group of plant viruses possess a lipid-containing envelope. Tubular capsids have helical symmetry, and their length depends on the length of the enclosed nucleic acid genome. Helical viruses may be either rigid rods or filamentous, and are most common among plant viruses. Helical viruses have CPs characterized by all α helical topology. Bacilliform particles are bullet-shaped and prevalent in rhabdoviruses and caulimoviruses. The capsids that are closed have quasi-spherical shells with icosahedral symmetry. The smallest capsids with icosahedral symmetry are constructed from 60 identical subunits. Larger capsids with icosahedral symmetry are made from 60T subunits, where T is the triangulation number. The three-dimensional structures of many of the icosahedral plant viruses are now available, and these structures have provided insights into their mode of assembly, movement, and vector transmission. The coat protein of isometric viruses display a common three-dimensional fold called a jellyroll β-barrel, which provides a rigid, brick-like structure, and such a shape is most suitable for the formation capsids. Geminate particle morphology is observed in geminiviruses where the particles appear twinned due to the fusion of two *T*=1 capsids. Most of the viral CPs form VLPs when expressed in heterologous systems, and can encapsidate nonspecific RNA. However, in vivo, the assembly is highly specific and only the cognate genomic

RNA is encapsidated. The origin of this specificity remains to be understood. High-resolution structures of the CPs of flexuous viruses, DNA viruses, and other complex viruses are yet to be determined. When such structures are available, they will provide in-depth insights into the assembly of DNA viruses.

REFERENCES

Abad-Zapatero C, Abdel-Meguid SS, Johnson JE, Leslie AGW, Rayment I, Rossmann MG, Suck D, Tsukihara T (1980). Structure of southern bean mosaic virus at 2.8 Å resolution. *Nature* 286: 33–39.

Acheson NH (2011). *Fundamentals of Molecular Virology*. Wiley, New York.

Adams MJ, Antoniw JF, Kreuze J (2009). Virgaviridae: A new family of rod-shaped plant viruses. *Arch Virol* 154(12): 1967–1972.

Agirrezabala X, Méndez-López E, Lasso G, Sánchez-Pina MA, Aranda M, Valle M (2015). The near-atomic cryoEM structure of a flexible filamentous plant virus shows homology of its coat protein with nucleoproteins of animal viruses. *eLife* 4: e11795.

Anindya R, Savithri HS (2003). Surface-exposed amino- and carboxy-terminal residues are crucial for the initiation of assembly in Pepper vein banding virus: A flexuous rod-shaped virus. *Virology* 316(2): 325–336.

Ban N, McPherson A (1995). The structure of satellite panicum mosaic virus at 1.9 Å resolution. *Nat Struct Biol* 2(10): 882–890.

Benson SD, Bamford JKH, Bamford DH and Burnett RM (2001). The X-ray crystal structure of P3, the major coat protein of the lipid-containing bacteriophage PRD1, at 1.65 Å resolution. *Acta Crystallog D Biol Crystallog* 58: 39–59.

Berman HM, Westbrook J, Feng Z, Gilliland G, Bhat TN, Weissig H, Shindyalov IN, Bourne PE (2000). *The Protein Data Bank Nucleic Acids Research* 28: 235–242. (www.rcsb.org)

Bernal JD, Fankuchen I (1941). X-ray and crystallographic studies of plant virus preparations. *J Gen Physiol* 25: 147–165.

Bhuvaneshwari M, Subramanya HS, Gopinath K, Savithri HS, Nayudu MV, Murthy MRN (1995). Structure of sesbania mosaic virus at 3 Å resolution. *Structure* 3(10): 1021–1030.

Bisby FA, Roskov YR, Orrell TM, Nicolson D, Paglinawan LE, Bailly N, Kirk PM, Bourgoin T, van Hertum J, eds. (2008). *Species 2000 & ITIS Catalogue of Life: Annual Checklist*. Reading, UK.

Borah BK, Sharma S, Kant R, Johnson AMA, Saigopal VR, Dasgupta I (2013). Bacilliform DNA-containing plant viruses in the tropics: Commonalities within a genetically diverse group. *Mol Plant Path* 14(8): 759–771.

Brunger AT, Adams PD, Clore GM, Delano WL, Gros P, Grosse-Kunstleve RW, Jiang JS, Kuszewski J, Nilges M, Pannu NS, Read RJ, Rice LM, Simonson T, Warren GL (1998). Crystallography & NMR System (CNS): A new software system for macromolecular structure determination. *Acta Crystallog D Biol Crystallog* 54: 905–921.

Canady MA, Larson SB, Day J, McPherson A (1983). Crystal structure of turnip yellow mosaic virus. *Nat Struct Biol* 3(9): 771–781.

Carrillo-Tripp M, Shepherd CM, Borelli IA, Venkataraman S, Lander G, Natarajan P, Johnson JE, Brooks III CL, Reddy VS (2009). VIPERdb2: An enhanced and web API enabled relational database for structural virology. *Nucl Acid Res* 37: D436–D442.

Caspar DL, Klug A (1962). Physical principles in the construction of regular viruses. *Cold Spring Harbor Symp Quant Biol* 27: 1–24.

Caspar DL, Namba K (1990). Switching in the self-assembly of tobacco mosaic virus. *Adv Biophys* 26: 157–185.

Chapman SN (2013). Plant viruses with rod shaped virions. In: *eLS*. Wiley, Chichester. (http://www.els.net.)

Chen Z, Stauffacher C, Schmidt Y, Li T, Bomu W, Kamer G, Shanks M, Lomonossoff G, Johnson JE (1989). Protein-RNA interactions in an icosahedral virus at 3.0 Å resolution. *Science* 245: 154–159.

Crick FH, Watson JD (1956). Structure of small viruses. *Nature* 177(4506): 473–475.

Cyrklaff M, Kühlbrandt W (1994). High-resolution electron microscopy of biological specimens in cubic ice. *Ultramicroscopy* 55: 141–153.

Dadonaite B, Vijayakrishnan S, Fodor E, Bhella D, Hutchinson EC (2016). Filamentous influenza viruses. *J Gen Virol* 97(8): 1755–1764.

DiMaio F, Chen C, Yu X, Frenz B, Hsu Y, Lin N, Egelman EH (2015). The molecular basis for flexibility in the flexible filamentous plant viruses. *Nat Struct Mol Biol* 22: 642–644.

Dolja VV, Boyko VP, Agranovsky AA, Koonin EV (1991). Phylogeny of capsid proteins of rod-shaped and filamentous RNA plant viruses: Two families with distinct patterns of sequence and probably structure conservation. *Virology* 184(1): 79–86.

Fauquet CM, Mayo MA, Maniloff J, Desselberger U, Ball LA (2005). *Virus taxonomy: Eighth report of the International Committee on Taxonomy of Viruses*. Elsevier/Academic Press, London, UK.

Franklin RE, Holmes KC (1956). The helical arrangement of the protein subunits in tobacco mosaic virus. *Biochim Biophys Acta* 21(2): 405–406.

Fujisawa H, Hayashi M (1977). Two infectious forms of bacteriophage phi X 174. *J Virol* 23(2): 439–442.

Gabrenaite-Verkhovskaya R, Andreev IA, Kalinina O, Torrance L, Taliansky ME, Mäkinen K (2008). Cylindrical inclusion protein of potato virus A is associated with a subpopulation of particles isolated from infected plants. *J Gen Virol* 89: 829–838.

Ge P, Zhou ZH (2011). Hydrogen-bonding networks and RNA bases revealed by cryo electron microscopy suggest a triggering mechanism for calcium switches. *Proc Natl Acad Sci U S A* 108: 9637–9642.

Godara S, Saini N, Paul Khurana SM, Biswas KK (2015). Lack of resistance in cotton against cotton leaf curl begomovirus disease complex and occurrence of natural virus sequence variants. *Indian Phytopathol* 68 (3): 326–333.

Godara S, Paul Khurana SM, Biswas KK (2017). Three variants of cotton leaf curl begomoviruses with their satellite molecules are associated with cotton leaf curl disease aggravation in New Delhi. *J Plant Biochem Biotech* 26(1): 97–105.

Goldsmith CS, Miller SE (2009). Modern uses of electron microscopy for detection of viruses. *Clin Microbiol Rev* 22(4): 552–563.

Golmohammadi R, Valegard K, Fridborg K, Liljas L (1993). The refined structure of bacteriophage MS2 at 2.8 Å resolution. *J Mol Biol* 234(3): 620–639.

Grimes JM, Burroughs JN, Gouet P, Diprose JM, Malby R, Zientara S, Mertens PP, Stuart DI (1998). The atomic structure of the bluetongue virus core. *Nature* 395(6701): 470–478.

Gulati A, Alapati K, Murthy A, Savithri HS, Murthy MR (2015). Structural studies on Tobacco streak virus coat protein: Insights into the pleomorphic nature of ilarviruses. *J Struc Biol* 193(2): 95–105.

Gulati A, Murthy A, Abraham A, Mohan K, Natraj U, Savithri HS, Murthy MR (2016). Structural studies on chimeric Sesbania mosaic virus coat protein: Revisiting SeMV assembly. *Virology* 489: 34–43.

Harrison SC, Olson AJ, Schutt CE, Winkler FK, Brigone G (1978). Tomato bushy stunt virus at 2.9 Å resolution. *Nature* 276: 368–372.

Hogle JM, Maeda A, Harrison SC (1986). Structure and assembly of turnip crinkle virus. I. X-ray crystallographic structure analysis at 3.2Å resolution. *J Mol Biol* 191: 625–638.

Hoh F, Uzest M, Drucker M, Plisson-Chastang C, Bron P, Blanc S, Dumas C (2010). Structural insights into the molecular mechanisms of cauliflower mosaic virus transmission by its insect vector. *J Virol* 84(9): 4706–4713.

Horne RW, Brenner S, Waterson AP, Wildy P (1959). The icosahedral form of an adenovirus. *J Mol Biol* 1: 84–86.

Hull R (2001). *Matthews' Plant Virology*. 4th ed. Academic Press, San Diego, CA.

Hulo C, de Castro E, Masson P, Bougueleret L, Bairoch A, Xenarios I, LeMercier P (2011). ViralZone: A knowledge resource to understand virus diversity. *Nucl Acids Res* 39(Database issue): D576–D582.

Huxley HE, Zubay G (1960a). Structure of the protein shell of turnip yellow mosaic virus. *J Mol Biol* 2: 10.

Jackson AO, Dietzgen RG, Goodin MM, Bragg JN, Deng M (2005). Biology of plant rhabdoviruses. *Annu Rev Phytopathol* 43: 623–660.

Joklik WK (1966). The poxviruses. *Bacteriol Rev* 30: 33–66.

Jones TA, Liljas L (1984). Structure of satellite tobacco necrosis virus after crystallographic refinement at 2.5 Å resolution. *J Mol Biol* 177(4): 735–767.

Ke J, Schmidt T, Chase E, Bozarth RF, Smith TJ (2004). Structure of Cowpea mottle virus: A consensus in the genus Carmovirus. *Virology* 321(2): 349–358.

Kendall A, McDonald M, Bian W, Bowles T, Baumgarten SC, Shi J, Stewart PL, Bullitt E, Gore D, Irving TC, Havens WM, Ghabrial SA, Wall JS, Stubbs G (2008). Structure of flexible filamentous plant viruses. *J Virol* 82(19): 9546–9554.

Kendall A, William D, Bian W, Stewart PL, Stubbs G (2013). Barley stripe mosaic virus: Structure and relationship to the tobamoviruses. *Virology* 443(2): 265–270.

Kirthi N, Savithri HS (2003). A conserved zinc finger motif in the coat protein of Tomato leaf curl Bangalore virus is responsible for binding to ssDNA. *Arch Virol* 148(12): 2369–2380.

Klug A (1999). The tobacco mosaic virus particle: Structure and assembly. *Philos Trans R Soc Lond B Biol Sci* 354(1383): 531–535.

Koonin EV, Dolja VV, Krupovic M (2015). Origins and evolution of viruses of eukaryotes: The ultimate modularity. *Virology* 479–480, 2–25.

Krishna SS, Hiremath CN, Munshi SK, Prahadeeswaran D, Sastri M, Savithri HS, Murthy MR (1999). Three-dimensional structure of physalis mottle virus: Implications for the viral assembly. *J Mol Biol* 289(4): 919–934.

Krol MA, Olson NH, Tate J, Johnson JE, Baker TS, Ahlquist P (1999). RNA-controlled polymorphism in the in vivo assembly of 180-subunit and 120-subunit virions from a single capsid protein. *Proc Natl Acad Sci U S A* 96: 13650–13655.

Kumar A, Reddy VS, Yusibov V, Chipman PR, Hata Y, Fita I, Fukuyama K, Rossmann MG, Loesch-Fries LS, Baker TS, Johnson JE (1997). The structure of alfalfa mosaic virus capsid protein assembled as T=1 icosahedral particle at 4.0-Å resolution. *J Virol* 71: 7911–7916.

Kunik T, Palanichelvam K, Czosnek H, Citovsky V, Gafni Y (1998). Nuclear import of the capsid protein of tomato yellow leaf curl virus (TYLCV) in plant and insect cells. *Plant J* 13(3): 393–399.

Lai-Kee-Him J, Schellenberger P, Dumas C, Richard E, Trapani S, Komar V, Demangeat G, Ritzenthaler C, Bron P (2013). The backbone model of the Arabis mosaicvirus reveals new insights into functional domains of Nepovirus capsid. *J Struct Biol* 182(1): 1–9.

Larson SB, Day J, Greenwood A, McPherson A (1998). Refined structure of satellite tobacco mosaic virus at 1.8 Å resolution. *J Mol Biol* 277(1): 37–59.

Larson SB, Day J, Canady MA, Greenwood A, McPherson A (2000). Refined structure of desmodium yellow mottle tymovirus at 2.7 Å resolution. *J Mol Biol* 301(3): 625–642.

Lin T, Chen Z, Usha R, Stauffacher CV, Dai JB, Schmidt T, Johnson JE (1999). The refined crystal structure of cowpea mosaic virus at 2.8 A resolution. *Virology* 265(1): 20–34.

Lin T, Clark AJ, Chen Z, Shanks M, Dai J, Li Y, Schmidt T, Oxelfelt P, Lomonossoff GP, Johnson JE (2000). Structural Fingerprinting: Subgrouping of comoviruses by structural studies of red clover mottle virus to 2.4-Å resolution and comparisons with other comoviruses. *J Virol* 74: 493–504.

Lucas RW, Larson SB, McPherson A (2002a). The crystallographic structure of brome mosaic virus. *J Mol Biol* 317(1): 95–108.

Lucas RW, Larson SB, Canady MA and McPherson A (2002b). The structure of tomato aspermy virus by X-ray crystallography. *J Struct Biol* 139(2): 90–102.

Mahy BWJ, Van Regenmortel MHV (2010). *Desk Encyclopedia of General Virology*. Academic Press, San Diego, CA.

Mathur C, Mohan K, Usha Rani TR, Reddy KM, Savithri HS (2014). The N-terminal region containing the zinc finger domain of tobacco streak virus coat protein is essential for the formation of virus-like particles. *Arch Virol* 159(3): 413–423.

Morgunova EYu, Dauter Z, Fry E, Stuart DI, Stel'mashchuk VYa, Mikhailov AM, Wilson KS, Vainshtein BK (1994). The atomic structure of Carnation Mottle Virus capsid protein. *FEBS Lett* 338(3): 267–271.

Moyer JW, German T, Sherwood JL and Ullman D (1999). An update on tomato spotted wilt virus and related tosposviruses. *APSnet Features*. Online. doi: 10.1094/APSnetFeatures-1999-0499.

Nakagawa A, Miyazaki N, Taka J, Naitow H, Ogawa A, Fujimoto Z, Mizuno H, Higashi T, Watanabe Y, Omura T, Cheng RH, Tsukihara T (2003). The atomic structure of rice dwarf virus reveals the self-assembly mechanism of component proteins. *Structure* 11(10): 1227–1238.

Namba K, Stubbs G (1986). Structure of Tobacco mosaic virus at 3.6 A resolution: Implications for assembly. *Science* 231: 1401–1406.

Namba K, Pattanayek R, Stubbs G (1989). Visualization of protein-nucleic acid interactions in a virus. Refined structure of intact tobacco mosaic virus at 2.9 A resolution by X-ray fiber diffraction. *J Mol Biol* 208(2): 307–325.

Oda Y, Saeki K, Takahashi Y, Maeda T, Naitow H, Tsukihara T, Fukuyama K (2000). Crystal structure of tobacco necrosis virus at 2.25 Å resolution. *J Mol Biol* 300(1): 153–169.

Olson AJ, Bricogne G, Harrison SC (1983). Structure of tomato busy stunt virus IV. The virus particle at 2.9 Å resolution. *J Mol Biol* 171(1): 61–93.

Pache L, Venkataraman S, Reddy VS, Nemerow GR (2008a). Structural variations in species B adenovirus fibers impact CD46 Association. *J Virol* 82(16): 7923–7931.

Pache L, Venkataraman S, Nemerow GR, Reddy VS (2008b). Conservation of fiber structure and CD46 usage by subgroup B2 adenoviruses. *Virology* 375: 573–579.

Palanichelvam K, Kunik T, Citovsky V, Gafni Y (1998). The capsid protein of tomato yellow leaf curl virus binds cooperatively to single-stranded DNA. *J Gen Virol* 79(Pt 11): 2829–2833.

Pappachan A, Chinnathambi S, Satheshkumar PS, Savithri HS, Murthy MRN (2009). A single point mutation disrupts the capsid assembly in Sesbania Mosaic Virus resulting in a stable isolated dimer. *Virology* 392 (2): 215–221.

Parker L, Kendall A, Stubbs G (2002). Surface features of potato virus X from fiber diffraction. *Virology* 300(2): 291–295.

Parker L, Kendall A, Berger PH, Shiel PJ, Stubbs G (2005). Wheat streak mosaic virus—Structural parameters for a Potyvirus. *Virology* 340(1):64–69.

Pettersen EF, Goddard TD, Huang CC, Couch GS, Greenblatt DM, Meng EC, Ferrin TE (2004). UCSF Chimera—A visualization system for exploratory research and analysis. *J Comput Chem* 25(13): 1605–1612.

Priyadarshini P, Savithri HS (2009). Kinetics of interaction of Cotton Leaf Curl Kokhran Virus-Dabawali (CLCuKV-Dab) coat protein and its mutants with ssDNA. *Virology* 386(2): 427–437.

Qu C, Liljas L, Opalka N, Brugidou C, Yeager M, Beachy RN, Fauquet CM, Johnson JE, Lin T (2000). 3D domain swapping modulates the stability of members of an icosahedral virus group. *Structure Fold Des* 8(10): 1095–1103.

Reddy VS, Natchiar SK, Stewart PL, Nemerow GR (2010). Crystal structure of human adenovirus at 3.5 A resolution. *Science* 329(5995): 1071–1075.

Reinisch KM, Nibert ML, Harrison SC (2000). Structure of the reovirus core at 3.6 Å resolution. *Nature* 404 (6781): 960–967.

Ribeiro D, Foresti O, Denecke J, Wellink J, Goldbach R, Kormelink RJ (2008). Tomato spotted wilt virus glycoproteins induce the formation of endoplasmic reticulum- and Golgi-derived pleomorphic membrane structures in plant cells. *J Gen Virol* 89(Pt 8):1811–1818.

Richardson JF, Tollin P, Bancroft JB (1981). The architecture of the potexviruses. *Virology* 112(1): 34–39.

Ruska, E (1986). "*Ernst Ruska autobiography*." Nobel Foundation. Retrieved 2010-01-31.

Sachse C, Chen JZ, Coureux PD, Stroupe ME, Fändrich M, Grigorieff N (2007). High-resolution electron microscopy of helical specimens: Aa fresh look at tobacco mosaic virus. *J Mol Biol* 371: 812–835.

Sangita V, Lokesh GL, Satheshkumar PS, Vijay CS, Saravanan V, Savithri HS, Murthy MR (2004). T=1 capsid structures of Sesbania mosaic virus coat protein mutants: determinants of T=3 and T=1 capsid assembly. *J Mol Biol* 342: 987–999.

Sangita V, Lokesh GL, Satheshkumar PS, Saravanan V, Vijay CS, Savithri HS, Murthy MR (2005a). Structural studies on recombinant T=3 capsids of Sesbania mosaic virus coat protein mutants. *Acta Crystallog D Biol Crystallog* 61: 1402–1405.

Sangita V, Satheshkumar PS, Savithri HS, Murthy MR (2005b). Structure of a mutant T=1 capsid of Sesbania mosaic virus: Role of water molecules in capsid architecture and integrity. *Acta Crystallog D Biol Crystallog* 61: 1406–1412.

Sastri M, Kekuda R, Gopinath K, Kumar CT, Jagath JR, Savithri HS (1997). Assembly of physalis mottle virus capsid protein in Escherichia coli and the role of amino and carboxy termini in the formation of the icosahedral particles. *J Mol Biol* 272(4): 541–552.

Satheshkumar PS, Lokesh GL, Sangita V, Saravanan V, Vijay CS, Murthy MR, Savithri HS (2004). Role of metal ion-mediated interactions in the assembly and stability of Sesbania mosaic virus T=3 and T=1 capsids. *J Mol Biol* 342: 1001–1014.

Savithri HS, Murthy MRN (2010). Structure and assembly of Sesbania mosaic virus. *Curr Sci* 98: 346–352.

Schellenberger P, Sauter C, Lorber B, Bron P, Trapani S, Bergdoll M, Marmonier A, Schmitt-Keichinger C, Lemaire O, Demangeat G, Ritzenthaler C (2011). Structural Insights into viral determinants of nematode mediated grapevine fanleaf virus transmission. *PLoS Pathogens* 7(5): e1002034.

Shanmugam G, Polavarapu PL, Kendall A, Stubbs G (2005). Structures of plant viruses from vibrational circular dichroism. *J Gen Virol* 86(8): 2371–2377.

Silva AM, Rossmann MG (1987). Refined structure of southern bean mosaic virus at 2.9 Å resolution. *J Mol Biol* 197(1): 69–87.

Smith TJ, Chase E, Schmidt T, Perry KL (2000). The structure of cucumber mosaic virus and comparison to cowpea chlorotic mottle virus. *J Virol* 74(16): 7578–7586.

Solovyev AG, Makarov VV (2016). Helical capsids of plant viruses: Architecture with structural lability. *J Gen Virol* 97(8): 1739–1754.

Speir JA, Munshi S, Wang G, Baker TS, Johnson JE (1995). Structures of the native and swollen forms of cowpea chlorotic mottle virus determined by X-ray crystallography and cryo-electron microscopy. *Structure* 3(1): 63–78.

Stehle T, Gamblin SJ, Yan Y, Harrison SC (1996). The structure of simian virus 40 refined at 3.1 Å resolution. *Structure* 4(2): 165–182.

Tamm T, Truve E (2000). Sobemoviruses. *J Virol* 74(14): 6231–6241.
Tars K, Zeltins A, Liljas L (2003). The three-dimensional structure of cocksfoot mottle virus at 2.7 Å resolution. *Virology* 310(2): 287–297.
Torrance L, Andreev IA, Gabrenaite-Verhovskaya R, Cowan G, Mäkinen K, Taliansky ME (2006). An unusual structure at one end of potato potyvirus particles. *J Mol Biol* 357: 1–8.
Umashankar M, Murthy MRN, Savithri HS (2006). Mutation of interfacial residues disrupts subunit folding and particle assembly of Physalis mottle tymovirus. *J Biol Chem* 278: 6145–6152.
Unseld S, Frischmuth T, Jeske H (2004). Short deletions in nuclear targeting sequences of African cassava mosaic virus coat protein prevent geminivirus twinned particle formation. *Virology* 318(1): 90–101.
Veesler D, Ng TS, Sendamarai AK, Eilers BJ, Lawrence CM, Lok SM, Young MJ, Johnson JE, Fu CY (2013). Atomic structure of the 75 MDa extremophile Sulfolobusturreted icosahedral virus determined by CryoEM and X-ray crystallography. *Proc Natl Acad Sci U S A* 110(14): 5504–5509.
Venkataraman S, Reddy SP, Loo J, Idamakanti N, Hallenbeck PL, Reddy VS (2008). Structure of Seneca Valley Virus-001: An oncolytic picornavirus representing a new genus. *Structure* 16(10): 1555–1561.
Wada Y, Tanaka H, Yamashita E, Kubo C, Ichiki-Uehara T, Nakazono-Nagaoka E, Omura T, Tsukihara T (2008). The structure of melon necrotic spot virus determined at 2.8 Å resolution. *Acta Cryst Sec F* 64(Pt 1): 8–13.
Wang H, Stubbs G (1994). The structure of cucumber green mottle mosaic virus at 3.4 Å resolution by X-ray fiber diffraction: Significance for the evolution of tobamoviruses. *J Mol Biol* 239: 371–384.
Wang H, Culver JN, Stubbs G (1997). Structure of ribgrass mosaic virus at 2.9 Å resolution. Evolution and taxonomy of tobamoviruses. *J Mol Biol* 269: 769–779.
Wikoff WR, Liljas L, Duda RL, Tsuruta H, Hendrix RW, Johnson JE (2000). Topologically linked protein rings in the bacteriophage HK97 capsid. *Science* 289(5487): 2129–2133.
Yang S, Wang T, Bohon J, Gagné MÈ, Bolduc M, Leclerc D, Li H (2012). Crystal structure of the coat protein of the flexible filamentous papaya mosaic virus. *J Mol Biol* 422(2): 263–273.
Yasaki M, Hirano Y, Uga H, Hanada K, Uehara-Ichiki T, Toda T, Furuya H, Fuji S (2017). Characterization of a new carmovirus isolated from an adonis plant. *Arch Virol* 162(2): 501–504.
Zhang W, Olson NH, Baker TS, Faulkner L, Agbandje-McKenna M, Boulton MI, Davies JW, McKenna R (2001). Structure of the Maize streak virus geminate particle. *Virology* 279: 471–477.
Zlotnick A, Aldrich R, Johnson JM, Ceres P, Young MJ (2000). Mechanism of capsid assembly for an icosahedral plant virus. *Virology* 277: 450–456.

2 Biotypes of *Cowpea Aphid-Borne Mosaic Virus* in Brazil

*José Albérsio de Araújo Lima,
Aline Kelly Queiroz do Nascimento, Laianny Morais Maia,
and Francisco de Assis Câmara Rabelo Filho*

CONTENTS

2.1 INTRODUCTION

The northeast of Brazil has a great potential for production of cowpea [*Vigna unguiculata* (L.) Walp. subsp. *unguiculata*] and several tropical fruit crops, including passion fruit (*Passiflora edulis* Sims) (Lima et al., 2005, 2015). Cowpea is a leguminous food with great importance for small growers in northeastern Brazil, and the State of Ceará is one of the highest producers in Brazil (Freire Filho et al., 2005). Many factors can affect the productivity of cowpea, especially the virus diseases, which have been responsible for great economic losses, considering that the crop has economic and social relevance for the region. More than 20 virus species can naturally infect cowpea around the world (Hampton et al., 1997; Lima et al., 2005, 2015), but *Cowpea aphid-borne mosaic virus* (CABMV), family *Potyviridae*, genus *Potyvirus*, which was first isolated in the State of Ceará in 1981, is considered the most important virus species that infect cowpea in northeastern Brazil (Lima et al., 2005, 2015).

Passion fruit is a fruit crop cultured in Brazil that is also affected by virus diseases that adversely affect its production and expansion (Nicolini et al., 2012). Several virus species have already been identified, causing disease in passion fruit in Brazil, and species from the genus *Potyvirus* have been considered responsible for the most important problems (Bezerra et al., 1995; Lima et al., 2015). A virus isolated from naturally infected passion fruit orchards in the State of Ceará was identified as *Passion fruit woodiness virus* (PWV), from genus *Potyvirus* based on its biological and serological properties (Bezerra et al., 1995; Lima et al., 2015).

Other virus isolates from the genus *Potyvirus* obtained from passion fruit with fruit woodiness symptoms in Brazil have been identified as CABMV based on molecular analyses (Nascimento et al., 2004, 2006; Barros et al., 2007; Nascimento, 2014). Some other virus species from the genus *Potyvirus* have been also identified causing infections in passion fruit, mainly *Passion fruit mottle virus* (PaMV) in Taiwan (Chang, 1992), *Soybean mosaic virus* (SMV) in Colombia (Benscher et al., 1996), *Passiflora virus Y* (PaVY) in Australia and Indonesia (Parry et al., 2004), and *East Asian passiflora virus* (EAPV) in Japan (Iwai et al., 2006).

Based on biological, serological, and mainly molecular results, it was proposed that the CABMV isolates which occur in northeast Brazil should be reclassified into biotype CABMV-C (C for cowpea) to include isolates obtained from cowpea that do not infect passion fruit, and biotype CABMV-P (P for passion fruit) to include the virus isolates obtained from passion fruit in Brazil that do not infect cowpea (Nascimento, 2014; Lima et al., 2015).

Two isolates of CABMV were originally obtained from naturally infected cowpea in Fortaleza (CABMV-CFor), causing mosaic, and in Bela Vista farm (CABMV-CBv) causing severe mosaic, both in the State of Ceará, and two isolates of CABMV-P, one obtained from naturally infected passion fruit orchards with severe symptoms in the State of Ceará (CABMV-PSevere) and an isolate which causes mild symptoms (CABMV-PMild) in passion fruit, were obtained in the State of Pernambuco (Bezerra et al., 1995; Lima et al., 2015). The virus isolates obtained from cowpea and from passion fruit have been maintained in *in vivo* conditions at the Plant Virus Laboratory at the Federal University of Ceará, by periodic mechanical inoculations into healthy young plants of their respective natural hosts: cowpea (CABMV-CFor and CABMV-CBv) and passion fruit (CABMV-PMild and CABMV-PSevere). CABMV-PMild and CABMV-PSevere infect and cause mosaic symptoms in passion fruit approximately 15 days after mechanical inoculation, but the symptoms induced by CABMV-PSevere are more severe than those caused by CABMV-PMild (Figure 2.1a and b). On the other hand, cowpea plants systemically infected with CABMV-CBv or with CABMV-CF present similar symptoms of severe mosaic, with leaf deformations (Figure 2.1c and d).

2.2 BIOLOGICAL PROPERTIES

Isolates of CABMV-C obtained from cowpea (Figure 2.1) present host range similarities, and there are also similarities in host range between CABMV isolates obtained from passion fruit (CABMV-P). Nevertheless, several differences can be observed among those isolates obtained from cowpea (CABMV-C), when compared with the isolates obtained from passion fruit (CABMV-P). Virus symptom reactions in different plant species demonstrate biological differences among the virus isolates obtained from cowpea and those obtained from passion fruit (Table 2.1). In all cases, the presence of the viruses is confirmed by plate-trapped antigen enzyme-linked immune absorbent assay (PTA-ELISA). All isolates of CABMV-C and CABMV-P cause local lesions in *Chenopodium amaranticolor* Coste & A. Reyn and in *C. quinoa* Willd, but the virus isolates do not infect *Gomphrena globosa* L. Plant species from the Fabaceae family, including *Canavalia ensiformis* (L.) DC, *Macroptilium lathyroides* (L.) Urb. and *V. unguiculata* are systemically infected by isolates of CABMV-C (CABMV-CBv and CABMV-CFort), causing severe symptoms, but isolates obtained from passion fruit (CABMV-PSevere and CABMV-PMild) do not infect cowpea.

Macroptilium lathyroides is also infected by all CABMV isolates, including those obtained from cowpea (CABMV-CBv and CABMV-CFor) and those obtained from passion fruit (CABMV-PSevere and CABMV-PMild), but *C. ensiformis* is not infected by CABMV-PMild (Table 2.1). All CABMV-C and CABMV-P isolates also systemically infect *Sesamum indicum* L., causing local clorotic lesions followed by mosaic (Table 2.1), indicating that these plant species could be a natural host for those virus isolates in the field. One CABMV-C isolate was obtained from *S. indicum* with symptoms of mosaic and leaf distortions in experimental cowpea fields in the State of Ceará, and the CABMV-C isolate infected several cowpea cultivars (Lima et al., 1991). Sreenivasulu et al. (1994) also isolated a virus from *S. indicum*, identified as a strain of CABMV-C by Pappu et al. (1997) in Georgia that also infected cowpea.

Biological differences can be observed among CABMV-P isolates (CABMV-PMild and CABMV-PSevere) and CABMV-C isolates (CABMV-CBv and CABMV-CFor). According to host range studies, *M. lathyroides* and *O. basilicum* 'Toscano' are common systemic hosts for all CABMV isolates, and *C. ensiformis* is infected by CABMV-PSevere, CABMV-CBv, and CABMV-CFor, but not by CABMV-PMild (Table 2.1). It has also been demonstrated that all CABMV-C and CABMV-P isolates systemically infect *S. indicum*, indicating that it could be a natural host for those virus isolates in the field.

FIGURE 2.1 Passion fruit (*Passiflora edulis*) and cowpea (*Vigna unguiculata*) systemically infected with isolates of *Cowpea aphid-borne mosaic virus* (CABMV). (a) Passion fruit infected with CABMV-PMild presenting mild symptoms. (b) Passion fruit infected with CABMV-PSevere showing severe symptoms. (c) Cowpea infected with CABMV-CFor. (d) Cowpea infected with CABMV-CBv.

Host range studies have also demonstrated that isolates of CABMV-P do not infect cowpea varieties, nor do CABMV-C isolates infect the following plant species from the family Passifloraceae: *Passiflora edulis* Sims cv. Macae*, P. setacea* D.C.*, P. cincinnata* Mast., and *P. gibertii* N.E. Brown (Table 2.1). According to Bock (1973), there are several distinct CABMV strains that can be differentiated by their biological properties, and according to Rodrigues et al. (2015), isolates of CABMV-P obtained from infected yellow passion fruit in four Brazilian states infect *M. atropurpureum* but do not infect cowpea.

This host range information confirms biological differences among isolates of CABMV-C and CABMV-P, showing that they can be differentiated according to their biological properties and that CABMV-CBv and CABMV-CFor cause severe mosaic in several cowpea genotypes, especially in cv. Pitiuba, an important cultivar produced in northeastern Brazil (Table 2.1). This agrees with the results of studies obtained by Nascimento (2014). Earlier virus infections, especially with CABMV in cowpea, cause the greatest losses in crop production (Taiwo and Akinjoguna, 2006).

Although Nascimento et al. (2006) demonstrated that CABMV isolates obtained from passion fruit caused systemic infections in cowpea, neither CABMV-PSevere nor CABMV-PMild infect cowpea (Table 2.1). On the other hand, Nascimento (2014) and Rodrigues et al. (2015) demonstrate that isolates of CABMV-P obtained from passion fruit do not infect cowpea. Bock (1973) reported

TABLE 2.1

Symptoms and Serological Results of Different Plant Species and Cultivars Mechanically Inoculated with Virus Isolates of *Cowpea aphid-borne mosaic virus* (CABMV) Obtained from *Passiflora edulis* (CABMV-PMild and CABMV-PSevere) and from *Vigna unguiculata* (CABMV-CBv and CABMV-CFor)

	CABMV-PSevere	CABMV-PMild	CABMV-CBv	CABMV-CFor
Family/Plant Species/Cultivar	**Symp. (Elisa)**	**Symp. (Elisa)**	**Symp. (Elisa)**	**Symp. (Elisa)**
Amaranthaceae				
Chenopodium amaranticolor	LL; NL (+)	LL; NL (+)	LL; NL (+)	LL; NL (+)
C. quinoa		LL; NL (+)	LL; NL (+)	LL; NL (+)
Gomphrena globosa	w/s (–)	w/s (–)	w/s (–)	w/s (–)
Fabacea				
Canavalia ensiformis	M (+)	w/s (–)	M (+)	M (+)
Macroptilium lathyroides	SM; Bl (+)	SM; Bl (+)	SM; Bl (+)	SM; Bl (+)
Pisum sativum	mM (+)	mM (+)	mM (+)	w/s (–)
Vigna mungo	w/s (–)	w/s (–)	w/s (–)	w/s (–)
Vigna unguiculata				
'Macaibo'	w/s (–)	w/s (–)	SM; Bl (+)	SM; Bl (+)
'Adzuki'	w/s (–)	w/s (–)	w/s (–)	w/s (–)
'Pitiuba'	w/s (–)	w/s (–)	SM; Bl, Lfd (+)	SM; Bl, Lfd (+)
'CE 566'	w/s (–)	w/s (–)	M (+)	M (+)
'CE 113'	w/s (–)	w/s (–)	SM (+)	SM (+)
'CE 189'	w/s (–)	w/s (–)	SM (+)	M (+)
'CE 524'	w/s (–)	w/s (–)	SM (+)	SM (+)
'Paulista-PB'	w/s (–)	w/s (–)	SM (+)	SM (+)
'Lizão'	w/s (–)	w/s (–)	SM (+)	SM (+)
'Costela-de-vaca'	w/s (–)	w/s (–)	SM (+)	SM (+)
'Manteigão'	w/s (–)	w/s (–)	SM (+)	M (+)
'Setentão'	w/s (–)	w/s (–)	SM (+)	SM (+)
'Pingo-de-ouro'	w/s (–)	w/s (–)	SM (+)	SM (+)
'Paulistinha'	w/s (–)	w/s (–)	SM (+)	SM (+)
'Sempre-verde'	w/s (–)	w/s (–)	SM (+)	SM (+)
'Clay'	w/s (–)	w/s (–)	SM (+)	SM (+)
Passifloraceae				
Passiflora edulis	SM (+)	mM (+)	w/s (–)	w/s (–)
P. edulis 'Macae'	SM (+)	mM (+)	w/s (–)	w/s (–)
P. setacea	SM (+)	mM (+)	w/s (–)	w/s (–)
P. cincinnata	SM (+)	mM (+)	w/s (–)	w/s (–)
P. gibertii	SM (+)	mM (+)	w/s (–)	w/s (–)
Pedaliaceae				
Sesamum indicum	M (+)	M (+)	M (+)	M (+)
Solanaceae				
Nicotiana benthamiana	mM (+)	mM (+)	w/s (–)	w/s (–)

Note: Bl: bubbles; Lfd: leaf deformations; LL: local lesion; mM: mild mosaic; NL: necrotic lesions; SM: severe mosaic; w/s: without symptoms.

the absence of symptoms in cowpea inoculated with virus isolates from the genus *Potyvirus* obtained from passion fruit, and that the CABMV-C isolates obtained from cowpea do not infect passion fruit, which is strong evidence that the causal agent of passion fruit woodiness does not infect cowpea (Nascimento, 2014; Lima et al., 2015).

Species from the family Passifloraceae can be systemically infected by all isolates of CABMV-P, including CABMV-PMild and CABMV-PSevere and isolates obtained from passion fruit from different regions in the State of Ceará (CABMV-PGua, CABMV-PSb, and CABMV-PUba). However, neither of the CABMV-C isolates obtained from cowpea, including CABMV-CBv and CABMV-CFor, infected any of more than 40 inoculated plants from the genus *Passiflora*.

Biological differences among isolates of CABMV-C and CABMV-P demonstrate that they can be differentiated according to their biological properties (host range) and that CABMV-C isolates do not infect passion fruit and CABMV-P isolates do not infect cowpea cultivars. According to Nascimento (2014), more than 50 cowpea genotypes evaluated by mechanical inoculations in greenhouse conditions confirmed those differences. On the other hand, it has been demonstrated that isolates of CABMV-C, including CABMV-CBv and CABMV-CFor, cause severe mosaic in several cowpea genotypes, especially in cv. Pitiuba, an important cultivar cultivated in northeastern Brazil (Nascimento et al., 2006). According to Taiwo and Akinjoguna (2006), earlier virus infection, especially with CABMV-C in cowpea, causes the greatest losses in crop production.

Although Lima and Nelson (1977) demonstrated that *Cowpea severe mosaic virus* (CPSMV) from the genus *Comovirus* was the most prevalent virus on cowpea in the State of Ceará, northeastern Brazil, during the 1970s, currently CABMV-C is responsible for the most common virus disease that occurs in northeast Brazil (Nascimento et al., 2006; Lima et al., 2015). Isolates of CABMV-C obtained from naturally infected cowpea, including CABMV-CBv and CABMV-CFor, infect several species from the botanical family Fabaceae, but did not cause infection in any one of more than 50 individual plants of *Passiflora edulis*, nor in plants of *P. edulis*, 'Macae,' *P. setacea*, *P. cincinnata*, and *P. gibertii* mechanically inoculated in greenhouse conditions (Table 2.1). Bock (1973) stated that strains of CABMV-C are responsible for the most important virus diseases of cowpea in Africa, but neither of the CABMV-C strains infected passion fruit.

On the other hand, CABMV isolates obtained from passion fruit, including CABMV-PMild and CABMV-PSevere, do not infect any of several cowpea genotypes, including Macaibo, Pitiuba, Otília, Quarentão, Quarenta Dias, Roxinho, Sempre Verde, Seridó, and Setentão (Table 2.1). This constitutes strong evidence that the causal agent of passion fruit woodiness disease does not naturally infect cowpea (Bezerra et al., 1995; Lima et al., 2003, 2005, 2015). Nascimento et al. (2006) identified 14 virus isolates from the genus *Potyvirus* obtained from passion fruit as CABMV, based on serological tests with antiserum for CABMV-C, without considering the strong serological relationship between CABMV and PWV and serological differences between CABMV-C and CABMV-P. Nevertheless, the virus isolates identified by Nascimento et al. (2004, 2006) and Zerbini et al. (2005) could be strains of the biotype CABMV-P.

2.3 UNILATERAL CROSS-PROTECTION AMONG BIOTYPES

Cross-protection experiments in passion fruit demonstrate that the mild virus isolate obtained from passion fruit (CABMV-PMild) does not protect passion fruit plants against the severe virus isolate (CABMV-PSevere). Passion fruit plants inoculated with CABMV-PMild and 15 days later inoculated with CABMV-PSevere developed severe symptoms (Figure 2.2a), indicating the presence of CABMV-PSevere. This absence of cross-protection was confirmed by inoculation of double-inoculated plants into *C. ensiformis*, which is not infected by CABMV-PMild (Table 2.1), but the plants develop symptoms and test positively for CABMV-PSevere by PTA-ELISA. These results, demonstrating the absence of cross-protection between these two virus isolates, are probably due to the low level of CABMV-PMild replication in passion fruit when compared with CABMV-PSevere replication level.

FIGURE 2.2 Cross-Protection Studies: (a) *Passiflora edullis* with severe symptoms, after inoculation with a mild biotype of *Cowpea aphid-borne mosaic virus* (CABMV-PMild) and further inoculated with a severe biotype (CABMV-PSevere). (b) *Canavalia ensiformis* inoculated with CABMV-PSevere.

Interaction studies between CABMV-CFor and CABMV-PSevere also provide evidence for the absence of cross-protection in *C. ensiformis*. Plants of *C. ensiformis* inoculated with CABMV-CFor and super-inoculated with CABMV-PSevere, showed the most severe symptoms 15 days later than plants inoculated with only CABMV-CFor. In addition, extracts from double-inoculated *C. ensiformis* plants caused infections in cowpea as well as in passion fruit plants, which exhibited virus symptoms. The presence of CABMV-PSevere in inoculated *C. ensiformis* was also confirmed by PTA-ELISA.

Canavalia ensiformis plants inoculated with CABMV-PSevere, and after 15 days super-inoculated with CABMV-CFor, did not present stronger symptoms when compared with plants inoculated only with CABMV-PSevere. Cowpea plants inoculated with extracts from the double-inoculated *C. ensiformis* plants were not infected, evidencing the absence of CABMV-CFor in the *C. ensiformis* plants double-inoculated with CABMV-PSevere and CABMV-CFor. Those results provide good evidence of cross-protection of CABMV-PSevere against CABMV-CFor in *C. ensiformis*, confirming unilateral cross-protection between those virus isolates, in function of the host plant used. However, passion fruit plants inoculated with extracts from *C. ensiformis* double-inoculated with CABMV-PSevere and CABMV-CFor showed symptoms (Figure 2.2b), indicating the presence of CABMV-PSevere, which was confirmed by PTA-ELISA.

The absence of cross-protection between CABMV-PMild and CABMV-PSevere is an indication that this mild strain (CABMV-PMild) is not recommended for use as a biological control for woodiness disease in passion fruit. The absence of cross-protection is probably due to the low level of CABMV-PMild replication in passion fruit when compared with the CABMV-PSevere replication level. Novaes and Rezende (2003) also observed the lack of cross-protection between mild and severe strains of virus from the genus *Potyvirus* in passion fruit plants and attributed the absence of cross-protection to apparent differences in the distribution of the mild virus strain and the possible low concentration of it in the infected tissues. The low concentration of mild virus strain in infected tissues will leave clusters of uninfected cells, permitting the replication and establishment of the severe strain. Nevertheless, cross-protection has commonly been observed between other virus strains (Novaes and Rezende, 2003; Lima et al., 2015).

In addition to being an alternative for controlling plant viruses in the field, the cross-protection phenomenon constitutes a biological criterion for strain definition and identification inside virus species, especially in those from the genus *Potyvirus*. The unilateral cross-protection observed between CABMV-PSevere and CABMV-CFor represents good evidence of the close relationship between these two virus isolates, which is confirmed by serological and molecular studies (Nascimento, 2014; Figure 2.3). Associated with cross-protection criterion between virus strains, the identity of nucleotide sequences from the coat protein gene (*cp*) and the non-translated 3′ region of the virus genome are biological and molecular parameters of great value for identification of virus species in the family *Potyviridae* (Van Regenmortel et al., 2000).

2.4 POLYCLONAL ANTISERUM AND SEROLOGICAL INTERACTIONS BETWEEN BIOTYPES

The plant species selected for virus propagation to be used for virus purification should not contain high concentrations of tannin components, latex, wax, or phenol compounds which could be caustic and with low solubility in water interfering with the virus infectivity. Those chemical compounds are normally associated with woody plants which, when possible, should be avoided for use in virus purification (Lima et al., 1979, 2015). As passion fruit plants are considered a sub-woody plant which contains some phenol compounds that could interfere with virus purification, *M. lathyroides* is selected for CABMV-P propagation. Although *M. lathyroides* has a small amount of leaf tissue per individual plant, it presents good qualities for virus propagation in an acceptable concentration.

The protocol for virus purification, using *n*-butanol, provides good plant leaf extract clarification without affecting the chemical and biological integrity of the virus isolates. This virus purification protocol is very efficient, permitting the production of purified virus preparations with concentrations varying from 20 to 25 mg of virus per Kg of infected plant tissue. Purified virus preparations present an ultraviolet (UV) absorption spectrum, with a maximum at 259 nm and a minimum at 250 nm. The absorption ratios between 260 and 280 nm (A_{260}/A_{280}) are approximately 1.25. Only one protein component of approximately 34 kDa is observed in polyacrylamide gel electrophoretic analyses of purified virus preparations treated with sodium dodecyl sulfate, indicating that the virus capsid protein is not degraded during the purification process or after storage at –20°C (Lima et al., 1979).

The antisera produced against purified CABMV isolates are very specific. The first blood collections, 15 days after the last rabbit immunizations, produce antisera, presenting good reactions with purified virus preparations and with extracts from infected plants in PTA-ELISA (Figure 2.3). The antiserum titers demonstrate that the best antiserum dilution to use in routine PTA-ELISA is 1:8000, when the absorption readings for the infected tissues are eight times the absorption readings obtained for healthy plants. After dilution to 1:1000, the difference between the absorption reading values of extracts from infected and healthy plants increase greatly (over four times), making possible its use for definitive identification of virus-infected plants.

The best CABMV antiserum dilutions to be used in routine PTA-ELISA around 1:8000 indicates a good quality for the antisera. Despite the impossibility of determining antibody concentration in the antiserum by its tittering, it is possible to conclude by the absorption reading values the levels of virus-specific antibodies present in the antiserum (Van Regenmortel and Dubs, 1993; Lima et al., 2015). The antiserum produced for CABMV-CFor also reacts with plant healthy extracts in PTA-ELISA only in dilutions up to 1:500, indicating that in dilutions over 1:1000, the difference between the absorption reading values of extracts from infected and healthy plants were over 2.5.

After the tenth antiserum collection, 150 days after the last rabbit immunization, the antisera for the virus isolates continues in a crescent phase, showing excellent absorption reading values at PTA-ELISA against extracts from samples of virus infected plants and low absorption readings with extracts from healthy plants. According to Van Regenmortel and Dubs (1993) and Lima et al. (2012), the production of high quality antiserum for plant viruses requires the use of purified virus

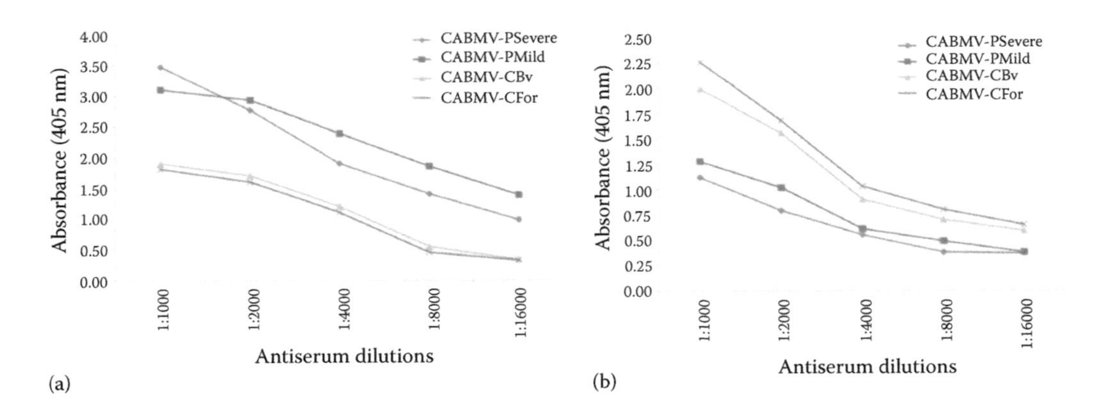

FIGURE 2.3 Results of PTA-ELISA with antiserum against: (a) an isolate of *Cowpea aphid-borne mosaic virus* (CABMV) obtained from *Passiflora edulis* (CABMV-PSevere) and (b) an isolate of CABMV obtained from *Vigna unguiculata* (CABMV-CFor), in different dilutions against extracts of *P. edulis* infected with CABMV-PMild and CABMV-PSevere and extracts from *V. unguiculata* infected with CABMV-CBv and CABMV-CFor. The results separate the virus isolates into two groups: passion fruit (CABMV-PSevere and CABMV-PMild) and cowpea (CABMV-CBv and CABMV-CFor).

preparations relatively free from antigenic plant proteins, which could induce antibodies that might interfere with the interpretation of serological tests. A high immune response depends on the concentration of the antigen used, its antigenicity, and the physical and healthy conditions of the animal (Van Regenmortel and Dubs, 1993; Lima et al., 2012). After an antigen injection, several reactions occur in the immune system of the immunized animal, which culminate with antibody production. The first injection produces a primary response, which is usually weak but keeps the animal organism on alert for subsequent exposure to the same antigen. The subsequent responses are faster and produce greater antibody concentrations, after which they start to decrease, the periodicity of those phases being dependent on the antigen immunization qualities (Almeida and Lima, 2001; Lima et al., 2012).

Reciprocal serological tests in PTA-ELISA with homologous and heterologous antisera for CABMV-PSevere isolated from passion fruit and for CABMV-CFor isolated from cowpea, in different dilutions, demonstrate that CABMV-PSevere and CABMV-PMild are serologically related to CABMV-CBv and CABMV-CFor, with some differences, which are indicated by the antiserum dilutions (Figure 2.3). Serological results are important to group the four virus isolates from the genus *Potyvirus* into the serologically related groups: Group Passion fruit and Group Cowpea (Figure 2.3). Lower cross-reactions are observed in PTA-ELISA between antiserum for CABMV-PSevere and extracts from cowpea plants infected with either CABMV-CBv or CABMV-CFor than with extracts from passion fruit plants infected with CABMV-PSevere or CABMV-PMild. The greatest serological differences are observed with antiserum samples collected in the first 6 weeks after the rabbit immunization. The antiserum obtained for CABMV-PSevere shows low cross-reaction with CABMV-CBv and CABMV-CFor, presenting indexes of 16% and 25% of cross-reactions, respectively (Table 2.2). According to Shukla and Ward (1989) and Shukla et al. (1993), the serological relationships between virus isolates from the genus *Potyvirus* are more clearly shown with antiserum obtained in the early collections, after the rabbit immunization. Similarly, the antiserum obtained in the first bleedings after the rabbit immunization shows higher specificity for CABMV-PSevere and CABMV-PMild, demonstrating that the antiserum can be used for detecting serological distinctions between virus species or serologically related strains.

Serological results with antiserum for CABMV-PSevere and with antiserum for CABMV-CFor (Figure 2.3) in PTA-ELISA group the virus isolates from the genus *Potyvirus* (CABMV-CBv, CABMV-CFor, CABMV-PMild, and CABMV-PSevere) into two serologically related groups:

TABLE 2.2
Percentage of Cross-Reactions with Antiserum for an Isolate of *Cowpea aphid-borne mosaic virus* (CABMV) Obtained from *Passiflora edulis* (CABMV-PSevere) against Plants Infected with the CABMV Isolates Obtained from *P. edulis* (CABMV-PMild and CABMV-PSevere) and from *Vigna unguiculata* (CABMV-CBv and CABMV-CFor)

Virus Isolates	Y = a + bx	% of Cross-Reaction $[(x_1/x_2)*100]$
CABMV-PSevere	Y = 3.825 – 0.1420x	100
CABMV-PMild	Y = 2.828 – 0.0050x	71
CABMV-CFor	Y = 0.578 – 0.0015x	16
CABMV-CBv	Y = 0.496 – 0.0250x	25

Group Passion fruit (CABMV-PSevere and CABMV-PMild) and Group Cowpea (CABMV-CBv and CABMV-CFor) (Figure 2.3).

2.5 MOLECULAR STUDIES WITH VIRUS ISOLATES

Virus RNA products from the immune-precipitated (IP) viruses are free from plant DNA, proteins, and plant RNA contaminants (Lima et al., 2013, 2015). The virus RNA extracts obtained from the IP products are excellent, using either small quantities of infected plant tissue (50 to 100 mg) or larger quantities of infected plant tissue (>1.0 g). The RNA extraction procedure is done in less than 1 h, indicating that a larger number of samples can be processed at the same time (Lima et al., 2013).

DNA fragments with approximately 1700 bp are amplified from the passion fruit samples infected with CABMV-PMild and CABMV-PSevere and from cowpea plants infected with CABMV-CFor and CABMV-CBv, using the primers M10/PY11 (Silva et al., 2013). The amplified DNA fragments, involving the 3′ terminal genomic portion of the NTR region of the 3′ terminus of the Nlb, where the entire *cp* is inserted.

Through the phylogenetic analyses (Figure 2.4) it is possible to observe that the virus isolates (CABMV-PMild, CABMV-PSevere, CABMV-CBv, and CABMV-CFor) grouped with species of CABMV isolated from cowpea, presenting the highest percentage of identity with them. These analyses demonstrate that the isolates obtained from naturally infected passion fruit (CABMV-PMild and CABMV-PSevere) group in the same branch separated from CABMV-CBv and CABMV-CFor obtained from cowpea, which group close to isolates from the genus *Potyvirus* CABMV (JF427592.1) and CABMV (GI4185863), obtained from cowpea, indicating a grouping according to the original host. The CABMV-PMild and CABMV-PSevere isolates group in the same branch produced by CABMV isolates, obtained from passion fruit and deposited in the GenBank, show identity percentages varying from 88.4% to 91.9%. On the other hand, CABMV-CBv and CABMV-CFor group in a branch produced by CABMV isolates obtained from cowpea and already deposited in the GenBank show percentage of identity varying from 77.69% to 80.44%. It is of great importance that the isolates CABMV-PSevere and CABMV-PMild group in different branches than the two isolates CABMV-CBv and CABMV-CFor, which are closest to the CABMV isolates (EU004070) and CABMV (JF833425) obtained from cowpea and deposited in the GenBank (Figure 2.4), indicating a grouping based on the original plant host. On the other hand, Barros et al. (2011), based on phylogenetic comparisons, stated that the CABMV isolates should be grouped according to their geographical regions.

Nucleotide sequences analyses by reverse transcription-polymerase chain reaction (RT-PCR) of a genomic portion corresponding to the partial *cp* also indicate that the isolates CABMV-PSevere, CABMV-PMild, CABMV-CBv, and CABMV-CFor group with CABMV, showing a higher percentage of identity with them (Figure 2.4). Phylogenetic analyses between those CABMV

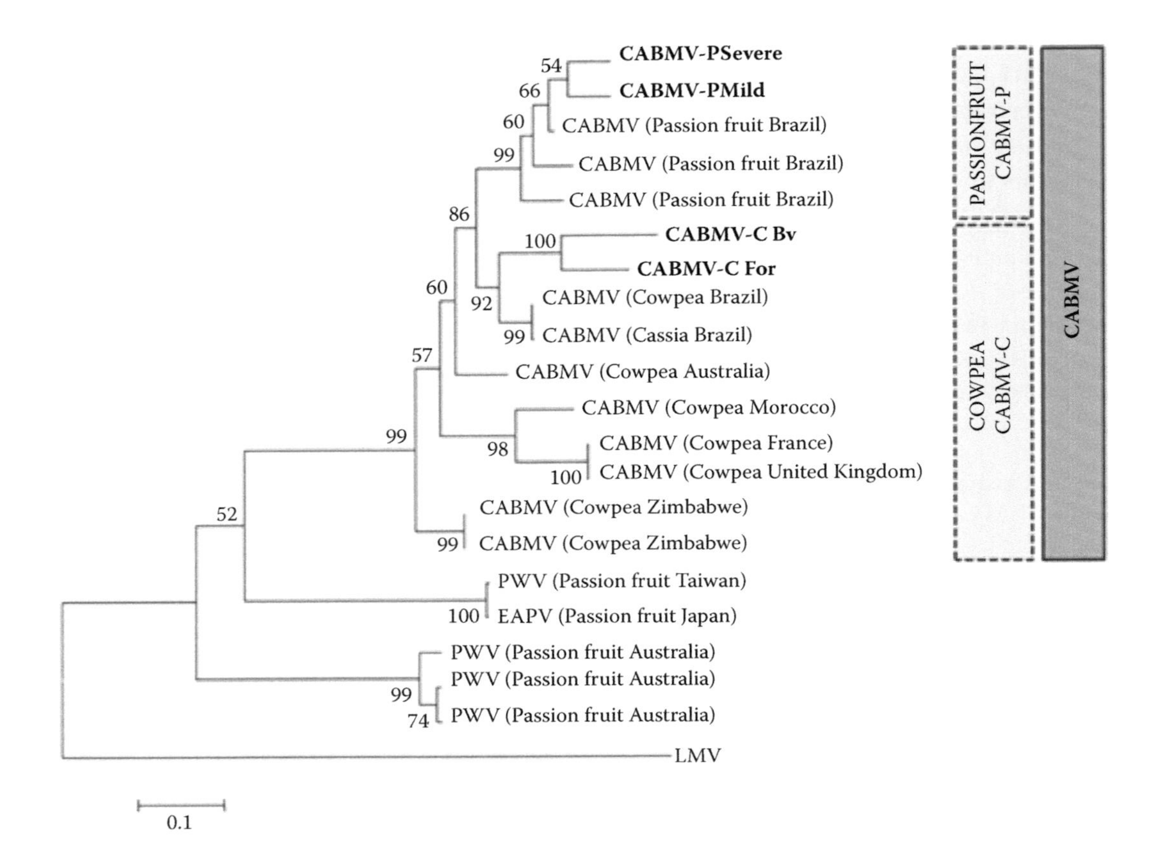

FIGURE 2.4 Phylogenetic tree obtained by MEGA 6 program, through the maximum likelihood method of multiple nucleotide, aligning of the genomic portion corresponding to coat protein gene (*cp*) of four isolates of *Cowpea aphid-borne mosaic virus* (CABMV) characterized in the present research, being two isolates obtained from naturally infected cowpea (CABMV-CBv and CABMV-CFor) and two obtained from naturally infected passion fruit (CABMV-PMild and CABMV-PSevere) with sequences of virus species from the genus *Potyvirus* deposited in the GenBank: CABMV (KC777406), CABMV (AY253907), CABMV (KC77407), CABMV (JF833425), CABMV (JF833424), CABMV (JF427592), CABMV (AF083558), CABMV (Y17822), CABMV (Y178221), CABMV (X82873), CABMV (NC004013), PWV (AF208662), *East Asian Passiflora virus* (EAPV) (AB185021.1), PWV (HQ122652), PWV (AJ430527), PWV (DQ8998218) and *Lettuce mosaic virus* (LMV) (EU247455). The numbers correspond to the values of bootstrap for 2000 replications. The bar represents the substitution number per site.

sequences present two distinct groups when compared with those sequences obtained for other CABMV isolates deposited in the GenBank. CABMV-PSevere and CABMV-PMild isolates group in the same branch of the CABMV isolates obtained from passion fruit already sequenced, with percentage of identities of 77.18% and 91.26%. On the other hand, CABMV-CBv and CABMV-CFor group in a branch produced by CABMV isolates obtained from cowpea and already deposited in the GenBank, with percentages of identities of 87.53% to 83.92% (Figure 2.4). The International Committee on Taxonomy of Viruses adopts as a criterion to designate new virus species in the genus *Potyvirus* less than 80% identity in the coat protein amino acid sequences and less than 85% identity in the nucleotide sequences in the complete genome (Fauquet et al., 2005). Nevertheless, Adams et al. (2005) suggest sequence identities of 77% and 76% for the *cp* nucleotides and complete genome nucleotides, respectively, as sufficient for designation of new virus species inside the genus *Potyvirus*.

In spite of the little variation in biological properties, CABMV isolates from São Paulo do not group consistently by origin or hosts in the maximum likelihood tree. The parent strains of CABMV isolates in São Paulo occupied a more basal position in the phylogenetic tree, suggesting the possible origin of Brazilian isolates. Isolates from São Paulo are the major parents of CABMV recombinant strains of cowpea, passion fruit, and peanut (*Arachis hypogaea* L.) compared to other Brazilian states. These results agree with the expansion pattern of passion fruit cultures in the country and suggest that the São Paulo region is the origin of CABMV in Brazil, from which isolates spread to other regions and evolved.

The genetic diversity of CABMV-P isolates obtained from infecting yellow passion fruit in four Brazilian states presents moderate genetic diversity between one another, infect *M. atropurpureum*, but do not infect cowpea (Rodrigues et al., 2015). Similarly, Nascimento (2014) and Rodrigues et al. (2015) demonstrated that isolates of CABMV-P obtained from passion fruit do not infect cowpea. As mentioned earlier, Bock (1973) reported absence of symptoms in cowpea inoculated with virus isolates from the genus *Potyvirus* obtained from passion fruit.

Results with Immunoprecipitation Reverse Transcription Polymerase Chain Reaction (IP-RT-PCR) for CABMV isolates have been showing to be a practical and sensitive technique for plant virus RNA amplification, indicating that it could be utilized as a diagnostic method for plant virus diseases, mainly because it presents the advantage of combining the virus serological specificity and the technical molecular properties in the same test (Lima et al., 2012, 2015). To avoid false negatives in molecular techniques, several strategies have been developed with the objective of simplifying the plant sample preparations and giving priority to the process of nucleic acid amplification, eliminating the necessity for their purification, especially for virus RNAs. In addition, the IP-RT-PCR approach reduces the risk of cross-contamination during the process of total RNA extraction from infected plant tissues, avoiding possible interference of plant RNAs without the necessity for expensive equipment and specific reagents. The use of this technique therefore represents a technical up-grade to the traditional RT-PCR for plant virus detection, providing plant virus RNA samples of good quality prepared in a simple way.

The extracted RNA products obtained from the IP free from plant DNA, proteins, and RNA contamination confirm that RNA extracted from IP virus for RT-PCR is very useful for virus RNA extraction, especially because its products are excellent, using either small quantities of infected plant tissue (50 to 100 mg) or larger quantities of infected plant tissue (>1.0 g). Extracts from some plant species contain high levels of polyphenols and polysaccharides (Melo et al., 2008), which could interfere with nucleic acid purification, especially RNA, interfering with the nucleic acid amplification techniques (Geuna et al., 1998). The viral RNA extraction for RT-PCR from IP virus particles overcame the problems caused by polyphenol actions.

2.6 CONCLUDING REMARKS

Host range, molecular analyses, and serological and biological results demonstrate the existence of two different biotypes of CABMV in northeastern Brazil, corresponding to Group C (C for cowpea), represented by CABMV-CBv and CABMV-CFor obtained from naturally infected cowpea, and Group P (P for Passion fruit) represented by CABMV-PMild and CABMV-PSevere obtained from naturally infected passion fruit. According to all technical and scientific evidence, it is proposed that a division of the virus from the genus *Potyvirus* found naturally infecting cowpea (CABMV-CBv and CABMV-CFor) and those found naturally infecting passion fruit (CABMV-PMild and CABMV-PSevere) into two biotypes: Biotype CABMV-C (C for cowpea) to include the virus isolates from the genus *Potyvirus* found infecting and causing disease in cowpea without infecting passion fruit and the biotype CABMV-P (P for passion fruit) to involve the virus isolates from the genus *Potyvirus* responsible for passion fruit woodiness disease and that do not infect cowpea.

Similarly, based on the serological similarity, *Watermelon mosaic virus*-1 (WMV-1), genus *Potyvirus* was considered as a biotype of *Papaya ringspot virus* (PRSV), also from the genus *Potyvirus*, which was divided into two biotypes: PRSV-P and PRSV-W. Biotype PRSV-P infects and causes disease in papaya (*Carica papaya* L.), whereas PRSV-W infects and causes disease in cucurbits but does not infect papaya (Purcifull and Hiebert, 1979; Purcifull et al., 1984). PRSV-P and PRSV-W are closely related serologically but differ from *Watermelon mosaic virus* (WMV), formerly known as WMV-2 (Purcifull and Hiebert, 1979). Molecular studies demonstrate Nlb amino acid similarities of approximately 98%, as well as *cp* and 3′ UTR of both PRSV-P and PRSV-W, confirming that they are two biotypes or strains from the same virus species (Quemada et al., 1990). Similarly, although biotypes CABMV-C and CABMV-P have very similar morphological and molecular properties, they can be differentiated by biological and serological methods and have specific ecological niches of host plants, with serological differences detectable by PTA-ELISA (Table 2.1; Figures 2.3 and 2.4). The isolates classified as CABMV-C are responsible for relevant virus disease in cowpea without causing infection in passion fruit, while the CBMV-P isolates cause the passion fruit woodiness, but do not infect cowpea.

REFERENCES

Adams MJ, Antoniw JF, Fauquet CM (2005). Molecular criteria for genus and species discrimination within the family *Potyviridae*. *Arch Virol* 150:459–479.

Almeida AMR, Lima JAA (2001). Técnicas sorológicas aplicadas à Fitovirologia. In: Almeida AMR, Lima JAA (Eds.) *Princípios e técnicas de diagnose aplicados em Fitovirologia*. Publicações SBF, Fortaleza, Brazil, pp. 33–62.

Barros DR, Alfenas-Zerbini P, Beserra JEA, Antunes TFS, Zerbini FM (2011). Comparative analysis of the genomes of two isolates of *Cowpea aphid-borne mosaic virus* (CABMV) obtained from different hosts. *Arch Virol* 156:1085–1091.

Barros DR, Beserra JEA, Alfenas-Zerbini P, Pio-Ribeiro G, Zerbini FM (2007). Complete genomic sequence of two isolates of *Cowpea aphid-borne mosaic virus* (CABMV) obtained from different hosts. *Virus Rev Res* 12:238–239.

Benscher D, Pappu SS, Niblett CL, Varón de Aguedlo F, Morales F, Hodson E, Alvarez E, Costa O, Lee RF (1996). A strain of *Soybean mosaic virus* infecting *Passiflora* spp. in Colombia. *Plant Dis* 80:258–262.

Bezerra DR, Lima JAA, Xavier-Filho J (1995). Purificação e caracterização de um isolado cearense do vírus do endurecimento dos frutos do maracujazeiro. *Fitopatologia Brasileira* 20:553–560.

Bock KR (1973). East African strains of *Cowpea aphid-borne mosaic virus*. *Ann Appl Biol* 74:75–83.

Chang CA (1992). Characterization and comparison of *Passion fruit mottle virus*, a newly recognized potyvirus, with *Passion fruit woodiness virus*. *Phytopathology* 82:1358–1363.

Fauquet CM, Mayo MA, Maniloff J, Desselberger U, Ball LA (2005). Virus taxonomy: Classification and nomenclature of viruses. *Eighth Report of the International Committee on Taxonomy of Viruses*. Elsevier, Academic Press, London.

Freire Filho FR, Lima JAA, Ribeiro VQ (2005). *Feijão—Caupi Avanços Tecnológicos*. Brasília: Embrapa Informação Tecnológica, vol. 1. p. 519.

Geuna F, Hartings H, Scienza A (1998). A new method for rapid extraction of high quality RNA from recalcitrant tissues of grapevine. *Plant Mol Biol Rep* 16:61–67.

Hampton RO, Thottappily G, Rossel HW (1997). Viral diseases of cowpea and their control by resistance-confering genes. In: Singh BB, Mohan Raj DR, Dashiell KE, Jackai LEN (Eds.) *Advances in Cowpea Research*. Ibadan, Nigéria: IITA, JIRCAS, pp. 159–175.

Iwai H, Yamashita Y, Nishi N, Nakamura M (2006). The potyvirus associated with the dappled fruit of *Passiflora edulis* in Kagoshima prefecture, Japan is the third strain of the proposed new species *East Asian Passiflora virus* (EAPV) phylogenetically distinguished from strains of *Passion fruit woodiness virus*. *Arch Virol* 151:811–818.

Lima JAA, Nelson MR (1977). Etiology and epidemiology of mosaic of cowpea in Ceará, Brasil. *Plant Dis Report* 61:864–867.

Lima JAA, Freire Filho FR, Rosal CJ de S, Lopes AC de A (2003). Resistência de genótipos de caupi (*Vigna unguiculata* L. Walp.) de tegumento branco à isolados de vírus das famílias *Bromoviridae, Comoviridae* e *Potyviridae*. *Revista Científica Rural* 8:85–92.

Lima JAA, Nascimento AKQ, Barbosa GS, Maia LM, Silva FR (2015). Etiologia, sintomatologia, distribuição geográfica e estratégias de controle de viroses em culturais tropicais. In: Lima JAA (Ed.). *Virologia Essencila & Viroses em Culturas Tropicais*. Fortaleza, Brazil: Edições UFC, pp. 303–527.

Lima JAA, Nascimento AKQ, Radaelli P, Purcifull DE (2012). Serology applied to plant virology. In: Al-Moslih M (Ed.) *Serological Diagnosis of Certain Human, Animal and Plant Diseases*. Rijeka, Croatia: InTech, pp. 71–94.

Lima JAA, Nascimento AKQ, Radaelli P, Silva AKF and Silva FR (2013). A technique combining immunoprecipitation and RT-PCR for RNA plant virus detection. *J Phytopathol* 162:426–433.
Lima JAA, Purcifull DE, Hiebert E (1979). Purification, partial characterization, and serology of *Blackeye cowpea mosaic virus*. *Phytopathology* 69:1252–1258.
Lima JAA, Silveira LFS, Santos CDG, Gonçalves MFB (1991). Infecção natural em gergelim ocasionada por um potyvirus. *Fitopatologia Brasileira* 16:60–62.
Lima JAA, Sittolin IM, Lima RCA (2005). Diagnose e estratégias de controle de doenças ocasionadas por vírus. In: Freire Filho FR, Lima JAA, Silva PHS, Ribeiro VQ (Eds.) *Feijão caupi: Avanços tecnológicos*. Embrapa Informação Tecnológica, pp. 404–459.
Melo EA, Maciel MIS, Lima VLAG, Nascimento RJ (2008). Capacidade antioxidante de frutas. *Rev Bras Cienc Farm* 44:193–201.
Nascimento AKQ (2014). *Inovações Tecnológicas Para Diagnose de Viroses de Plantas e Caracterização de Biótipos de Cowpea aphid borne mosaic virus*. Doctoral dissertation, Universidade Federal do Ceará.
Nascimento AVS, Santana NE, Braz ASK, Alfenas PF, Pio-Ribeiro G, Andardae GP, Caravlho MG, Zerbini FM (2006). *Cowpea aphid-borne mosaic virus* (CABMV) is widespread in passionfruit in Brazil and causes passionfruit woodiness disease. *Arch Virol* 151:1797–1809.
Nascimento AVS, Souza ARR, Alfenas PF, Andrade GP, Carvalho MG, Pio-Ribeiro G (2004). Análise filogenética do Potyvirus causando endurecimento dos frutos do maracujazeiro no Nordeste do Brasil. *Fitopatologia Brasileira* 29:378–383.
Nicolini C, Rabelo Filho FAC, Resende RO, Andrade GP, Kitajima EW, Pio-Ribeiro G, Gand Nagata T (2012). Possible host adaptation as an evolution factor of *Cowpea aphid-borne mosaic virus* deduced by coat protein gene analysis. *J Phytopathol* 160:82–87.
Novaes QS, Rezende JAM (2003). Selected mild strains of *Passion fruit woodiness virus* (PWV) fail to protect pre-immunized vines in Brazil. *Scientia Agricola* 60:699–708.
Pappu HR, Pappu SS, Sreenivasulu P (1997). Molecular characterization and interviral homologies of potyviruses infecting sesame (*Sesamum indicum*) in Georgia. *Arch Virol* 142:1919–1927.
Parry JN, Davis RI, Thomas JE (2004). Passiflora virus Y: A novel virus infecting Passiflora spp. In Australia and the Indonesian Province of Papua. *Austral Plant Pathol* 33:423–427.
Purcifull D, Edwardson J, Hiebert JE, Gonsalves D (1984). *Papaya ringspot virus*. CMI/AAB Descriptions of plant viruses, No. 292. (No. 84 revised, July 1984.) Wallingford, UK: CAB International
Purcifull DE, Hiebert E (1979). Serological distinction of *Watermelon mosaic virus isolates*. *Phytopathology* 19:116–122.
Quemada H, L'Hostis B, Gonsalves D, Reardon IM, Heinrikson R, Hiebert EL, Sieu LC, Slightom JL (1990). The nucleotide sequences of the 3′-terminal regions of *Papaya ringspot virus* strains W and P. *J Gen Virol* 71:203–210.
Rodrigues LK, Silva LA, Garcez RM, Chaves ALR, Duarte LML, Giampani JS, Colariccio A, Harakava R, Eiras M (2015). Phylogeny and recombination analysis of Brazilian yellow passion fruit isolates of *Cowpea aphid-borne mosaic virus*: Origin and relationship with hosts. *Austral Plant Pathol* 44:31–41.
Shukla DD, Ward CW (1989). Identification and classification of potyviruses on the basis of coat protein sequence data and serology. *Arch Virol* 106:171–200.
Shukla DD, Ward CW, Brant AA (1993). *The Potyviridae*. Wallingford, UK: CAB International.
Silva KN, Nicolini C, Silva MS, Fernandes CD, Nagata T, Resende RG (2013). First report of Johnsongrass mosaic virus (JGMV) infecting *Pennisetum purpureum* in Brazil. *Plant Dis* 97:1003–1003.
Sreenivasulu P, Demski JW, Purcifull DE, Christie RG, Lovell GR (1994). A potyvirus causing mosaic disease of sesame (*Sesamum indicum*). *Plant Dis* 78:95–99.
Taiwo MA, Akinjogunla OJ (2006). Cowpea viruses: Quantitative and qualitative effects of single and mixed viral infections. *Afr J Biotech* 5:1749–1756.
Van Regenmortel MHV, Dubs HC (1993). Serological procedures. In: Matthews REF (Ed.) *Diagnoses of Plant Virus Diseases*. Boca Raton, FL: CRC Press, pp. 159–214.
Van Regenmortel MHV, Fauquet CM, Bishop DHL, Castens E, Estes MK, Lemon S, Maniloff J, Mayo JA, McGeoch DJ, Pringle CR, Wickner R (Eds.) (2000). *Virus Taxonomy. Seventh Report of the International Committee on the Taxonomy of Viruses*. New York: Academic Press, p. 1121.
Zerbini FM, Nascimento AVS, Alfenas PF, Torres LB, Braz ASK, Santana EN, Otoni WC, Carvalho MG (2005). Análise de plantas transgênicas de maracujá-amarelo (*Passiflora edulis* f. *Clavicarpa*) resistente ao endurecimento dos frutos: Efeito de dosagem gênica no espectro da resistência. *Summa Phytopathologica* 31:147–149.

3 Characterization and Control of Potato Virus Y in Crops

Nikolay M. Petrov, Mariya I. Stoyanova*, and R.K. Gaur*

CONTENTS

3.1 INTRODUCTION

The genus *Potyvirus* covers almost one-third of the known plant viruses. Potato virus Y (PVY) is most economically significant, is spread worldwide, and causes damage to potato, tobacco, tomato, pepper,

* These two authors contributed equally to the chapter.

and many other crops. PVY is characterized by a variety of strains and recombinants which make the virus evolutionary flexible and hard to control.

3.2 GENERAL CHARACTERISTICS OF POTYVIRUSES

3.2.1 Taxonomy of Potyviruses

Family *Potyviridae* have filamentous particles and positive-sense ssRNA genomes of about 10 kb that are translated to form a single polyprotein. The family was ratified in 1993 including three genera: *Potyvirus* (with type species Potato virus Y), *Bymovirus* (with type species Barley yellow mosaic virus), and Rymovirus (with type species Ryegrass mosaic virus). The three genera were initially distinguished on the basis of their transmitting vector: *Potyvirus* were transmitted by aphids, *Rymovirus* by mites, and *Bymovirus* by fungi infecting plant roots (Brunt 1992; Murphy et al. 1995). In 1998 another genus—*Ipomovirus*, transmitted by whiteflies with type species Sweet potato mild mottle virus, was officially ratified by the International Committee on Taxonomy of Viruses (ICTV). Further molecular characterization was the reason for formation of another genus transmitted by aphids but distinct from the viruses in the other recognized genera: *Macluravirus* (with type species Maclura mosaic virus) (Badge et al. 1997). Wheat streak mosaic virus (WSMV) was found to be significantly different from Ryegrass mosaic virus in gene sequences and was placed as the type species of the new genus *Tritimovirus* (Salm et al. 1996; Pringle 1999). Genus *Brambyvirus* (with type species Blackberry virus Y) and genus *Poacevirus* (with type species Triticum mosaic virus) were ratified in 2009 and 2010, respectively. In the last release of ICTV, family *Potyviridae* included eight genera: *Brambyvirus* (including one species), *Bymovirus* (including six species), *Ipomovirus* (including six species), *Macluravirus* (including six species), *Poacevirus* (including two species), *Potyvirus* (including 158 species), *Rymovirus* (including three species), *Tritimovirus* (including six species), and two unassigned viruses—Rose yellow mosaic virus and Spartina mottle virus.

3.2.2 Epidemiology, Symptoms, Spread, and Hosts of PVY

PVY is the type species of genus *Potyvirus* and one of the earliest described viruses. It was first reported in potatoes in England (Smith 1931). PVY infects a number of hosts—495 plant species from 72 genera and 36 families. Forty aphid species from 20 genera act as vectors (Kerlan 2006). Transmission is carried out in a non-persistent manner. The main vector is *Myzus persicae* (Kennedy et al. 1962) but in warm climates other species such as *Aphis fabae*, *Aphis gossypii*, *Aphis nasturtii*, and *Macrosiphum euphorbiae* are also important because of their early appearance and prolonged colonization on potato crops (Ragsdale et al. 2001). Virus aphid transmission depends on temperature and humidity. Transmission is 30–35%, more effective at relative humidity 80–90% and temperatures 25–30°C, and is reduced by 50% at moderate realtive humidity (50%) (Singh et al. 1988). Mechanical transmission is possible but of minor importance in the field (Ragsdale et al. 2001).

The main host range includes plants from families *Solanaceae*, *Chenopodiaceae*, and *Commelinaceae*: *Solanum tuberosum*, *S. lycopersicon*, *S. demissum*, *S. chacoense*, *Nicotiana tabacum*, *N. glutinosa*, *Physalis floridana*, *Chenopodium quinoa*, *C. amaranticolor*, *Capsicum annuum*, *C. frutescens*, *Lycium* sp., *Tinantia erecta* and etc. *Datura stramonium* and *S. demissum* 'A' are not infected with PVY (Buchen-Osmond 1987).

PVY is spread worldwide but predominantly in countries with temperate and warm climates growing potatoes, tobacco, tomatoes, and peppers (Buchen-Osmond 1987). Symptoms induced after infection with PVY vary significantly according to the virus strain, plant host species and variety, geographic region, and climatic conditions. Local and system symptoms can be observed. Most common symptoms on potatoes include mosaic and leaf rumpling, weak or distinguished leaf spots leading to deformation and folding, chlorosis and necrosis, vein necrosis, necrotic spots and rings of tubers, leaf wilt, and defoliation. Plants affected by secondary infection are stunted and fragile.

Tobacco, tomato, and pepper develop leaf spots and mosaic, leaf rumpling and vein necrosis, and defoliation. Petunias' flowers change color (Buchen-Osmond 1987). PVY can cause severe damage through mixed infections together with Tomato mosaic virus (Petrov 2014a) and Cucumber mosaic virus (CMV) (Petrov 2015a) in tomato, with Potato leaf roll virus, Potato virus X, Potato virus S, Potato virus M, and Potato virus A (Petrov 2015b).

PVY is considered the main problem in seed potato production worldwide (Jeffries et al. 2006; Khurana and Garg 1992; Valkonen 2007), causing losses of up to 85% when potato plants are grown from infected seed tubers (Whitworth et al. 2006). PVY is now ranked at fifth place of the top ten plant viruses based on scientific and economic importance (Scholthof et al. 2011).

3.3 GENOME ORGANIZATION

Potyviral virions are non-enveloped, flexuous filamentous particles, 680–900 nm in length and 11–15 nm in diameter, with a helical symmetry of the capsid. Potyviruses can be classified on the basis of the length of the gene for the capsid protein and the variables in the length of the N-end (Shukla et al. 1988; Shukla and Ward 1989). Differences among potyviruses range between 263 and 330 amino acids of the protein, and sequence homology is from 38 to 71%. Different strains of a particular virus share the same gene length of the capsid protein and high gene homology (90–99%) (Shukla and Ward 1989). On the basis of this bimodal division, the gene for the capsid protein serves as a marker for distinguishing potyviruses and potyvirus strains.

Virion of PVY is 740 nm in length and 11 nm in diameter. The coat protein (CP) is composed of 267 amino acids, with a total of 30 kDa (Shukla et al. 1994). The single virion includes approximately 2000 copies of CP, which is 95% of the virion weight (Hollings and Brunt 1981; Talbot 2004). The genome is monopartite, linear, single-strand positive sense (ssRNA(+)) and with 9704 nucleotides. The 5′ terminus is a 144 nucleotide-non-translating region (5′-NTR) rich in adenine residues and associated with a genome-linked protein (VPg) (Hollings and Brunt 1981; Carrington and Freed 1990). The highly conservative motive (UCAACACAACAU) in the 5′-NTR is called "Potybox" and is specific for potyviruses (Tordo et al. 1995). The 3′ terminus has a poly (A) tract (Van der Vlugt et al. 1989).

The replication cycle begins when the viral genomic RNA enters a host cell from a neighboring cell or by inoculation by a vector and undergoes decapsidation, translation, and formation of mature proteins. The ssRNA component encodes one large (350 kDa) polyprotein which undergoes co- and posttranslational cleavage by three different proteases (NIa, P1, and HC-Pro) to yield ten different mature proteins—P1 (P1 Protein), HC-Pro (Helper Component Proteinase), P3 (P3 Protein), 6K1 (6-kDa Protein 1), CI (Cylindrical Inclusion body), 6K2 (6-kDa Protein 2), VPg (Viral Genome-linked Protein), NIa (Nuclear Inclusion Protein a, Proteinase domain), NIb (Nuclear Inclusion Protein b), and the CP (Coat Protein) (Talbot 2004; Chung et al. 2008; Cuevas et al. 2012). An additional peptide, P3N-PIPO, is translated from an overlapping ORF after +2 frameshifting of the P3 cistron (Delgado-Sanchez and Grogan 1970; Cuevas et al. 2012) (Figure 3.1).

Initial processing of the viral polyprotein is autocatalytic and cotranslational. NIa is the major potyvirus protease, cleaving at all proteolytic sites except the first two in the N-terminal region (Dougherty and Carrington 1988; Hellmann et al. 1988; Garcia et al. 1989; Riechmann et al. 1992)

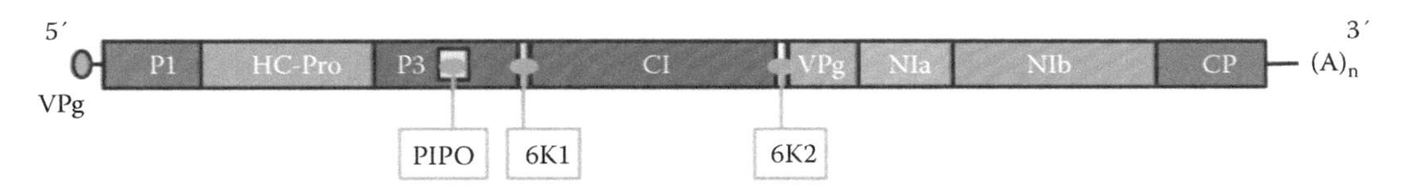

FIGURE 3.1 Organization of PVY genome. (From Cuevas JM et al., *Plos ONE* 7(5):e37853, Figure 1, 2012. doi:10.1371/journal.pone.0037853.g001. With permission.)

(Figure 3.2). P1 is a serine-type proteinase which catalyses autoproteolytic cleavage at a Tyr-Ser dipeptide in the conservative C-terminal at the P1/HC-Pro border (Adams et al. 2005a; Verchot et al. 1992; Yang et al. 1998). HC-Pro recognizes a consensus sequence YXVGG and cleaves a Gly-Gly dipeptide at this site. The structure represents a postcleavage state in which the cleaved C-terminus remains tightly bound at the active site cleft to prevent trans-activity (Guo et al. 2011).

P1 is the most variable protein of PVY and reaches 77% lowest amino-acid sequence similarity (Tordo et al. 1995). The C-terminal of the gene region is conservative and encodes 147 amino acids which form the complete functional proteinase, and the N-terminal encodes 157 amino acids which are not compulsory for proteinase activity (Verchot et al. 1992). P1 protein exhibits nonspecific RNA-binding activity (Brantley and Hunt 1993; Soumounou and Laliberte 1994) similar to that described for known movement proteins of plant viruses (Citovsky et al. 1991, 1992; Osman et al. 1992, 1993; Schoumacher et al. 1992) and it was initially suggested that P1 could be involved in cell-to-cell transport of a virus in plants (Atabekov and Taliansky 1990; Brantley and Hunt 1993; Dougherty and Semler 1993; Riechmann et al. 1992). However, immunogold labeling showed localization of P1 both in association with cytoplasmic inclusion bodies and in the cytoplasm of infected plant cells, and absence of P1 from the cell wall and the plasmodesmata of tobacco leaves (Arbatova et al. 1998). Previously, Verchot and Carrington (1995a,b) showed that deletion of the entire P1 coding sequence of another potyvirus, tobacco etch virus (TEV), had only minor effects on cell-to-cell and long-distance transport but considerably reduced genome amplification. These investigations suggest that P1 is not involved in virus transport but contributes to virus replication (Arbatova et al. 1998).

P1 takes part in the suppression of the antiviral post-transcriptional gene silencing (PTGS) as a natural response of the host plant (Kasschau and Carrington 1998; Maki-Valkama et al. 2000).

HC-Pro is a multifunctional protein needed for virus spread and transfer. HC-Pro increases viral pathogenicity through suppression of PTGS in the host plant. In the absence of functional HC-Pro, the

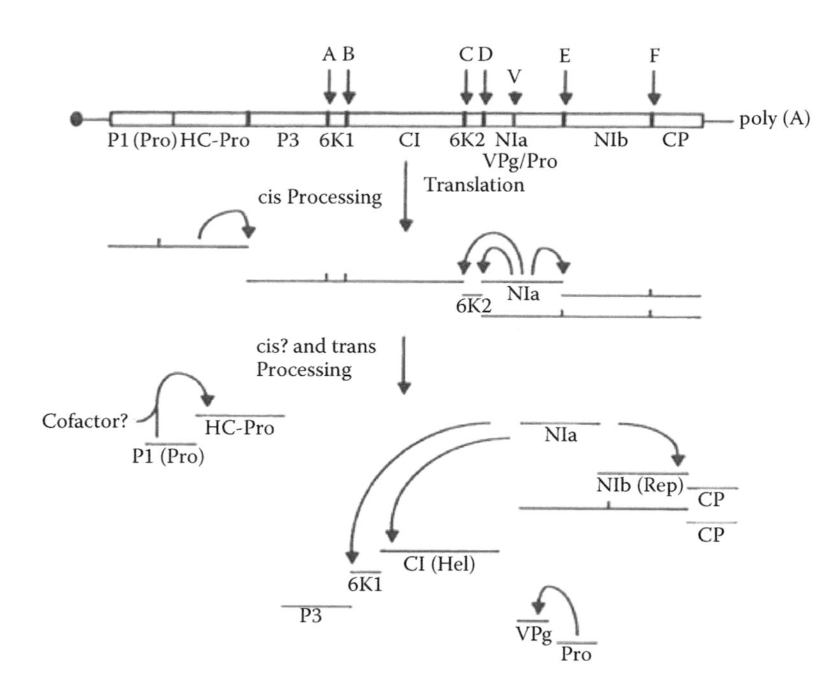

FIGURE 3.2 Processing of the viral polyprotein. Arrows A to F and V show the cleaving sites of NIa. (From Riechmann JL et al. *J Gen Virol* 73:1–16, 1992. With permission.)

viral RNA is targeted to the natural response of gene silencing of the host plant (Kasschau and Carrington 1998). The protein contains highly conservative KITC and PTK motives, necessary for the protein-protein interaction during aphid-mediated transmission (Blanc et al. 1998; Peng et al. 1998). HC-Pro acts as a dimer: the domain KITC at the N-terminal interacts with the aphid stilet, and PTK at the central part is necessary for the binding with the CP of the virus (Pirone and Blanc 1996; Peng et al. 1998; Wang et al. 1998). The central part also contains another conservative CC/CS domain essential for the systemic spread of the virus in the host plant (Cronin et al. 1995) and IGN domain participating in the viral replication (Kasschau et al. 1997). Two other independent domains in the central part are related to the binding of HC-Pro with the viral RNA (Urcuqui-Inchima et al. 2000). C-terminal part is a cysteine proteinase which autocatalytically cleaves the site between HC-Pro and P3 (Carrington and Herndon 1992) and is responsible for the modification of plasmodesmata to ensure the cell-to-cell transport of the viral RNA (Rojas et al. 1997). HC-Pro ensures entry and exit of the virus from the vascular tissues of the host plant (Cronin et al. 1995).

VPg is the only protein covalently binded to the 5′-end of viral RNA by a tyrosine residual (Murphy et al. 1991). Mutation in Tyr_{1860} leads to lack of infectiousness in protoplasts (Murphy et al. 1996). VPg participates in the replication of the viral genome (Anindya et al. 2005) and interacts with the plant translation initiation factors eIF4E and eIF(iso)4E (Leonard et al. 2000, 2004; Wittmann et al. 1997). Uridinated VPg functions as a primer for the viral RNA synthesis (Anindya et al. 2005). Its N-terminal interacts with cell components to help the movement at long distances, which leads to symptom expression in the host plant (Dunoyer et al. 2004).

P3 is a highly variable protein (Adams et al. 2005b) which participates in the virus replication (Merits et al. 1999), accumulation (Klein et al. 1994), symptom formation (Chu et al. 1997; Saenz et al. 2000), overcoming of plant resistance to virus (Johansen et al. 2001; Jenner et al. 2002; Desbiez et al. 2003), and cell-to-cell movement (Johansen et al. 2001). C-terminal is responsible for the virus ability to infect different hosts (Suehiro 2004), and the complex C-terminal of P3/6K1 determines the pathogenicity of the virus (Saenz et al. 2000).

The protein of cylindrical inclusion (CI) bodies aggregates in the cytoplasm of the infected plant cells to form bundle-like inclusions associated with P1 protein (Arbatova et al. 1998). The protein has seven conservative domains: I, Ia, II, III, IV, V, and VI (Kadare and Haenni 1997) in the N-terminal which express nucleoside-triphosphate (NTP)-binding, RNA-binding, and RNA-helicase activities (Lain et al. 1990; Eagles et al. 1994; Fernandez and Garcia 1996; Fernandez et al. 1997). CI is a major component in the replication complex (Shukla et al. 1998). It interacts with the plasmodesmata and the viral ribonucleotide complex to help cell-to-cell movement of the virus (Carrington et al. 1998). It also possesses adenosine-triphosphatase (ATP-ase) activity to provide energy for modification of plasmodesmata (Chen et al. 1994).

NIa forms nuclear inclusions and consists of two domains—VPg and a protease which cleaves all proteins at the C-end of the precursor (Hong and Hunt 1996).

NIb codes the viral RNA-dependent-RNA polymerase (Allison et al. 1986; Hong and Hunt 1996). The polymerase recognizes the polyadenilated RNA-loops at 3′UTR which act as promoters (Hsue and Masters 1997; Haldeman-Cahill et al. 1998).

CP is a structural protein which constructs the virus capsid and consists of three domains: N-terminal, a central part, and C-terminal (Shukla et al. 1998). N-terminal and C-terminal are expressed on the surface of the virion particle (Shukla et al. 1991). The N-terminal contains the basic viral-specific antigene determinants. This part is highly variable with the exception of the DAG motive, which participates in the aphid-mediated transmission and is highly conservative (Atreya et al. 1995; Lopez-Moya et al. 1999). The specific interaction between DAG domaine of CP and KITC domaine of HC-Pro is a prerequisite for the aphid-based transmission (Blanc et al. 1997; Flasinski and Cassidy 1998). N-terminal and C-terminal are necessary for the movement of the virus at long distances while the central part, together with HC-Pro, modifies plasmodesmata to ensure the cell-to-cell transport of the viral RNA (Dolja et al. 1994, 1995; Rojas et al. 1997). CP interacts with GDD domain of NIb and regulates viral RNA synthesis (Hong et al. 1995).

6K protein modifies the membrane of the endoplasmic reticulum in the cytoplasm of the infected cell, and by its hydrophobic domaine ensures the bond between the viral genomic RNA and the intercell membranes (Restrepo-Hartwig and Carrington 1994; Schaad et al. 1997).

Replication complex is formed by NIb, CI, VPg, 6K2, Nia, and other probable cell factors. The genomic ssRNA(+), or the "sense" chain, is replicated into ssRNA(–) or "antisense" chain, which is used as a matrice for generating new viral genomic RNAs (Kadare and Haenni 1997). After replication, viral descendants move to other plant cells through the plasmodesmata with the help of HC-Pro, CI, VPg, and CP (Lopez-Moya and Garcia 2008), and at long distances with the help of CP, HC-Pro, and VPg following the assimilates (Carrington et al. 1996; Revers et al. 1999; Santa Cruz 1999; Ruiz-Medrano et al. 2001).

3.4 STRAIN DIVERSITY

PVY was initially accepted as a complex of isolates (Smith 1931). Potato virus C (PVC) reported in 1930 (Salaman 1930; Bawden 1936; Dykstra 1936), was later classified as a PVY isolate (Bawden 1943), and is the first from the PVY^C strain group. Another strain group PVY^N (De Bokx 1961) was established in 1935 in tobacco near potato fields (Smith and Dennis 1940), and later in potatoes (Nobrega and Silberschmidt 1944; Silberschmidt 1960).

The primary classification of PVY was based on the host plant the virus was isolated from. This led to the grouping of PVY into potato, pepper, tobacco, and tomato strains. Isolates in each group are further characterized on the basis of biological (symptoms and resistance response), serological, and molecular properties (Jacquot et al. 2005; Rolland et al. 2008; Tribodet et al. 2005). In 1950 PVY caused tomato yield losses in Australia (Sturgess 1956), South and North America (Silberschmidt 1956, 1957; Simons 1959), and pepper losses in Florida (Simons 1959) and Israel (Nitzany 1962). In 1970s and 1980s due to the emergence of necrotic isolates PVY has raised as a major threat to these crops (Tomlinson 1987; Legnani 1995; Marchoux et al. 1995).

With the development of technologies, PVY was divided into strains on the basis of the sequences for the capsid protein. The identity of the coding regions is 93.99% and of the non-coding regions is 83–98% (Van der Vlugt et al. 1993). Five major groups of strains have been identified: PVY^O, PVY^C, PVY^Z, PVY^N, and PVY^E (Singh et al. 2008) but many recombinant strains exist. In terms of virus host specificity, epidemiology, and control, PVY isolates are still considered as potato, tobacco, tomato, and pepper strains.

3.4.1 POTATO STRAINS

PVY infecting potato is the most variable group of strains. It consists of three main strain groups, known as O, C, and N on the basis of the local and systemic symptoms induced in specific varieties of tobacco and potato—*Nicotiana tabacum* cv. Samsun and *Solanum tuberosum* subsp. *tuberosum* (Kerlan and Moury 2008). Some strains do not belong to any of these three groups (Delgado-Sanchez and Grogan 1970). Unlike PVY^O and PVY^N, some PVY^C isolates are not transmitted with aphids (Watson 1956; De Bokx and Piron 1978; Blanco-Urgoiti et al. 1998a).

3.4.1.1 PVY^O

The O strain group is the most widely-spread group (Singh et al. 2008) and induces hypersensitive response in potato cultivars harbouring the *Ny* genes. Symptoms generally include mosaic, crinkle, and leaf and stem necrosis in potato, and mottling and mosaic in tobacco. The strain is present in Africa, Europe, New Zealand, and South America (Lorenzen et al. 2006; Rigotti and Gugerli 2007; Rolland et al. 2008).

3.4.1.2 PVY^C

PVY^C was first reported in 1930 as Potato virus C (PVC) (Salaman 1930; Bawden 1936; Dykstra 1936) and later classified as PVY^C isolate (Bawden 1943) based on the hypersensitive resistance response it produced on potato cultivars having the *Nc* gene. The main symptom it induces in susceptible potato is stipple streak (Rolland et al. 2008; Singh et al. 2008).

3.4.1.3 PVY^N

PVY^N was reported for the first time in 1935 in a tobacco field near experimental potato plants (Smith and Dennis 1940; De Bokx 1961) and shortly after that in potato cultivars in Peru and Bolivia (Nobrega and Silberschmidt 1944; Silberschmidt 1960). This strain group caused severe epidemics on potato and tobacco in Europe in the 1950s (Klinkowski 1960; Horvath 1967a). PVY^N induces systemic veinal necrosis symptoms in tobacco, necrosis in *P. floridana*, and only mild mottling in potato (Van der Vlugt et al. 1993).

3.4.1.4 PVY^{NTN}

PVY^{NTN} was first described in Hungary in 1978 (Beczner et al. 1984; Le Romancer et al. 1994) and later reported in many European potato-producing countries (Weidemann and Maiss 1996). PVY ^{NTN} is serologically close to PVY^N but it causes characteristic necrotic ring spots on potato tubers (van den Heuvel et al. 1994; Glais et al. 1996). On tobacco, it causes veinal necrosis symptoms similar to PVY^N (Rolland et al. 2008). Today PVY^{NTN} is considered as a subgoup of PVY^N (van den Heuvel et al. 1994; Glais et al. 1996; Ogawa et al. 2008).

3.4.1.5 Wilga Strain

PVY$^{N\text{-}Wilga}$ is a subgroup of PVY^N (Ogawa et al. 2008) that was first described in Poland in 1984. This strain differs in virulence and aggressiveness from the earlier reported PVY^N isolates and is serologically close to PVY^O (Chrzanowska 1991). Its genome is a recombinantion of PVY^O and PVY^N sequences (McDonald and Singh 1996; Shukla et al. 1994; Ogawa et al. 2008). PVY$^{N\text{-}Wilga}$ causes vein necrosis in tobacco like PVY^N but the symptoms are less severe (Shukla et al. 1994; Kogovsek et al. 2008). Isolates from North America induce lethal necrosis in *Solanum brachycarpum* (Schubert et al. 2007).

Isolates with characteristics of PVY$^{N\text{-}Wilga}$ were reported in Canada (McDonald and Singh 1996), Spain (Blanco-Urgoiti et al. 1998a), and France (Kerlan et al. 1999). Later it was established that PVY^{Wilga}, PVY$^{N\text{-}Wilga}$, and PVY$^{N:O}$ are the same recombinant strain (Singh et al. 2008).

3.4.1.6 PVY^Z and PVY^E

PVY^Z was described in the United Kingdom in 1984 as serologically related to PVY^O but able to overcome the resistance genes against both PVY^O and PVY^C—*Ny* and *Nc* (Jones 1990; Kerlan et al. 1999). It does not induce necrosis in tobacco (Kerlan et al. 1999; Singh et al. 2008). PVY^E is a variant of PVY^Z strain reported in Spain which overcomes the resistant *Nz* gene (Jones 1990; Kerlan et al. 1999; Singh et al. 2008).

3.4.2 Tobacco Strains

Tobacco PVY isolates can be determined as "strong" or "moderate" on the basis of their ability to induce necrosis in certain tobacco genotypes (Gooding 1985). However, historically they are pathotypically classified in three groups: MsNr, MsNr, and NsNr, according to the reaction of tobacco cultivars to root nematode *Meloidogyne incognita.* A group of strains—VAM-B—overcoming the resistance of tobacco cultivars was also found (Gooding 1985; Latore and Flores 1985; Reddick and Miller 1991; Blancard et al. 1994; Kerlan and Moury 2008; Singh et al. 2008). Latore and Flores (1985) suggest that the tobacco genotype VAM can serve as an additional host for genotyping of PVY tobacco isolates. Blancard et al. (1998) distinguish six pathotypes according to the reaction of the virus to four tobacco genotypes, including VAM and genotype NG TG52, which is resistant to VAM-B. In this classification MsNr and MsMr strain groups of Gooding and Tolin vary in the group of mosaic-inducing strains, and the necrotic strains are divided in four pathotypes: (0), (1), (2), and (1,2). The (0) pathotype includes NsNr strain group (Latore and Flores 1985). Generally, tobacco strains are referred to as mosaic or veinal necrosis type (Ali et al. 2008; Singh et al. 2008).

3.4.3 Pepper Strains

Pepper isolates of PVY are classified in three pathotypes according to their ability to overcome the resistance genes *pvr1* and *pvr2* in pepper cultivars: (0), (0,1), and (0,1,2). *Pvr* alleles code the eIF4E translation-initiating factor. Resistant genes in homozygotic state possess point mutations which interfere with the interaction of eIF4E, with the viral protein VPg blocking viral accumulation and movement from cell to cell and at long distance (Kang et al. 2005). Pathotype 0 is unable to overcome both resistant genes, pathotype (0,1) infects plants having only *pvr2* gene, and pathotype (0,1,2) overcomes both *pvr1* and *pvr2* (Kerlan and Moury 2008; Singh et al. 2008). Pathotypes do not show significant differences in their coat protein sequences (Romero et al. 2001). In these groups, the isolates are additionally determined as "common" (PVY^C) or "necrotic" (PVY^N) (D'Aquino et al. 1995). Most pepper isolates do not infect potato (Singh et al. 2008), but few potato isolates show limited ability to infect pepper when inoculated with aphids (Romero et al. 2001). Serological studies show relatedness between most pepper isolates and potato PVY^O and PVY^C strains (Aramburu et al. 2006), but monoclonal antibodies for potato strains do not detect pepper isolates (Romero et al. 2001).

3.4.4 Tomato Strains

Tomato PVY isolates possesses great biological diversity (Walter 1967) and lack a defined classification. Gebre Selassie et al. (1985) and Marchoux et al. (1995) distinguish the isolates which induce mosaic and necrosis from the ones which induce only mosaic. Legnani (1995) differentiates seven tomato PVY pathotypes according to their virulence to genotypes of *Solanum lycopersicon*. The most widely spread pathotype includes both necrotic and non-necrotic isolates. The wild species *Lycopersicon hirsutum* possesses *pot1* gene which confers resistance to PVY but has not yet been used for classification purposes (Moury et al. 2004; Singh et al. 2008).

3.4.5 Other Strains

Reports of emergence of new or variant strains have become a common feature with PVY, especially from isolates infecting potato. The NE-11 PVY isolate, previously classified as a North American NTN strain, was reclassified as a new strain variant class based on its genome sequence (Lorenzen et al. 2008). This phenomenon has started raising the question of defining an efficient system of PVY classification.

Some isolates do no belong to PVY^O, PVY^N, and PVY^C strain groups. A previously described fourth group called PVY^{An} (Horvath 1967b) comprises potato and tomato isolates, including isolates with the properties of PVY^O and PVY^N. Many virus isolates from potatoes were characterized to varying degrees (De Bokx et al. 1975; Thompson et al. 1987; Chrzanowska 1994), including the non-transmitted with aphids PVY^C, which is considered the first to have arisen evolutionarily (Cockerham 1943; Bawden and Kassanis 1947). Two isolates assigned as PVY^C but serologically distinct (De Bokx et al. 1975; Calvert et al. 1980) were separated from PVY into a new separate virus, called potato virus V-PVV (Fribourg and Nakashima 1984).

Frequent and almost simultaneous appearance of new necrotic strains of potato, tobacco, tomato, or pepper in the same area, for example, in 1990s in Canada (McDonald and Kristjansson 1993; Stobbs et al. 1994) or in Mediterranean countries such as Slovenia (Pepelnjak 1993), led to questions about whether these emerging strains are able to migrate from one crop to another.

Host specificity of PVY is highly variable. Some studies show a high level of specificity of PVY to the host (Gebre Selassie et al. 1985). McDonald and Kristjansson (1993) described a pepper isolate infecting tomato and tobacco but not potatoes. They also found many inconsistencies in typing a few PVY isolates, and reported that some isolates causing necrosis in tobacco (Gooding and Tolin 1973) can not be classified within PVY^{N} group of strains, as they fail to cause systemic infection in some potato cultivars. D'Aquino et al. (1995) emphasized the specificity of pepper isolates and suggested a separate group for pepper strains. Blanco-Urgoiti et al. (1996) proposed a classification based on the restriction fragment length polymorphism (RFLP) assay pattern of the coat protein gene which divided the strains into three main clusters: potato PVY^{O}, potato PVY^{C}, and non-potato PVY (PVY^{NP}) including pepper, tobacco, and *Datura* spp isolates.

However, separate PVY isolates can infect different crops. Isolates of potato and tobacco can infect pepper (Horvath 1966, 1967c; Marte et al. 1991; McDonald and Kristjansson 1993; Le Romancer et al. 1994). Tomato varieties are susceptible to a non-tomato PVY^{N} isolate (Stobbs et al. 1994) and to the PVY isolates from different hosts (Legnani 1995). A necrotic tomato isolate (Kerlan and Tribodet 1996) and a decolorizing PVY isolate from petunia (Boonham 1999) cause necrosis of potato tubers. Tomato (Legnani 1995) and tobacco (Blancard et al. 1998) are susceptible to the most PVY isolates, including isolates from potato and pepper.

Determination of host specificity of PVY is essential for the epidemiology studies and control of the virus. In Canada, it is accepted that tomato crops can serve as a reservoir for seasonal PVY^{N}, which could infect neighboring potato or tobacco fields (Stobbs et al. 1994) and on the other hand influence the evolution of PVY.

Comparing the results of studies using different strains is difficult mainly because the initial host is not always specified or the strain is not identified (Stobbs et al. 1994). The ratio between the potato, tobacco, tomato, and pepper strains is also unknown. For example, the relationship between the PVY^{N} group of strains isolated from potato and tobacco is not yet clarified, although PVY^{N} was identified almost simultaneously in both species (as the Tobacco veinal necrosis virus, TVNV) (De Bokx and Huttinga 1981).

3.5 EVOLUTION OF PVY

Due to the constant monitoring of different hosts, there is reason to conclude that new PVY isolates were identified soon after their occurrence, which makes PVY particularly interesting in terms of evolution.

Viruses use two main paths for alteration—mutation and recombination. Unlike DNA-dependent polymerases, RNA-dependent polymerases contribute to a high degree of mutation (or one error in 10–100 kb) due to the low level of accuracy and lack of 3′-5′-exonuclease activity (Domingo et al. 1995; Ramirez et al. 1995). As a result, the replicating virus does not correspond to a genome but is a population of varying sequences, leading to the concept of "quasi-species" (Smith et al. 1997), a term that refers to a population of different chemical species, subjected to a biological selection (Domingo et al. 1995). Recombinations can be classified as homologous and non-homologous. Crossing-over areas can occur between homologous RNA sequences in corresponding or non-corresponding locations, the latter leading to insertions or deletions in recombinant progeny (Simon and Bujarski 1994). Non-homologous or illegitimate recombination occurs between two independent RNA molecules with complementary sequences in the crossing-over positions (Mayo and Jolly 1991; Meyers et al. 1991). Recombinations occur during the process of copying of viral RNA, when the viral RNA templates are exchanged (Nagy and Bujarski 1997).

PVY evolution was observed for the first time in the 1950s, when the biological properties and the genetic studies of new isolates established the formation of new genomes by recombination. For example, a recombination between PVY^{O} and PVY^{N} strains is clearly visible in potatoes and triggers the appearance of PVY^{NTN} and $PVY^{N:O}$ strains. Historically, PVY^{NTN} strains are initial, but genetic studies present three recombinations in PVY^{NTN} strains and only two in $PVY^{N:O}$ strains, suggesting the opposite (Piche 2004).

Since there is a good correlation between genotype and phenotype of potato PVY^{N} and PVY^{O} strain groups, genetic analyses suggest the existence of a third and possibly a fourth group of strains (Blanco-Urgoiti et al. 1998a). Although PVY strains have a broad host range, an interesting observation is that several pepper isolates cannot infect potato. This may mean that two different PVY clones develop independently in these hosts.

Cluster analysis of CP and 5′-NTR places PVY^{NTN} strains in a homogenic group connected with PVY^{N} strains (Revers et al. 1996; Blanco-Urgoiti et al. 1996, 1998a; Chachulska et al. 1997). However, the same analysis of CP and 3′-NTR places PVY^{NTN} together with PVY^{O} strains according to the geographic region (Revers et al. 1996). Another two recombinations in PVY^{NTN} genomes have been established between HC-Pro and P3 and between 6K2 and Nia, which makes PVY^{NTN} genetically related to PVY^{N} from 5′-NTR to HC-Pro and from NIa to the middle of CP, and genetically related to PVY^{O} from P3 to 6K2 and from the middle of CP to 3′NTR (Glais et al. 1999).

Analysis of CP of $PVY^{N\text{-}W}$ isolates indicates a high degree of homogeneity within this strain (97.8%) (McDonald et al. 1997) and suggests that these isolates are more closely related to PVY^{O} (96% homology) than to PVY^{N} (92% homology) (Chachulska et al. 1997). Isolates of $PVY^{N\text{-}W}$ from Poland ($PVY^{N\text{-}Wi}$) (Chrzanowska 1991) and Canada (I-136 and I-L56) (McDonald et al. 1997) are remotely related. The relationship between PVY^{N} and Canadian isolates is closer than that observed between PVY^{O} isolates and $PVY^{N\text{-}W}$ from Poland (Tordo et al. 1995). The conservative motif UUUCA, located at position 124-128 of the 5′-NTR and connected with belonging to subgroup PVY^{NTN} (Tordo 1993) is also present in the Canadian $PVY^{N\text{-}W}$ isolates (McDonald et al. 1997). At 5′-NTR region $PVY^{N\text{-}W}$ isolates a high genetic variability is observed. Glais et al. (1999) confirms the division of $PVY^{N\text{-}W}$ isolates in two subgroups. Other findings show that the last 5670 nucleotides are related to PVY^{O}, unlike the first 4063 nucleotides (Glais et al. 1998), which suggests a recombination. In this meaning, $PVY^{N\text{-}W}$ isolates are a heterogeneous group with genomes resulting from one or more recombinations. Potential crossing-over points are between HC-Pro and P3, as in PVY^{NTN} and in P1. $PVY^{N\text{-}W}$ genomes can be related to PVY^{O} or PVY^{N} from 5′-NTR to the middle of P1, to PVY^{N} from the middle of P1 to HC-Pro, and the rest is connected to PVY^{O} (Glais et al. 1999). Other recombinant strains between PVY^{N} and PVY^{O} have also been established in potato (Revers et al. 1996).

The genome of PVY^{Z} is still not well understood, but an initial RFLP analysis of the CP gene of this isolate shows it belonging to PVY^{O} (Blanco-Urgoiti et al. 1998b). RFLP analysis of the 5′-NTR of the two PVYZ isolates indicates that they are related to PVY^{NTN} (Blanco-Urgoiti et al. 1998b). Homologous recombination is also found in pepper strains of PVY (Fakhfakh et al. 1995).

There is growing evidence amending viral evolution as a result of changes in the environment from natural and anthropogenic factors (Murphy 1999). Populations of plant viruses are in constant communication with the hosts, intercontinental transport of plants, agricultural products, and viral vectors change due to the increasing use of pesticides. Moreover, the effect of increasing concentrations of carbon dioxide in the atmosphere can change the response of plants to viral infection (Malmström and Field 1997). All these environmental reasons modify the selection pressure and can lead to a rapid evolution of viral genomes, especially in RNA-viruses (Holland and Domingo 1998).

3.6 CONTROL OF PVY

PVY control is complex, and it can be achieved by two strategies: control of the vector and control by plant resistance to the virus. Control of the vector is accomplished by chemical (Van Toor et al. 2009), biological (Cabral et al. 2009; Rashki et al. 2009), and cultural methods (Hooks and Fereres 2006).

3.6.1 Chemical Control

Chemical control includes treatment of tubers and foliar applications with insecticides. However, measures are limited due to non-persistent transmission of PVY. The short time period sufficient for transmission grants a dominant role of the non-colonizing vectors which compete with the need of time for action of most chemicals (Schepers et al. 1977; Perring et al. 1999; Radcliffe and Ragsdale 2002). Only the fast-acting pyrethroids have reduced the spread of PVY (Gibson et al. 1982; Rice et al. 1983). Studies with insecticides in New Zealand, including combined seed and foliar treatments, managed to maintain low aphid populations. However, multiple uses of the same or related insecticides leads to insecticide resistance (Van Toor et al. 2009).

A more environmentally friendly method was suggested by Bradley (1963), who showed that mineral oil interferes with aphid transmission of stylet-borne viruses. Vegetable oil sprays also could control potyviruses aphid spread (Bhargava and Khurana 1969) in papaya. Reductions in the spread of PVY were achieved by mineral oil treatments of seed potatoes; however, weekly applications were necessary (Vanderveken 1977; Sharma and Varma 1982; Kirchner et al. 2014). A better effect was achieved by mixed treatments of chemicals and mineral oil, which increased the chemical deposits on the plants and allowed the use of low application volumes (Raccah et al. 1983; Gibson and Cayley 1984).

3.6.2 Biological Control

Biological control exploits natural enemies to control a pathogen or pest. Good perspectives for biological control of the vector of PVY (aphids) bring the use of parasitoid, predator population or fungal enthomopathogen, but these have been tested only in laboratory conditions (Cabral et al. 2009; Rashki et al. 2009).

3.6.3 Field Practices

Field practices aim mainly at reduction of aphid infestation. Control of planting time is intended to avoid peaks in aphid migration, and barrier/border plants, which are generally taller than the primary crop, act as a physical barrier (Hooks and Fereres 2006). According to the virus-sink hypothesis, the virus gets lost from the vector during probing on border crop plants and by reduction of the edge-effect (DiFonzo et al. 1996; Fereres 2000; Hooks and Fereres 2006; Boiteau et al. 2009). Mechanical barriers such as polypropylene sheets or straw mulches have also been shown to reduce aphid spread, and thus PVY spread (Ragsdale et al. 2001; Summers et al. 2005). The ability of straw mulches to reduce virus incidence was achieved in potatoes (Saucke and Doring 2004) and vegetables (Ragsdale et al. 2001; Summers et al. 2004, 2005). The efficiency of straw mulch in reducing PVY appears to be greatest when vector flight activity peaks early in the season (Saucke and Doring 2004).

A precondition for the effectiveness of all field practices is the use of virus-free seeds.

3.6.4 Meristem Cultures

In tropical and subtropical areas it is difficult to produce potato seed tubers due to lack of appropriate storage facilities and transport. Plant tissue cultures are an alternative method for production and rapid propagation of planting material. Meristem cultures also allow the production of pathogen-free seeds; however, sizes of meristem and explant have a critical role in virus elimination (Ali et al. 2013; Khurana and Sane 1998). PVY is easily freed from 3–4 primordia meristems (approx. 0.6–0.8 mm). Prior heat treatment of tubers increases the rate of successful virus elimination (Khurana and Sane 1998).

3.6.5 Resistant Cultivars

Despite the use of chemicals and cultural measures for vector control, the control of PVY achieved in the field is usually still insufficient. Using resistant cultivars is one of the most efficient approaches

(Nemecek et al. 1995; Valkonen 2015). Resistant cultivars are produced either by breeding or by plant transformation.

Breeding is the oldest method used. The major drawbacks are the long time needed to produce a resistant cultivar and the probability of the resistance to be overcome by new virus strains. Most cultivars are not resistant to all PVY strains (Tian and Valkonen 2013; Zimnoch-Guzowska et al. 2013).

There are two types of resistance to infection among potato cultivars: partial or field resistance, based on polygenes, and extreme resistance, based on a single major gene. Partial resistance occurs at different degrees in cultivars and is nearly the same under different epidemiological conditions, only the numbers of infected plants vary (Rohloff 1979).

Wild potato species possess genes for hypersensitive resistance (HR) to PVY in potato, some of which have been used in resistance breeding (Cockerham 1970; Ross 1986; Zimnoch-Guzowska et al. 2013). Unfortunately, most HR genes such as *Ny*, *Nc*, or *Nz* (Jones 1990) are virus strain-specific and cannot recognize new variants of the virus which overcomes resistance. Strain typing of local PVY strains is essential for selection of cultivars carrying the appropriate HR genes. PVY strains recognized by *Ny*, *Nc*, and *Nz* are placed to strain groups PVY^O, PVY^C, and PVY^Z (Singh et al. 2008). The most problematic PVY strains overcome all three genes. Most of them belong to strain group PVY^N but also to a large group of recombinant PVY strains designated as PVY^E (Jones 1990; Kerlan et al. 1999; Singh et al. 2008). HR efficiency is temperature-dependent and may fail to protect the crop (Adams et al. 1986; Valkonen 2015). HR to PVY^O in potato cv. Pito is effective at the temperature of 16–18°C, but at higher temperatures (19–24°C) the virus spreads and causes leaf-drop and mosaic symptoms in upper parts of the plant (Valkonen 2015). The HR genes *Ny-1* and *Ny-2* protect potato cultivars at 20°C from systemic infection with the $PVY^{N\text{-}W}$ (Chrzanowska 1991; Glais et al. 2002) but not at 28°C (Szajko et al. 2014).

Extreme resistance (ER) is conferred by *Ry* genes derived from wild *Solanum* species, which initiates an extreme hypersensitive reaction restricting the virus at or close to the inoculation site (Wiersema 1972; Davidson 1979; Ross 1986). This leads to inhibition of virus multiplication and protects potato plants against all strains of the virus, which allows control of the severe recombinant PVY strains (Ross 1986). Although it is considered that ER ensures complete protection against infections in the field, PVY^N strains can invade ER cultivars under experimental conditions (Wiersema 1972; Davidson 1979, 1980; Ross 1986) and limited systemic necrosis which may develop after graft inoculation (Jones 1990; Valkonen et al. 1996; Vidal et al. 2002).

ER to PVY is epistatic to HR (Valkonen 2015), and a potato genotype carrying *Ry* and *N* genes remains symptomless. These observations indicate that the genes for ER act earlier and more efficiently than the genes for HR (Valkonen 2015). However, allelic variation of additional genes involved in virus recognition and defense signaling may interfere with ER reaction and cause the expression of HR rather than ER (Ross 1986, Valkonen 2015). The genes Ry_{adg}, Ry_{sto}, and *Ry* derived from *Solanum tuberosum* subsp. *andigena*, *Solanum stoloniferum*, and *Solanum chacoense*, respectively, are used in potato breeding programs (Hämäläinen et al. 1997; Chrzanowska et al. 1998; Hosaka et al. 2001). *Ry* gene at present is carried by about 20 European cultivars.

3.6.6 Transgenic Plants

Transgenic plants have been developed to resist viral infections based on the expression of a viral component (pathogen-derived resistance) on the expression of a recombinant antibody directed against viral protein or on PTGS (RNA-derived resistance) (Ghosh et al. 2002; Bouaziz et al. 2009; Zhu et al. 2009). The viral coat protein gene has been the preferred viral component used in pathogen-derived resistance (Zhu et al. 2009). Transgenic plant expressing antibodies showed higher levels of resistance when specific against functional protein rather than specific against the coat protein. However, the expression of recombinant antibodies in the plants needs to be improved (Bouaziz et al. 2009), and the possibility of heterogeneous recombination with other viruses in nature remained the major disadvantage of this method (Zhu et al. 2009).

RNA-derived resistance involves only viral sequences and allows the production of transgenic plants with multiple virus resistance (Zhu et al. 2009). It is a more recent technology for transformation of plants following the discovery of PTGS in plants as a defense mechanism against invaders. The process is initiated by the double-stranded RNAs (dsRNAs) produced during viral replication. The dsRNAs are recognized by the plant as a "non-own" and subsequently cut by Dicer-like cellular enzymes to form small interfering RNAs (siRNAs) of 21–25 bp in length (Hammond et al. 2000; Denli and Hannon 2003). The siRNAs initiate the formation of a multicomponent cell complex (RISC) which destroys complementary-specific viral mRNAs (Martinez et al. 2002). RNA gene silencing spreads from cell to cell, inducing a systemic signal in the whole plant (Mlotshwa et al. 2002). Naturally, plant viruses encode proteins capable of suppressing RNA gene silencing as a response like HC-Pro in Potyviruses (Voinnet and Baulcombe 1997; Kasschau and Carrington 1998; Mlotshwa et al. 2002). Attacking viral suppressor proteins holds the key to unlocking host plant self-defense mechanisms.

The first cases of directed use of PTGS were transgenic plants bearing artificially introduced genes. Virus coat protein-mediated resistance has been successfully employed for PVY (Smith et al. 1994; Ghosh et al. 2002; Marwal et al. 2016). Potato plants engineered to express dsRNAs derived from the highly conserved 3′ terminal part of the coat protein gene of PVY were highly resistant to three strains of PVY, belonging to PVY^{N}, PVY^{O}, and PVY^{NTN} (Missiou et al. 2004). These mechamisms can offer a universal means of induced resistance to one or many viruses through stacked individual or multiple functional transgenes (Marwal et al. 2016).

3.6.7 Induction of Resistance

Use of transgenic plants has raised concerns about possible unexpected changes in genome regulation. Acquired resistance is safer than use of transgenic plants and insecticides for control and does not require time for breeding, but it cannot be inherited. Systemic acquired resistance (SAR) can be induced by bioagents or chemicals that mimic the biological activation of SAR. Treatment of infected potato tubers with *Pseudomonas fluorescens* and *Rhodotorula* sp. suspensions before sowing in the field accelerated plant emergence, improved plant growth, and reduced disease severity caused by PVY (Al-Ani et al. 2013).

Baebler et al. (2011) showed that sensitivity to PVY is related to different endogenous salicylic acid (SA). High SA levels result in low or no PVY symptom development. SA reduces viral accumulation and replication, and although further spread of the virus is unaffected, symptom appearance is delayed (Naylor et al. 1998; Nie 2006; Baebler et al. 2011). Two classes of chemicals are considered as functional analogues of SA: 2,6-dichloro isonicotinic acid (INA) and its derivatives (Metraux et al. 1991) and the benzo [1,2,3] thiadiazole derivatives (Kunz et al. 1997). S-methylbenzo [1,2,3] thiadiazole-7-carbothiate (acibenzolar-S-methyl (ASM)) was the first commercial product marketed under the trade names BION, ACTIGARD, and BOOST (Lawton et al. 1996; Oostendorp et al. 2001). SAR against PVY was successfully induced by treatments with BION and EXIN of tomato cv. Ideal. Treatment with aqueous solution of 3 mM BION three days before PVY inoculation resulted in 86% protection. When combining BION with EXIN, the result was 92% protection (Petrov and Andonova 2012). Similar results were achieved in pepper plants. The combination treatment with EXIN and BION three days before PVY inoculation of pepper plants cv. Kurtovska kapia induced resistance in 86% of the plants (Petrov 2014b).

A novel technology is the use of PTGS for induction of resistance without transformation of plants. Specific dsRNAs and pool of siRNAs covering conservative regions of PVY can be produced in vitro using the polymerase complex of bacteriophage phi6 and applied in the plant without affecting the plant genome and genome regulation pathways. Using this strategy, PTGS was achieved in potato plants cv. Agria by silencing the HC-Pro gene of PVY^{NTN} in newly grown leaves of potato plants (Petrov et al. 2015a). Similar results were established in potatoes cv. Arinda against PVY^{N}. Newly grown leaves of the treated plants were virus free, while the old leaves remained infected but later defoliated (Petrov et al. 2015c).

Recovery from infection with PVY was observed also in tobacco and tomato plants. Regardless of the virus strain, reduction in PVY RNA levels was observed in the upper leaves of the plants. Removal of the first three leaves above the inoculated leaves interfered with the occurrence of recovery, suggesting that the signal(s) mediating the recovery is likely generated in these leaves. Recovery in tobacco was not obvious only in plants infected with PVY^O, which induced mild symptoms (Nie and Molen 2015).

3.6.8 Thermotherapy and Electrotherapy

Other methods for control such as thermotherapy and electrotherapy of meristem cultures and potato tubers have been investigated.

Significant differences were found among the varieties for percent survival of meristem and production of percent virus-free plantlets derived from meristem of PVY-infected potato. In addition, results were not very promising. Thermotherapy with 30°C freed about 38–40% of the meristem plantlets, but survival of plantlets was 18–26%. A hormone combination of 3.0 mg L^{-1} BAP and 0.2 mg L^{-1} GA3 was able to free 34–63% of meristem derived plantlets at a survival rate of 15–46% (Ali et al. 2013).

Thermotherapy had no applicable effect on potato seed tubers. Electrotherapy at 25mA for 12 min eliminates PVY in 70% of the tubers cv. Agria, and germination rate was reduced to 50%. The best effect was achieved by a combination of electrotherapy at 15mA for 10 min followed by thermotherapy at 55°C for 15 min, which eliminated PVY in 93% of potato tubers at a germination rate of 60% (Petrov and Lyubenova 2011).

3.6.9 Application of Plant Extracts

Only a limited number of investigations concerning management through plant extracts have been carried out. According to Theoming et al. (2003), plant extracts have a significant role in controlling vector population, and may trigger the host defense system. Plant extracts from onion, ginger, and garlic at 1% concentrations were effective in reducing the disease incidence caused by PVY in three potato cultivars (Ahmad et al. 2011). Sprays on tobacco with extract from St. John's wort at 5% concentration and tansy at 20% concentration reduced the virus titer to the level of healthy plants (Petrov et al. 2015b, 2016).

3.7 CONCLUSION

PVY is a virus comprising a variety of strains and recombinants. Despite the number of available measures for limiting its spread, the virus is a rising threat to crops. Identification of the proportions of strains in local populations and investigations on new techniques are essential to develop adequate control.

REFERENCES

Adams M, Antoniw J, Beaudoin F (2005a) Overview and analysis of the polyprotein cleavage sites in the family Potyviridae. *Mol Pl Pathol* 6:471–487.

Adams M, Antoniw J, Fauquet C (2005b) Molecular criteria for genus and species discrimination within the family Potyviridae. *Arch Virol* 150:459–479.

Adams SE, Jones RA, Coutts RH (1986) Effect of temperature on potato virus X infection in potato cultivars carrying different combinations of hypersensitivity genes. *Plant Pathol* 35:517–526.

Ahmad N, Khan MA, Ali S et al. (2011) Epidemiological studies and management of potato germplasm against PVX and PVY. *Pak J Phytopathol* 23:159–165.

Al-Ani RA, Athab MA, Matny ON (2013) Management of potato virus Y (PVY) in potato by some biocontrol agents under field conditions. *Adv Environ Biol* 7(3):441–444.

Ali MA, Nasiruddin KM, Haque MS, Faisal SM (2013) Virus elimination in potato through meristem culture followed by thermotherapy. *SAARC J Agri* 11:71–80.

Ali MC, Maoka T, Natsuaki KT (2008) Whole genome sequence and characterization of a novel isolate of PVY inducing tuber necrotic ringspot in potato and leaf mosaic in tobacco. *J Phytopathol* 156:413–418.

Allison R, Johnston RE, Dougherty WG (1986) The nucleotide sequence of the coding region of tobacco etch virus genomic RNA: Evidence for the synthesis of a single polyprotein. *Virology* 154:9–20.

Anindya R, Chittori S, Savithri H (2005) Tyrosine 66 of Pepper vein banding virus genome-linked protein is uridylylated by RNA-dependent RNA polymerase. *Virology* 336:154–162.

Aramburu J, Galipienso L, Matas M (2006) Characterization of potato virus Y isolates from tomato crops in northeast Spain. *Eur J Pl Pathol* 115:247–258.

Arbatova J, Lehto K, Pehu E, Pehu T (1998) Localization of the P1 protein of potato Y potyvirus in association with cytoplasmic inclusion bodies and in the cytoplasm of infected cells. *J Gen Virol* 79:2319–2323.

Atabekov JG, Tallansky ME (1990) Expression of a plant virus-coded transport function by different viral genomes. *Adv Virus Res* 38:201–240.

Atreya P, Lopez-Moya J, Chu M et al. (1995) Mutational analysis of the coat protein N-terminal amino acids involved in potyvirus transmission by aphids. *J Gen Virol* 76:265–270.

Badge J, Robinson DJ, Brunt AA, Foster GD (1997) 3′-Terminal sequences of the RNA genomes of narcissus latent and maclura mosaic viruses suggest that they represent a new genus of the Potyviridae. *J Gen Virol* 78:253–257.

Baebler S, Stare K, Kovac M et al. (2011) Dynamics of responses in compatible potato—Potato virus Y interaction are modulated by salicylic acid. *PLoS ONE* 2011 6:e29009.

Bawden FC (1936) The viruses causing top necrosis (acronecrosis) of the potato. *Ann Appl Biol* 23:487–497.

Bawden FC (1943) Some properties of the potato viruses. *Ann Appl Biol* 30:82–83.

Bawden FC, Kassanis B (1947) The behaviour of some naturally occurring strains of potato virus Y. *Ann Appl Biol* 34:503–516.

Beczner L, Horvath J, Romhanyi I, Forster H (1984) Studies on the etiology of tuber necrotic ringspot disease in potato. *Potato Res* 27:339–352.

Bhargava KS, Khurana SM Paul (1969) Papaya mosaic control by oil sprays. *Phytopath Z* 64:338–343.

Blanc S, Ammar E, Garcia-Lampasona S et al. (1998) Mutations in the potyvirus helper component protein: Effects on interactions with virions and aphid stylets. *J Gen Virol* 79:3119–3122.

Blanc S, Lopez-Moya J, Wang R et al. (1997) A specific interaction between coat protein and helper component correlates with aphid transmission of a potyvirus. *Virology* 231:141–147.

Blancard D, Ano G, Cailleteau B (1994) Principaux virus affectant le tabac en France. *Ann Tabac* 2:39–50.

Blancard D, Cailleteau B, Ano G (1998) *Maladies du tabac: Observer, identifier, lutter.* INRA Editions, Paris.

Blanco-Urgoiti B, Sanchez F, Dopazo J, Ponz F (1996) A strain-type clustering of potato virus Y based on the genetic distance between isolates calculated by RFLP analysis of the amplified coat protein gene. *Arch Virol* 141:2425–2442.

Blanco-Urgoiti B, Sanchez F, Perez de San Roman C et al. (1998a) Potato virus Y group C isolates are a homogeneous pathotype but two different genetic strains. *J Gen Virol* 79:2037–2042.

Blanco-Urgoiti B, Tribodet M, Leclere S, Ponz F, Perez de San Roman C, Legorburu FJ, Kerlan C (1998b) Characterization of potato Potyvirus Y (PVY) isolates from seed potato batches. Situation of the NTN, Wilga and Z isolates. *Eur J Pl Pathol* 104:811–819.

Boiteau G, Singh M, Lavoie J (2009) Crop border and mineral oil sprays used in combination as physical control methods of the aphid-transmitted potato virus Y in potato. *Pest Manag Sci* 65:255–259.

Boonham N, Hims M, Barker I, Spence N (1999) Potato virus Y from petunia can cause symptoms of potato tuber necrotic ringspot disease (PTNRD). *Eur J Plant Pathol* 105:617–621.

Bouaziz D, Ayadi M, Bidani A et al. (2009) A stable cytosolic expression of VH antibody fragment directed against PVY NIa protein in transgenic potato plant confers partial protection against the virus. *Plant Sci* 176:489–496.

Bradley RH (1963) Some ways in which a paraffin oil impedes aphid transmission of potato virus Y. *Can J Microbiol* 9:369–380.

Brantley JD, Hunt AG (1993) The N-terminal protein of the polyprotein encoded by the potyvirus tobacco vein mottling virus is an RNA binding protein. *J Gen Virol* 74:1157–1162.

Brunt AA (1992) The general properties of Potyviruses. *Arch Virol* 5 (Suppl):3–16.

Buchen-Osmond C (1987) Potato virus Y. In: Descriptions and Lists from the VIDE Database. Online http://www.agls.uidaho.edu/ebi/vdie/descr652.htm

Cabral S, Soares AO, Garcia P (2009) Predation by *Coccinella undecimpunctata* L. (*Coleoptera: Coccinellidae*) on *Myzus persicae sulzer* (*homoptera: Aphididae*): Effect of prey density. *Biol Control* 50:25–29.

Calvert EL, Cooper P, Mc Clure J (1980) An aphid transmitted strain of PVYC recorded in potatoes in Northern Ireland. *Rec Agricult Res* 28:63–74.

Carrington JC, Freed DD (1990) Cap-independent enhancement of translation by a plant potyvirus 5′ nontranslated region. *J Virol* 64:1590–1597.

Carrington JC, Herndon K (1992) Characterization of the potyviral HC-Pro autoproteolytic cleavage site. *Virology* 187:308–315.

Carrington JC, Jensen P, Schaad M (1998) Genetic evidence for an essential role for potyvirus CI protein in cell-to-cell movement. *Plant J* 14:393–400.

Carrington JC, Kasschau KD, Mahajan SK, Schaad MC (1996) Cell-to-cell and long-distance transport of viruses in plant. *Plant Cell* 8:1669–1681.

Chachulska AM, Chrzanowska M, Robaglia C, Zagorski W (1997) Tobacco veinal necrosis determinants are unlikely to be located within the 5′ and 3′ terminal sequences of the potato virus Y genome. *Arch Virol* 142:765–779.

Chen S, Das P, Hari V (1994) In situ localization of ATPase activity in cells of plants infected by Maize dwarf mosaic potyvirus. *Arch Virol* 134:433–439.

Chrzanowska M (1991) New isolates of the necrotic strain of Potato virus Y (PVYN) found recently in Poland. *Potato Res* 34:179–182.

Chu M, Lopez-Moya JJ, Llave-Correas C, Pirone TP (1997) Two separate regions in the genome of the tobacco etch virus contain determinants of the wilting response of tabasco pepper. *Mol Plant Microbe Interact* 10:472–480.

Chung BY, Miller WA, Atkins JF, Firth AE (2008) An overlapping essential gene in the *Potyviridae. Proc Natl Acad Sci U S A* 105:5897–5902.

Citovsky V, Knorr D, Zambryski P (1991) Gene I, a potential cell-to-cell movement locus of cauliflower mosaic virus, encodes an RNA-binding protein. *Proc Natl Acad Sci U S A* 88:2476–2480.

Citovsky V, Wong ML, Shaw AL et al. (1992) Visualization and characterization of tobacco mosaic virus movement protein binding to single-stranded nucleic acids. *Plant Cell* 4:397–411.

Cockerham G (1943) The reaction of potato varieties to viruses X, A, B and C. *Ann Appl Biol* 30:338–344.

Cockerham G (1970) Genetical studies on resistance to potato viruses X and Y. *Heredity* 25:309–348.

Cronin S, Verchot J, Haldeman-Cahill R et al. (1995) Long-distance movement factor: A transport function of the potyvirus helper component. *Plant Cell* 7:549–559.

Cuevas JM, Delaunay A, Visser JC et al. (2012) Schematic representation of the PVY genome, including UTR regions and gene distribution. Figure 1. *Plos ONE* 7(5):e37853.

D'Aquino L, Dalmay T, Burgyan J (1995) Host range and sequence analysis of an isolate of potato virus Y inducing veinal necrosis in pepper. *Plant Dis* 79:1046–1050.

Davidson TM (1979) Breeding for resistance to virus diseases of the potato (*Solanum tuberosum*) at the Scottish Plant Breeding Station. *59th Annual Report of the Scottish Plant Breeding Station* pp. 100–108.

De Bokx JA (1961) Waardplanten van het aardappel-YN-virus. *Tijdschrift over Plantenziekten* 67:273–277.

De Bokx JA, Huttinga H (1981) Potato Virus Y. In: *CMI/AAB Descriptions of Plant Viruses* No 242 (No 37 rev), CMI/AAB, Kew, Surrey, England.

De Bokx JA, Kratchanova B, Maat ZD (1975) Some properties of a deviating strain of potato virus Y. *Potato Res* 18:38–51.

De Bokx JA, Piron PG (1978) Transmission of potato virus YC by aphids. In: *Abstracts of Conference Papers of the 7th Triennial Conference of the European Association for Potato Research*, Ziemnaka Institute, Warsaw, Poland, pp. 244–245.

Delgado-Sanchez S, Grogan RG (1970) Potato virus Y. In: *CMI/AAB Descriptions of Plant Viruses* No 37, CMI/AAB, Kew, Surrey, England.

Denli AM, Hannon GJ (2003) RNAi: An ever-growing puzzle. *Trends Biochem Sci* 284:196–201.

Desbiez C, Gal-On A, Girard M et al. (2003) Increase in zucchini yellow mosaic virus symptom severity in tolerant zucchini cultivars is related to a point mutation in P3 protein and is associated with a loss of relative fitness on susceptible plants. *Phytopathology* 93:1478–1484.

DiFonzo CD, Ragsdale DW, Radcliffe EB et al. (1996) Crop borders reduce potato virus Y incidence in seed potato. *Ann Appl Biol* 129:289–302.

Dolja V, Haldeman R, Robertson N et al. (1994) Distinct functions of capsid protein in assembly and movement of tobacco etch potyvirus in plants. *EMBO J* 13:1482–1491.

Dolja V, Haldeman-Cahill R, Montgomery A et al. (1995) Capsid protein determinants involved in cell-to-cell and long distance movement of tobacco etch potyvirus. *Virology* 206:1007–1016.

Domingo E, Holland JJ, Briebricher C, Eigen M (1995) Quasi-species: The concept and the word. In: Gibbs AJ, Calisher CH, Garcia-Arenal F (eds). *Molecular Basis of Virus Evolution*, Cambridge University Press, Cambridge, pp. 181–191.

Dougherty WG, Carrington JC (1988) Expression and function of potyviral gene products. *Ann Rev Phytopathol* 26:123–143.

Dougherty WG, Semler BL (1993) Expression of virus-encoded proteinases: Functional and structural similarities with cellular enzymes. *Microbiol Rev* 57:781–822.

Dunoyer P, Thomas C, Harrison S et al. (2004) A cysteine-rich plant protein potentiates potyvirus movement through an interaction with the virus genome-linked protein VPg. *J Virol* 78:2301–2309.

Dykstra TP (1936) Comparative studies of some European and American potato viruses. *Phytopathology* 26:597–606.

Eagles RM, Balmorimelian E, Beck DL et al. (1994) Characterization of NTPase, RNA-binding and RNA-helicase activities of the cytoplasmic inclusion protein of tamarillo mosaic potyvirus. *Eur J Biochem* 224: 677–684.

Fakhfakh H, Makni M, Robaglia C, Elgaaied A, Marrakchi M (1995) Polymorphisme des régions capside et 3′ NTR de 3 isolats tunisiens du virus Y de la pomme de terre (PVY). *Agronomie* 15:569–579.

Fereres A (2000) Barrier crops as a cultural control measure of non-persistently transmitted aphid-borne viruses. *Virus Res* 71:221–231.

Fernandez A, Garcia J (1996) The RNA helicase CI from plum pox potyvirus has two regions involved in binding to RNA. *FEBS Lett* 388:206–210.

Fernandez A, Guo H, Saenz P et al. (1997) The motif V of plum pox potyvirus CI RNA helicase is involved in NTP hydrolysis and is essential for virus RNA replication. *Nucl Acids Res* 25:4474–4480.

Flasinski S, Cassidy B (1998) Potyvirus aphid transmission requires helper component and homologous coat protein for maximal efficiency. *Arch Virol* 143:2159–2172.

Fribourg CE, Nakashima J (1984) Characterization of a new potyvirus from potato. *Phytopathology* 74:1363–1369.

Garcia JA, Reichmann JL, Lain S (1989) Proteolytic activity of the plum pox potyvirus NI_a-like protein in *Escherichia coli*. *Virology* 170:362–369.

Gebre Selassie K, Marchoux G, Delecolle B, Pochard E (1985) Variabilité naturelle des souches du virusY de la pomme de terre dans les cultures de piment du sud-est de la France. Caractérisation et classification en pathotypes. *Agronomie* 5:621–630.

Ghosh SB, Nagi LHS, Ganapathi TR, Khurana SM Paul, Bapat VA (2002) Cloning and sequencing of potato virus Y coat protein gene from an Indian isolate and development of transgenic tobacco for PVY resistance. *Curr Sci* 82:855–859.

Gibson RW, Cayley GR (1984) Improved control of potato virus Y by mineral oil plus pyrethroid cypermethrin applied electrostatically. *Crop Prot* 3:469–478.

Gibson RW, Rice AD, Sawicki RM (1982) Effects of pyrethroid deltamethrin on the acquisition and inoculation of viruses by *Myzus persicae*. *Ann Appl Biol* 100:49–59.

Glais L, Kerlan C, Tribodet M et al. (1996) Molecular characterization of potato virusYN isolates by PCR-RFLP. *Eur J Pl Pathol* 102:655–662.

Glais L, Tribodet M, Gauthier JP, Astier-Manifacier S, Robaglia C, Kerlan C (1998) RFLP mapping of the whole genome of ten viral isolates representative of different biological groups of potato virus Y. *Arch Virol* 143:2077–2091.

Glais L, Tribodet M, Kerlan C (1999) Le phénomène de recombinaison observe chez le virus Y de la pomme de terre (PVY) est—il impliqué dans l'apparition des nécroses sur tabac et sur les tubercules de pomme de terre? In: *Rencontres de Virologie Végétales*, Aussois, France, p. 81.

Glais L, Tribodet M, Kerlan C (2002) Genomic variability in *Potato potyvirus Y* (PVY): Evidence that PVYNW and PVYNTN variants are single to multiple recombinants between PVYO and PVYN isolates. *Arch Virol* 147:363–378.

Gooding GV Jr (1985) Relationship between strains of potato virus Y and breeding for resistance, cross-protection and interference. *Tobacco Sci* 29:99–104.

Gooding GV Jr, Tolin SA (1973) Strains of potato virus Y affecting flue-cured tobacco in the southeastern United States. *Plant Dis Rep* 57:200–204.

Guo BI, Lin J, Ye K (2011) Structure of the autocatalytic cysteine protease domain of potyvirus helper-component proteinase. *J Biol Chem* 286:21937–21943.

Haldeman-Cahill R, Daros JA, Carrington JC (1998) Secondary structures in the capsid protein coding sequence and 3′-non translated region involved in amplification of the tobacco etch virus genome. *Virology* 72:4072–4079.

Hämäläinen JH, Watanabe KN, Valkonen JP et al. (1997) Mapping and marker-assisted selection for a gene for extreme resistance to potato virus Y. *Theor Appl Gen* 94:192–197.
Hammond SM, Bernstein E, Beach D, Hannon GJ (2000) An RNA-directed nuclease mediates post-transcriptional gene silencing in *Drosophila cells. Nature* 404:293–296.
Hellmann GM, Shaw JG, Rhoads RE (1988) *In vitro* analysis of tobacco vein mottling virus NI_a cistron: Evidence for a virus-encoded protease. *Virology* 163:554–562.
Holland J, Domingo E (1998) Origin and evolution of viruses. *Virus Genes* 16:13–21.
Hollings M, Brunt AA (1981) Potyvirus group. In: *CMI/AAB Descriptions of Plant Viruses* No 245, CMI/AAB, Kew, Surrey, England.
Hong Y, Hunt A (1996) RNA polymerase activity catalyzed by a potyvirus-encoded RNA-dependent RNA polymerase. *Virology* 226:146–151.
Hong Y, Levay K, Murphy J et al. (1995) A potyvirus polymerase interacts with the viral coat protein and VPg in yeast cells. *Virology* 214:159–166.
Hooks CR, Fereres A (2006) Protecting crops from non-persistently aphid-transmitted viruses: A review on the use of barrier plants as a management tool. *Virus Res* 120:1–16.
Horvath J (1966) Studies on strains of potato virus Y. 2. Normal strain. *Acta Phytopathol Acad Sci Hung* 1:334–352.
Horvath J (1967a) Studies on strains of potato virus Y. 3. Strain causing browning of midribs in tobacco. *Acta Phytopathol Acad Sci Hung* 2:95–108.
Horvath J (1967b) Studies on strains of Potato virus Y. 4. Anomalous strain. *Acta Phytopathol Acad Sci Hung* 2:195–210.
Horvath J (1967c). Virulenz differenzen verschiedener Stamme und Isolate des Kartoffel-Y-virus an Capsicum-Arten und Varietaten. *Acta Phytopathol Acad Sci Hung* 2:17–37.
Hosaka K, Hosaka Y, Mori M et al. (2001) Detection of a simplex RAPD marker linked to resistance to potato virus Y in a tetraploid potato. *Am J Potato Res* 78:191–196.
Hsue B, Masters PS (1997) A bulged stem-loop structure in the 3′ untranslated region of the genome of the coronavirus mouse hepatitis virus is essential for replication. *J Virol* 71:7567–7578.
ICTV Online. *Virus Taxonomy*: 2014 Release. http://www.ictvonline.org/virusTaxonomy.asp
Jacquot E, Tribodet M, Croizat F et al. (2005) A single nucleotide polymorphism-based technique for specific characterization of PVYO and PVYN isolates of Potato virus Y (PVY). *J Virol Methods* 125:83–93.
Jeffries C, Barker H, Khurana SM Paul (2006) Viruses & viroids. In: Gopal J, Paul SM Khurana (eds). *Handbook of Potato Production, Improvement and Post Harvest Management*. Haworth Press, Philadelphia, PA, pp. 387–448.
Jenner CE, Tomimura K, Ohshima K et al. (2002) Mutations in turnip mosaic virus P3 and cylindrical inclusion proteins are separately required to overcome two Brassica napus resistance genes. *Virology* 300:50–59.
Johansen IE, Lund OS, Hjulsager CK, Laursen J (2001) Recessive resistance in *Pisum sativum* and potyvirus pathotype resolved in a gene-for-cistron correspondence between host and virus. *J Virol* 75:6609–6614.
Jones RA (1990) Strain group specific and virus specific hypersensitive reactions to infection with potyviruses in potato cultivars. *Ann Appl Biol* 117:93–105.
Kadare G, Haenni A (1997) Virus-encoded RNA helicase. *J Virol* 71:2583–2590.
Kang B, Yeam I, Jahn M (2005) Genetics of plant virus resistance. *Ann Rev Phytopathol* 43:581–621.
Kasschau K, Carrington J (1998) A counterdefensive strategy of plant viruses: Suppression of posttranscriptional gene silencing. *Cell* 95:461–470.
Kasschau K, Cronin S, Carrington J (1997) Genome amplification and long-distance movement functions associated with the central domain of tobacco etch potyvirus helper component-proteinase. *Virology* 228:251–262.
Kennedy J, Day M, Eastop V (1962) A conspectus of aphids as vectors of plant viruses. *Commonwealth Institute of Entomology*. London.
Kerlan C (2006) Potato Virus Y. In: *CMI/AAB Descriptions of Plant Viruses* No 414, CMI/AAB, Kew, Surrey, England.
Kerlan C, Moury B (2008) Potato Virus Y. In: Mahy BWJ, Van Regenmortel MHV (eds). *Encyclopedia of Virology*. 3rd edn. Elsevier, Oxford.
Kerlan C, Tribodet M (1996) Are all PVYN isolates able to induce potato tuber necrosis ringspot disease? In: *Proceedings of the 13th Triennial Conference of the European Association of Potato Research*. European Association of Potato Research, Veldhoven, The Netherlands, pp. 65–66.
Kerlan C, Tribodet M, Glais L, Guillet M (1999) Variability of potato virus Y in potato crops in France. *J Phytopathol* 147:643–651.

Khurana SM Paul, Garg ID (1992) Potato mosaics. pp. 148–164 In: HS Chaube, J Kumar, AN Mukhopadhyay, US Singh (eds). *Plant Diseases of International Importance*, Vol. II. Prentice Hall, Englewood Cliffs, NJ, pp. 65–66.

Khurana SM Paul, Sane A (1998) Apical meristem culture—A tool for virus elimination. In: SM Paul Khurana, R Chandra, MD Upadhya (eds). *Comprehensive Potato Biotechnology*. Malhotra Publishing House, Delhi, India, pp. 207–232.

Kirchner SM, Hiltunen LH, Santala J et al. (2014) Comparison of straw mulch, insecticides, mineral oil, and birch extract for control of transmission of potato virus y in seed potato crops. *Potato Res* 57:59–75.

Klein PG, Klein RR, Rodríguez-Cerezo E et al. (1994) Mutational analysis of the tobacco vein mottling virus genome. *Virology* 204:759–769.

Klinkowski M, Schmelzer K (1960) A necrotic type of potato virus Y. *Am Potato J* 37:221–227.

Kogovsek P, Gow L, Pompe-Novak M et al. (2008) Single-step RT real-time PCR for sensitive detection and discrimination of Potato virus Y isolates. *J Virol Methods* 149:1–11.

Kunz W, Schurter R, Maetzke T (1997) The chemistry of benzothiadiazole plant activators. *Pesticide Sci* 50: 275–282.

Lain S, Riechmann JL, Garcia JA (1990) RNA helicase: A novel activity associated with a protein encoded by a positive strand RNA virus. *Nucleic Acids Res* 18:7003–7006.

Latore BA, Flores V (1985) Strain identification and cross-protection of potato virus Y affecting tobacco in Chile. *Plant Dis* 69:930–932.

Lawton KA, Friedrich L, Hunt M et al. (1996) Benzothiadizole induces disease resistance in *Arabidopsis* by activation of the systemic acquired resistance signal transduction pathway. *Plant J* 10:71–82.

Le Romancer M, Kerlan C, Nedellec M (1994) Biological characterization of various geographical isolates of potato virus Y inducing superficial necrosis on potato tubers. *Plant Pathol* 43:138–144.

Legnani R (1995) Evaluation and inheritance of the *Lycopersicon hirsutum* resistance against Potato virus Y. *Euphytica* 86:219–226.

Leonard S, Plante D, Wittmann S et al. (2000) Complex formation between potyvirus VPg and translation eukaryotic initiation factor 4E correlates with virus infectivity. *J Virol* 74:7730–7737.

Leonard S, Viel C, Beauchemin C et al. (2004) Interaction of VPg-Pro of turnip mosaic virus with the translation initiation factor 4E and the poly(A)-binding protein in planta. *J Gen Virol* 85:1055–1063.

Lopez-Moya JJ, Wang R, Pirone T (1999) Context of the coat protein DAG motif affects potyvirus transmissibility by aphids. *J Gen Virol* 80:3281–3288.

Lopez-Moya JJ, Garcia JA (2008) Potyviruses. In: Mahy BW, Van Regenmortel MH (eds). *Encyclopedia of Virology*. Vol 4, 3rd ed. Elsevier, Oxford.

Lorenzen J, Piche L, Gudmestad N et al. (2006) A multiplex PCR assay to characterize Potato virus Y isolates and identify strain mixtures. *Plant Dis* 90:935–940.

Maki-Valkama T, Valkonen J, Kreuze J, Pehu E (2000) Transgenic resistance to PVY(O) associated with post-transcriptional silencing of P1 transgene is overcome by PVY(N) strains that carry highly homologous P1 sequence and recover transgene expression at infection. *Mol Plant Micr Interact* 13:366–373.

Malmström CM, Field CB (1997) Virus-induced differences in the response of oat plants to elevated carbon dioxide. *Plant Cell Envir* 20:178–188.

Marchoux G, Palloix P, Gebre Selassie K et al. (1995) Variabilité du virus Y de la pomme de terre et des Potyvirus voisins. Diversité des sources de résistance chez le piment (*Capsicum* sp.). *Ann Tabac* 2:25–34.

Marte M, Belleza G, Polverari A (1991) Infective behaviour and aphid-transmissibility of Italian isolates of potato virus Y in tobacco and peppers. *Ann Appl Biol* 118:309–317.

Martinez J, Patkaniowska A, Urlaub H et al. (2002) Single-stranded antisense siRNAs guide target RNA cleavage in RNAi. *Cell* 110:563–574.

Marwal A, Gaur RK, Khurana SM Paul (2016) RNAi mediated gene silencing against plant viruses. In: P Chowdappa, P Sharma, D Singh, AK Misra (eds). *Perspectives of Plant Pathology in Genomic Era*. Today & Tomorrow Printers and Publishers, Delhi, India, pp. 235–254.

Mayo MA, Jolly CA (1991) The 5′-terminal sequence of potato leafroll virus RNA: Evidence for recombination between virus and host RNA. *J Gen Virol* 72:2591–2595.

McDonald JG, Kristjansson GT (1993) Properties of strains of *Potato virus YN* in North America. *Plant Dis* 77:87–89.

McDonald JG, Singh RP (1996) Host range, symptomology, and serology of isolates of *Potato virus Y* (PVY) that share properties with both the PVYN and PVYO strain groups. *Am Potato J* 73: 309–315.

McDonald JG, Wong E, Henning D, Tao T (1997) Coat protein and 5′ non-translated region of a variant of Potato virus Y. *Can J Plant Pathol* 19:138–144.

Merits A, Guo D, Järvekülg L, Saarma M (1999) Biochemical and genetic evidence for interactions between potato A potyvirus-encoded proteins P1 and P3 and proteins of the putative replication complex. *Virology* 263:15–22.

Metraux JP, Ahl-Goy P, Staub T et al. (1991) Induced systemic resistance in cucumber in response to 2,6-dichloro-isonicotinic acid and pathogens. In: Hennecke H and Verma DPS (eds). *Advances in Molecular Genetics of Plant-Microbe Interactions*. Vol 1, Kluwer Academic Publishers, Dordrecht, pp. 432–439.

Meyers G, Tautz N, Dubovi EJ, Thiel HJ (1991) Viral cytopathogenicity correlated with integration of ubiquitin-coding sequences. *Virology* 180:602–616.

Missiou A, Kalantidis K, Boutla A et al. (2004) Generation of transgenic potato plants highly resistant to potato virus Y (PVY) through RNA silencing. *Mol Breed* 14:185–197.

Mlotshwa S, Voinnet O, Mette MF et al. (2002) RNA silencing and the mobile silencing signal. *Plant Cell* 14: S289–S301.

Moury B, Morel C, Johansen E et al. (2004) Mutations in *Potato virus Y* genome-linked protein determine virulence towards recessive resistances in *Capsicum annuum* and *Lycopersicon hirsutum*. *Mol Plant Micr Interact* 3:322–329.

Murphy FA (1999) The evolution of viruses, the emergence of viral diseases: A synthesis that Martinus Beijerinck might enjoy. *Arch Virol* 15 Suppl:73–85.

Murphy FA, Fauquet CM, Bishop DH et al. (eds) (1995) Virus taxonomy—The classification and nomenclature of viruses. *Sixth Report of the International Committee on Taxonomy of Viruses*. Springer-Verlag, Vienna.

Murphy J, Klein P, Hunt A, Shaw J (1996) Replacement of the tyrosine residue that links a potyviral VPg to the viral RNA is lethal. *Virology* 220:535–538.

Murphy J, Rychlik W, Rhoads R et al. (1991) A tyrosine residue in the small nuclear inclusion protein of tobacco vein mottling virus links the VPg to the viral RNA. *J Virol* 65:511–513.

Nagy PD, Bujarski JJ (1997) Engineering of homologous recombination hotspots with AU-rich sequences in brome mosaic virus. *J Virol* 71:3799–3810.

Naylor M, Murphy AM, Berry JO, Carr JP (1998) Salicylic acid can induce resistance to plant virus movement. *Mol Plant Microbe Interact* 11:860–868.

Nemecek T, Schwärzel R, Derron JO (1995) Quels facteurs déterminent l'infection virale des pommes de terre? Une analyse de systéme de la production de plants. *Revue suisse d'agriculture* 27:73–77.

Nie X (2006) Salicylic acid suppresses Potato virus Y Isolate N:O-induced symptoms in tobacco plants. *Phytopathology* 96:255–263.

Nie X, Molen A (2015) Host recovery and reduced virus level in the upper leaves after Potato virus Y infection occur in tobacco and tomato but not in potato plants. *Viruses* 7:680–698.

Nitzany FE, Tanne E (1962) Virus diseases of peppers in Israël. *Phytopathol Mediterr* 4:180–182.

Nobrega NR, Silberschmidt K (1944) Sobre una provavel variante do virus "Y" da batatinha (Solanum virus 2, Orton) que tem a peculiaridade de provocar necroses em plantas de fumo. *Arq Inst Biol (Sao Paulo)* 15:307–330.

Ogawa T, Tomitaka Y, Nakagawa A, Ohshima K (2008) Genetic structure of a population of *Potato virus Y* inducing potato tuber necrotic ringspot disease in Japan; comparison with North American and European populations. *Virus Res* 131:199–212.

Oostendorp M, Kunz W, Dietrich B, Staub T (2001) Induced disease resistance in plants by chemicals. *Eur J Plant Pathol* 107:19–28.

Osman TA, Hayes RJ, Buck KW (1992) Cooperative binding of the red clover necrotic mosaic virus movement protein to single stranded nucleic acids. *J Gen Virol* 73:223–227.

Osman TA, Thommes P, Buck KW (1993) Localization of a single-stranded RNA-binding domain in the movement protein of Red clover necrotic mosaic dianthovirus. *J Gen Virol* 74:2453–2457.

Peng Y, Kadoury D, Gal-On A et al. (1998) Mutations in the HC-Pro gene of zucchini yellow mosaic potyvirus: Effects on aphid transmission and binding to purified virions. *J Gen Virol* 79:897–904.

Pepelnjak M (1993) Potato virus YNTN on tomato. In: *Proceedings of the 12th Triennial Conference of the European Association of Potato Research*, European Association of Potato Research, Paris.

Perring TM, Gruenhagen NM, Farrar CA (1999) Management of plant viral diseases through chemical control of insect vectors. *Ann Rev Entomol* 44:457–481.

Petrov N (2014a) Damaging effects of Tomato mosaic virus and Potato virus Y on tomato plants. *Sci Technol* 4:56–60.

Petrov N (2014b) Induction of resistance in pepper to Potato virus Y by activation of defense mechanisms of the host plant. *In: Proceedings of Seminar of Ecology*, 24–25 Apr. 2014, IBER, Sofia, Bulgaria, pp. 134–139.

Petrov N (2015a) Mixed viral infections in tomato as a precondition for economic loss. *Agri Sci Tech* 7:124–128.
Petrov N (2015b) Sensitivity of potato cultivars grown in Bulgaria to plant viruses PVY and PLRV. In: *Proceedings of the Union of Scientists–Ruse* 7:233–236.
Petrov N, Andonova R (2012) Bion and exin as SAR elicitors against Potato virus y infection in tomato. *Sci Technol* 2:46–49.
Petrov N, Lyubenova V (2011) Thermotherapy and electrotherapy of potato tubers infected with potato virus Y – PVY. In: Scientific Papers of the Jubilee National Scientific Conference with International Participation "The Man and the Universe," 6–8 Oct. 2011, Smolyan, Bulgaria, Part II. pp. 678–685.
Petrov N, Stoyanova M, Andonova R, Teneva A (2015a). Induction of resistance to potato virus Y strain NTN in potato plants through RNAi. *Biotech Biotechnol Eq* 29:21–26.
Petrov N, Stoyanova M, Valkova M (2015b) The antiviral activity of extract from St. John's wort against Potato virus Y. In: *Proceedings of the Union of Scientists–Ruse* 7:229–232.
Petrov N, Stoyanova M, Valkova M (2016) Antiviral activity of plant extract from *Tanacetum vulgare* against Cucumber Mosaic Virus and Potato Virus Y. *J BioSci Biotechnol* 5(2):189–194.
Petrov N, Teneva A, Stoyanova M, Andonova R, Denev I, Tomlekova N (2015c) Blocking the systemic spread of potato virus Y in the tissues of potatoes by posttranscriptional gene silencing. *Bul J Agri Sci* 21:288–294.
Piche LM, Singh RP, Nie X, Gudmestad NC (2004) Diversity among Potato virus Y isolates obtained from potatoes grown in the United States. *Phytopathology* 94:1368–1375.
Pirone T, Blanc S (1996) Helper-dependent vector transmission of plant viruses. *Ann Rev Phytopathol* 34:227–247.
Pringle CR (1999) The universal system of virus taxonomy, updated to include the new proposals ratified by the International Committee on Taxonomy of Viruses during 1998. *Arch Virol* 144:421–429.
Raccah B, Antignus Y, Cohen-Braun M (1983) Effect of a combination of mineral oil and pyrethroid on the transmission of CMV in laboratory and on the natural infection of MDMV in a corn field. In: Proceedings of the 4th International Congress of Plant Pathology, Melbourne, Australia.
Radcliffe EB, Ragsdale DW (2002) Aphid-transmitted potato viruses: The importance of understanding vector biology. *Am Potato J* 79:353–386.
Ragsdale D, Radcliffe E, Difonzo C (2001) Epidemiology and field control of PVY and PLRV. In: G Loebenstein, Berger P, Brunt A, Lawson R (eds). *Virus and Virus-Like Diseases of Potatoes and Production of Seed-Potatoes*. Springer, Dordrecht, Netherlands, pp. 237–270.
Ramirez BC, Barbier P, Seron K et al. (1995) Molecular mechanisms of point mutations in RNA viruses. In: Gibbs AJ, Calisher CH, Garcia-Arenal F (eds). *Molecular Basis of Virus Evolution*. Cambridge University Press, Cambridge, UK, pp. 105–118.
Rashki M, Kharazi-pakdel A, Allahyari H, van Alphen JJ (2009) Interactions among the entomopathogenic fungus, *beauveria bassiana* (*Ascomycota: Hypocreales*), the parasitoid, *Aphidius matricariae* (*Hymenoptera: Braconidae*), and its host, *Myzus persicae* (*Homoptera: Aphididae*). *Biol Control* 50:324–328.
Reddick BB, Miller RD (1991) Identification of virus population from TN 86 vs. non-TN 86 Burley tobacco field in East Tennessee. *Tobacco Abstr* 35:650.
Restrepo-Hartwig MA, Carrington JC (1994) The tobacco etch potyvirus 6-kilodalton protein is membrane associated and involved in viral replication. *J Virol* 68:2388–2397.
Revers F, Le Gall O, Candresse T et al. (1996) Frequent occurrence of recombinant potyvirus isolates. *J Gen Virol* 77:1953–1965.
Revers F, Le Gall O, Candresse T, Maule AJ (1999) New advances in understanding the molecular biology of plant/potyvirus interactions. *Mol Plant Micr Interact* 12:367–376.
Rice AD, Gibson RW, Stribley ME (1983) Effects of deltamethrin on walking, flight, and potato virus Y-transmission by pyrethroid-resistant Myzuspersicae. *Ann Appl Biol* 102:229–236.
Riechmann JL, Lain S, Garcia JA (1992) Highlights and prospects of potyvirus molecular biology. *J Gen Virol* 73:1–16.
Rigotti S, Gugerli P (2007) Rapid identification of Potato virus Y strains by by one-step triplex RT-PCR. *J Virol Methods* 140:90–94.
Rohloff H (1979) Beitrag zur Analyse der Kartoffet-Y-Virus Epidemie in 1976. *Gesunde Pflanzen* 31:296–299.
Rojas M, Zerbini F, Allison R et al. (1997) Capsid protein and helper component-proteinase function as potyvirus cell-to-cell movement proteins. *Virology* 237:283–295.
Rolland M, Glais L, Kerlan C, Jacquot E (2008) A nucleotide Polymorphisms interrogation assay for reliable Potato virus Y group and variant characterization. *J Virol Methods* 147:108–117.

Romero A, Blanco-Urgoiti B, Soto MJ et al. (2001) Characterization of typical pepper-isolates of PVY reveals multiple pathotypes within a single genetic strain. *Virus Res* 79:71–80.
Ross H (1986) Potato breeding—Problems and perspectives. *J Plant Breed* Suppl 13.
Ruiz-Medrano R, Xoconostle-Cazares B, Lucas WJ (2001) The phloem as a conduit for inter-organ communication. *Curr Opin Plant Biol* 4:202–209.
Saenz P, Cervera M, Dallot S et al. (2000) Identification of a pathogenicity determinant of plum pox virus in the sequence encoding the C-terminal region of protein P3+6K(1). *J Gen Virol* 81:557–566.
Salaman RN (1930) Virus diseases of potato: Streak. *Nature* 126:241.
Salm SN, Rey ME, Rybicki EP (1996) Phylogenetic justification for splitting the Rymovirus genus of the taxonomic family Potyviridae. *Arch Virol* 141:2237–2242.
Santa Cruz S (1999) Perspective: Phloem transport of viruses and macromolecules—What goes in must come out. *Trends Microbiol* 7:237–241.
Saucke H, Döring TF (2004) *Potato virus Y* reduction by straw mulch in organic potatoes. *Ann Appl Biol* 144:347–355.
Schaad M, Jensen P, Carrington J (1997) Formation of plant RNA virus replication complex on membranes: Role of an endoplasmic reticulum-targeted viral protein. *EMBO J* 16:4049–4059.
Schepers A, Bus CB, de Bokx JA, Cuperus C (1977) De verspreiding van YN-Virus in aardappelen. *Landbouwkundig Tijdschrift* 89:123–128.
Scholthof KB, Adkins S, Czosnek H et al. (2011) Top 10 plant viruses in molecular plant pathology. *Mol Plant Pathol* 12:938–954.
Schoumacher F, Erny C, Berna A et al. (1992) Nucleic acid binding properties of the alfalfa mosaic virus movement protein produced in yeast. *Virology* 188:896–899.
Schubert J, Fomitcheva V, Sztangret-Wisniewskab J (2007) Differentiation of *Potato virus Y* strains using improved sets of diagnostic PCR-primers. *J Virol Methods* 140:66–74.
Sharma SR, Varma V (1982) Control of virus diseases by oil sprays. *Zentralbl Mikrobiol* 137:329–347.
Shukla DD, Frenkel M, Ward C (1991) Structure and function of the potyvirus genome with special reference to the coat protein coding region. *Can J Plant Pathol* 13:178–191.
Shukla DD, Strike PM, Tracy SL et al. (1988) The N and C termini of the coat proteins of potyviruses are surface-located and the N-terminus contains the major virus-specific epitopes. *J Gen Virol* 69:1497–1508.
Shukla DD, Ward CW (1989) Identification and classification of potyviruses on the basis of coat protein sequence data and serology: Brief review. *Arch Virol* 106:171–200.
Shukla DD, Ward CW, Brunt AA (1994) *The Potyviridae*, CAB International, Wallingford, UK.
Shukla DD, Ward C, Brunt A, Berger P (1998) Potyviridae family. In: *CMI/AAB Descriptions of Plant Viruses* No 366, CMI/AAB, Kew, Surrey, England.
Silberschmidt K (1956) Una doenca do tomateiro em piedade causada pelo virus Y da batatinha. *Arq Inst Biol (Sao Paulo)* 23:125–150.
Silberschmidt K (1957) Cross-protection ("premunity") tests with two strains of potato virus Y in tomatoes. *Turrialba* 7:34–43.
Silberschmidt K (1960) Types of potato virus Y necrotic to tobacco: History and recent observation. *Am Potato J* 37:151–159.
Simon AE, Bujarski JJ (1994) RNA-RNA recombination and evolution in virus-infected plants. *Ann Rev Phytopathol* 35:337–362.
Simons JN (1959) Potato virus Y appears in additional areas of peppers and tomato production in South Florida. *Plant Dis Rep* 43:710–711.
Singh MN, Khurana SM Paul, Nagaich BB. et al. (1988) Environmental factors influencing aphid transmission of potato virus Y and potato leafroll virus. *Potato Res* 31:501–509.
Singh RP, Valkonen JP, Gray SM et al. (2008) Discussion paper: The naming of Potato virus Y strains infecting potato. *Arch Virol* 153:1–13.
Smith DB, McAllister J, Casino C, Simmonds P (1997) Virus "quasispecies": Making a mountain out of a molehill? *J Gen Virol* 78:1511–1519.
Smith HA, Swaney SL, Parks TD, Wernsman EA, Dougherty WG (1994) Transgenic plant virus resistance mediated by untranslatable sense RNAs: Expression, regulation and fate of nonessential RNAs. *Plant Cell* 6:1441–1453.
Smith KM (1931) On the composite nature of certain potato virus diseases of the mosaic group. *Nature* 127:702.
Smith KM, Dennis RW (1940) Some notes on a suspected variant of *Solanum virus 2 (Potato virus Y)*. *Ann Appl Biol* 27:65–70.
Soumounou Y, Laliberte J-F (1994) Nucleic acid-binding properties of the P1 protein of turnip mosaic potyvirus produced in *Escherichia coli*. *J Gen Virol* 75:2567–2573.

Stobbs LW, Poysa V, Van Schagen JG (1994) Susceptibility of cultivars of tomato and pepper to a necrotic strain of potato virus Y. *Can J Plant Pathol* 16:43–48.
Sturgess OW (1956) Leaf shrivelling virus diseases of the tomato. *Queensl J Agric Sci* 13:175–220.
Suehiro N, Natsuaki T, Watanabe T, Okuda S (2004) An important determinant of the ability of turnip mosaic virus to infect *Brassica* spp. And/or *Raphanus sativus* is in its P3 protein. *J Gen Virol* 85:2087–2098.
Summers CG, Mitchell JP, Stapleton JJ (2004) Management of aphid-borne viruses and *Bemisia argentifolii* (Homoptera: Aleurodidae) in zucchini squash by using UV reflective plastic and wheat straw mulches. *Envir Entomol* 33:1447–1457.
Summers CG, Mitchell JP, Stapleton JJ (2005) Mulches reduce aphid-borne viruses and whiteflies in cantaloupe. *Calif Agric* 59:90–94.
Szajko K, Strzelczyk-Zyta D, Marczewski W (2014) *Ny-1* and *Ny-2* genes conferring hypersensitive response to potato virus Y (PVY) in cultivated potatoes: Mapping and marker-assisted selection validation for PVY resistance in potato breeding. *Mol Breed* 34:267–271.
Talbot NJ (2004) *Plant-Pathogen Interaction*. Blackwell Publishing, Oxford, UK.
Theoming U (2003) *Potato Atlas and Compendium of Pakistan*. Pak Agri Res Council, Islamabad, Pakistan.
Thompson GJ, Hoffman DC, Prins PJ (1987) A deviant strain of potato virus Y infecting potatoes in South Africa. *Potato Res* 30:219–228.
Tian YP, Valkonen JP (2013) Genetic determinants of *Potato virus Y* required to overcome or trigger hypersensitive resistance to PVY strain group O controlled by the gene *Ny* in potato. *Mol Plant Micr Interact* 26:297–305.
Tomlinson JA (1987) Epidemiology and control of virus diseases of vegetables. *Ann Appl Biol* 110:661–681.
Tordo VM (1993) *Région 5'du génome du virus Y de la pomme de terre: Etude par séquençage de son polymorphisme et des motifs conservés. Contribution à l'étude de la protéine P1. PhD thesis*, University of Paris.
Tordo VM, Chachulska AM, Fakhfakh H et al. (1995) Sequence polymorphism in the 5′ NTR and in the P1 coding region of potato virus Y genomic RNA. *J Gen Virol* 76:939–949.
Tribodet M, Glais L, Kerlan C, Jacquot E (2005) Characterization of Potato virus Y (PVY) molecular determinants involved in the vein necrosis symptom induced by PVYN isolates in infected Nicotiana tabacum cv. xanthi. *J Gen Virol* 86:2101–2105.
Urcuqui-Inchima S, Maia I, Arruda P et al. (2000) Deletion mapping of the potyviral helper component-proteinase reveals two regions involved in RNA binding. *Virology* 268:104–111.
Valkonen JP (2007) Potato viruses: Economical losses and biotechnological potential. In: Vreugdenhil D, Bradshaw J, Gebhardt C et al. (eds). *Potato Biology and Biotechnology*, Elsevier, Amsterdam, Netherlands, pp. 619–641.
Valkonen JP (2015) Elucidation of virus-host interactions to enhance resistance breeding for control of virus diseases in potato. *Breeding Science* 65:69–76.
Valkonen JP, Jones RA, Slack SA, Watanabe KN (1996) Resistance specificities to viruses in potato: Standardization of nomenclature. *Plant Breed* 115:433–438.
van den Heuvel JF, Verbeek M, Van der Wilk F (1994) Endosymbiotic bacteria associated with circulative transmission of potato leafroll virus by *Myzus persicae*. *J Gen Virol* 75:2559–2565.
Van der Vlugt R, Allefs S, De Haan P, Goldbach R (1989) Nucleotide sequence of the 3′-terminal region of potato virus YN RNA. *J Gen Virol* 70:229–233.
Van der Vlugt R, Leunissen J, Goldbach R (1993) Taxonomic relationships between distinct potato virus Y isolates based on detailed comparisons of the viral coat proteins and 3′ non-translated regions. *Arch Virol* 131:361–375.
van Toor RF, Drayton GM, Lister RA, Teulon DA (2009) Targeted insecticide regimes perform as well as a calendar regime for control of aphids that vector viruses in seed potatoes in New Zealand. *Crop Prot* 28:599–607.
Vanderveken JJ (1977) Oils and other inhibitors of nonpersistent virus transmission. In: Harris KF, Maramorosch K (eds). *Aphids as Virus Vectors*. Academic Press, New York, pp. 435–454.
Verchot J, Carrington JC (1995a) Debilitation of plant potyvirus infectivity by P1 proteinase-inactivating mutations and restoration by second-site modifications. *J Virol* 69:1582–1590.
Verchot J, Carrington JC (1995b) Evidence that the potyvirus P1 proteinase functions in trans as an accessory factor for genome amplification. *J Virol* 69:3668–3674.
Verchot J, Herndon K, Carrington J (1992) Mutational analysis of the tobacco etch potyviral 35-kDa proteinase: Identification of essential residues and requirements for autoproteolysis. *Virology* 190:298–306.

Vidal S, H Cabrera, RA Andersson, A Fredriksson, JPT Valkonen (2002) Potato gene *Y-1* is an *N* gene homolog that confers cell death upon infection with *Potato virus Y*. *Mol Plant Micr Interact* 15:717–727.

Voinnet O, Baulcombe DC (1997) Systemic signalling in gene silencing. *Nature* 389:553.

Walter JM (1967) Hereditary resistance to disease in tomato. *Ann Rev Phytopathol* 5:131–144.

Wang R, Powell G, Hardie J, Pirone T (1998) Role of the helper component in vector-specific transmission of potyviruses. *J Gen Virol* 79:1519–1524.

Watson MA (1956) The effect of different host plants of potato virus C in determining its transmission by aphids. *Ann Appl Biol* 44:599–607.

Weidemann HL, Maiss E (1996) Detection of the potato tuber necrotic ringspot strain of potato virus Y (PVYNTN) by reverse transcription and immunocapture polymerase chain reaction. *J Plant Dis Prot* 103:337–345.

Whitworth JL, Nolte P, McIntosh C, Davidson R (2006) Effect of potato virus Y on yield of three potato cultivars grown under different nitrogen levels. *Plant Disease* 90:73–76.

Wiersema HT (1972) Breeding for resistance. In: JA de Bokx (ed.). *Viruses of Potatoes and Seed-Potato Production*. Pudoc, Wageningen, Netherlands, pp. 174–187.

Wittmann S, Chatel H, Fortin M, Laliberte J (1997) Interaction of the viral protein genome linked of turnip mosaic potyvirus with the translational eukaryotic initiation factor (iso) 4E of *Arabidopsis thaliana* using the yeast two-hybrid system. *Virology* 234:84–92.

Yang L, Hidaka M, Masaki H, Uozumi T (1998) Detection of potato virus Y P1 protein in infected cells and analysis of its cleavage site. *Biosci Biotech Biochem* 62:380–382.

Zhu C, Song Y, Yin G, Wen F (2009) Induction of RNA-mediated multiple virus resistance to *Potato virus Y*, *Tobacco mosaic virus* and *Cucumber mosaic virus*. *J Phytopathol* 157:101–107.

Zimnoch-Guzowska E, Yin Z, Chrzanowska M, Flis B (2013) Sources and effectiveness of potato PVY resistance in IHAR's breeding research. *Am J Potato Res* 90:21–27.

4 Tobamoviruses and Their Diversity

Yuri L. Dorokhov, Ekaterina V. Sheshukova, and Tatiana V. Komarova

CONTENTS

4.1 INTRODUCTION

Tobacco mosaic virus (TMV), which is a filterable agent identified by Dmitri Ivanovsky and Martinus Willem Beijerinck, was the first virus described and later had profound significance in addressing fundamental questions regarding the nature of viruses in general. Why did TMV play a major role in defining what a virus is? Three circumstances contributed to the fact that it was TMV from which the history of virology began. First, there was an economic interest in tobacco as a profitable crop in the second half of the nineteenth century. For example, Adolf Mayer, a German who became the director of the Agricultural Experiment Station at Wageningen in The Netherlands in 1886, described the decline in tobacco yields from curling and brittleness of their leaves, which made tobacco unsuitable for the production of cigars. Dimitrii Ivanovsky, while still a student, was sent to Bessarabia, Ukraine and Crimea, provinces of the Russian Empire, by the Ministry of Agriculture to study the tobacco disease that was greatly damaging tobacco plantations in 1887, and again in 1890. Second, the high level of TMV production: the yield of purified TMV reaches 10 g/kg of leaf fresh weight (Komarova et al. 2011). Finally, TMV virion is extremely stable (Caspar 1963). All of these circumstances predetermined TMV to become the first identified virus and an object of virology and molecular biology (Scholthof et al. 2011; Creager et al. 1999). TMV was the first virus to be chemically purified and visualized in an electron microscope. Its RNA was used in the first decisive experiments that showed that nucleic acid alone is sufficient for viral infectivity. TMV coat protein (CP) was the first viral protein whose amino acid sequence was determined. Today, TMV is still an object of virology and molecular biology but has also gained great importance in solving practical problems of biotechnology. Vectors based on the genome of Brassicaceae-infecting tobamovirus allowed the first production of anti-cancer antibodies and vaccines in plants (Gleba et al. 2014).

In this chapter, we consider the *Tobamovirus* genus of the *Virgaviridae* family as well as its classification, origin, and evolution. We review current knowledge of TMV genome diversity, symptomatology, replication, and cell-to-cell movement.

4.2 TRANSMISSION AND SYMPTOMATOLOGY

Unlike most plant viruses, tobamoviruses are not transmitted by insects. Instead, TMV is spread mechanically through wounding. Damage to the plant epidermis caused by abiotic (wind, hail, and rain) and biotic (insects and animals) factors may allow the penetration of TMV. TMV is very easily transmitted when an infected leaf rubs against the leaf of a healthy plant. Tobamoviruses are also carried on seeds, leading to the infection of germinating seedlings. Tobamoviruses are easily spread by contact, including touching infected and healthy plants during operations such as transplanting, pruning, tying, cultivation, spraying, and harvesting. Contaminated tools and the hands of workers after smoking cigarettes may also serve as a source of infection.

Mechanical damage leads to the emission of methanol, which is quickly perceived by intact leaves of damaged plants and neighboring plants. Exposure to methanol may result in a "priming" effect on intact leaves that promotes cell-to-cell communication, facilitating the spread of TMV in neighboring plants (Dorokhov et al. 2012; Komarova et al. 2014).

Symptoms induced by TMV are dependent on the host plant species and are described as mosaic (Figure 4.1), including mottling, necrosis, stunting, leaf curling, and yellowing of plant tissues. The symptoms are dependent on the virus strain, age of the infected plant, and environmental conditions. Plants infected as seedlings are usually stunted and pale.

FIGURE 4.1 Mosaic symptoms on *N. tabacum* cv. Samsun leaves systemically infected with tobacco mosaic virus.

4.3 VIRION STRUCTURE AND STABILITY

Tobacco mosaic virus is a 300-nm rigid rod-like particle that is 18 nm in diameter (Figure 4.2). TMV is composed of 2130 subunits of a CP, which has a molecular weight of approximately 17,500 kDa, and one molecule of genomic single strand RNA, which is approximately 6400 bases long (Stubbs and Kendall 2012). Virions contain no lipids and no carbohydrates. A central hole that is approximately 4 nm in diameter runs down the viral axis. A single RNA strand follows the viral helix, with three nucleotides binding to each coat protein subunit and 16.34 protein subunits per turn of the viral helix. The complete structure of the TMV U1 strain was determined by X-ray fiber diffraction at a 2.9 Å resolution (Namba et al. 1989). The CP N- and C-termini are located on the surface of the virion. The interactions between neighboring CP subunits make up a mosaic of hydrophobic and hydrophilic sites, stabilizing the overall TMV structure. Moreover, U1 strain structures have also been determined for several other tobamoviruses (Holmes 2010), including tobacco mild green mosaic virus (TMGMV, also called TMV-U2), cucumber green mottle mosaic virus (CGMMV), ribgrass mosaic virus (RMV), Hibiscus latent Fort Pierce virus (HLFPV), and odontoglossum ringspot virus (ORSV).

Features of the TMV structure make it one of the most stable viruses. TMV is extremely stable: native particles are stable after heating over to 60°C for 30 min, at pH values ranging from 2 to 10, and in organic solvents of up to 80% by volume. In alkaline media (pH >9), the particles sequentially lose protein subunits from the 5′-end of the RNA. TMV remains intact in water-rich tetrahydrofuran and dimethylsulfoxide (DMSO) mixtures. At high DMSO concentrations (72% v/v), a significant portion of the TMV rods disassemble in a unique polar fashion starting at the 5′-end of the viral RNA (Nicolaïeff and Lebeurier 1979). Furthermore, TMV is very stable; for example, *in vitro* longevity in infected sap is 3000 days. Purified TMV is infectious even after 50 years of storage in the laboratory at 4°C. In addition, it has a very wide survival range. TMV can be detected in cigarettes and in the

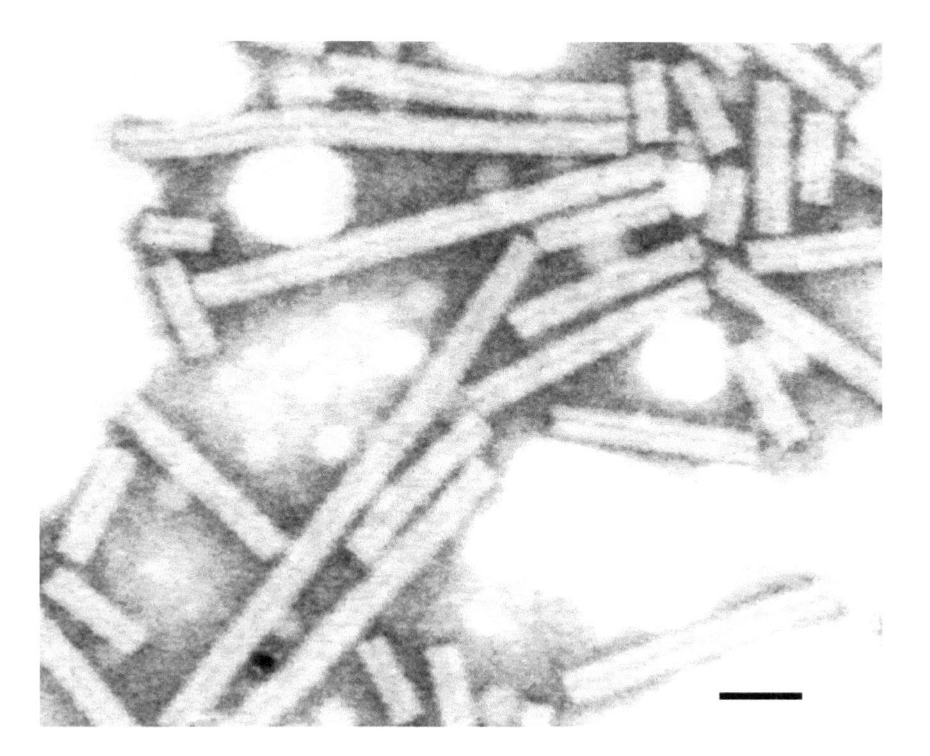

FIGURE 4.2 Electron microscope images of tobacco mosaic virus particles. Bar = 60 nm.

saliva of smokers (Balique et al. 2012). TMV has also been detected in the environment, including soil, water, clouds, and even in Greenland glacial ice that was approximately 140,000 years old (Castello et al. 1999).

4.4 GENOME ORGANIZATION

Tobamoviruses are rod-shaped (Figure 4.2) plant viruses with a single component Sindbis-like (+)-RNA genome that is approximately 6400 nucleotides in length. The complete nucleotide sequence of TMV (6395 nts) was determined in 1982 (Goelet et al. 1982) and placed in the National Center for Biotechnology Information as reference sequence for TMV (NC_001367.1). Interestingly, if a cDNA is created following this sequence, as Bret Cooper tried to do, it would be noninfectious, but to turn it into an infectious copy only two nucleotide substitutions need to be made (Cooper 2014). Infectious RNA from full-length cDNA clones was obtained in 1986 (Dawson et al. 1986; Meshi et al. 1986). TMV was the second plant virus after Brome mosaic virus (Ahlquist and Janda 1984) for which an infectious copy was obtained.

Genomic RNA serves as a template for protein synthesis and has a 7-methylguanosine 5′ triphosphate [$m^7G(5')ppp$] cap structure at the 5′ terminus. The 5′ untranslated region (UTR) is 70 nucleotides in length, with a (CAA)n motif and direct ACAAUUAC repeat. The 3′ UTR is ~200 nucleotides in length and contains a tRNA-like structure and an upstream region with three consecutive pseudoknot structures. The genome contains six ORFs and encodes six proteins: a 126-kDa protein (ORF2), a read-through derivative of 183 kDa (ORF1), a 30-kDa protein (ORF4), and a 17.4-kDa CP (ORF5) (Figure 4.3, Table 4.1). A fifth 54-kDa protein (ORF3) has an undefined virological role (if any). A sixth, 4.8-kDa protein encoded by ORF6 that overlaps the ORF4 and ORF5 genes influences symptomology, but is not found in all tobamovirus species (Canto et al. 2004). The 126-kDa and 183-kDa proteins participate in viral RNA replication and are synthesized by translation of genomic RNA (Ishibashi and Ishikawa 2016), whereas MP and CP are synthesized from respective subgenomic RNAs that are 5′-capped and 3′-co-terminal to genomic RNA. The 30-kDa protein participates in viral spread from infected cells to neighboring uninfected cells in host plant tissues (Atabekov and Dorokhov 1984).

The genome organizations of Brassicaceae-infecting tobamoviruses differ from those of other tobamoviruses, as there is a 77 nucleotide overlap of the MP and CP ORFs, which results in the loss of ORF6 (Figure 4.4 and Table 4.1). Moreover, due to the internal ribosome entry site (IRES) upstream of the CP gene, the Brassicaceae-infecting tobamovirus crTMV (crucifer-infecting TMV) is able to provide early synthesis of CP directly from genomic RNA (Dorokhov et al. 2002).

4.5 TAXONOMY AND HOST RANGE

Tobamoviruses are usually named after the host plant from which they have been isolated, although many of them have a wider host range. For example, crucifer-infecting tobamoviruses infect not only Brassicaceae but also Solanaceae. By the end of 2016, the genus *Tobamovirus* consisted of 35 species (ICTV, virus taxonomy: 2016 release), of which TMV is the type species. The natural host ranges of the members of the *Tobamovirus* genus include Solanaceae, Brassicaceae, Cucurbitaceae, Malvaceae, Cactaceae, Passifloraceae, Fabaceae, Apocynaceae, and Orchidaceae (Table 4.1). Initially, virus classification was built on the basis of their serological behavior or the amino acid sequence of their coat proteins (Gibbs 1999). Now, species demarcation criteria in the genus is based on the nucleotide sequence similarity. A less than 10% overall nt sequence difference is considered necessary to characterize strains of the same species, although most of the sequenced species have considerably less than 90% sequence identity. A variant of this approach is the use of nucleic acid signatures and specific combinations of nucleotides (nucleotide combination [NC] motifs) as criteria. The procedure to search for the "4404–4450 NC motif" in a convenient and phylogenetically conserved region of the tobamovirus RNA polymerase gene was previously described (Gibbs et al. 2004). Now *Tobamovirus*

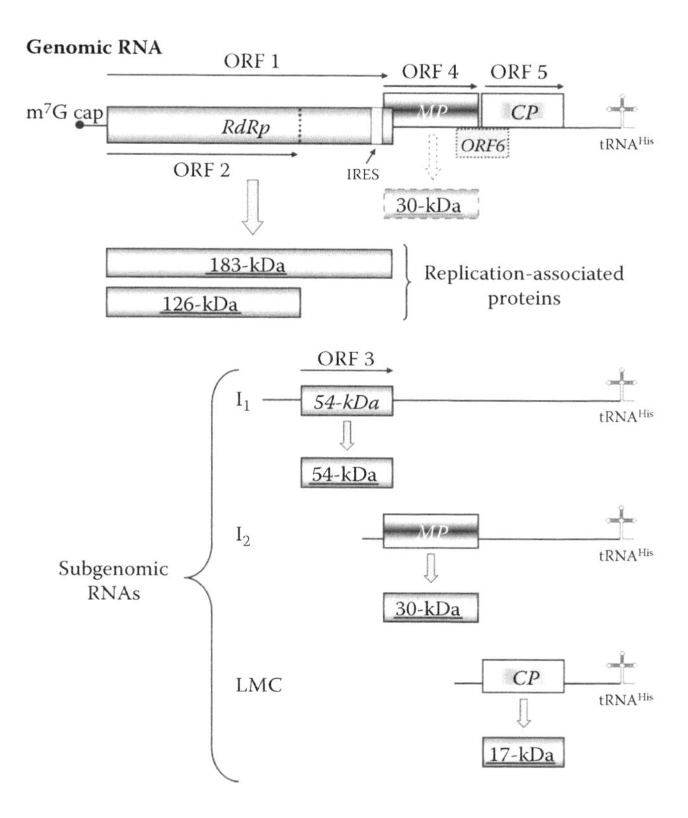

FIGURE 4.3 A schematic representation of tobacco mosaic virus genomic organization and expression. Tobacco mosaic virus strain U1 genomic RNA is 6395 nt long, 5′-capped and contains tRNA-like structure at the 3′-end. There are six ORFs 183- and 126-kDa replication proteins are translated directly from the genomic RNA in a 5′-dependent manner while MP could be synthesized via internal ribosome entry site (IRES). However, *MP* and *CP* are mainly expressed from the subgenomic RNAs (sgRNAs) I_2 and LMC, respectively.

genus subgroups are defined according to their genome organization, sequence homologies, and host range. Tobamoviruses can be divided into at least nine subgroups: Solanaceae, Brassicaceae, Cucurbitaceae, Malvaceae, Cactaceae, Passifloraceae, Fabaceae, Apocynaceae, and Orchidaceae-infecting tobamoviruses, according to the amino acid composition of their CPs. Such grouping agrees with the serological specificity, CP amino acid sequence similarity, and host from which the viruses were originally isolated. Initially, tobamoviruses were classified into two subgroups (Fukuda et al. 1981), which had different genomic locations of their origin of virion assembly (OA). The OA is located within the MP ORF of the representatives of subgroup 1, which includes Solanaceae-infecting tobamoviruses, with the exception of the ORSV that infects orchids. The Solanaceae family comprises 75 genera with over 2000 species, including peppers, potatoes, tomatoes, and tobacco. The tobamoviruses of subgroup 2 include Cucurbitacea- and Fabaceae-infecting tobamoviruses and the OA site is within the CP gene. Once the complete nucleotide sequences of the first four tobamoviruses isolated from Brassicaceae (former Cruciferaceae) became available (Dorokhov et al. 1994; Lartey et al. 1995; Aguilar et al. 1996), a third group was proposed (Lartey et al. 1996). This group also includes the first tobamovirus that was shown to be pathogenic to Brassicaceae, Holmes' RMV, which was originally known as the ribgrass strain of TMV (Chavan and Pearson 2016; Wetzel et al. 2006). The OA of the viruses from subgroup 3 is again located in the MP ORF; however, the genome organization of crucifer-pathogenic tobamoviruses is different from those of other tobamoviruses, and their MP and CP ORFs overlap by 77 nucleotides. Isolation of new tobamoviruses, such as Frangipani mosaic virus (FrMV) (Lim et al. 2010), Maracuja mosaic virus (MarMV) (Song et al. 2006), Passion

TABLE 4.1
Tobamovirus Taxonomy and Genome Structure

#	Virus	Host Range	EMBL ID	Genome (nts)	Open Reading Frames and Intergenic Junctions, nts Position						
					ORF1 180K	ORF2 130K	ORF1-4 Junction	ORF4 MP	ORF4-5 Junction	ORF5 CP	ORF6
1	Brugmansia mild mottle virus (BMMV)		AM398436	6381	61-4902	61-3408	11-nt overlap	4892-5707	3-nt spacer	5709-6188	–
2	Bell pepper mottle virus (BpeMV)		DQ355023	6375	72-4868	72-3422	13-nt overlap	4856-5692	4-nt spacer	5695-6177	–
3	Obuda pepper virus (ObPV)		D13438	6507	69-4919	69-3414	9-nt spacer	4927-5751	4-nt spacer	5754-6239	–
4	Paprika mild mottle virus (PaMMV)		KX187305	6522	69-4919	69-3416	18-nt spacer	4936-5745	4-nt spacer	5748-6233	+
5	Pepper mild mottle virus (PMMV)		AB276030	6356	70-4908	70-3423	2-nt overlap	4909-5682	4-nt spacer	5685-6158	–
6	Rehmannia mosaic virus (ReMV)	Solanaceae	AB628188	6395	72-4922	72-3422	17-nt overlap	4906-5709	4-nt spacer	5712-6191	+
7	Tobacco mosaic virus (TMV)		V01408	6395	69-4917	69-3419	15-nt overlap	4903-5710	3-nt spacer	5712-6189	+
8	Tobacco mild green mosaic virus (TMGMV)		M34077	6355	71-4900	3407-4900	11-nt overlap	4890-5660	7-nt spacer	5666-6145	+
9	Tomato mosaic virus (ToMV)		X02144	6384	72-4922	72-3422	17-nt overlap	4906-5634	70-nt spacer	5703-6182	+
10	Tropical soda apple mosaic virus (TSAMV)		KU659022	6350	70-4908	70-3423	3-nt spacer	4910-5680	4-nt spacer	5683..6165	+
11	Yellow tailflower mild mottle virus (YTMMV)		KF495565	6379	71-4915	71-3430	10-nt spacer	4924-5724	4-nt spacer	5727-6203	–
12	Odontoglossum ringspot virus (ORSV)	Orchidaceae	AY571290	6612	64-4902	64-3402	95-nt overlap	4808-5719	4-nt spacer	5722-6198	+
13	Cucumber fruit mottle mosaic virus (CFMMV)		AF321057	6562	63-5063	63-3561	2-nt spacer	5064-5840	7-nt spacer	5846-6334	–
14	Cucumber green mottle mosaic virus (CGMMV)	Cucurbitaceae	AB015146	6423	61-5007	61-3495	14-nt overlap	4994-5788	26-nt overlap	5763-6248	–
15	Cucumber mottle virus (CuMoV)		AB261167	6485	61-5079	61-3561	14-nt overlap	5066-5854	29-nt overlap	5826-6314	–
16	Kyuri green mottle mosaic virus (KGMMV)		AB015145	6515	58-5070	58-3558	4-nt spacer	5073-5861	6-nt spacer	5866-6351	–
17	Zucchini green mottle mosaic virus (ZGMMV)		AJ252189	6513	60-5072	60-3560	4-nt spacer	5075-5860	6-nt spacer	5865-6350	–
18	Hibiscus latent Fort Pierce virus (HLFPV)	Malvaceae	KP828049	6453	49-4989	49-3465	20-nt overlap	4970-5824	29-nt overlap	5796-6272	–
19	Hibiscus latent Singapore virus (HLSV)		AF395898	6485	59-4978	59-3463	5-nt spacer	4968-5816	14-nt overlap	5803-6294	–

(Continued)

TABLE 4.1 (CONTINUED)
Tobamovirus Taxonomy and Genome Structure

#	Virus	Host Range	EMBL ID	Genome (nts)	Open Reading Frames and Intergenic Junctions, nts Position						
					ORF1 180K	ORF2 130K	ORF1-4 Junction	ORF4 MP	ORF4-5 Junction	ORF5 CP	ORF6
20	Clitoria yellow mottle virus (CYMV)	Cactaceae	JN566124	6514	76-4977	76-3477	8-nt overlap	4970-5860	26-nt overlap	5835-6326	–
21	Sunn-hemp mosaic virus (SHMV) [Tobacco mosaic virus (cowpea strain)]	Cactaceae	U47034 J02413	N/A	75-4962	75-3467	8-nt overlap	4955-5805	29-nt overlap	5777-6485	–
22	Crucifer-infecting tobacco mosaic virus (Cr-TMV)[a]	Brassicaceae	Z29370	6312	69-4874	69-3392	4-nt spacer	4877-5680	77-nt overlap	5604-6077	–
23	Turnip vein-clearing virus (TVCV)	Brassicaceae	U03387	6311	68-4873	68-3391	4-nt spacer	4876-5679	77-nt overlap	5603-6076	–
24	Ribgrass mosaic virus (RMV)	Brassicaceae	HQ667979	6311	68-4873	68-3391	4-nt spacer	4876-5679	77-nt overlap	5603-6076	–
25	Wasabi mottle virus (WMV)	Brassicaceae	AB017503	6298	67-4860	67-3378	4-nt spacer	4863-5666	77-nt overlap	5590-6063	–
26	Youcai mosaic virus (YoMV)[b]	Brassicaceae	AF254924	6301	69-4862	69-3380	13-nt spacer	4874-5668	77-nt overlap	5592-6065	–
27	Streptocarpus flower break virus (SFBV)	Gesneriaceae	AM040955	6279	59-4879	59-3376	20-nt overlap	4860-5726	77-nt overlap	5668-6072	–
28	Frangipani mosaic virus (FrMV)	Apocynaceae	HM026454	6643	78-5030	78-3518	14-nt overlap	5017-5787	59-nt spacer	5845-6369	–
29	Passion fruit mosaic virus (PafMV)	Passifloraceae	HQ389540	6791	57-4892	57-3359	375-nt spacer	5266-6198	77-nt overlap	6122-6613	–
30	Maracuja mosaic virus (MarMV)	Passifloraceae	DQ356949	6794	57-4900	55-3366	371-nt spacer	5270-6202	77-nt overlap	6126-6617	–
31	Cactus mild mottle virus (CMMoV)	Cactaceae	EU043335	6449	57-5006	57-3485	47-nt overlap	4960-5880	146-nt overlap	5735-6220	–
32	Rattail cactus necrosis-associated virus (RCNaV)	Cactaceae	JF729471	6506	67-4995	67-3507	7-nt spacer	5001-6005	212-nt overlap	5794-6306	–

Source: Aguilar I, Sánchez F, Martín AM, Martínez-Herrera D, Ponz F (1996). Nucleotide sequence of Chinese rape mosaic virus (oilseed rape mosaic virus), a crucifer tobamovirus infectious on Arabidopsis thaliana. *Plant Mol Biol* 30:191–197; Lartey RT, Voss TC, Melcher U (1996). Tobamovirus evolution: Gene overlaps, recombination, and taxonomic implications. *Mol Biol Evol* 13:1327–1338.

[a] Strain of TVCV.

[b] Synonymous with Chinese rape mosaic virus or Oilseed rape mosaic virus.

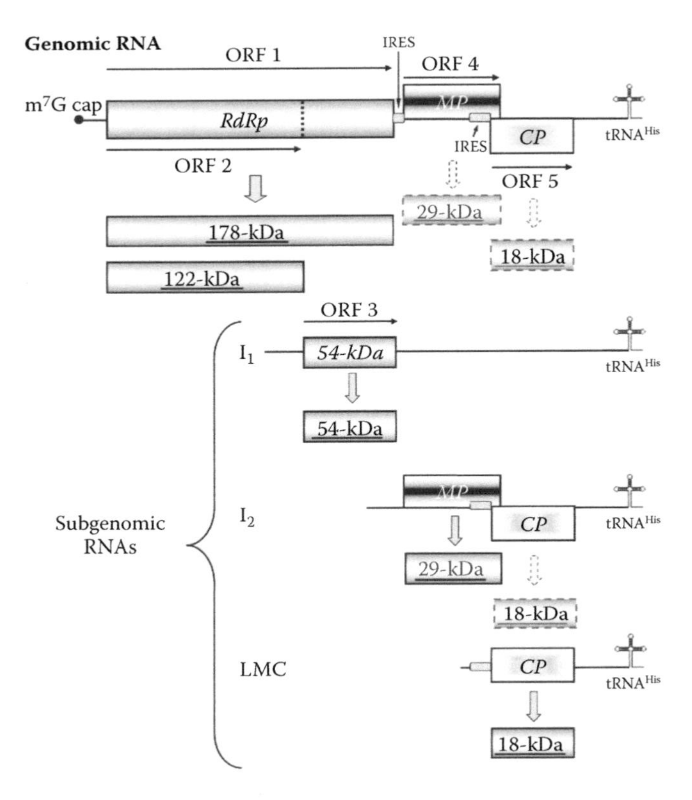

FIGURE 4.4 A schematic representation of Crucifer-infecting tobacco mosaic virus (crTMV) genomic organization and expression. crTMV genomic RNA is 5′-capped and contains tRNA-like structure at the 3′-end. 178- and 122-kDa replication proteins are translated directly from the genomic RNA in a 5′-dependent manner while MP and CP could be synthesized via internal initiation of translation from IRES. However, *MP* and *CP* are mainly expressed from the subgenomic RNAs (sgRNAs) I_2 and LMC, respectively.

fruit mosaic virus (PafMV) (Song and Ryu 2011), Cactus mild mottle virus (CMMoV) (Min et al. 2009; Song et al. 2006), and Rattail cactus necrosis-associated virus (RCNaV) (Kim et al. 2012), and complete sequencing of their genomes revealed that the MP-CP overlap is not unique to Brassicaceae-infecting tobamoviruses, as Apocynaceae-, Passifloracea-, and Cactaceae-infecting tobamoviruses also have this feature. Using the MP-CP overlap as a classifier is rather arbitrary because some representatives of tobamoviruses assigned to group 2 also contain a small MP-CP overlap, although it is no more than 29-nucleotides in length.

We believe that the tobamovirus grouping shown in Table 4.1 should be based on the following criteria: the host from which the virus was originally isolated, MP-CP overlap, and CP amino acid primary structure (Figure 4.5).

Molecular phylogenetic analysis of tobamoviruses calculated from the amino acid sequences of their MPs (Figure 4.6) and RNA-dependent RNA polymerases (RdRp) (Figure 4.7) generally confirms the conclusions drawn on the basis of the CP amino acid primary structure. The minor differences between these dendrograms probably reflect differences in the origin of the tobamoviruses.

4.6 ORIGIN AND EVOLUTION

The origin of tobamoviruses is ancient, and it is believed that tobamoviruses coevolved with their hosts (Stobbe et al. 2012; Gibbs et al. 2015). In 1999, Adrian Gibbs assessed the age of origin

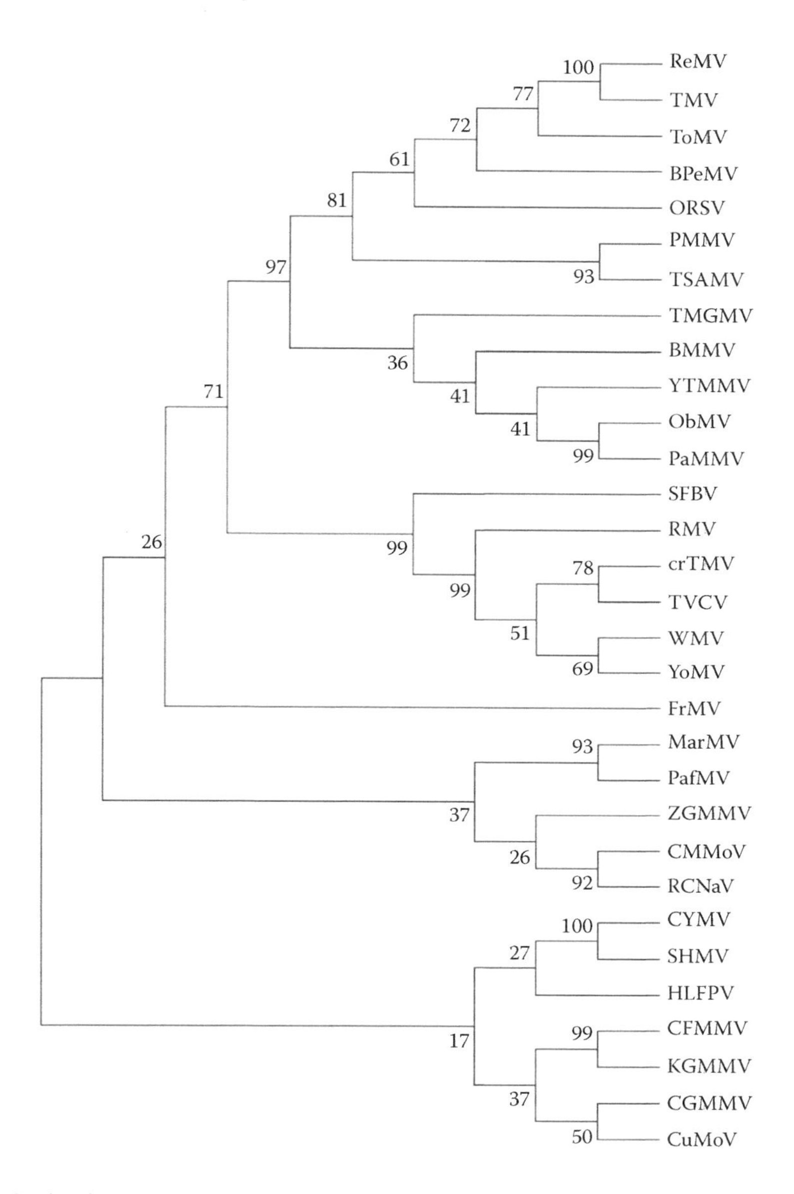

FIGURE 4.5 Molecular phylogenetic analysis of tobamoviruses calculated from the amino acid sequences of their CPs by the maximum likelihood method and based on the JTT matrix-based model. (Jones DT, Taylor WR, Thornton JM (1992). The rapid generation of mutation data matrices from protein sequences. *Comput Appl Biosci CABIOS* 8:275–282.) The evolutionary analyses were conducted in MEGA7. (Kumar S, Stecher G, Tamura K (2016). MEGA7: Molecular evolutionary genetics analysis version 7.0 for bigger datasets. *Mol Biol Evol* 33:1870–1874.).

of tobamovirus by extrapolation and calculated that "If they were evolving at a rate of 1% every 2–10 million years, like many other proteins (Wilson et al. 1977), then the proto-tobamovirus probably arose about 120–600 million years ago." Later, he concluded that "all the tobamoviruses, except those found in brassicas, have probably co-diverged with their eudicotyledonous hosts, since they formed the asterid, rosid, and caryophyllid lineages around 112.9 million years ago" (Gibbs et al. 2015). The New World, in particular some parts of Peru, Bolivia, or Brazil, is the place of origin for tobamoviruses, and their appearance in Australia is a recent event associated with the breakup of Gondwana.

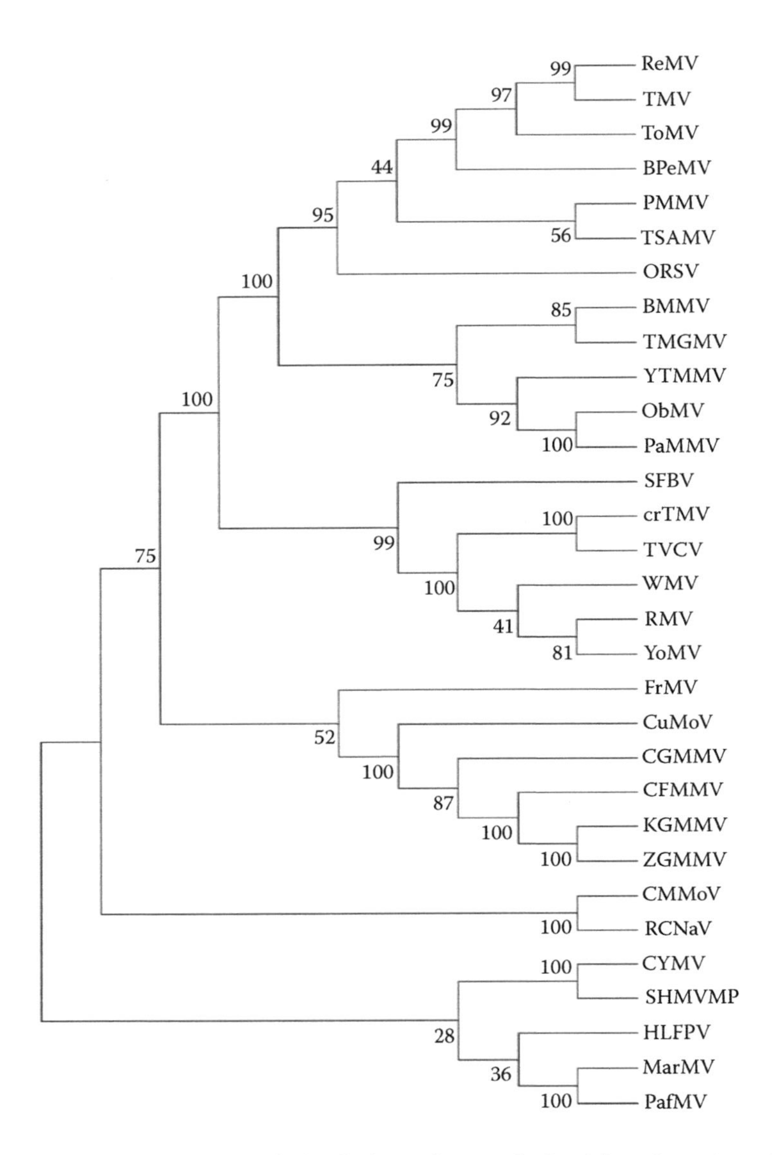

FIGURE 4.6 Molecular phylogenetic analysis of tobamoviruses calculated from the amino acid sequences of their MPs by a maximum likelihood method based on the JTT matrix-based model. (Jones DT, Taylor WR, Thornton JM (1992). The rapid generation of mutation data matrices from protein sequences. *Comput Appl Biosci CABIOS* 8:275–282.) The evolutionary analyses were conducted in MEGA7. (Kumar S, Stecher G, Tamura K (2016). MEGA7: Molecular evolutionary genetics analysis version 7.0 for bigger datasets. *Mol Biol Evol* 33:1870–1874.)

TMV, as an RNA-virus, is capable of genetic variability that is observed both in nature and in the laboratory. Adrian Gibbs estimated the rate of tobamovirus evolution as 2.2×10^{-8} nucleotide substitutions per site as calculated from a maximum-likelihood analysis of 320 tobamovirus CP genes representing approximately 25 species (Gibbs et al. 2010). A major role in the evolution of the virus was played by agriculture when human activity appeared 8000–13000 years ago in at least nine geographic areas around the world. Cultivation of plants and their distribution throughout the world contributed to the spread of viruses as well as evolution, in particular for tobamoviruses (Gibbs et al. 2010). The origin of tobamovirus genes is well described by the theory of modular evolution, which

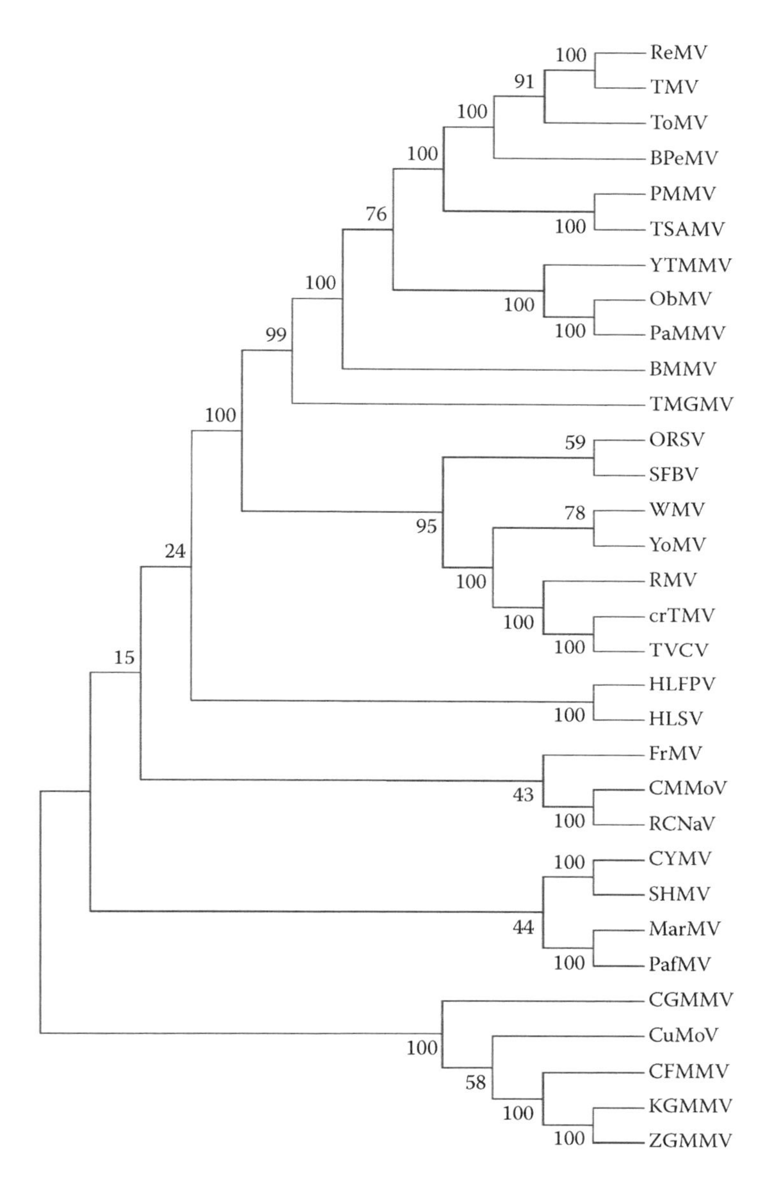

FIGURE 4.7 Molecular phylogenetic analysis of tobamoviruses calculated from the amino acid sequences of their RdRps by the maximum likelihood method and based on the JTT matrix-based model. (Jones DT, Taylor WR, Thornton JM (1992). The rapid generation of mutation data matrices from protein sequences. *Comput Appl Biosci CABIOS* 8:275–282.) The evolutionary analyses were conducted in MEGA7. (Kumar S, Stecher G, Tamura K (2016). MEGA7: Molecular evolutionary genetics analysis version 7.0 for bigger datasets. *Mol Biol Evol* 33:1870–1874.)

was detected earlier in bacteriophages and emerged as a major feature of the evolution of all viruses (Hull 2014). TMV replicase likely has its origin from eukaryotic RdRp (Koonin and Dolja 1993; Mushegian et al. 2016). The cell-to-cell movement protein of tobamoviruses forms a 30K superfamily and is thought to originate from integrated pararetroviruses (Mushegian and Elena 2015). The tobamovirus coat proteins are clearly related in sequence and structure to those of other viruses with rod-shaped and filamentous virions and have close relationships with chymotrypsin-like proteinases (Koonin and Dolja 1993).

4.7 REPLICATION OF TOBAMOVIRUS RNA

TMV infection starts when the virus enters a primary infected cell after mechanical damage of the cell wall and plasma membrane. Within a few minutes after entry, the TMV CP begins to disassemble from the capsid with ribosome participation (cotranslational disassembly of virus particles) (Wilson 1984). Genomic 5′ ORFs (126 kDa and 183 kDa) are immediately translated to form the replicase complex, which initiates replication of the viral genome (Shaw 1999). The tobamovirus replication machinery or viral factory may form on vacuolar and endoplasmic reticulum (ER) membranes. The 126-kDa and 183-kDa proteins have methyltransferase and guanylyltransferase activity, which is responsible for the formation of the RNA 5′ cap (Merits et al. 1999), as well as RNA helicase and RNA polymerase activities (Nishikiori et al. 2006). The 183-kDa protein is necessary for viral RNA replication as mutants carrying defects in the 183-kDa protein genes were unable to multiply in tobacco plants.

In addition to the 126-kDa and 183-kDa proteins, the following host proteins that were identified in a study of mutants of *A. thaliana* are involved in virus replication (Ishibashi and Ishikawa 2016):

a. The transmembrane protein TOM1, encoded by the *TOM1* gene and interacts with the helicase domain of replicase
b. Transmembrane protein TOM2, encoded by the *TOM1A* gene and interacts with the transmembrane protein TOM1
c. ADP-ribosylation factor 8 (ARF8) and translation elongation factors eEF1A and eEF1B

Thus the main steps of tobamovirus replication include (i) synthesis of viral replication proteins by translation of genomic RNA; (ii) binding of replication proteins to the 5′-terminal region of genomic RNA to inhibit further translation; (iii) formation of a replicative complex and recruitment of genomic RNA by replication proteins onto membranes interacting with the host proteins TOM1 and ARF8; and (iv) synthesis of complementary (negative-strand) RNA in the complex followed by synthesis of progeny genomic RNA.

The negative-strand RNA serves as a template for the synthesis of subgenomic RNAs (sgRNAs): dicistronic and monocistronic mRNAs encoding MP and CP, respectively. The localization of the MP and CP sgRNA promoters was determined (Grdzelishvili et al. 2000). It turned out that for MP and CP sgRNAs promoter activity, their stem-loop structures but not their particular sequences are essential.

Some tobamoviruses are capable of expressing the 3′ proximal genes not only through mechanism of subgenomization but also by an internal translation initiation directly from genomic RNA. The internal initiation of translation was shown for *MP* gene expression from genomic RNA of both TMV U1 and crTMV (Zvereva et al. 2004). In contrast to TMV strain U1, the sequence of the crTMV genomic RNA upstream of the *CP* gene includes a 148-nt internal ribosome entry site ($IRES^{CR}_{148,CP}$) (Ivanov et al. 1997) containing a stem-loop hairpin structure surrounded from both sides by a polypurin-rich region (PPR): the long 33-nt PPR33 before the hairpin and short 11-nt PPR11 after the hairpin. The PPR33 sequence is enough for $IRES^{CR}_{CP,148}$ translational activity. Moreover, a simplified version of such a sequence, which consists of only 16 (GAAA) repeats, also provides the high cross-kingdom efficiency of cap-independent translation initiation in plant and mammalian cells (Dorokhov et al. 2002). The contribution of $IRES^{CR}_{CP,148}$ to the total synthesis of the CP during viral infection is at least 3%. It has been suggested that internal translation initiation is important for the synthesis of CP and provides the early formation of the virus replication complexes and enhanced systemic movement of the virus (Dorokhov et al. 2006).

4.8 MOVEMENT

Mature virions of TMV do not participate in intercellular transport. Nevertheless, cell-to-cell movement of the viral genetic material requires a specific viral protein, MP. Modern understanding of intercellular plant virus spreading began with development of the concept of the TMV "transport protein" (Atabekov and Dorokhov 1984) or "translocation protein" (Leonard and Zaitlin 1982) or the now generally accepted term "movement protein" (Deom et al. 1987). Suggestion that TMV non-structural 30-kDa protein is able to perform the function of the MP (Atabekov and Morozov 1979) was soon confirmed by the study of the temperature-sensitive mutant Ls1 (Deom et al. 1987; Leonard and Zaitlin 1982; Taliansky et al. 1982). To explain the function of the MP, two mechanisms were originally suggested (Atabekov and Dorokhov 1984). According to the first mechanism, plasmodesmata (PD), which were initially closed to the virus, are modified by MPs to become open, allowing the transfer of viral genetic material. Thus the MP function is PD gating. The second mechanism assumed that the function the MP is in the host nucleus where MP mediates overcoming plant cell resistance to the virus.

To date, the prevailing concept is that the function of the TMV MP is associated with intracellular movement of viral RNA to the PD and PD gating, as MP has the ability to non-specifically bind to TMV RNA and interact with the ER membrane (Heinlein 2015; Tilsner et al. 2014; Ueki and Citovsky 2014). By carrying out its functions, the TMV MP interacts with cellular proteins and the structural components of PD whose ultimate role in PD gating is not clear.

TMV cell-to-cell movement and long-distance transport are distinct processes. The long-distance transport of TMV, similar to other plant viruses, is conducted through the plant vascular tissue (usually the phloem sieve tubes). PD between the sieve elements of the phloem, that is, the sieve pores, is nonselective and allow for the free passage of large macromolecular complexes. TMV requires a CP for long-distance transport, but the form in which the infection entity is transported is unknown as well as the mechanism by which the CP contributes to TMV systemic movement (Conti et al. 2017).

4.9 CONCLUSION

Currently, the genus *Tobamovirus* comprises no less than 35 species, and there is confidence that this number will increase in the future because of its adaptability and co-evolution with the host plant. In addition, the role of the agricultural activities of man cannot be ruled out for driving this increase. Tobamoviruses demonstrate high stability and resistance to external factors and have high flexibility in the organization of their genome and expression mechanisms. All of these factors allow tobamoviruses to expand their host range. The relatively simple organization, excellence in gene expression, and ability to exploit the maximal biosynthetic activity of a plant cell makes tobamoviruses a perfect tool for biotechnology. Despite the abundance of data, there are still many unknowns for TMV regarding its mechanisms of gene expression, interactions with the host, and intercellular transport of viral genetic material.

ACKNOWLEDGMENTS

Yuri Dorokhov acknowledges the support of the Russian Science Foundation (project No. 16-14-00002). Ekaterina Sheshukova would like to acknowledge the support of the Russian Foundation for Basic Research (project No. 16-34-00062_mol_a). Tatiana Komarova acknowledges the support of the President of Russian Federation grant (MD-5697.2016.4).

REFERENCES

Aguilar I, Sánchez F, Martín AM, Martínez-Herrera D, Ponz F (1996). Nucleotide sequence of Chinese rape mosaic virus (oilseed rape mosaic virus), a crucifer tobamovirus infectious on Arabidopsis thaliana. *Plant Mol Biol* 30:191–197.

Ahlquist P, Janda M (1984). cDNA cloning and in vitro transcription of the complete brome mosaic virus genome. *Mol Cell Biol* 4:2876–2882.

Atabekov JG, Dorokhov YL (1984). Plant virus-specific transport function and resistance of plants to viruses. *Adv Virus Res* 29:313–364.

Atabekov JG, Morozov SY (1979). Translation of plant virus messenger RNAs. *Adv Virus Res* 25:1–91.
Balique F, Colson P, Raoult D (2012). Tobacco mosaic virus in cigarettes and saliva of smokers. *J Clin Virol Off Publ Pan Am Soc Clin Virol* 55:374–376.
Canto T, MacFarlane SA, Palukaitis P (2004). ORF6 of Tobacco mosaic virus is a determinant of viral pathogenicity in Nicotiana benthamiana. *J Gen Virol* 85:3123–3133.
Caspar DL (1963). Assembly and stability of the tobacco mosaic virus particle. *Adv Protein Chem* 18:37–121.
Castello JD, Rogers SO, Starmer WT, Catranis CM, Ma L, Bachand GD, Zhao Y, Smith JE (1999). Detection of tomato mosaic tobamovirus RNA in ancient glacial ice. *Polar Biol* 22:207–212.
Chavan RR, Pearson MN (2016). Molecular characterisation of a novel recombinant Ribgrass mosaic virus strain FSHS. *Virol J* 13:29.
Conti G, Rodriguez MC, Venturuzzi AL, Asurmendi S (2017). Modulation of host plant immunity by Tobamovirus proteins. *Ann Bot* 119:737–747.
Cooper B (2014). Proof by synthesis of Tobacco mosaic virus. *Genome Biol* 15:R67.
Creager ANH, Scholthof K-BG, Citovsky V, Scholthof HB (1999). Tobacco mosaic virus: Pioneering research for a century. *Plant Cell* 11:301–308.
Dawson WO, Beck DL, Knorr DA, Grantham GL (1986). cDNA cloning of the complete genome of tobacco mosaic virus and production of infectious transcripts. *Proc Natl Acad Sci U S A* 83:1832–1836.
Deom CM, Oliver MJ, Beachy RN (1987). The 30-kilodalton gene product of tobacco mosaic virus potentiates virus movement. *Science* 237:389–394.
Dorokhov YL, Ivanov PA, Komarova TV, Skulachev MV, Atabekov JG (2006). An internal ribosome entry site located upstream of the crucifer-infecting tobamovirus coat protein (CP) gene can be used for CP synthesis in vivo. *J Gen Virol* 87:2693–2697.
Dorokhov YL, Ivanov PA, Novikov VK, Agranovsky AA, Morozov SY, Efimov VA, Casper R, Atabekov JG (1994). Complete nucleotide sequence and genome organization of a tobamovirus infecting cruciferae plants. *FEBS Lett* 350:5–8.
Dorokhov YL, Komarova TV, Petrunia IV, Frolova OY, Pozdyshev DV, Gleba YY (2012). Airborne signals from a wounded leaf facilitate viral spreading and induce antibacterial resistance in neighboring plants. *PLoS Pathog* 8:e1002640.
Dorokhov YL, Skulachev MV, Ivanov PA, Zvereva SD, Tjulkina LG, Merits A, Gleba YY, Hohn T, Atabekov JG (2002). Polypurine (A)-rich sequences promote cross-kingdom conservation of internal ribosome entry. *Proc Natl Acad Sci U S A* 99:5301–5306.
Fukuda M, Meshi T, Okada Y, Otsuki Y, Takebe I (1981). Correlation between particle multiplicity and location on virion RNA of the assembly initiation site for viruses of the tobacco mosaic virus group. *Proc Natl Acad Sci* 78:4231–4235.
Gibbs A (1999). Evolution and origins of tobamoviruses. *Philos Trans R Soc Lond B Biol Sci* 354:593–602.
Gibbs AJ, Armstrong JS, Gibbs MJ (2004). A type of nucleotide motif that distinguishes tobamovirus species more efficiently than nucleotide signatures. *Arch Virol* 149:1941–1954.
Gibbs AJ, Fargette D, García-Arenal F, Gibbs MJ (2010). Time—The emerging dimension of plant virus studies. *J Gen Virol* 91:13–22.
Gibbs AJ, Wood J, Garcia-Arenal F, Ohshima K, Armstrong JS (2015). Tobamoviruses have probably co-diverged with their eudicotyledonous hosts for at least 110 million years. *Virus Evol* 1:019.
Gleba YY, Tusé D, Giritch A (2014). Plant viral vectors for delivery by Agrobacterium. *Curr Top Microbiol Immunol* 375:155–192.
Goelet P, Lomonossoff GP, Butler PJ, Akam ME, Gait MJ, Karn J (1982). Nucleotide sequence of tobacco mosaic virus RNA. *Proc Natl Acad Sci U S A* 79:5818–5822.
Grdzelishvili VZ, Chapman SN, Dawson WO, Lewandowski DJ (2000). Mapping of the Tobacco mosaic virus movement protein and coat protein subgenomic RNA promoters in vivo. *Virology* 275:177–192.
Heinlein M (2015). Plant virus replication and movement. *Virology* 479–480:657–671.
Holmes KC (2010). 50 years of fiber diffraction. *J Struct Biol* 170:184–191.
Hull R (2014). Origins and evolution of plant viruses. *Plant Virology* (Fifth Edition). Academic Press, Boston, MA, pp. 423–476.
ICTV. *Virus Taxonomy*: 2016 Release. https://talk.ictvonline.org/taxonomy/ (accessed September 28, 2017).
Ishibashi K, Ishikawa M (2016). Replication of Tobamovirus RNA. *Ann Rev Phytopathol* 54:55–78.
Ivanov PA, Karpova OV, Skulachev MV, Tomashevskaya OL, Rodionova NP, Dorokhov YL, Atabekov JG (1997). A tobamovirus genome that contains an internal ribosome entry site functional in vitro. *Virology* 232:32–43.
Jones DT, Taylor WR, Thornton JM (1992). The rapid generation of mutation data matrices from protein sequences. *Comput Appl Biosci CABIOS* 8:275–282.

Kim NR, Hong JS, Song YS, Chung BN, Park JW, Ryu KH (2012). The complete genome sequence of a member of a new species of tobamovirus (rattail cactus necrosis-associated virus) isolated from Aporcactus flagelliformis. *Arch Virol* 157:185–187.

Komarova TV, Petrunia IV, Dorokhov YL (2011). Vaccine peptide display on recombinant TMV particles. In: FM Buonaguro (ed.) *Plant-derived Vaccines: Technologies and Applications*. Future Medicine, London, pp. 44–54.

Komarova TV, Sheshukova EV, Dorokhov YL (2014). Cell wall methanol as a signal in plant immunity. *Front Plant Sci* 5:101.

Koonin EV, Dolja VV (1993). Evolution and taxonomy of positive-strand RNA viruses: Implications of comparative analysis of amino acid sequences. *Crit Rev Biochem Mol Biol* 28:375–430.

Kumar S, Stecher G, Tamura K (2016). MEGA7: Molecular evolutionary genetics analysis version 7.0 for bigger datasets. *Mol Biol Evol* 33:1870–1874.

Lartey RT, Voss TC, Melcher U (1995). Completion of a cDNA sequence from a tobamovirus pathogenic to crucifers. *Gene* 166:331–332.

Lartey RT, Voss TC, Melcher U (1996). Tobamovirus evolution: Gene overlaps, recombination, and taxonomic implications. *Mol Biol Evol* 13:1327–1338.

Leonard DA, Zaitlin M (1982). A temperature-sensitive strain of tobacco mosaic virus defective in cell-to-cell movement generates an altered viral-coded protein. *Virology* 117:416–424.

Lim MA, Hong JS, Song YS, Ryu KH (2010). The complete genome sequence and genome structure of frangipani mosaic virus. *Arch Virol* 155:1543–1546.

Merits A, Kettunen R, Mäkinen K, Lampio A, Auvinen P, Kääriäinen L, Ahola T (1999). Virus-specific capping of tobacco mosaic virus RNA: Methylation of GTP prior to formation of covalent complex p126-m7GMP. *FEBS Lett* 455:45–48.

Meshi T, Ishikawa M, Motoyoshi F, Semba K, Okada Y (1986). In vitro transcription of infectious RNAs from full-length cDNAs of tobacco mosaic virus. *Proc Natl Acad Sci U S A* 83:5043–5047.

Min BE, Song YS, Ryu KH (2009). Complete sequence and genome structure of cactus mild mottle virus. *Arch Virol* 154:1371–1374.

Mushegian AR, Elena SF (2015). Evolution of plant virus movement proteins from the 30K superfamily and of their homologs integrated in plant genomes. *Virology* 476:304–315.

Mushegian AR, Shipunov A, Elena SF (2016). Changes in the composition of the RNA virome mark evolutionary transitions in green plants. *BMC Biol* 14:68.

Namba K, Pattanayek R, Stubbs G (1989). Visualization of protein-nucleic acid interactions in a virus. Refined structure of intact tobacco mosaic virus at 2.9 A resolution by X-ray fiber diffraction. *J Mol Biol* 208:307–325.

Nicolaïeff A, Lebeurier G (1979). Polar uncoating of tobacco mosaic virus (TMV) with dimethylsulfoxide (DMSO) and subsequent reassembly of partially stripped TMV. *Mol Gen Genet MGG* 171:327–333.

Nishikiori M, Dohi K, Mori M, Meshi T, Naito S, Ishikawa M (2006). Membrane-bound tomato mosaic virus replication proteins participate in RNA synthesis and are associated with host proteins in a pattern distinct from those that are not membrane bound. *J Virol* 80:8459–8468.

Scholthof K-BG, Adkins S, Czosnek H, Palukaitis P, Jacquot E, Hohn T, Hohn B, Saunders K, Candresse T, Ahlquist P, Hemenway C, Foster GD (2011). Top 10 plant viruses in molecular plant pathology. *Mol Plant Pathol* 12:938–954.

Shaw JG (1999). Tobacco mosaic virus and the study of early events in virus infections. *Philos Trans R Soc Lond B Biol Sci* 354:603–611.

Song YS, Min BE, Hong JS, Rhie MJ, Kim MJ, Ryu KH (2006). Molecular evidence supporting the confirmation of maracuja mosaic virus as a species of the genus Tobamovirus and production of an infectious cDNA transcript. *Arch Virol* 151:2337–2348.

Song YS, Ryu KH (2011). The complete genome sequence and genome structure of passion fruit mosaic virus. *Arch Virol* 156:1093–1095.

Stobbe AH, Melcher U, Palmer MW, Roossinck MJ, Shen G (2012). Co-divergence and host-switching in the evolution of tobamoviruses. *J Gen Virol* 93:408–418.

Stubbs G, Kendall A (2012). Helical viruses. *Adv Exp Med Biol* 726:631–658.

Taliansky ME, Malyshenko SI, Pshennikova ES, Kaplan IB, Ulanova EF, Atabekov JG (1982). Plant virus-specific transport function. I. Virus genetic control required for systemic spread. *Virology* 122:318–326.

Tilsner J, Taliansky ME, Torrance L (2014). Plant virus movement, In: eLS Wiley, Chichester. (http://www.els.net.)

Ueki S, Citovsky V (2014). Plasmodesmata-associated proteins: Can we see the whole elephant? *Plant Signal Behav* 9:e27899.

Wetzel T, Njapo Ngangom HO, Chotewutmontri S, Krczal G (2006). Nucleotide sequence of a new isolate of ribgrass mosaic tobamovirus infecting Impatiens New Guinea. *Arch Virol* 151:787–791.

Wilson AC, Carlson SS, White TJ (1977). Biochemical evolution. *Ann Rev Biochem* 46:573–639.
Wilson TM (1984). Cotranslational disassembly of tobacco mosaic virus in vitro. *Virology* 137:255–265.
Zvereva SD, Ivanov PA, Skulachev MV, Klyushin AG, Dorokhov YL, Atabekov JG (2004). Evidence for contribution of an internal ribosome entry site to intercellular transport of a tobamovirus. *J Gen Virol* 85:1739–1744.

5 Plum Pox (Sharka Disease)
Pest, Challenge, and Curiosity

Antoniy Stoev

CONTENTS

5.1 REVEALING PLUM POX AS A NEW VIRAL DISEASE

The viral disease sharka on plums (plum pox) attracted the attention of scientists in the fourth decade of twentieth century. It was described for the first time as a new viral disease on plums in Bulgaria by Atanasoff (1933). He named the disease sharka by analogy with the human disease varicella (chicken pox). In English language literature, the name plum pox prevailed, while other sources refer to it as šarka (Šutić 1980), scharka (Grüntzig et al. 1986; Kegler and Kleinhempel 1987), ospa (Vlasov and Larina 1982), szarka and ospowatošč (Basak 1968), and vaiolatuara and anello (Canova 1960).

The first scientific report was followed by other reports on sharka in Bulgaria and other countries including Yugoslavia (Josifović 1937; Pobegajlo 1939), Romania—1942 according to Savulescu and Macovei (1967), Germany (Schuch 1957), and others. These specialized sources created the impression of an explosive epidemiological expansion of sharka from a single primary source—Bulgaria—to the neighboring countries (Ravelonandro et al. 2007) as well as to most countries of the continent.*

Clarification of the real phytosanitary situation concerning plum pox and its international spread required the analysis of a number of conditions that can be grouped into objective (environmental, natural) (Polischuk 2005) and subjective (anthropogenic) (Stoev 2007; Thresh 2005) (Table 5.1).

The establishment of PPV on other continents and the results of detailed studies of viral isolates on site (Kegler and Hartmann 1998; Nemchinov et al. 1998) suggested a different explanation of plum pox appearance and spread. The analysis of the subjective factors should account for the condition of the country, type and size of farms, and use of farm produce.

* See also www.cabi.org/isc/datasheet/42203 and *Bulletin OEPP/EPPO Bulletin* (2006).

TABLE 5.1
Conditions to be Taken into Account during Evaluation of the Phytosanitary Situation

Natural Conditions		
Abiotic	**Biotic**	**Anthropogenic**
Location, altitude, climate, temperature regime, rainfall, soil (as mineral structure), water basins, landscape, ground relief (lay), etc.	Soil (as an organized matter), flora (species as virus hosts) and fauna (insects as virus vectors), parasite sand pathogens including viruses (in hand as a living organisms)	Social and political regime, traditions in agriculture, size and type of farms, migration, age structure of human population, state policy to agriculture, level of knowledge and activities concerning plant health

In the first half of the twentieth century, farming land in Bulgaria was cultivated by a large number of owners. Many of them owned less than one hectare. The cultivated land in the areas with developed fruit growing was not always consolidated. Plum and other cultivated plants often were grown in semi-mountainous areas in small orchards. Traditionally, part of the harvest was intended for family consumption (Birnikoff 1942).

After World War I, Bulgaria was a defeated country that fell into isolation and had to pay reparations to the victorious countries. The defeated country was not in a position to impose conditions. The reparations were in convertible currency, coal, and cattle and not in sapling material. Besides the reparations, Bulgaria lost territories along its west and south borders that were rendered to Greece and the former Yugoslavia.

The economic uplift between the two world wars stimulated the farmers to create new orchards. The period marked a growing demand for sapling material and the growth of its production. Such quantities could not possibly come only from the area where sharka was established for the first time in Bulgaria. There were other areas with highly developed fruit growing in the country, including a nursery for planting material. Purchasing saplings for new orchards from a nearby nursery was more profitable because of savings on transportation costs.

The lack of reliable means for detecting plant viruses increased the possibility of spreading the infection over greater distances by means of the planting material. Taking that into account, the competent Bulgarian authorities issued measures for limiting sharka disease. According to them, the infected trees had to be destroyed. In some areas, the farmers resisted those measures (Kovachevski 1945).

The contemporary analysis of the situation created by this (Stoev 2007) leads to the following conclusions:

- The protests of some farmers were reasonable, because the trees they owned were giving fruit of sufficient quantity and quality.
- In those times, no database was available about the reaction of plum cultivars and forms to PPV.
- In those times, sharka had not yet spread widely in the inspected regions.

The infected trees in some areas were cut instead of being eradicated, hence the possibility of keeping the infection in the suckers and shoots from cut trees (Kovachevski 1945).

The attempts for PPV control by imposition of domestic quarantine were also unsuccessful. This was a result of:

- Insufficient knowledge of the phylogenic specialization of the pathogen and its epidemiology
- Insufficient number of specialists with qualification in plant pathology and virology
- Lack of means for mass and reliable testing of orchards and nurseries
- Lack of systemic control to vectors of the infection

In the early decades of the twentieth century, plant virology was still in the process of becoming a separate branch of phytopathology. That is why it was only much later that sharka was identified as a separate disease from the so-called mosaic diseases (Atanasoff 1934).

Bulgaria was ruled by a totalitarian communist regime after World War II. It imposed a collective farming system by force, where private initiative was suppressed by administrative orders. Large-scale orchards were created, plum orchards among them. The lack of reliable phytosanitary control of viral infections increased the risk of PPV spread in remote and large areas (Stoev et al. 2006; Stoev 2007).

The prevailing cultivar in the new orchards was Kyustendilska sinya, that was susceptible to the plum pox virus and reacted with deformations, fruit drop, and deteriorated fruit quality. Sharka was one of the reasons for the decline of plum production in Bulgaria in the last 30 years of the twentieth century (Stoev 1996).

The possibilities of PPV transmission between the East European countries increased only after the creation of the Council for Mutual Economic Assistance (COMECON, CMEA) in 1949 within the frames of the former socialist countries. There were countries with similar natural conditions, established traditions in fruit growing, and valuable local cultivars in the new economic union. It would be impossible for those countries to rely on Bulgaria only to satisfy their demands for planting material. In addition, it would be impossible for Bulgaria to provide large quantities of planting material in a short period of time if requested by the CMEA partners.

The analysis of the data on the appearance of sharka in the first half of the twentieth century pointed to an area with specific geographical coordinates, where the disease existed. Its identification as a new disease in plum as well as other drupaceous fruit species came at a time when there were trained specialists in the field of plant virology, which was in the process of becoming a branch of phytopathology.

Atanasoff (1933) pointed out in the first scientific report on sharka that he had seen fruit with the disease symptoms on the market in Brno (Czech Republic). In Bulgaria, the symptoms were observed on the leaves and fruit of Kyustendilska sinya sliva (Prunus domestica subsp. domestica L.). This is not a completely stable cultivar but a cultivar type. By pomology characteristics, it relates to the famous Požegača in Serbia and Hauszwetsche or Bauernpflaume (Germany).

According to Christoff (1947), plum pox disease was observed on plum trees in Macedonia during World War I. It would hardly be possible for the infection to spread in the territory of Bulgaria in such a short time and penetrate into former Yugoslavia, where the natural conditions, as given in Table 5.1, were the same, in addition to the established traditions in plum growing.

The interest in sharka increased with the increase of the losses it caused. As of 1933, not all plum tree farmers suffered losses. The reason was that some forms within the cultivar type Kyustendilska sinya responded differently to the plum pox pathogen and fruit quality was preserved better in cases of infection. Until World War I, most farms in Bulgaria were small to medium in size and they developed in relative isolation. The owners of such farms were able to use their own planting material (seedlings and root scions) and be independent of the market.

The development of the diagnostic methods in the field of virology in the second half of the twentieth (Hansen et al. 1986) and beginning of the twenty-first century (Cambra et al. 2005) made possible not only virus detection but also strain characterization within the virus population. The strain differentation of PPV in the areas contaminated with PPV is proof of an evolutionary process in time in certain conditions.

One more piece of evidence was the spread of the disease in the major fruit species. For example, sharka was established in peach trees in Bulgaria and former Yugoslavia only at the end of the last century (Dulić and Sarić 1986; Yankulova et al. 1990). We have to emphasize that from an epidemiological point of view, the natural spread of sharka in peach trees did not coincide with that of plum trees (Rankovic and Sutic 1980). Jordović (1985) found that sharka virus from plum was not able to infect any examined peach cultivar but only plum cultivars and tested indicator plants.

The late identification of sharka in sour cherry and cherry trees should also be taken into consideration (Kalašan and Bilkej 1989; Nemchinov et al. 1998).

5.2 CURRENT SPREAD OF PLUM POX

At the beginning of the twenty-first century, in addition to Europe, the disease became known in Asia, Africa, and North and South America. It was widespread in some countries (Albania, Bulgaria, Germany, Greece, Hungary, Serbia, Slovakia, Egypt, and Chile) on the above continents, while in others (Italy, Portugal, Norway, and Canada) the distribution was restricted or there were few occurrences. It was found both in agricultural lands (orchards) and in places with self-sown host plants of the sharka causal agent (yards, parks, and neglected zones, etc.).

According to Roy and Smith (1994), there were three zones with different levels of infection in Europe during the last decade of the twentieth century. The level of infection was generally high in Central and East European countries (Bosnia-Herzegovina, Bulgaria, Croatia, Czech Republic, Hungary, Moldova, Poland, Romania, Serbia, Slovakia, Slovenia, and Ukraine).

The disease is new to the countries of the Mediterranean but there is a high risk of further spread (Albania, Cyprus, Egypt, Greece, Italy, Portugal, Spain, Syria, and Turkey). The spread of the disease to the north and west is very irregular.*

Following the reports on sharka in Asia (Thakur et al. 1994; Navratil et al. 2005; Fujiwara et al. 2011) and North and South America (Levy et al. 2000; Thompson et al. 2001; Herrera 1994; Zotto et al. 2006), the disease became known worldwide.

5.3 ECONOMIC IMPORTANCE OF PLUM POX

Plum pox is a serious risk to production of plum cultivars in the countries of the Balkan Peninsula and Central Europe. The infection on the trees of some cultivars compromises the yield due to premature fruit drop and deteriorated quality. Bark cracking is another harmful consequence (Zavadzka 1980) as well as the drying of separate skeleton branches and even whole trees of cultivars that respond to the pathogen with hypersensitivity (Trifonov 1972).

The yield of other sensitive Prunus cultivars can be compromised in terms of quality and quantity as well (Kegler and Hartmann 1998). The pathogenicity of PPV variations in species and cultivars was characterized with different resistances to the pathogen (Rankovic and Sutic 1986).

5.4 CAUSAL AGENT OF PLUM POX

After the differentiation of sharka from the so-called mosaic diseases, the pathogen was described according to the system of Johnson and Hoggan (1935) as Prunus virus 7 Christhoff, and was listed as such in the older specialized sources.

According to the modern classification, PPV belongs to the Group RNA viruses, family *Potyviridae*, genus *Potyvirus*, species *Plum pox virus*. The virus has filamentous particles 750 nm long and 15 nm in diameter (Kegler et al. 1964; Kegler and Hartmann 1998). It contains single-stranded RNA. Protein inclusion type pinwheel is formed in the cytoplasm of leaf tissue cells (Van Bakel and Van Oosten 1972). The presence of these inclusions is important for the diagnostics of plum pox. Van Oosten (1970) observed numerous bundles or needles as well as granular inclusions in the cytoplasm of parenchyma cells of plum fruits. The inclusion bodies in plum fruits were identical to those found in sharka virus-infected Nicotiana clevelandii.

The phylogenic specialization of PPV is connected with the family *Rosaceae*, subfamily *Prunoideae*, genus *Prunus*. The plums (*P. domestica*), apricots (*P. armeniaca*) and peaches (*P. persica*)

* See also CABI/EPPO Distribution Maps of Plant Diseases: 1998, 1999, 2007.

are the most endangered. Cases of infection in natural conditions were reported in sour cherry (*P. cerasus*) (Kalašan and Bilkej 1989) and cherry (*P. avium*) (Nemchinov et al. 1998). Almond (*P. dulcis*) was found receptive to PPV after transplanting onto infected rootstock (Petrov and Stoev 2011).

The spreading of the virus in the infected trees depends of the method of infection. In the case of a permanent source of infection such as infected rootstock, for example, the probability for the infection to penetrate faster and spread evenly in the transplant is greater. It should be noted that this is possible when the transplant, representing an infectious recipient, does not respond with hypersensitivity. Otherwise, a necrotic line could be formed between the transplant and rootstock and cause the transplant to perish. Genotype K 4 after graft inoculation with PPV-M is an example of hypersensitivity (Kegler and Hartmann 1998).

Polak (2001) reported natural infection with PPV in *Ligustrum vulgare* and *Euonymus europaea*. During a walnut mosaic study, Hristov (1976) assumed that sharka in plum trees could be transmitted to walnut trees (*Juglandaceae*).

5.5 EPIDEMIOLOGY

The natural spread of PPV occurs by leaf lice (aphids) (Atanasoff 1933). Species *Brachycaudus helichrysi*, *B. cardui*, *Myzus persicae*, and *Phorodon humuli* transmit the pathogen more efficiently. Other species—*Myzodes varians*, *Brachycaudus persicae*, and *Hyalopterus amygdali*—also transmit the infection. PPV is a non-persistently transmissible virus. The acquisition time amounts to 5–10 minutes, and the vectors remain infectious only for a few hours. This circumstance allows the spread of the infection over long distances in the case of passive flight (Kegler and Hartmann 1998).

The virophority of the above indicated species can be proved both by infecting indicator plants and by laboratory tests, e.g., real-time RT–PCR (Kamenova 2015).

The spread of the infection by means of aphids depends on the natural conditions (wind, temperature, and precipitation), in which the reproduction cycle of the vectors takes place.

The life cycle of the separate species, their seasonal migration, in which they pass from woody to herbaceous species and back, as well as the dynamics of the population density have been studied (Kunze and Krczal 1971; Grigorov 1980). Weed species are possible reservoirs of infection (Milusheva and Rankova 2002).

The infection also depends on the forms in the population of the vector and the specific interaction between vector and plant as a potential infectious recipient. The resistance to the vectors and infection transmission are of practical importance for disease control in orchard conditions (Massonié and Maison 1985).

The spread of the disease could result from human activity, when people use infected vegetative plant material (trees, cuttings, and buds) for the creation of new orchards (Trifonov 1972). The generative posterity (seedlings) remains virus free (Christoff 1947; Schimanski et al. 1988; Pasquini and Barba 2006).

5.6 DIAGNOSTICS AND STRAIN CHARACTERIZATION

Sharka diagnostics is a complex undertaking that includes visual checkup of the plants, corroboration of the infection, identification of the pathogen, and its strain characterization.

The results of these diagnostic activities can be complemented for the full clarification of the disease etiology and epidemiology. Modern virology has suitable methods for rapid tests both in natural and laboratory conditions. The isolation, maintenance, and amplification of viral isolates are also parts of the diagnostic process. These activities take place in cultivation chambers and laboratories. The observation should be focused on the symptoms characteristic of the disease (Atanasoff 1933; Christoff 1947; Van Oosten 1972; Zavadzka 1980) (Table 5.2).

TABLE 5.2
Plum Pox Symptoms to be Taken into Account during Observation

Symptoms on the Leaves	Symptoms on the Fruits	Other Symptoms
Chlorotic spots with different size and shapes	Dark purple arcs, stripes, and spots on the fruit skin	Violated coloration of petals (mottling of petals)
Ring spots or spots with irregular shapes	Internal darkening in the pulp just below the purple spots	Premature dropping of fruits
Deformation of lamina		
Arcs, stripes and chlorotic zones between leaf lamina ribs, mottling	Cavities in the fruit flesh (mesocarp), which could be resinous	Bark splitting
Reversal pox—green spots onto chlorotic background of leaf lamina	Conglomeration between pulp and pit; occurrence of spot on the pit surface	Drying of branches of some cultivars

The symptoms on leaves are found either on separate branches or the whole canopy. The manifestation rate depends on the following:

- The species and cultivar of the host plant
- The vegetative phase and temperature regime
- Other reasons of infectious and non-infectious nature

In Central and East European countries, the symptoms are most clearly defined on the leaves at the end of spring and the first half of summer. The symptoms become less obvious with the heat wave under the influence of other biotic and abiotic factors (pests and physiological disturbances, etc.).

After the visual evaluation, the diagnostic process continues with confirming the infectious etiology, identification of the causal agent, and strain differentiation in its population. Currently, specialists use a number of methods, applied depending on the phytosanitary conditions and the requirements for crop biosecurity (Cambra et al. 2005; Cooper 2005; El-Hammady et al. 1995; Hansen et al. 1986).

5.6.1 Biological Testing

The herbaceous indicators are most often inoculated by mechanical treatment of their leaves with sap from an infected plant. The representatives of the botany families *Chenopodiaceae* (*Chenopodium foetidum*) and *Solanaceae* (*Nicotiana benthamiana, N. Clevelandii, N. occidentalis*) are suitable as herbaceous indicators (Basak 1967; Kamenova 1987).

The biological test procedure includes preparation of the inoculum by grinding the leaves in a mortar with added buffer solution. Preliminary dusting of the leaves of the indicator plant with carborundum enhances the opening of micro wounds and facilitates virus pervasion in the plant tissue (Kassanis and Šutić 1965).

The indicator can manifest symptoms systemically or locally (Basak 1967). The damage on plant tissue can be chlorotic, necrotic, and chlorotic-necrotic (Kassanis and Šutić 1965).

Woody indicators of the *Prunus* genus have also been used for sharka diagnostics (Nemeth and Kölber 1980; Šutić 1964) The inoculation of the woody indicators is done by chip budding. In the trials conducted by Kamenova (2015), there were differences in the response to PPV on the leaves of peach seedling GF 305 depending on the strain PPV—rec (peach isolate) and PPV M (plum isolate). Tip necrosis can be observed in the woody indicators as well (Kegler and Hartmann 1998). The indicator plants can be also inoculated by natural carriers in controlled conditions.

The woody indicators should be grown in special isolators that do not allow the penetration of aphids—carriers of PPV (Stoev 1996). It is recommended to use insecticides for better protection against external infection.

5.6.2 Laboratory Investigations

The laboratory tests give a much faster reaction for identifying the pathogen. A number of methods are known in virology practice, including agar gel diffusion and agglutination of the biological inert bodies (bentonite and latex, etc.) that are preliminarily loaded with antibodies.

5.6.3 Serological Testing

Plant virology uses tests that are based on the reaction of specific antibodies, formed in the animal organisms against the virus, injected into the animal.* There are different methods applied in diagnostics that are based on the specific interaction between virus and antibody (Hood et al. 1985; Hansen et al. 1986). For example, viral particles diffused in agar gel meet the antibodies, and a precipitation zone is formed. The use of monoclonal antibodies has an advantage over other sera because of the lack of non-specific reaction to plant antigens.

PPV immunological activity varies. Antiserum production is not feasible for every strain, due to the different serological aptitude of the strains. Obtaining of antiserum with higher titer depends on viral accumulation in the host plant. From this point of view, *Nicotiana clevelandii* and *N. occidentalis* are the most suitable. The agglutinate that appears during the reaction of the viral antigen with its homologous antibodies is indicative of agglutination method.

Enzyme-linked immunosorbent assay (ELISA) based on the interaction between the pathogen (viral protein) and the pathogen-specific antibodies is applied for plum pox diagnostics (Adams 1978). If a pathogen is present, it connects with the antibodies, some marked with an enzyme. The enzyme disintegrates a specific substrate. A color reaction takes place.

Immune specific electron microscopy (ISEM) is a reliable method for pathogen identification that reveals viral particles, bound with specific antibodies (Derrick 1973; Kaytazova 1983).

5.6.4 Structural Investigation on Molecular Level

Detection of genomic PPV is another method for assay of the pathogen. The procedure includes virus purification, RNA extraction and characterization, complementary DNA synthesis and characterization, and hybridization of complementary DNA to RNAs. The final result is a diagnostic possibility especially useful for strain differentiation and for cases of non-specific reactions in serology (Hansen et al. 1986).

Polymerase chain reaction (PCR) is a technique used in molecular biology to amplify a single copy or a few copies of a piece of DNA across several orders of magnitude, generating thousands to millions of copies of a particular DNA sequence (Bartlett and Stirling 2003).

Reverse transcription polymerase chain reaction (RT-PCR)—a variant of PCR—is a technique commonly used in molecular biology to detect RNA expression.

Immunocapture reverse transcription polymerase chain reaction (IC-RT-PCR) is more sensitive than ELISA and RT-PCR alone. It allows detection of the virus without isolating the RNA.

Strain characterization of the pathogen is the next level in the diagnostic process. The information obtained is important both for theory and practice, as follows:

- First-time identification of the pathogen in a certain area
- Identification of new host plant species of the virus

* Plants do not form antibodies as immune response against pathogens.

- Geographic mapping of the strains and evaluation of damage caused by the disease in agricultural species, known as hosts of PPV
- Elaboration of specific reagents for diagnostic purposes, etc.

At the end of the twentieth/beginning of the twenty-first century, the PPV strains in the PPV population were differentiated as conventional (Kegler and Hartmann 1998) and PPV-Cherry subgroup including PPV-Sour cherry and PPV-Sweet cherry (Nemchinov et al. 1998). Complete and partial genome sequences of the unusual PPV isolates from sour cherry in Russia suggested their classification to a new PPV strain (Glasa et al. 2012).

The strains M and D belong to the conventional group. Viral isolates from occidental Europe normally corresponded to D-strains, whereas those from oriental Europe showed a profile similar to M-strains (Kegler and Hartmann 1998). Results obtained after virological screening by using monoclonal antibodies for DAS ELISA showed the presence and distribution of PPV-M strain in monitored plum orchards in Bulgaria. The prevalence of PPV-M was confirmed in other areas of the country along with PPV-D. Mixed M+D infections were also found (Kamenova et al. 2003). The comparison of amino acid sequence of the RT-PCR-amplified capsid protein (CP) genes differentiated four isolates as PPV-rec (Candresse and Cambra 2006; James and Glasa 2006).

Based on the amino acid sequence comparison of the RT-PCR-amplified part of CP gene isolates from Bulgaria could be classified as belonging to PPV-rec, and another isolate belonging to PPV-D (Szathmary et al. 2006; Kamenova et al. 2015).

The results of the study of the amino acid consequences of CP of Bulgarian PPV-M, PPV-D and PPV-rec isolates showed that they contained amino triplet DAG that participated in the transmission of potyviruses by aphids. The divergences in RNA sequence separated a strain originating from Egypt (Wetzel et al. 1991). This strain is known as El Amar. In Canada, a strain of PPV was distinguished as Winona (James et al. 2003). The near-complete (99.7%) genome sequence of a novel Russian *Plum pox virus* (PPV) isolate Pk, belonging to the strain Winona (W), has been determined by 454 pyrosequencing with the exception of the 31 5′-terminal nucleotides (Sheveleva et al. 2013).

5.7 AGGRESSIVENESS OF PPV

The complete evaluation of virus aggressiveness, PPV included, can be done through their organotropic and histotropic specialization (Stoev 2016).

Kamenova and Stoev (1987) confirmed the presence of PPV in leaf and blossom samples taken from plum trees through DAS ELISA (Adams 1978). Stoev and Kamenova (1995) registered positive reaction for fruit samples of plum cultivar Zhulta butilkovidna. The sample originated from the spotted zone on the exocarp and adjoined mesocarp. The reaction for another sample from the opposite side without symptoms was negative.

Petrov (2014) proved the presence of both pathogens in the exocarp during the examination of plant organ and tissue tropism of PPV and prunus necrotic ringspot virus (PNRV). Only PPV penetrated into the seed coat but the embryo remained virus free.

Kamenova (2015) confirmed the presence of PPV in cotyledons and embryo of apricot through IC-RT-PCR but the generative posterity remained virus free. The seedlings remained without symptoms of sharka. The results from the realized twofold serological and molecular assays during the next two years were negative.

The tracking of PPV virus in the vegetative organs showed that the infection was not evenly distributed in the crown of trees, growing in natural conditions (Ranković and Vuksanović 1985; Fuchs et al. 1989). As determined by ELISA, there were differences in the relative concentration of PPV among the infected cultivars. Based on the relative virus concentration, determined through DAS ELISA, Albrechtova (1989) found a correlation between low PPV concentration and fruit damage indexes, but there was no correlation between high virus concentration and fruit damage indexes of the cultivars. According to the PPV concentration in flowers and fruits, the tested cultivars were distributed into three groups: very low absorbance values, medium, and high PPV concentration in the tissue.

The infection did not affect the top parts of vigorously growing shoots. The non-infected part may serve as a source for the production of propagative material for restoration of cultivars, possessing valuable qualities (Trifonov 1972).

5.8 REACTION OF FRUIT HOST SPECIES TO PPV

Plant species are divided into immune and receptive, according to their affinity to the pathogen. There is no infection in immune species. The virus is not present in the cells of the plant organism.

In the receptive plants, the virus finds its way to the cell and can continue to reproduce at the expense of the cell. The cultivars of the receptive species are tolerant or susceptible.

The tolerant plants respond to infection with symptoms on the leaves, but the fruits remain healthy. The sensitive ones manifest symptoms on both leaves and fruits. It is possible that the symptoms on fruits are more clearly defined, whereas leaf symptoms appear only in some years or remain localized in part of the canopy (Stoev 1996; Trifonov 1972).

Plums may manifest qualitative resistance, characterized with virus rejection by hypersensitive reaction in the point of inoculation (Kegler and Verderevskaya 1985). The necrosis and gummosis are manifestations of hypersensitivity to the pathogen.

Other types of reaction such as mild symptoms, low viral concentration in the host plant, and partial infection are regarded as quantitative resistance. The different virulence of PPV strains must be taken into consideration in assessing resistance (Kegler and Kleinhempel 1987).

Comparing the visual assessment and the results obtained through DAS ELSA, Topchiiska (1995) classified the investigated peach and nectarine cultivars, rootstocks, and hybrids as tolerant, poorly susceptible, moderately susceptible, and susceptible.

The manifestations of hypersensitivity to the virus in the case of leaf inoculation prevent the spread of infection in the plant organism. Such plants cannot be infected by aphids in natural conditions because of the local necrosis developing on the inoculation spot. The non-inoculated plants have qualitative resistance.

Iliev and Stoev (2002) suggested two scales for the assessment of the reaction to PPV. The first one alludes to the spread and distribution of symptoms in the tree crown.

- 0—no symptoms
- 1—symptoms in one part of the crown (skeleton branch or its bias)
- 2—Symptoms on several branches
- 3—Symptoms manifested overall

The second is targeted at fruits status:

- 0—no symptoms and changes from typical cultivar properties
- 1—Superficial symptoms without deterioration of fruit quality
- 2—Up to 10% of the fruits have the typical external and internal symptoms of plum pox, obvious fruit deformation and low consumer quality
- 3—Over 10% are with pathological changes mentioned in the preceding level

The above scales enabled the visual screening of proven cultivars and elites. The juxtaposition of data from the visual evaluation to those from ELISA DAS confirmed the possibility of employing more cultivars with the aim of reducing the losses caused by plum pox disease as well as restoring and developing plum production in Bulgaria (Iliev et al. 1999).

Stoev (2000) suggested a scale for the distribution of the pathological changes on fruits of plum (*P. domestica*) and wild plum (*P. cerasifera*) caused by PPV:

- 0—smooth skin with normal color
- 1—Skin surface mottling—purple rings and arcs
- 2—Initial concavity in the mottled zone
- 3—Several concavities in the mottled zone
- 4—Concavities merge in larger area in ring, arc, or indefinite shape
- 5—Several concavity zones in ring, arc, or indefinite shape
- 6—Considerable part of the exocarp is caved and wrinkled—impaired fruit morphological characteristics, specific to the cultivar

The collected data on damage distribution allowed for the calculation of the disease index (DI) as suggested by Kegler and Hartmann (1998):

$$DI = \frac{100\sum_{i}^{n} n_i m_i}{N\ m},$$

where

- n_i = number of injured fruits in the respective fraction
- m_i = degree of injury for the respective fraction
- m = maximal injury
- N = number of observed fruits

Stoev (1996) recommended increasing the share of the Bulgarian cultivar Izobilie and Nancy Mirabelle in new fruit plantations. Iliev and Stoev (2002) found that the plum form Besztercei MM 122, introduced for testing in Bulgaria, was tolerant to PPV in natural conditions. This kind of reaction was confirmed with Black-skinned plum cultivars (Prunus domestica), Vengerka donetskaya rannaya, V. kavkazkaya, and V. krupnaya sladkaya (Stoev and Iliev 2009). The pomological features of the mentioned form and cultivars were close to those of cultivar type Kyustendilska sinya. In instances of tree infection with PPV, the fruits stay healthy as in Čačanska najbolja and Stanley, included in the study for the sake of comparison. The assessment of the cultivars should take into consideration the strain aggressiveness of PPV in association with other properties which characterize plant pathogens (Stoev 2016). The assessment is a complex activity that engages specialists in different fields including fruit growing, plant pathology and virology, statistics, and others. The interaction with pathogens is part of the overall agrobiological characteristics of the cultivars (Iliev et al. 1985). Along with the viral pathogens PPV, prune dwarf virus (PDV), PNRV, and apple chlorotic leafspot virus (ACLV), fungal pathogens such as *Polystigma rubrum*, *Monilia fructigena* and *M. laxa*, and *Stygmina carpophila* present a hazard as well (Stoev 1996; Milusheva and Kamenova 2006; Iliev et al. 2011; Mitre et al. 2015).

The consistent selection of drupaceous fruit species at the end of the twentieth/beginning of twenty-first century allowed for the introduction of cultivars with valuable economic characteristics and such that are not endangered by possible PPV infection. The cultivar variety allows farmers to set profitable production in areas where there are conditions for the appearance and spread of sharka (Iliev et al. 1999). In the regions contaminated with PPV, only cultivars that are tolerant to the pathogen should be planted (Trifonov 1965; Stoev and Iliev 2007).

Virological screening in the region of Plovdiv (Bulgaria) showed tolerance to PPV of plum cultivars with valued and market qualities (Milusheva and Kamenova 2006). In spite of the severe symptoms, manifested on the leaves, there were mild symptoms on the fruits of Bluefre, Čačanska rana, Čačanska najblolja, Čačanska lepotica, Tuleu timpuriu, and Althan's gage.

TABLE 5.3
Recommended List of Cultivars of Different Drupaceous Fruit Species Connected with PPV Control

Fruit Species	Cultivars	Properties
Plum	Czar	Tolerant
	Čačanska najbolja	Tolerant
	Elena	Tolerant
	Herman	Tolerant
	Hanita	Tolerant
	Jo jo	Resistant
	Katinka	Tolerant
	Nancy mirabelle	Resistant
	Opal	Tolerant
	Oullins Reine Claude	Tolerant
	President	Tolerant
	Stanley	Tolerant
Apricot	Gold rich	Resistant
	Harlayne	Resistant
	Kuresia	Resistant
	Orange red	Resistant

It became the responsibility of the government to provide specialized information about the suitable cultivars through the state-funded scientific institutes and respective government organizations (Table 5.3) (Nußbaum and Maring 2005). International targeted projects are one more source of valuable information (SharCo).*

Diversification of the production assortment can be also achieved with cultivars produced by genetic technologies. The HoneySweet plum is one of them. It was originally selected *in vitro* from a hypocotyl slice that was transferred with Agrobacterium tumefaciens EHA 101 carrying plasmid pGA482GG/PPV-CP-33. The regenerated transgenic shoot coded as C5 was rooted *in vitro* and transferred to a greenhouse. It was tested in the greenhouse using graft and aphid inoculation with M and D strains of PPV and later patented as HoneySweet (Scorza et al. 2016). Generally, no systemic infection was detected and if it was, it was limited and with mild symptoms that decreased over multiple growing seasons (Polak et al. 2012).

5.9 CONTROL OF SHARKA (PLUM POX) ON DRUPACEOUS FRUIT SPECIES

Historically, the decrease of losses to sharka can be summarized as follows:

- Quarantine measures for prevention of the spread of plum pox
- Creation of cultivars resistant to the pathogen
- Creation of cultivars resistant to infectious vectors
- Diversification of the production assortment
- Application of technologies for growing PPV-susceptible cultivars with valuable economic parameters

* See also http://sharco.eu Sharka Containment in View of EU Expansion. Seventh Framework Programme.

The choice of measures and actions depends mainly on the natural conditions and type of agriculture. Suitable quarantine measures are recommended for areas that are sharka-free but with favorable conditions for its development. Eventual sources of infection should be eradicated immediately (Fujiwara et al. 2011). The primary task should be to identify the density of the infectious background, species variety, and population dynamics of PPV vectors (Milusheva and Kamenova 2006).

Pathogen tolerant, slightly susceptible, and resistant cultivars should prevail in the commercial plantations in natural infectious background, large number of leaf aphid vectors, and probable existence of virulent strains as well as reduced control of the virus status of the planting material (Topchiiska 1995). The growing of sensitive cultivars is admissible on small non-commercial farms for family use.

The creation of large orchards of susceptible cultivars is possible with available virus-free plant material, planted in isolated locations far from potential infectious sources. Regardless of distance, the new plantations should be inspected for sharka on an annual basis and the trees should be treated with insecticides for prevention. If the disease is found on any trees, they should be eradicated or destroyed with systemic herbicides. Along with visual observation, mother plants should be checked periodically by DAS ELISA and IC-RT-PCR as an extra control against latent infection (Dragoyski et al. 1990; Dragoyski 2011; Petrov 2014).

The tests should also cover other viruses such as PNRV and PDV, which cause dangerous diseases in drupaceous fruit species in single or mixed infections (Stoev 1996). The results are also important because of eventual pseudo pox (Van Oosten and Van Bakel 1970; Schmid 1981).

The fungal pathogens *Monilinia laxa*, *Monilia fructigena*, *Polystigma rubrum*, and *Stigmina carpophila* present a risk to some of the PPV-resistant or -tolerant plum cultivars. The diseases they cause—blossom blight, brown rot, red spot, and shot hole—require tree treatment with fungicides (Mitre et al. 2015).

5.10 CHALLENGES AND FUTURE RESEARCH ISSUES

In the second decade of the twenty-first century, fruit farmers have a sufficient number of cultivars and technological solutions at their disposal for successful fruit production in spite of the presence of PPV in the infectious background. Elimination of the problems caused by sharka in the past is the outcome of long-term research activity of plant pathologists, agronomists, plant breeders, and so forth. The research is not going to stop here. The sharka pathogen has become known worldwide just like the tobacco mosaic virus (TMV). The accumulated diverse information on PPV is a solid and reliable landmark for future investigations in the field of virology. The interaction between virus and host plant, the variability of the pathogen, which is fully integrated with the cells of the host plant during its reproduction, and also the viral aggressiveness will continue to intrigue scientists. They have to establish how stable the strain differences are.

The discovery of naturally infected plant species outside the botany family of agricultural plants is important for the identification of the infectious cycle in plum pox.

The question of the extent of infection in the sexual process is still open. Are the embryo and seed generation virus free? PPV could be the target of experiments with new technologies for the suppression of virus reproduction and disease prevention in plants.

REFERENCES

Adams A (1978). The detection of plum pox virus in Prunus species by enzyme-linked immunosorbent assay (ELISA). *Ann Appl Biol* 90:215–221.

Albrechtova L (1989). Zur Bewertung der Resistenz von Pflaumensorten und–hybriden gegenüber dem Sharka-Virus (plum pox virus). *Zeitschrift für Pflanzenkrankheiten und Pflanzenschutz* 96(5):455–463.

Atanasoff D (1933). The sharka on plums—A new viral disease. *Yearbook of the University of Sofia*, Faculty of Agriculture, 11:49–70.

Atanasoff D (1934). Mosaic disease of drupaceous fruit trees. *Yearbook of the University of Sofia*, Faculty of Agriculture, 13:9–42.
Bartlett JMS, Stirling D (2003). A short history of the polymerase chain reaction. *Methods Mol Biol* 226:3–6.
Basak W (1967). Herbaceous host ranges of two trunus line-patren viruses. *Bull Acad Polonaise Sci C.V.* 15 (5):313–319.
Basak W (1968). Chororby Wirusowe Roślin Sadowniczych. Panstwowe Wydawnictwo Rolnicze i Leśne, Warszawa.
Birnikoff D (1942). Lebensproblemen der bulgarischen Landbevölkerung und Massnahmen für ihre Lösung, Zemedelsko Obrazowanie, Jahrb. 1, 3/4:2
Cambra M, Bertolini E, Olmos A, Nieves C (2005). Molecular methods for detection and quantitation of viruses in aphids. In: *Virus Diseases and Crop Biosecurity*, Cooper I, Kühne T, Polischuk V (eds). Springer in cooperation with NATO Public Diplomacy Division, 81–88.
Candresse T and Cambra M (2006). Causal agent of sharka disease: Historical perspective and current status of Plum pox virus strains. *Bulletin OEPP/EPPO Bulletin* 36:239–246.
Canova A (1960). Le virosi delle planto de frutto. *Edizioni Agricole Bologna.*
Christoff A (1947). Plum Pox. News of the People Chamber of Culture, ser. Biology, Agriculture and Forestry, 1, 2:261–296.
Cooper I (2005). Genomic approaches in virus diagnostics: A personal assessment of realities when faced with viruses in a plant biosecurity control. In: *Virus Diseases and Crop Biosecurity*, Cooper I, Kühne T, Polischuk V (eds). Springer in cooperation with NATO Public Diplomacy Division, 71–80.
Derrick KS (1973). Quantitative assay for plant viruses using serologically specific electron microscopy. *Virology* 56(2):652–653.
Dragoyski K (2011). The experience in establishment of plum pox virus free scion orchards in RIMSA Troyan. *Acta Hort* 899:159–163.
Dragoyski K, Minev I, Mondeshka P, Dinkova H, Michovska B (1990). Work related to the Plum pox at the Institute of Mountain Animal Breeding and Land Management. *Plant Science* 24, 4:10–15.
Dulic I, Saric A (1986). Outbreak of Plum pox virus in Yugoslavia. *Acta Hort* 183:161–165.
El-Hammady M, El-Abbas A F, Mazyed H, El-Ela AA (1995). Studies on the Egyptian isolate of Plum pox virus (isolation and identification). *Plant Science* 32, 4:47–54.
Fuchs E, Grüntzig M, Otto F, Kad B (1989). The distribution of fruit viruses in individual trees. *Proceedings of the 10th Conference of the Czechoslovak Plant Virologists*, Prague, 27–30 June.
Fujiwara Y, Saito N, Kasugai K, Tsukamoto T, Aihara F (2011). Occurrence and eradication of Plum pox virus in Japan. *Acta Hort* 899:165–174.
Glasa M, Prichodko Y, Zhivaeva T et al. (2016). Complete and partial genome sequences of the unusual PPV isolates from sour cherry in Russia suggest their classification to a new PPX strain. Conference Paper. https://www.researchgate.net/publication/262822114.
Grigorov S (1980). Leaf lice and fight with them. *ZEMIZDAT*, Sofia.
Grüntzig M, Fuchs E and Kegler H (1986). Untersuchungen zum Nachweis des Scharka-Virus (plum pox virus) in Pflaumenbäumen. *Arch Phytopathol Pflanzenschutz,* Berlin 22(6):441–449.
Hansen AJ, Hamilton R, Martin R, Stace-Smith R (1986). Improved methods for tree friut virus detection. *Acta Hort* 193:229–231.
Herrera G (1994). Detección de la enfermedad de sharka (plum pox virus) en una vieja collection de carozos en la subestración experimental Los Tilos (INIA), Chile. *Agric Téc (Chile)* 187–191.
Hood LE, Weisman IL, Wood WB (1985). *Immunology*. Science and Art, Sofia. (Translation from English.)
Hristov A (1976). La mosaïcue du noyer. *Horticultural and Viticultural Science* 13(2):50–56.
Iliev I, Vasilev V, Georgiev V et al. (1985). *Small Pomology—Drupaceous Fruit Species*. Hristo G. Danov, Plovdiv.
Iliev P, Dragoiski K, Stoev A, Kamenova I (1999). The introduction of more varieties in plum plantations as a condition for lowering the damages caused by Plum Pox virus. *Agricultural Science* 6:14–15.
Iliev P, Stoev A (2002). Behavior of three forms of Kyustendilska sinya (Prunus donestica L.) under the conditions of natural infection with sharka disease (PPV). *Acta Hort* 577:269–274.
Iliev P, Stoev A, Petrov N (2011). Between sharka and monilia. *Acta Hort* 899:171–174.
James D, Glasa M (2006). Causal agent of sharka disease: New and emerging events associated with Plum pox virus characterization. *Bulletin OEPP/EPPO Bulletin* 36(2):247–250.
James D, Varga A, Thompson D, Hayes S (2003). Detection of a new and unusual isolate of Plum pox virus in plum (Prunus domestica). *Plant Disease* 87(9):1119–1124.
Johnson J, Hoggan A (1935). A descriptive key for plant viruses. *Phytopathology* 25:328–343.

Jordović M (1985). Prilog proučavanju šarke šljive i breskve. *Zaštita bilja* 172:155–159.
Josifović M (1937). Mosaik na šljivi. *Arhiv Ministarstva Pljoprivrede* 4(7):31–143.
Kalašan J, Bilkej N (1989). Identifikace Viru Šarky Švestki na Višni. Plant Virology. *Proceedings of the 10th Conference of the Czechoslovak Plant Virologists*, Prague, 27–30 June.
Kamenova I (1987). Sharka Disease on Plums. Methods for Diagnostics. PhD diss., Agricultural Academy, Sofia, Institute of Plant Protection.
Kamenova I (2015). A Research of Epidemiology of Plum Pox Virus on Drupaceous Fruit Species in Bulgaria. Doctor Sc. Diss., Agricultural Academy, AgroBioInstitute, Sofia.
Kamenova I, Borisova A, Dragoyski K et al. (2015). Plum Pox virus strains in Bulgaria. *Acta Hort* 1063:47–54.
Kamenova I, Stoev A (1987). Application of the ELISA-test for detecting the Plum pox virus in plums. *Soil Science Agrochem Plant Protect* 22(2):96–101.
Kamenova V, Milusheva S, Borisova A, Stoev A, Myrta A (2003). Plum pox virus strains in Bulgaria: Options Méditerranéennes, Serie B/n° 45 Virus and Virus-like Diseases on Stone Fruits, with Particular Reference to the Mediterranean Region. Myrta A, Terlizzi di B, and Savino V (eds).
Kassanis BM, Šutić D (1965). Some results of recent investigation on sharka (plum pox) virus disease. *Zaštita bilja* 85–86:335–340.
Kaytazova P (1983). Immuno-electron microscopy—High sensible and specific method for virus diagnostics. *Plant Science* 5:74–75.
Kegler H, Hartmann W (1998). Present status of controlling conventional strains of plum pox virus. In: *Plant Virus Disease Control*, Hadidi A, Khetarpal RK, Koganezawa H (eds). APS Press, St. Paul, MN, 616–628.
Kegler H, Kleinhempel H (1987). *Virusresistenz der Pdlanzen*. Akademie-Verlag, Berlin.
Kegler H, Schmidt HB, Trifonov D (1964). Identifizirung, Nachweis und Eigenschaften des Scharkavirus der Pflaume (Plum Pox Virus). *Phytopathologische Zeitschrift*, 50(2):97–11.
Kegler H, Verderevskaya T (1985). Virusresistenz bei Obstgehölzen. *Arch Gartenbau* 33(7/8):453–461.
Kovachevski I (1945). Ten Years Plant Protection. Report for Activity of the Institute of Plant Protection in Ministry of Agriculture and State estates, Sofia (in Bulgarian).
Kunze L and Krczal H (1971). Transmission of sharka virus by aphids. In: *Proceedings of the 8th European Symposium on Fruit Tree Virus Diseases*. INRA, Paris, 255–260.
Levy L, Damsteegt V, Welliver R (2000). First report of Plum pox (sharka disease) in Prunus persica in USA. *Plant Disease* 84(2):202.
Massonié G, Maison P (1985). Investigation on the resistance of peach varieties to aphid transmission of plum virus. *Acta Hort* 193:207–211.
Milusheva S, Kamenova I (2006). Plum pox virus sanitary status in Plovdiv Region of Bulgaria. *Voćarstvo* 40, 155(3):199–208.
Milusheva S, Rankova Z. (2002). Plum pox poty virus detection in Weed Species under Field Conditions. Proc. 7th International Symposium on Plum and Prune Genetics, Breeding and Pomology. *Acta Horticulturae* 577:283–287.
Mitre I Jr, Tripon A, Mitre I, Mitre V (2015). The response of several plum cultuvars to natural infection with *Monilinia laxa*, *Polystigma rubrum* and *Stigmina carpophila*. *Nat Sci Biol* 7(1):136–139.
Navratil M, Safarova A, Karesova R, Petrzik K (2005). First incidence of Plum pox virus on apricot trees in China. *Plant Disease* 89(3):338.
Nemchinov L, Crescenzi A, Hadidi A, Piazzolla P, Verderevskaya T (1998). Present status of the new cherry subgroup of plum pox virus (PPV-C). In: *Plant Virus Disease Control*, Hadidi A, Khetarpal RK, Koganezawa H (eds). APS Press, St. Paul, MN, 629–638.
Nemeth M, Kölber M (1980). GF 31 a reliable myrobalan field indicator for rapid detection of plum pox virus. *Acta Phytopath Acad Sc Hung* 15:207–212.
Nußbaum R-P, Maring E (2005). Merkblatt: Scharkakrankheit am Steinobst. Thüringer Ministerium für Landwirtschaft, Naturschutz und Umwelt. Thüringer Landesanstalt für Landwirtschaft, Erfurt-Kühnhausen, Germany.
Pasquini G, Barba M (2006). The question of seed transmissibility of Plum pox virus. *Bulletin OEPP/EPPO Bulletin* 36(2):287–292.
Petrov N (2014). Tracking of plant organ and tissue tropism of plum pox virus (PPV) and Prunus necrotic ring spot virus (PNRV) in plums and apricots. *J Mountain Agriculture Balkans* 17(2):433–445.
Petrov N, Stoev A (2011). Almond (Prunus dulcis) as a host of Plum Pox. *Acta Hort* 899:73–78.
Pobegajlo I (1939). Plum sharka (Plum mosaic). Agricultural Experimental and Control Station Beograd—Topchider, 3:1–40.

Polak J (2001). European spindle tree and common privet a new natural hosts of plum pox virus. ISHS *Acta Hort* 550:125–128.
Polak J, Kumar J, Krska B, Ravelonandro M (2012). Biotech GM crops in horticulture: Plum Cv. honey sweet resistant to plum pox virus. *Plant Protect Sci* 48:543–548.
Polischuk V (2005). Abiotic environmental factors: Effects on epidemiology of plant virus infections. In: *Virus Diseases and Crop Biosecurity*, Cooper I, Kühne T, Polischuk V (eds). Springer in cooperation with NATO Public Diplomacy Division, 120–132.
Ranković M, Šutić D (1980). Investigation of peach as a host of sharka (plum pox) virus. *Acta Phytopath Acad Sc Hung* 15(1–4):215–221.
Ranković M, Šutić D (1986). Resistance of some peach cultivars and variable pathogenicity of the sharka (plum pox) virus. *Acta Hort* 183:193–200.
Ranković M, Vuksanović S (1985). Mogućnosti I problemi dijagnostiek virusa šarke u kajsii prmenom ELISA tehnike. *Zaštita bilja* 36(2):161–166.
Ravelonandro M, Kundu J, Briard P, Milusheva S, Minoiu N, Ranković M (2007). Updated characterization of plum pox virus isolates from the Balkan area to Western Europe. *Acta Hort* 734:275–279.
Roy AS, Smith IM (1994). Plum Pox situation in Europe. *Bulletin OEPP/EPPO Bulletin* 24(3):515–523.
Savulescu A, Macovei A (1967). Comparative experimental data concerning plum pox and plum line pattern virus in Romania. VII Europäisches Symposium über Virukrankheiten der Obstbäume, Berlin, 10–16 July 1967. *Tagungsberichte Nr.* 97:241–249.
Schimanski H–H, Grüntzig M, Fuchs E (1988). Non-transmission of the plum pox virus in plum and apricot seed sourced clones. *Zentralbl Mikrobiol* 143:121–123.
Schmid G (1981). Transmission experiments with pseudo-pox and similar disorders of plum. *Acta Hort* 94:159–166.
Schuch K. (1957) Viruskrankheiten und Ähnliche Erscheinungen bei Obstgewächsen. Mitt. Bild. Bundesanst. Land- und Forstwirtsch Berlin–Dahlem, 88:1–96.
Scorza R, Ravelonandro M, Callahan A et al. (2016). Honey sweet (C5), the first genetically engineered plum pox virus–resistant plum (Runus domestiac L.) cultivar. *Hotscience* (5):601–603.
Sheveleva A, Kudryavtseva A, Speranskaya A, Belenikin M, Melnikova N, Chirkov S (2013). Complete genome sequence of a novel plum pox virus strain w isolate determined by 454 pyrosequencing. *Virus Genes* 47(2):385–388.
Stoev A (1996). Mixed Viral Infections on Plum. PhD Diss., Ministry of Agriculture and Forestry, Institute of Plant Protection, Bulgaria.
Stoev A (2000). A scale for pathological changes in fruit of plum (Prunus domestica L.) and myrobalan (Prunus cerasifera Ehrh.) infected by Plum Pox virus. *J Mountain Agriculture Balkans* 3(2):278–285.
Stoev A (2007). Development of the plum pox (sharka disease) control in Bulgaria. *Ecology Future* 6(1):14–16.
Stoev A (2016). Aggressiveness of phytopathogenic viruses: Trace investigation and evaluation. *Acta Microbiologica Bulgarica* 32(2):121–125.
Stoev A, Iliev P (2007). Manifestation of tolerance to sharka (plum pox virus) of plum cultivars imported in Bulgaria. *Agricultural Science* 40(5):34–37.
Stoev A, Iliev P (2009). *Testing of Black Skinned Plum Cultivars (Prunus Domestica L.) for Reaction to Plum Pox Virus in Natural Conditions*. Fotonika, Sofia.
Stoev A, Iliev P, Milenkov M (2006). Sharka (Plum Pox) disease—A permanent challenge? *Ecology Future* 5(4):21–24.
Stoev A, Kamenova I (1995). Studies on the relationship between plum cultivar Zhulta Butilkovidna and Plum Pox virus. *Plant Science* 32(4):44–145.
Šutić D (1964). Sejanci kajsije kao indikatora virusa šarke šljive. *Zaštita bilja* 77:87–91.
Šutić D (1980) Viroze biljaka. Publisher Nolit, Beograd.
Szathmary E, Dragoyski K, Hristova D, Stoev A, Tobias I, Palkovics L (2006). Molecular characterization of some plum pox virus isolates originated from Troyan Region, Bulgaria. *Plant Science* 43:407–412.
Thakur PD, Bhardwaj SV, Garg ID, Kishore K, Sharma DR (1994). Plum pox virus on stone fruits from India—A new record. *Plant Dis Res* 9(1):100–102.
Thompson D, McCann M, McLeod M, Lye D, Green M, James D (2001). First report of plum pox potyvirus in Ontario, Canada. *Plant Disease* 85(1):97.
Thresh JM (2005). Crop viruses and virus diseases: A global perspective. In: *Virus Diseases and Crop Biosecurity*, Cooper I, Kühne T, Polischuk V (eds). Springer with NATO Public Diplomacy Division, 9–31.
Topchiiska M (1995). Susceptibility of peach cultivars, roorstocks and hybrids to plum pox virus. *Plant Science* 32(4):19–23.

Trifonov D (1965). Plum pox virus infection rate of some varieties of plums in heavily contaminated regions of Bulgaria. *Zaštita bilja* 85–88:375–378.

Trifonov D (1972). Viral diseases on fruit trees. *ZEMIZDAT*, Sofia, p. 159.

Van Bakel CHJ, van Oosten H (1972). Additional data on the ultrastrucure of inclusion bodies evoked by sharka (plum pox) virus. *Neth J Pl Path* 78:160–16.

Van Oosten HJ (1970). The Diagnostic Value of Inclusion Bodies in the Fruits of Plum Trees Infected with Sharka Virus. Further Information about the Herbaceous Host Range of Sharka (Plum Pox Virus). *VIII-e Symposium Européen sur les Maladies a Virus des Arbres Frutiers*, Bordeaux, 24–30 June, 1970, 211–219.

Van Oosten HJ (1972). Diagnosis of sharka (plum pox) by internal and external fruit symptoms. *Neth J Pl Path* 78:99–106.

Van Oosten HJ, van Bakel CHJ (1970). Inclusion bodies in plants infected with sharka (plum pox) virus. *Neth J Pl Path* 76:313–319.

Vlasov Y, Larina E (1982). Agricultural Virology. *Kolos*, Moscow.

Wetzel T, Candresse T, Ravelonandro M et al. (1991). Nucleotide sequence of the 3'-terminal region of the RNA of the E1 Amar strain of plum pox potyvirus. *J Gen Virol* 72(7):1741–1746.

Yankulova M, Kamenova I, Stoev A, Gabova R (1990). Plum pox virus on peach in Bulgaria. *Plant Science* 27(4):42–47.

Zavadzka B (1980). The response of several plum cultivars to infection with plum pox virus. *Acta Phytopath Acad Sc Hung* 15(1–4):215–221.

Zotto A, Ortego JM, Raigón JM, Callogero S, Rossini M, Ducasse DA (2006). First report in Argentina of Plum pox virus causing sharka disease in Prunus. *Plant Disease* 90(4):523.

6 Evolution and Emergence of Geminiviruses

Reasons and Consequences

R. Vinoth Kumar and S. Chakraborty

CONTENTS

6.1 INTRODUCTION

The *Geminiviridae* family includes a group of plant-infecting circular ssDNA viruses of ~2.8 kb in size. It includes a group of twinned icosahedral particles containing circular single-stranded genomic component(s). Based on the nature of virions, the term "Geminivirus" was coined several decades ago (Harrison et al. 1977). The International Committee on the Taxonomy of Viruses (ICTV) has classified geminiviruses based on insect vector, genome organization, host range, and genome-wide pairwise sequence identities into seven genera: *Becurtovirus*, *Begomovirus*, *Curtovirus*, *Eragrovirus*,

Mastrevirus, *Topocuvirus*, and *Turncurtovirus* (Varsani et al. 2014). Among them, begomoviruses are geographically distributed throughout the world and are transmitted by the arthropod vector, *Bemisia tabaci* Genn. They infect and cause substantial crop losses in a large number of economically important vegetable, fiber, and food crops, and ornamental plants worldwide (Rocha et al. 2013; Varma and Malathi 2003).

6.2 GENOME ORGANIZATION OF BEGOMOVIRUSES

Based on the requirement of genomic components for successful pathogenesis, begomoviruses are regarded as either bipartite (both DNA-A and DNA-B) or monopartite (only DNA-A molecule) (Brown et al. 2012). Figure 6.1 depicts the genome organization of both the monopartite and bipartite begomoviruses. The presence of monopartite and some bipartite begomoviruses is mainly documented from the "Old World" ("OW"), while the "New World" ("NW") is considered the primary source of bipartite begomoviruses, with a few notable exceptions of monopartite begomoviruses (Rojas et al. 2005). Members of monopartite and a few bipartite viruses have been found along with ssDNA satellite molecules named alphasatellite and betasatellite (Nawaz-ul-Rehman and Fauquet 2009). Generally, DNA-A/DNA-A-like genomic component encodes six ORFs coding for replication-associated protein (Rep or *AC1*), transcriptional activator protein (TrAP or *AC2*), replication enhancer protein (REn or *AC3*), silencing suppressor protein (*AC4*) in the complementary strand, and coat protein (CP or *AV1*) and pre-coat protein (PCP or *AV2*) in the virion strand (Figure 6.1). The DNA-B component codes for movement protein (MP or *BC1*) and nuclear shuttle protein (NSP or *BV1*) which aids in intra- and intercellular virus movement. In bipartite begomoviruses, both of these DNA components are needed for systemic infection and for typical symptom induction (Brown et al. 2012; Hanley-Bowdoin et al. 2013).

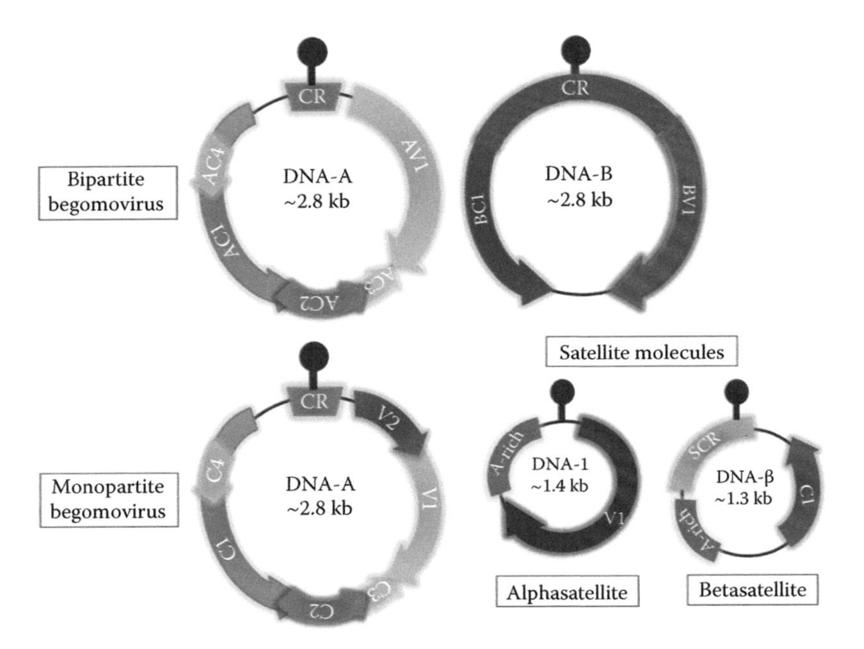

FIGURE 6.1 Genome organization of the begomovirus components such as bipartite (DNA-A and DNA-B) and monopartite (DNA-A only). DNA-A encodes two ORFs in virion strand (V1 and V2) and four ORFs in complementary strand (C1, C2, C3, and C4), whereas DNA-B encodes one ORF each in virion strand (BV1) and complementary strand (BC1). Both DNA-A and DNA-B share a conserved region (CR). In addition, alphasatellite and betasatellite encode a single ORF along with adenine rich (A-rich) region and satellite conserved region (SCR).

The majority of 'OW' begomoviruses are identified along with satellite molecules, called betasatellites, of ~1350 nucleotides in length (Nawaz-ul-Rehman and Fauquet 2009). These molecules contain a satellite conserved region (SCR), an adenine rich (A-rich) region, and a single ORF called βC1 in the complementary sense strand (Figure 6.1). However, the aspect of transreplication of these betasatellite molecules by various begomoviruses has been reported to be host specific (Ranjan et al. 2014). In addition to assisting in virus movement, these βC1 proteins have been demonstrated to counter RNAi silencing, and in monopartite begomoviruses they have been shown as a pathogenicity determinant and help in disease development (Saunders et al. 2004; Hanley-Bowdoin et al. 2013).

Alphasatellites (satellite-like molecules) are circular ss-DNA molecules of 1.4 kb in size which are mainly found in association with monopartite begomovirus-batasatellite complexes in the "OW" (Mansoor et al. 1999; Kumar et al. 2017b). However, few alphasatellites are also reported to be associated with "NW" begomoviruses (Paprotka et al. 2010; Jeske et al. 2014). These alphasatellites encode a single ORF in the virion strand, A-rich region, and SCR (Figure 6.1). They can replicate independently in the infected cells; however, they depend on the helper viruses for their insect transmission and systemic spread. These alphasatellites require helper begomoviruses for their intra- and intercellular movement, and also have been shown to ameliorate symptoms (Kumar et al. 2017b).

6.3 BEGOMOVIRUS TAXONOMY

On the basis of pairwise sequence identities of the complete nucleotide sequences, the Geminivirus study group of ICTV has devised various criteria for the nomenclature of begomovirus components. According to the guidelines, if a sequence possesses <91% pairwise identity to any of the known begomovirus species, it should be considered as a member of a new species (Brown et al. 2015). However, if the sequence shares 91–94% identity with all isolates of a particular begomovirus species, then a strain name should be proposed (Brown et al. 2015). Similarly, the demarcation threshold for betasatellites was recommended to be <78% (Briddon et al. 2008). No demarcation threshold for alphasatellites is recommended by the ICTV's Geminivirus study group, but a recent study has proposed <84% for their demarcation (Kumar et al. 2017b).

6.4 BEGOMOVIRUS DISEASES IN INDIA

In India, begomoviruses cause several diseases, such as leaf curl and yellow mosaic, in a wide variety of plant species belonging to *Brassicaceae*, *Caricaceae*, *Cucurbitaceae*, *Euphorbiaceae*, *Fabaceae*, *Leguminosae*, *Malvaceae*, and *Solanaceae*. The initial symptoms include leaf curling (upward and/or downward), leaf crinkling, leaf blistering, leaf rolling, yellowing of leaves, mosaic, enations, and interveinal chlorosis. However, at the severe stage of infection they cause stunting which eventually leads to the death of infected plants. Table 6.1 contains the complete list of the begomoviruses and their associated satellite molecules infecting various crop plants in India.

6.4.1 Mosaic and Leaf Curl Diseases of Amaranthus

The incidence of begomovirus infection causing leaf curl and yellow mosaic diseases of *Amaranthus* species have been documented in India. *Ageratum enation virus* (AEnV), a betasatellite and an alphasatellite, has been identified from an *Amaranthus* plants bearing yellow mosaic symptom in the Lucknow region (Srivastava et al. 2013a). Similarly, *Chilli leaf curl virus* (ChiLCV), *Tomato yellow leaf curl Thailand betasatellite* (TYLCTHB), and Chilli leaf curl alphasatellite (ChiLCA) were found to be associated with the leaf curl disease of amaranthus in the Banswara region of Rajasthan (George et al. 2014). In addition, association of *Papaya leaf curl virus* with the leaf curl disease of amaranth has been reported (Srivastava et al. 2015a). These studies suggest that amaranthus plants could act as an alternate host for begomoviruses.

TABLE 6.1
Begomovirus Genomic Components Infecting Various Crops in India

Crop (Family Name)	Associated Disease	Indentified Begomovirus Genomic Components[a]			Nature of Viral Genome
		Begomovirus(es)	Betasatellite(s)	Alphasatellite(s)	
Amaranthus (*Amaranthaceae*)	Mosaic	AEnV	ALCB	ALCA	Monopartite
	Leaf curl	ChiLCV PaLCuV	TYLCTHB	ChiLCA	Monopartite
Bhendi (*Malvaceae*)	Mosaic	BYVMV BYVBhV CLCuAlV CLCuBaV MeYVMV	BYVB	OLCuA	Monopartite
	Leaf curl	ChiLCV OEnLCV RaLCV	ToLCBDB	BYVMA CLCuBurA GMusSLA OLCuA OEnA	Monopartite
Bittergourd (*Cucurbitaceae*)	Mosaic	BGYMV PepLCBV	–	–	Monopartite
		ICMV	–	–	Bipartite
Cassava (*Euphorbiaceae*)	Mosaic	ICMV SLCMV	–	–	Bipartite
Chilli (*Solanaceae*)	Leaf curl	ChiLCV ChiLCAhmV ChiLCGdV ChiLCIV ChiLCSV ChiLCVeV PaLCuV PepLCBV PepLCLaV ToLCJoV ToLCV ToLCNDV	ChiLCB CroYVMB RaLCB ToLCBDB ToLCJoB ToLCRnB	ChiLCA GDarSLA	Monopartite
Cotton (*Malvaceae*)	Leaf curl	CLCuBaV CLCuBuV CLCuKoV-Lu CLCuMuV-Ra	CLCuMuB	–	Monopartite
Jatropha (*Euphorbiaceae*)	Mosaic	CYVMV JMINV JYMINV	CroYVMB	–	Monopartite
		ICMV	–	–	Bipartite
	Leaf curl	JLCV	–	–	Monopartite
Horsegram (*Leguminosae*)	Mosaic	HgYMV	–	–	Bipartite

(*Continued*)

TABLE 6.1 (CONTINUED)
Begomovirus Genomic Components Infecting Various Crops in India

Crop (Family Name)	Associated Disease	Indentified Begomovirus Genomic Components[a]			Nature of Viral Genome
		Begomovirus(es)	Betasatellite(s)	Alphasatellite(s)	
Mungbean (*Leguminosae*)	Mosaic	MYMV MYMIV	–	–	Bipartite
Soybean (*Leguminosae*)	Mosaic	MYMV MYMIV	–	–	Bipartite
		AEnV PaLCuV-Soy	–	AEnA	Monopartite
Frenchbean (*Leguminosae*)	Leaf curl	CYVMV FbLCV	CroYVMB PaLCuB	–	Monopartite
Velvet bean (*Leguminosae*)	Mosaic	VBSMV	–	–	Bipartite
Blackgram (*Leguminosae*)	Mosaic	MYMV MYMIV	–	–	Bipartite
Papaya (*Caricaceae*)	Leaf curl	ChiLCV-IN CYVMV PaLCrV PaLCuV-Luc ToLCNDV	ChiLCB CroYVMB PaLCB ToLCB	GMusSLA	Monopartite
Potato (*Solanaceae*)	Leaf curl	ToLCNDV	–	–	Bipartite
Pumpkin (*Cucurbitaceae*)	Mosaic	PYVMV SLCCNV	–	–	Bipartite
		ToLCPalV ToLCNDV	PeLCB ToLCBDB	–	Monopartite
Radish (*Brassicaceae*)	Leaf curl	CYVMV RaLCV	CroYVMB RaLCB	GDarSLA OLCuA	Monopartite
Tobacco (*Solanaceae*)	Leaf curl	PaLCuV-Tob TbLCPuV	ToLCBDB	ToLCuA	Monopartite
Tomato (*Solanaceae*)	Leaf curl	ToLCBV ToLCGV ToLCNDV ToLCPalV	–	GDarSLA	Bipartite
		AEnV CYVMV ToLCGV ToLCJoV ToLCKeV ToLCPaV ToLCRnV ToLCV	CroYVMB ToLCB ToLCPaB ToLCRnB TYLCTHB	–	Monopartite

[a] The names of the viral genomic components are abbreviated in the text.

6.4.2 Mosaic and Leaf Curl Diseases of Bhendi/Okra

Bhendi yellow vein mosaic disease (BYVMD) and Okra leaf curl disease (OLCD) are two economically important diseases of bhendi (*Abelmoschus esculentus*), hindering its productivity. Initially it was demonstrated that monopartite begomovirus (BYVMV) and betasatellite (*Bhendi yellow vein betasatellite* [BYVB]) cause typical yellow vein symptoms on the inoculated bhendi plants (Jose and Usha 2003; Venkataravanappa et al. 2011). In addition to BYVMV, several monopartite begomoviruses such as *Bhendi yellow vein Bhubhaneswar virus* (BYVBhV), *Cotton leaf curl Alabad virus* (CLCuAlV), *Cotton leaf curl Bangalore virus* (CLCuBaV), and *Mesta yellow vein mosaic virus* (MeYVMV) are also identified from bhendi plants affected with BYVMD in India (Kumar et al. 2017a; Venkataravanappa et al. 2012a, 2013; Zaffalon et al. 2012). Furthermore, ChiLCV, *Okra enation leaf curl virus* (OEnLCV), and *Radish leaf curl virus* (RaLCV) are the monopartite begomoviruses found to be associated with OLCD in India (Kumar et al. 2012a; Venkataravanappa et al. 2015, 2016). ToLCBDB was the only betasatellite group isolated from the okra plants bearing leaf curl symptoms. Similarly, the members of the alphasatellite groups i.e., *Bhendi yellow vein mosaic alphasatellite* (BYVMA), *Cotton leaf curl Burewala alphasatellite* (CLCuBurA), *Gossypium mustilinum symptomless alphasatellite* (GMusSLA), *Okra leaf curl alphasatellite* (OLCuA), and *Okra enation alphasatellite* (OEnA) were cloned from begomovirus-infected bhendi plants in India (Chandran et al. 2013; Zaffalon et al. 2012; Kumar et al. 2017b). In addition, a distinct bipartite begomovirus has been reported to be infecting bhendi plants (Venkataravanappa et al. 2012b). Altogether, the available literature underscores that monopartite begomovirus-betasatellite-alphasatellite complexes are found to be associated with begomoviral disease of bhendi in India.

6.4.3 Mosaic Disease of Bitter Gourd

Symptoms of yellow mosaic disease (YMD) on bitter gourd were first noticed in 2005, and subsequently a begomovirus named *Bitter gourd yellow mosaic virus* (BGYMV) has been reported from the diseased plant samples (Raj et al. 2005). The whitefly transmissibility of this begomovirus was also demonstrated (Rajinimala et al. 2005). Later, the association of the isolates of *Indian cassava mosaic virus* (ICMV) and *Pepper leaf curl Bangladesh virus* (PepLCBV) with YMD of bitter gourd have been documented in Tamil Nadu and Uttar Pradesh, respectively (Rajinimala and Rabindran 2007; Raj et al. 2010).

6.4.4 Cassava Mosaic Disease (CMD)

In India, the incidence of CMD was first reported by Alagianagalingam and Ramakrishnan in 1966. Initially, *Indian cassava mosaic virus* (ICMV) and *Sri Lankan cassava mosaic virus* (SLCMV) were identified as two bipartite begomoviruses found in association with CMD. Subsequently, SLCMV has been demonstrated to infect *Arabidopsis*, *N. tabacum*, and cassava, and SLCMV has been suggested to spread throughout southern India (Dutt et al. 2005; Mittal et al. 2008). Further, inoculated *N. benthamiana* plants were shown to produce DNA-B-derived defective DNA (def-DNA), which leads to altered symptoms and reduced DNA-B levels (Patil et al. 2007).

6.4.5 Chilli Leaf Curl Disease (ChiLCD)

ChiLCD was first documented in the 1960s (Mishra et al. 1963), and since then ChiLCD has emerged as a major limiting factor for chilli cultivation throughout the major chilli growing regions in India. Though the association of viruses with ChiLCD was reported several decades ago (Dhanraj and Seth 1968), *Chilli leaf curl virus* (ChiLCV) from Varanasi was the first complete begomovirus genome characterized in association with this disease (Chattopadhyay et al. 2008). Subsequently, *Chilli leaf curl India virus* (ChiLCIV), *Chilli leaf curl Vellanad virus* (ChiLCVeV), *Pepper leaf curl*

Lahore virus (PepLCLaV), and *Tomato leaf curl Joydebpur virus* (ToLCJoV) were identified along with betasatellite molecules including *Chilli leaf curl betasatellite* (ChiLCB) in India (Kumar et al. 2011, 2012b; Shih et al. 2007). However, a comprehensive analysis on the distribution of begomoviruses and satellites associated with ChiLCD in India was carried out recently (Kumar et al. 2015a). This report highlighted presence of various begomovirus species such as ChiLCV, *Chilli leaf curl Salem virus* (ChiLCSV), *Papaya leaf curl virus* (PaLCuV), PepLCBV, *Tomato leaf curl virus* (ToLCV), and *Tomato leaf curl New Delhi virus* (ToLCNDV) with ChiLCD-affected samples collected throughout the country (Kumar et al. 2015a). In addition, several members of betasatellites such as *Croton yellow vein mosaic betasatellite* (CroYVMB), *Radish leaf curl betasatellite* (RaLCB), *Tomato leaf curl Bangladesh betasatellite* (ToLCBDB), *Tomato leaf curl Joydebpur betasatellite* (ToLCJoB), and *Tomato leaf curl Ranchi betasatellite* (ToLCRnB) were reported in association with these samples (Kumar et al. 2015a). These begomoviruses and their associated betasatellites have been demonstrated to cause leaf curl disease in *N. benthamiana* and *C. annuum* plants, thereby fulfilling Koch's postulates for the first time (Chattopadhyay et al. 2008; Kumar et al. 2015a). The infectivity analysis emphasizes that these monopartite begomovirus-associated betasatellites are indispensable for symptom induction in *C. annuum* plants (Kumar et al. 2015a). Further, synergistic interaction of multiple begomovirus components was observed in the resistant chilli plants in India that lead to breakdown of resistance (Singh et al. 2016). Two new begomovirus species, Chilli leaf curl Ahmedabad virus and Chilli leaf curl Gonda virus, along with ToLCBDB, have been reported with ChiLCD from Gujarat and Uttar Pradesh, respectively (Bhatt et al. 2016; Khan and Khan 2017). Recently, members of the alphasatellites belonging to the groups ChiLCA and Gossypium darwinii symptomless alphasatellite (GDarSLA) were cloned from chilli plants affected with ChiLCD in India. Unlike betasatellites, results from the agro-inoculation studies using *N. benthamiana* plants suggest the dispensable role of ChiLCA in leaf curl disease development (Kumar et al. 2017b). Collectively, these reports suggest the increased association of diverse begomovirus-betasatellite-alphasatellite complex with ChiLCD at an alarming rate.

6.4.6 Cotton Leaf Curl Disease (CLCuD)

The appearance of leaf curl disease of cotton in northern India was first noticed by Rishi and Chauhan (1994). The first begomovirus identified in association with this disease was the 'Rajasthan' strain of *Cotton leaf curl Multan virus* (CLCuMuV-Ra) in 1994 from India. Subsequently, this disease has continuously spread to other cotton growing areas of the country. A new begomovirus species named CLCuBaV was isolated along with a DNA-β from *G. barbadense* in southern India (Reddy et al. 2005). Further, the association of *Cotton leaf curl Burewala virus* (CLCuBuV) and a "Lucknow" strain of *Cotton leaf curl Kokhran virus* (CLCuKoV-Lu) have been reported with CLCuD-affected samples (Kirthi et al. 2004; Kumar et al. 2010a; Zaffalon et al. 2012). *Cotton leaf curl Multan betasatellite* (CLCuMuB) is the only betasatellite group found to be associated with the above-mentioned monopartite begomoviruses associated with this disease (Reddy et al. 2005; Zaffalon et al. 2012). Recent studies have identified association of three distinct variants of begomoviruses and their associated satellites that lead to increasing CLCuD incidence in cotton over the years (Godara et al., 2015, 2017).

6.4.7 Mosaic and Leaf Curl Diseases of Jatropha

Several members of *Jatropha* spp have been reported to be affected by mosaic and leaf curl diseases caused by begomoviruses. Various members of begomoviruses infecting *Jatropha* spp are *Jatropha leaf curl virus* (JLCV), *Jatropha mosaic India virus* (JMINV), and *Jatropha yellow mosaic India virus* (JYMINV) (Snehi et al. 2011a; Srivastava et al. 2015b). In addition, a monopartite begomovirus, *Croton yellow vein mosaic virus* (CYVMV), along with DNA-β was reported to be associated with the YMD of *J. gossypifolia* in India (Snehi et al. 2011b). However, the only bipartite begomovirus known to be associated with YMD was 'Jatropha' strain of ICMV (Snehi et al. 2012).

6.4.8 Yellow Mosaic Disease of Legumes

Yellow mosaic disease (YMD) on mungbean was first reported in India by Nariani (1960). Currently, seven distinct begomovirus species are known to be associated with YMD of blackgram (*Vigna mungo*), French bean (*Phaseolus vulgaris*), horsegram (*Macrotyloma uniflorum*), mungbean (*Vigna radiata*), soybean (*Glycine max*), and velvet bean (*Mucuna pruriens*) (Table 6.1). *Mungbean yellow mosaic virus* (MYMV) and *Mungbean yellow mosaic India virus* (MYMIV) are the two important begomoviruses known to infect mungbean and several members of *Leguminoseae* family. A distinct bipartite begomovirus, *Velvet bean severe mosaic virus* (VBSMV), was reported from Lucknow, and this VBSMV has also been demonstrated to cause disease in the inoculated velvet bean plants (Zaim et al. 2011). Similarly, from Tamil Nadu, a novel bipartite begomovirus, *Horsegram yellow mosaic virus* (HgYMV), has been experimentally shown to cause typical yellow mosaic symptoms on inoculated horsegram plants (Barnabas et al. 2010). Recently, a new begomovirus species named *French bean leaf curl virus* (FbLCV) and an isolate of *Papaya leaf curl betasatellite* (PaLCuB) have been identified from leaf curl disease-affected French bean plants in the Kanpur region (Kamaal et al. 2013). A blackgram isolate of MYMV (DNA-A) and MYMIV (DNA-B) complex have been shown to produce symptom expression on the inoculated plants (Haq et al. 2011; Mandal et al. 1997). Pramesh et al. (2013) subsequently demonstrated infection of CYVMV and CroYVMB on French bean plants through whitefly transmission. In addition, a member of HgYMV has been shown to infect this crop (Haq et al. 2011; Malathi and John 2008). The DNA-A of MYVMV and DNA-B of MYVIV have been identified from the diseased blackgram plants, and the inoculated plants did not produce any visible symptoms, however the viral DNAs were detected by PCR (Haq et al. 2011). The association of MYMV with YMD of soybean has been identified in 2005 and the soybean plants inoculated with these cloned viral components developed yellow mosaic symptoms (Girish and Usha 2005). Subsequently, yellow mosaic virus infecting soybean in northern India has been found distinct from the species infecting soybean in southern and western India (Usharani et al. 2004). In addition, a 'Soybean' strain of PaLCuV was identified from the Lucknow region in 2011 (Brown et al. 2015). Collectively these studies suggest that both monopartite and bipartite begomovirus complexes infect various legumes leading to severe crop loss.

6.4.9 Papaya Leaf Curl Disease (PaLCuD)

Papaya (*Carica papaya*) is a fruit rich in various essential vitamins and minerals. In India, Saxena et al. (1998) predicted the association of bipartite begomovirus with PaLCuD in Uttar Pradesh, India. An analysis of begomoviruses infecting papaya suggests the presence of ChiLCV, *Papaya leaf crumple virus* (PaLCrV), and ToLCNDV in the infected plant samples (Singh-Pant et al. 2012). The association of CYVMV and PaLCuV with PaLCuD in India has also been documented (Sinha et al. 2016). ChiLCB, CroYVMB, *Papaya leaf curl betasatellite* (PaLCB), and *Tomato leaf curl betasatellite* (ToLCB) are the various groups of betasatellites reported in association with this disease in India (Singh-Pant et al. 2012). An isolate of GMusSLA was isolated from a papaya sample from Varanasi region (Kumar et al. 2017b).

6.4.10 Potato Apical Leaf Curl Disease (PoALCD)

In 2001, Garg et al. (2001) first reported a high incidence of apical leaf curl disease in early planted potato in northern India, and the associated begomovirus with PoALCD was tentatively named Potato apical leaf curl virus. However, it was later found that these viruses belong to ToLCNDV (Usharani et al. 2003). In addition, the presence of DNA-B component was shown in few potato samples affected with PoALCD (Venkatasalam et al. 2011), suggesting a possible association of bipartite begomovirus.

6.4.11 Pumpkin Yellow Mosaic Disease

Pumpkin (*Cucurbita moschata*) cultivation is affected largely due to begomovirus infections causing YMD, leading to severe crop loss. *Pumpkin yellow vein mosaic virus* (PYVMV) is a distinct bipartite begomovirus associated with YMD of pumpkins (Maruthi et al. 2003). A bipartite begomovirus, *Squash leaf curl China virus* (SLCCNV), from Coimbatore and Varanasi regions, has been demonstrated to cause typical yellow mosaic symptoms on pumpkin plants (Singh et al. 2009). Further, *Tomato leaf curl Palampur virus* (ToLCPalV) and ToLCBDB have been isolated from a pumpkin plant, which was followed by a report on the association of ToLCNDV without DNA-B or DNA-β (Namrata et al. 2010; Phaneendra et al. 2012). These data suggest the association of both monopartite and bipartite begomoviruses with pumpkin YMD in India.

6.4.12 Radish Leaf Curl Disease (RaLCD)

In 2012, the association of begomoviruses with RaLCD was reported for the first time in India. A new begomovirus-betasatellite complex, RaLCV and *Radish leaf curl betasatellite* (RaLCB), has been identified from Varanasi region, and similarly, CYVMV and CroYVMB were reported from Pataudi (Singh et al. 2012). All these cloned viral components have been demonstrated to infect *N. benthamiana* and *R. sativus* plants. Differential interaction of these begomoviral components with various tomato-infecting begomoviruses has also been demonstrated using *N. benthamiana*, *S. lycopersicum*, and *R. sativus* plants (Singh et al. 2012). GDarSLA and Okra leaf curl alphasatellite (OLCuA) were also cloned and characterized from the Pataudi sample (Kumar et al. 2017b). Collectively, in India, RaLCD is mainly caused by monopartite begomovirus-betasatellite-alphasatellite complexes.

6.4.13 Tobacco Leaf Curl Disease (TbLCD)

Leaf curl disease of tobacco is also prevalent throughout India. In 2009, a distinct begomovirus, *Tobacco leaf curl Pusa virus* (TbLCPuV), was identified along with ToLCBDB and an alphasatellite (Tomato leaf curl alphasatellite [ToLCuA]) from a tobacco sample in India (Singh et al. 2011). Another report suggests the presence of a 'tobacco' strain of PaLCuV with TbLCD (Kumar et al. 2009).

6.4.14 Tomato Leaf Curl Disease (ToLCD)

ToLCD is one of the major diseases infecting tomato and causing severe crop loss, thereby considered as a major limitation for tomato cultivation in India. This disease was first documented from northern India and then further reported from other parts of the country (Vasudeva and Raj 1948). Padidam et al. (1995) characterized a new bipartite begomovirus associated with ToLCD which was named as ToLCNDV. In addition, it was identified that ToLCNDV isolates from northern India have bipartite genome, whereas from southern India, only DNA-A component of ToLCNDV was detected. Subsequently, *Tomato leaf curl Bangalore virus* (ToLCBV) and *Tomato leaf curl Gujarat virus* (ToLCGV) are the two new begomoviruses identified from Karnataka and Gujarat, respectively (Kirthi et al. 2002; Chakraborty et al. 2003). Four distinct strains were recognized by ICTV within ToLCBV (A to D) and ToLCNDV (1 to 4) species (Brown et al. 2015). Later, several new begomovirus species such as ToLCJoV, *Tomato leaf curl Kerala virus* (ToLCKeV), *Tomato leaf curl Palampur virus* (ToLCPalV), *Tomato leaf curl Patna virus* (ToLCPaV), and *Tomato leaf curl Ranchi virus* (ToLCRnV) were found to be associated with ToLCD in India (Brown et al. 2015;

Pandey et al. 2010; Kumar et al. 2008; Kumari et al. 2010, 2011). In addition, the association of AEnV with ToLCD was reported from the Pantnagar region in Uttarakhand state (Swarnalatha et al. 2013). Recently, Tomato leaf curl Gandhinagar virus was reported from a tomato plant bearing ToLCD-like symptoms in Gujarat (Rathore et al. 2014). Among the begomoviruses associated with ToLCD, ToLCPalV and ToLCNDV are bipartite begomoviruses, while the other begomoviruses are monopartite in nature. Kanakala et al. (2013a) demonstrated the asymmetric synergism and hetero-encapsidation between ToLCPalV and ToLCNDV using tomato as an experimental host. However, recent reports suggest the predominant association of ToLCGV DNA-A component with TYLCTHBs rather than its cognate DNA-B component (Jyothsna et al. 2013a). Similarly, differential pathogenicity of ToLCGV isolates along with TYLCTHB has also been demonstrated (Ranjan et al. 2013). The diverse betasatellites were transreplicated by ToLCBV, which in turn enhanced the pathogenicity of ToLCBV to infect *N. benthamiana* and tomato plants (Tiwari et al. 2010). Ranjan et al. (2014) demonstrated the role of host-specific adaptation in the transreplication of diverse betasatellites by various tomato-infecting begomoviruses. Association of betasatellites has resulted in the enhanced level of helper virus components (ToLCNDV), but these betasatellites antagonistically interacted with DNA-B component (Jyothsna et al. 2013b). Further, in the absence of CroYVMB, tomato plants inoculated with CYVMV did not develop any symptoms, thereby suggesting the necessity of betasatellites in the symptom induction (Pramesh et al. 2013). The diversity of betasatellites associated with ToLCD in India was investigated by Sivalingam et al. (2010). This report emphasizes that betasatellites identified from central and southern parts of India were closer to each other than of that of betasatellites reported from north India. Collectively, several distinct begomoviruses (bipartite and monopartite) along with diverse betasatellites were found to be responsible for causing leaf curl disease of tomato in India.

6.4.15 Weeds and Ornamental Plant-Associated Begomoviruses

Weeds, the non-cultivated plants distributed throughout the world, have a high potential of environmental adaptability. The various begomovirus components associated with weeds or ornamental plants are summarized in Table 6.2. During off-season periods, the viruses can be maintained by these weeds and subsequently transmitted to cultivated crops by insect vectors. CYVMV is a widely known begomovirus species infecting various weeds, including *Croton bonplandianum*, in the Indian subcontinent (Pramesh et al. 2013). In 2007, the association of a novel monopartite begomovirus, MeYVMV, and cotton-associated betasatellites were reported with YMD in mesta (*Hibiscus cannabinus*) plants (Prajapat et al. 2014). Subsequently, a novel monopartite begomovirus named *Rhynchosia yellow mosaic India virus* was identified in association with YMD of leguminous weed *Rhynchosia minima* in India (Jyothsna et al. 2011). An 'OW' begomovirus-betasatellite-alphasatellite complex associated with leaf curl disease of guar (*Cyamopsis tetragonoloba*) plants was characterized by Kumar et al (2010b). Likewise, a variant of ToLCNDV has been demonstrated to cause YMD in *Solanum melongena* plants in India (Pratap et al. 2011). In winter, 2011–2012, the presence of weed-associated begomovirus-betasatellite (CYVMV and CroYVMB) complex was detected among the accessions of rapeseed-mustard germplasm displaying leaf curl disease symptoms (Roy et al. 2013). Furthermore, several ornamental plants (*Althea rosea*, *Cosmos bipinnata*, *Petunia hybrid*, and *Zinnia elegans*) and weeds (*Acanthospermum hispidum*, *Chenopodium amaranticolor*, *Datura stramonium*, *Nicandra physaloides*, *Phyllanthus niruri*, *Sonchus torvum*, and *Synedrella nodiflora)* cause disease symptoms upon transmission of CYVMV by whiteflies (Pramesh et al. 2013). The first report of association of a begomovirus with the leaf curl diseases of *Datura inoxia* and *Helianthus* spp was reported by Marwal et al. (2012) and Vanitha et al. (2013) (Table 6.2). The presence of ChiLCV and ChiLCIV with leaf curl disease of *Petunia hybrida* and *Mentha spicata* plants, respectively, was also documented from India (Nehra and Gaur 2015; Saeed et al. 2014). The AEnV has been found to be associated with leaf curl disease of zinnia and YMD of *Crassocephalum*

TABLE 6.2
Indian Weed or Ornamental Plants-Associated Begomoviral Genomic Components

Weed or Ornamental Plant(s)	Associated Begomoviral Genomic Components			Nature of Viral Component
	Begomovirus	Betasatellite	Alphasatellite	
Abutilon pictum	*Abutilon mosaic virus*	–	–	Bipartite
Coccinia grandis	*Coccinia mosaic virus*	–	–	Bipartite
Solanum melongena	*Tomato leaf curl New Delhi virus*	–	–	Bipartite
Rhynchosia minima	*Rhynchosia yellow mosaic India virus*	–	–	Bipartite
Calendula officinalis	*Ageratum enation virus*	*Ageratum leaf curl betasatellite*	–	Monopartite
Clerodendron inerme	*Clerodendron yellow mosaic virus*	–	–	Monopartite
Cyamopsis tetragonoloba	*Cyamopsis tetragonoloba leaf curl virus*	*Cotton leaf curl Multan betasatellite*	Cyamopsis tetragonolobaleaf curl alphasatellite	Monopartite
Datura inoxia	*Chilli leaf curl virus/ Tomato leaf curl virus*	–	–	Monopartite
Helianthus spp	*Tomato leaf curl Karnataka virus*	–	–	Monopartite
Hemidesmus indicus	*Hemidesmus yellow mosaic virus*	–	–	Monopartite
Hibiscus cannabinus	*Mesta yellow vein mosaic virus*	*Cotton leaf curl betasatellite*	–	Monopartite
Hibiscus rosa-sinensis	*Cotton leaf curl Multan virus*	*Cotton leaf curl Multan betasatellite*	–	Monopartite
Kalimeris indica	*Papaya leaf curl virus*	*Ageratum leaf curl betasatellite*	–	Monopartite
Lens culinaris	*Tomato leaf curl New Delhi virus - Lentil*	–	–	Monopartite
Mentha spicata	*Chilli leaf curl India virus*	*Ageratum yellow vein betasatellite*	–	Monopartite
Mirabilis jalapa	*Chilli leaf curl India virus*	–	–	Monopartite
Parthenium hysterophorus	*Tomato leaf curl virus*	*Papaya leaf curl betasatellite*	Ageratum yellow vein alphasatellite	Monopartite
Petunia hybrida	*Chilli leaf curl virus*	*Chilli leaf curl betasatellite*	–	Monopartite
Rapeseed-mustard	*Croton yellow vein mosaic virus*	*Croton yellow vein mosaic betasatellite*	–	Monopartite
Rosa indica	*Rose leaf curl virus*	*Digera leaf curl betasatellite*	–	Monopartite
Senna occidentalis	*Senna leaf curl virus*	*Papaya leaf curl betasatellite*	Ageratum yellow vein alphasatellite	Monopartite
Tagetes patula	*Ageratum enation virus*	*Ageratum leaf curl betasatellite*	Ageratum enation alphasatellite	Monopartite
Cestrum nocturnum and *Tabernaemontana coronaria*	*Pedilanthus leaf curl virus*	–	–	Monopartite

crepidioides and *Ageratum conyzoides* (Prajapat et al. 2014). Further, AEnV and ALCB was found infecting *Calendula officinalis* in India (Jaidi et al. 2015). Association of *Pedilanthus leaf curl virus* with yellow mottling of crape jasmine (*Tabernaemontana coronaria*) and leaf curl of night-blooming jasmine (*Cestrum nocturnum*) was documented by Srivastava et al. (2014). However, the presence of *Rose leaf curl virus* (RoLCuV) and *Digera leaf curl betasatellite* was obtained from the leaf curl disease that affected *Rosa indica* plants in Rajasthan (Table 6.2). By whitefly transmission, RoLCuV has been demonstrated to infect several cultivated crops such as *Abelmoschus esculentus*, *Capsicum annuum*, *Gossypium arboretum*, *Solanum lycopersicum*, and *Raphanus sativus* (Sahu et al. 2014). Similarly, PaLCuV and ALCB were isolated from aster plants (*Kalimeris indica*) displaying yellow vein disease symptoms in India (Srivastava et al. 2013b). In addition, the first molecular evidence of the presence of *Abutilon mosaic virus* ('NW' begomovirus) was observed from *Abutilon* plants showing bright yellow mosaic symptoms (Jyothsna et al. 2013c). A recent study also identifies the association of novel and unreported begomoviruses along with multiple satellite components with yellow mosaic and leaf curl diseases of hollyhock plants (Kumar et al. 2016). Altogether, the available data suggest that non-cultivated plants were infected by both bipartite and monopartite begomoviruses, and satellite molecules in India. These studies also indicate that non-cultivated plants act as source of primary inoculums of viruses by serving as alternate hosts for the virus spread.

6.5 DISEASES CAUSED BY MASTREVIRUSES IN INDIA

The members of *Mastrevirus* genus are known to infect chickpea and wheat crops in India. Nene and Reddy (1976) first recorded the incidence of chickpea stunt disease (CSD) in India. This is the most important disease affecting the productivity of chickpea worldwide. Association of *Chickpea chlorotic dwarf virus* (CpCDV) with CSD was reported in 1993, and the yield loss of chickpea due to CpCDV has subsequently been assessed (Horn et al. 1993; 1995). The presence of CpCDV strains C and D in India has recently been reported (Kraberger et al. 2015). The CpCDV resistance genotypes were screened through *Agrobacterium*-mediated delivery of this virus, and this screening might help to manage CSD in a sustainable manner (Kanakala et al. 2013b). In 2012, a new mastrevirus, *Wheat dwarf India virus* (WDIV), was reported from India, and the infectious clone of WDIV resulted in dwarfing of inoculated plants, thereby confirming it as a causal agent of wheat dwarf disease (Kumar et al. 2012c). The spread of this virus throughout the wheat growing regions of the country has been recorded (Kumar et al. 2015b). Subsequently, without the presence of any begomovirus, two different alphasatellites and a betasatellite have been found along with WDIV in India (Kumar et al. 2014a). Further, βC1 of the cloned betasatellite has been demonstrated to be a pathogenicity factor by functionally interacting with WDIV, a mastrevirus (Kumar et al. 2014b). This is the first report of an interaction between the components of begomoviruses and mastrevirus to cause wheat dwarf disease in India.

6.6 EVOLUTION OF INDIAN BEGOMOVIRUS POPULATION

Major factors driving the emergence and evolution of these viruses are described in the following sections.

6.6.1 Recombination

Genetic recombination is demonstrated to play pivotal roles in the emergence of geminiviruses, most devastating phytopathogens (George et al. 2015; Rocha et al. 2013). As these plant-infecting viruses replicate through recombination-driven processes, it is imperative to mention that the recombination event plausibly facilitates the exchange of genetic fragments between same and/or diverse begomoviruses. These recombination processes can be described as intraspecific (exchanging of fragments between the isolates of a single begomovirus species) and interspecific (genetic exchanges between diverse begomovirus species). Certainly, mixed infections play a major role in the emergence of new recombinant molecules capable of infecting new crop species. The majority of the recombination events detected in the begomoviruses infecting chilli crops belong to the begomoviruses infecting tomato (Kumar et al. 2015a). However, non-random distribution of recombination events has been reported from various tomato-infecting begomoviruses. In addition, the possibility of genetic exchanges between bhendi and cotton-infecting begomoviruses has been reported recently (Kumar et al. 2017a). Therefore it is apparent that solanaceous crop-infecting begomoviruses (chilli and tomato) and malvaceous crop-infecting begomoviruses (cotton and bhendi) in India have a complex history of recombination events by sharing different common ancestors. The multiple interspecific recombination events have been reported to be resulted in the emergence of distinct begomoviruses such as RaLCV and TbLCPuV (Singh et al. 2011, 2012). Similar to the helper begomoviruses, genetic exchange of fragments has been reported to be common among bhendi, chilli, and tomato-associated betasatellites (Kumar et al. 2015a, 2017b). A comprehensive study on recombination among alphasatellites suggests that alphasatellites associated with solanaceous and malvaceous crops in the 'OW' have a complex history of recombination, sharing a common ancestor (Kumar et al. 2017b). The existence of a plausible link between recombination and microsatellite has also been suggested (George et al. 2015). It has been speculated that the microsatellites could provide a platform for the host-encoded recombinase to bind and exchange the viral fragments. Hence it is worth mentioning that recombination in begomovirus and satellite components acts as a major driving force in the emergence of more pathogenic virus species and/or severe disease complex of begomoviruses and betasatellites.

6.6.2 Nucleotide Diversity and Nucleotide Substitutions

The ultimate source of genetic variation is mutation/substitution, which along with recombination determines the genetic makeup of pathogens, ultimately leading to either compatible or incompatible interaction with the host. The nucleotide substitution rate of ChiLCV and BYVMV was found to be similar to ssRNA viruses (10^{-3} to 10^{-5} nt/site/generation) rather than dsDNA viruses (10^{-8} nt/site/generation) (Kumar et al. 2015a, 2017a). This high substitution rate detected could be attributed to the involvement of error-prone DNA polymerases in the replication of begomovirus molecules (Richter et al. 2016). In addition, a high degree of genetic variability in the Indian begomoviruses (such as bhendi- and chilli-infecting begomoviruses) has been documented than in the begomoviruses reported from the "NW." The genetic variability among various begomoviruses identified from both "OW" and "NW" are shown in Table 6.3. Among these begomoviruses, a higher number of amino acids has been reported to be under negative selection, thereby suggesting the role of strong purifying selection in their evolution (Kumar et al. 2015a, 2017a). These data suggest that genetic variation and nucleotide substitution rate are significant factors involved in the evolution of Geminiviruses.

6.6.3 Pseudo-Recombination

The exchange of viral components by pseudo-recombination occurs among begomoviruses, which enables them to infect multiple hosts or increase the virulence. The pseudo-recombination between the components of ToLCGV and ToLCNDV (both bipartite begomoviruses) has been demonstrated to cause more severe disease in tomato than their counterparts (Chakraborty et al. 2008). On the other hand, this same pseudo-recombination event in chilli has been reported to be involved in breakdown of virus resistance (Singh et al. 2016). Importantly, pseudo-recombination between a monopartite and a bipartite begomovirus has been reported to result in the acquisition of DNA-B component by this monopartite begomovirus from the bipartite begomovirus (Saunders et al. 2002). The biological functions of genes encoded by the captured DNA-B component might enable this monopartite

TABLE 6.3
Comparison of Genetic Variability Detected among the Begomovirus Population Identified Worldwide

Viral Population	ToCmMV[a]	ToCMoV[a]	ToSRV[a]	ToYVSV[a]	BYVMV[b]	ChiLCV[c]	PepLCBV[c]
Number of sequences	22	22	27	26	67	19	7
Genome size	2560	2619	2588	2562	2587	2790	2743
Total number of segregating sites	103	135	148	49	1130	708	576
Total number of mutations	104	138	159	49	1473	849	690
Average number of nucleotide differences between sequences	36.645	18.4	26.5	5.381	208.02	253.5	263.9
Nucleotide diversity	0.0143	0.0070	0.0102	0.0021	0.0800	0.0934	0.0962

[a] ToCmMV, *Tomato common mosaic virus*; ToCMoV, *Tomato chlorotic mottle virus*; ToSRV, *Tomato severe rugose virus*; ToYVSV, *Tomato yellow vein streak virus* (Rocha et al. 2013)
[b] BYVMV, *Bhendi yellow vein mosaic virus* (Kumar et al. 2017a)
[c] ChiLCV, *Chilli leaf curl virus*; PepLCBV, *Pepper leaf curl Bangladesh virus* (Kumar et al. 2015a)
ToCmMV, ToCMoV, ToSRV, and ToYVSV are 'NW' begomoviruses, whereas BYVMV, ChiLCV, and PepLCBV are 'OW' begomoviruses

begomovirus to expand its host range of infection. Further, the expression of the disease can be modulated by pseudo-recombination of betasatellites in a host-dependent manner (Ranjan et al. 2014).

6.7 CONCLUSION

Geminiviridae is one of the largest families of plant-infecting viruses, infecting a wide range of the most economically important plants throughout the world and leading to significant crop loss. A large number of viruses have been identified from the Indian subcontinent, which might be due to its tropical climatic conditions or round-the-year crop cultivation aiding the spread of diseases. The rapid emergence of diverse begomoviruses and their associated satellite molecules in India renders it a center of origin of the evolution of these viruses. The main aspect of the evolution of Indian begomoviruses is their nature of having a broad host range, i.e., a begomovirus species (such as ChiLCV or ToLCNDV) infecting members of various hosts across many plant families. Mixed infection of begomoviruses and their associated satellite components has been reported to be common among Indian viruses, thereby providing a platform to undergo interspecific recombination. Furthermore, mixed infections have resulted in the interaction of unrelated viral components through complementation and synergism. However, the mechanistic details of these synergistic interactions are insufficiently understood, and in-depth analysis on these aspects could be an important breakthrough. The interaction of begomovirus molecules (alphasatellite and betasatellite) with a member of mastrevirus (WDIV), the acquisition of DNA-B component by a monopartite begomovirus (ICMV), and increased incidences of the presence of betasatellite (instead of DNA-B) with DNA-A component of a bipartite begomoviruses (ToLCGV and TYLCTHB) have resulted in enhanced spread of the disease and/or invading alternate hosts. Importantly, several weeds and ornamental plants have been reported to act as reservoirs that can be an important factor in the emergence of new disease complexes. However, studies on the distribution of viruses infecting non-cultivated plants such as weeds in India are still lacking, and hence comprehensive research is needed to fill the knowledge gap for this vital information. The Indian begomovirus components have also been found to have a high degree of genetic variability—deamination, lack of repair by host exonucleases, and virus replication by error-prone DNA polymerases are some of the plausible reasons. Together, mutation and recombination might lead to evolution of several variants/species, of which the best armed ones will prevail and may cause severe pandemics. A major reason for the successful emergence of several new disease complexes in India could be the polyphagous nature of insect vector that carries these viruses, continued crop cultivation throughout the year, or the presence of weeds in fields during off seasons. There is a strong research interest on studying the genetic distribution of begomoviruses in India; however, no significant improvement has been made in studying the virus-vector relationship, which is a crucial aspect in understanding the spread of diseases caused by these viruses. Studying the nature and distribution of insect vector, host population dynamics, and modulations in the virus-vector relationship may help predict the emergence of diseases in a particular area before it appears. The virus- and host-derived methods and biological control of viruses have not yielded effective solutions in tackling these viral diseases. Hence there is an urgent need to attempt various insect vector-based virus resistance strategies along with the implementation of efficient integrated pest or disease management approaches to protect various economically important crops from geminivirus infections.

ACKNOWLEDGMENT

The authors are thankful to DRS-1 grant of the University Grants Commission, Ministry of Human Resource and Development, Goverment of India, New Delhi for providing financial support.

REFERENCES

Alagianagalingam MN, Ramakrishnan K (1966). Cassava mosaic in India. *S Indian Hort* 14:71–72.

Barnabas AD, Radhakrishnan GK, Usha R (2010). Characterization of a begomovirus causing horsegram yellow mosaic disease in India. *Eur J Plant Pathol* 127:41–51.

Bhatt BS, Chahwala FD, Rathod S, Singh AK (2016). Identification and molecular characterization of a new recombinant begomovirus and associated betasatellite DNA infecting *Capsicum annuum* in India. *Arch Virol* 161(5):1389–1394.

Briddon RW, Brown JK, Morione E, Stanley J et al. (2008). Recommendations for the classification and nomenclature of the DNA-β satellites of begomoviruses. *Arch Virol* 153:763–781.

Brown JK, Fauquet CM, Briddon RW, Zerbini M et al. (2012). Family *Geminiviridae*. In: *Virus Taxonomy: Classification and Nomenclature of Viruses. Ninth Report of the International Committee on Taxonomy of Viruses*. King AMQ, Adams MJ, Carstens EB, Lefkowitz EJ (eds), 351–373. London: Elsevier Academic Press.

Brown JK, Zerbini FM, Navas-Castillo J, Moriones E et al. (2015). Revision of begomovirus taxonomy based on pairwise sequence comparisons. *Arch Virol* 160:1593–1619.

Chakraborty S, Pandey PK, Banerjee MK, Kalloo G, Fauquet CM (2003). *Tomato leaf curl Gujarat virus*, a new begomovirus species causing a severe leaf curl disease of tomato in Varanasi, India. *Phytopathology* 93:1485–1495.

Chakraborty S, Vanitharani R, Chattopadhyay B, Fauquet CM (2008). Supervirulent pseudo-recombination and asymmetric synergism between genomic components of two distinct species of begomovirus associated with severe tomato leaf curl disease in India. *J Gen Virol* 89:818–828.

Chandran SA, Packialakshmi RM, Subhalakshmi K, Prakash C et al. (2013). First report of an alphasatellite associated with *Okra enation leaf curl virus*. *Virus Genes* 46(3):585–587.

Chattopadhyay B, Singh AK, Yadav T, Fauquet CM et al. (2008). Infectivity of the cloned components of a begomovirus: DNA beta complex causing chilli leaf curl disease in India. *Arch Virol* 153:533–539.

Dhanraj KS, Seth ML (1968). Enation in *Capsicum annum* L (Chili) caused by a new strain leaf curl virus. *Indian J Hort* 25:70–71.

Dutt N, Briddon RW, Dasgupta I (2005). Identification of a second begomovirus, *Sri Lankan cassava mosaic virus*, causing cassava mosaic disease in India. *Arch Virol* 150:2101–2108.

Garg ID, Paul-Khurana SM, Kumar S, Lakra BS (2001). Association of a geminivirus with potato apical leaf curl in India and its immunoelectron microscopic detection. *J Indian Potato Assoc* 28:227–232.

George B, Kumar RV, Chakraborty S (2014). Molecular characterization of *Chilli leaf curl virus* and satellite molecules associated with leaf curl disease of *Amaranthus* spp. *Virus Genes* 48:397–401.

George B, Alam CM, Kumar RV, Gnanasekaran P, Chakraborty S (2015). Potential linkage between compound microsatellites and recombination in geminiviruses: Evidence from comparative analysis. *Virology* 482:41–50.

Girish KR, Usha R (2005). Molecular characterization of two soybean-infecting begomoviruses from India and evidence for recombination among legume-infecting begomoviruses from South-East Asia. *Virus Res* 108:167–176.

Godara S, Saini N, Paul Khurana SM, Biswas KK (2015). Lack of resistance in cotton against cotton leaf curl begomovirus disease complex and occurrence of natural virus sequence variants. *Indian Phytopath* 68 (3):326–333.

Godara S, Paul-Khurana SM, Biswas KK (2017). Three variants of cotton leaf curl begomoviruses with their satellite molecules are associated with cotton leaf curl disease aggravation in New Delhi. *J Plant Biochem Biotech*. 26(1):97–105.

Hanley-Bowdoin L, Bejarano ER, Robertson D, Mansoor S (2013). Geminiviruses: Masters at redirecting and reprogramming plant processes. *Nat Rev Microbiol* 11(11):777–788.

Haq QMI, Rouhibakhsh A, Ali A, Malathi VG (2011). Infectivity analysis of a blackgram isolate of *Mungbean yellow mosaic virus* and genetic assortment with MYMIV in selective hosts. *Virus Genes* 42:429–439.

Harrison BD, Barker H, Bock KR, Guthrie EJ et al. (1977). Plant viruses with circular single-stranded DNA. *Nature* 270:760–762.

Horn NM, Reddy SV, Roberts IM, Reddy DVR (1993). *Chickpea chlorotic dwarf virus*, a new leaf hopper transmitted geminivirus of chickpea in India. *Ann App Biol* 122(3):467–479.

Horn NM, Reddy SV, Reddy DVR (1995). Assessment of yield losses caused by chickpea chlorotic dwarf geminivirus in chickpea (*Cicer arietinum*) in India. *Eur J Plant Pathol* 101:221–224.

Jaidi M, Kumar S, Srivastava A, Raj SK (2015). First report of *Ageratum enation virus* and *Ageratum leaf curl betasatellite* infecting *Calendula officinalis* in India. *New Dis Rep* 32:6.

Jeske H, Kober S, Schäfer B, Strohmeier S (2014). Circomics of Cuban geminiviruses reveals the first alphasatellite DNA in the Caribbean. *Virus Genes* 49(2):312–324.

Jose J, Usha R (2003). Bhendi yellow vein mosaic disease in India caused by association of a DNA β satellite with a begomovirus. *Virology* 305:310–317.

Jyothsna P, Rawat R, Malathi VG (2011). Molecular characterization of a new begomovirus infecting a leguminous weed *Rhynchosia minima* in India. *Virus Genes* 42:407–414.

Jyothsna P, Rawat R, Malathi VG (2013a). Predominance of *tomato leaf curl Gujarat virus* as a monopartite begomovirus: Association with *tomato yellow leaf curl Thailand betasatellite*. *Arch Virol* 158:217–224.

Jyothsna P, Haq QM, Singh P, Sumiya KV et al. (2013b). Infection of *tomato leaf curl New Delhi virus* (ToLCNDV), a bipartite begomovirus with betasatellites, results in enhanced level of helper virus components and antagonistic interaction between DNA B and betasatellites. *Appl Microbiol Biotechnol* 97(12):5457–5471.

Jyothsna P, Haq QMI, Jayaprakash P, Malathi VG (2013c). Molecular evidence for the occurrence of *Abutilon mosaic virus*, a New World begomovirus in India. *Indian J Virol* 24(2):284–288.

Kamaal N, Akram M, Pratap A, Yadav P (2013). Characterization of a new begomovirus and a betasatellite associated with the leaf curl disease of French bean in northern India. *Virus Genes* 46(1):120–127.

Kanakala S, Jyothsna P, Shukla R, Tiwari N et al. (2013a). Asymmetric synergism and heteroencapsidation between two bipartite begomoviruses, *tomato leaf curl New Delhi virus* and *tomato leaf curl Palampur virus*. *Virus Res* 174(1–2):126–136.

Kanakala S, Verma HN, Vijay P, Saxena DR, Malathi VG (2013b). Response of chickpea genotypes to *Agrobacterium*-mediated delivery of *Chickpea chlorotic dwarf virus* (CpCDV) genome and identification of resistance source. *Appl Microbiol Biotechnol* 97(21):9491–9501.

Khan ZA, Khan JA (2017). Characterization of a new begomovirus and betasatellite associated with chilli leaf curl disease in India. *Arch Virol* 162(2):561–565.

Kirthi N, Maiya SP, Murthy MRN, Savithri HS (2002). Evidence for recombination among the tomato leaf curl virus strains/species from Bangalore, India. *Arch Virol* 147:255–272.

Kirthi N, Priyadarshini CGP, Sharma P, Maiya SP et al. (2004). Genetic variability of begomoviruses associated with cotton leaf curl disease originating from India. *Arch Virol* 149:2047–2057.

Kraberger S, Kumari SG, Hamed AA, Gronenborn B et al. (2015). Molecular diversity of *Chickpea chlorotic dwarf virus* in Sudan: High rates of intra-species recombination – a driving force in the emergence of new strains. *Infect Genet Evol* 29:203–215.

Kumar Y, Hallan V, Zaidi AA (2008). Molecular characterization of a distinct bipartite begomovirus species infecting tomato in India. *Virus Genes* 37:425–431.

Kumar J, Kumar A, Khan JA, Aminuddin JA (2009). First report of papaya leaf curl virus naturally infecting tobacco in India. *J Plant Pathol* 91:107.

Kumar A, Kumar J, Khan JA (2010a). Sequence characterization of cotton leaf curl virus from Rajasthan: Phylogenetic relationship with other members of geminiviruses and detection of recombination. *Virus Genes* 40:282–289.

Kumar J, Kumar A, Roy JK, Tuli R, Khan JA (2010b). Identification and molecular characterization of begomovirus and associated satellite DNA molecules infecting *Cyamopsis tetragonoloba*. *Virus Genes* 41:118–125.

Kumar Y, Hallan V, Zaidi AA (2011). *Chilli leaf curl Palampur virus* is a distinct begomovirus species associated with a betasatellite. *Plant Pathol* 60:1040–1047.

Kumar J, Kumar A, Singh SP, Roy JK et al. (2012a). First report of *Radish leaf curl virus* infecting okra in India. *New Dis Rep* 25:9.

Kumar RV, Singh AK, Chakraborty S (2012b). A new monopartite begomovirus species, *Chilli leaf curl Vellanad virus*, and associated betasatellites infecting chilli in the Vellanad region of Kerala, India. *New Dis Rep* 25:20.

Kumar J, Singh SP, Kumar J, Tuli R (2012c). A novel mastrevirus infecting wheat in India. *Arch Virol* 157:2031–2034.

Kumar J, Kumar J, Singh SP, Tuli R (2014a). Association of satellites with a mastrevirus in natural infection: Complexity of wheat dwarf India virus disease. *J Virol* 88(12):7093–7104.

Kumar J, Kumar J, Singh SP, Tuli R (2014b). βC1 is a pathogenicity determinant: Not only for begomoviruses but also for a mastrevirus. *Arch Virol* 159(11):3071–3076.

Kumar RV, Singh AK, Singh AK, Yadav T et al. (2015a). Complexity of begomovirus and betasatellite populations associated with chilli leaf curl disease in India. *J Gen Virol* 96(10):3143–3158.

Kumar J, Kumar J, Singh S, Shukla V et al. (2015b). Prevalence of *Wheat dwarf India virus* in wheat in India. *Cur Sci* 108(2):260–265.

Kumar M, Kumar RV, Ragunathan D, Chakraborty S (2016). Identification of begomovirus and associated satellite components in *Alcea rosea* L. in India. In: 8th International Geminivirus Symposium and 6th International ssDNA Comparative Virology Workshop, New Delhi, India, November 7–10.

Kumar RV, Prasanna HC, Singh AK, Ragunathan D et al. (2017a). Molecular genetic analysis and evolution of begomoviruses and betasatellites causing yellow mosaic disease of bhendi. *Virus Genes* 53(2):275–285.

Kumar RV, Singh D, Singh AK, Chakraborty S (2017b). Molecular diversity, recombination and population structure of alphasatellites associated with begomovirus disease complexes. *Infect Genet Evol* 49:39–47.

Kumari P, Singh AK, Chattopadhyay B, Chakraborty S (2010). Molecular characterization of a new species of begomovirus and betasatellite causing leaf curl disease of tomato in India. *Virus Res* 152(1–2):19–29.

Kumari P, Singh AK, Sharma VK, Chattopadhyay B, Chakraborty S (2011). A novel recombinant tomato-infecting begomovirus capable of transcomplementing heterologous DNA-B components. *Arch Virol* 156:769–783.

Malathi VG, John P (2008). Geminiviruses infecting legumes. In: *Characterization, Diagnosis and Management of Plant Viruses*. (Vol. 3). Rao GP, Kumar PL, Holguin–Peña RJ (eds), 97–123. New Delhi: Studium Press.

Mandal B, Varma A, Malathi VG (1997). Systemic infection of *Vigna mungo* using the cloned DNAs of mungbean yellow mosaic geminivirus through agroinoculation and transmission of the progeny virus through whiteflies. *J Phytopathol* 145:505–510.

Mansoor S, Khan SH, Bashir A, Saeed M et al. (1999). Identification of a novel circular single-stranded DNA associated with cotton leaf curl disease in Pakistan. *Virology* 259:190–199.

Maruthi MN, Colvin J, Briddon RW, Bull SE, Muniyappa V (2003). *Pumpkin yellow vein mosaic virus*: A novel begomovirus infecting cucurbits. *J Plant Pathol* 85(1):64–65.

Marwal A, Sahu A, Prajapat R, Choudhary DK, Gaur RK (2012). First report of association of a begomovirus with the leaf curl disease of a common weed, *Datura inoxia*. *Indian J Virol* 23(1):83–84.

Mishra MD, Raychaudhri SP, Jha A (1963). Virus causing leaf curl of chilli (*Capsicum annum* L.). *Indian J Microbiol* 3:73–76.

Mittal D, Borah BK, Dasgupta I (2008). Agroinfection of cloned *Sri Lankan cassava mosaic virus* DNA to *Arabidopsis thaliana*, *Nicotiana tabacum* and cassava. *Arch Virol* 153:2149–2155.

Namrata J, Saritha RK, Datta D, Singh M et al. (2010). Molecular Characterization of *Tomato leaf curl Palampur virus* and Pepper leaf curl betasatellite naturally infecting pumpkin (*Cucurbita moschata*) in India. *Indian J Virol* 21(2):128–132.

Nariani TK (1960). Yellow mosaic of mung (*Phaseolus aureus* L.). *Indian Phytopathol* 13:24–29.

Nawaz-ul-Rehman MS, Fauquet CM (2009). Evolution of geminiviruses and their satellites. *FEBS Lett* 583 (12):1825–1832.

Nehra C, Gaur RK (2015). Molecular characterization of *Chilli leaf curl viruses* infecting new host plant *Petunia hybrida* in India. *Virus Genes* 50(1):58–62.

Nene YL, Reddy MV (1976). Preliminary information on chickpea stunt. *Trop Grain Legume Bull* 5:31–32.

Padidam M, Beachy RN, Fauquet CM (1995). Tomato leaf curl geminivirus from India has a bipartite genome and coat protein is not essential for infectivity. *J Gen Virology* 76:25–35.

Pandey P, Mukhopadhyay S, Naqvi AR, Mukherjee SK et al. (2010). Molecular characterization of two distinct monopartite begomoviruses infecting tomato in India. *Virol J* 7:337.

Paprotka T, Metzler V, Jeske H (2010). The first DNA 1-like alpha satellites in association with New World begomoviruses in natural infections. *Virology* 404:148–157.

Patil BL, Dutt N, Briddon RW, Bull SE et al. (2007). Deletion and recombination events between the DNA-A and DNA-B components of Indian cassava-infecting geminiviruses generate defective molecules in *Nicotiana benthamiana*. *Virus Res* 124:59–67.

Phaneendra C, Rao KRSS, Jain RK, Mandal B (2012). *Tomato leaf curl New Delhi virus* is associated with pumpkin leaf curl: A new disease in northern India. *Indian J Virol* 23(1):42–45.

Prajapat R, Marwal A, Gaur RK (2014). Begomovirus associated with alternative host weeds: A critical appraisal. *Arch Phytopathol Plant Prot* 47:157–170.

Pramesh D, Mandal B, Phaneendra, C, Muniyappa V (2013). Host range and genetic diversity of *croton yellow vein mosaic virus*, a weed-infecting monopartite begomovirus causing leaf curl disease in tomato. *Arch Virol* 158(3):531–542.

Pratap D, Kashikar AR, Mukherjee SK (2011). Molecular characterization and infectivity of a *Tomato leaf curl New Delhi virus* variant associated with newly emerging yellow mosaic disease of eggplant in India. *Virology J* 8:305.

Raj SK, Khan MS, Singh R, Kumari N, Prakash D (2005). Occurrence of yellow mosaic geminiviral disease on bitter gourd (*Momordica charantia*) and its impact on phytochemical contents. *Intl J Food Sci Nutr* 56:185–192.

Raj SK, Snehi SK, Khan MS, Tiwari AK, Rao GP (2010). First report of *Pepper leaf curl Bangladesh virus* strain associated with bitter gourd (*Momordica charantia* L.) yellow mosaic disease in India. *Australas Plant Dis Notes* 5:14–16.

Rajinimala N, Rabindran R (2007). First report of *Indian cassava mosaic virus* on bittergourd (*Momordica charantia*) in Tamil Nadu, India. *Australas Plant Dis Notes* 2:81–82.

Rajinimala N, Rabindran R, Ramiah M, Kamalakannan A, Mareeswari P (2005). Virus–vector relationship of Bittergourd yellow mosaic virus and the whitefly *Bemisia tabaci* Genn. *Acta Phytopathol Hun* 40:23–30.

Ranjan P, Kumar RV, Chakraborty S (2013). Differential pathogenicity among *Tomato leaf curl Gujarat virus* isolates from India. *Virus Genes* 47:524–531.

Ranjan P, Singh AK, Kumar RV, Basu S, Chakraborty S (2014). Host-specific adaptation of diverse betasatellites associated with distinct Indian tomato-infecting begomoviruses. *Virus Genes* 48:334–342.

Rathore S, Bhatt BS, Yadav BK, Kale RK, Singh AK (2014). A new begomovirus species in association with betasatellite causing tomato leaf curl disease in Gandhinagar, India. *Plant Dis* 98(3):428.

Reddy RVC, Muniyappa V, Colvin J, Seal S (2005). A new begomovirus isolated from *Gossypium barbadense* in southern India. *Plant Pathol* 54:570.

Richter KS, Götz M, Winter S, Jeske H (2016). The contribution of translation synthesis polymerases on geminiviral replication. *Virology* 488:137–148.

Rishi N, Chauhan MS (1994). Appearance of leaf curl disease of cotton in Northern India. *J Cotton Res Devel* 8:179–180.

Rocha CS, Castillo-Urquiza GP, Lima ATM, Silva FN et al. (2013). Brazilian begomovirus populations are highly recombinant, rapidly evolving, and segregated based on geographical location. *J Virol* 87 (10):5784–5799.

Rojas MR, Hagen C, Lucas WJ, Gilbertson RL (2005). Exploiting chinks in the plant's armor: Evolution and emergence of geminiviruses. *Annu Rev Phytopathol* 43:361–394.

Roy A, Spoorthi P, Bag MK, Prasad TV et al. (2013). A leaf curl disease in germplasm of rapeseed-mustard in India: Molecular evidence of a weed-infecting begomovirus–betasatellite complex emerging in a new crop. *J Phytopathol* 161:522–535.

Saeed ST, Khan A, Kumar B, Ajaya Kumar PV, Samad A (2014). First report of *Chilli leaf curl India virus* infecting *Mentha spicata* (Neera) in India. *Plant Dis* 98(1):164–165.

Sahu AK, Marwal A, Shahid MS, Nehra C, Gaur RK (2014). First report of a begomovirus and associated betasatellite in *Rosa indica* and in India. *Australas Plant Dis Notes* 9:147.

Saunders K, Nazeera S, Mali VR, Malathi VG et al. (2002). Characterization of *Sri Lankan cassava mosaic virus* and *Indian cassava mosaic virus*: Evidence for acquisition of a DNA B component by a monopartite begomovirus. *Virology* 293:63–74.

Saunders K, Norman A, Gucciardo S, Stanley J (2004). The DNA β satellite component associated with ageratum yellow vein disease encodes an essential pathogenicity protein (βC1). *Virology* 324:37–47.

Saxena S, Hallan V, Singh BP, Sane PV (1998). Nucleotide sequence and inter geminiviral homologies of the DNA-A of papaya leaf curl geminivirus from India. *Biochem Mol Biol Intl* 45:101–113.

Shih SL, Tsai WS, Green SK, Singh D (2007). First report of *Tomato leaf curl Joydebpur virus* infecting chilli in India. *Plant Pathol* 56:341.

Singh AK, Mishra KK, Chattopadhyay B, Chakraborty S (2009). Biological and molecular characterization of a begomovirus associated with yellow mosaic vein mosaic disease of pumpkin from Northern India. *Virus Genes* 39:359–370.

Singh MK, Singh K, Haq QM, Mandal B, Varma A (2011). Molecular characterization of *Tobacco leaf curl Pusa virus*, a new monopartite Begomovirus associated with tobacco leaf curl disease in India. *Virus Genes* 43(2):296–306.

Singh AK, Chattopadhyay B, Chakraborty S (2012). Biology and interactions of two distinct monopartite begomoviruses and betasatellites associated with radish leaf curl disease in India. *Virology J* 9:43.

Singh AK, Kushwaha N, Chakraborty S (2016). Synergistic interaction among begomoviruses leads to the suppression of host defense-related gene expression and breakdown of resistance in chilli. *Appl Microbiol Biotechnol* 100(9):4035–4049.

Singh-Pant P, Pant P, Mukherjee SK, Mazumdar-Leighton S (2012). Spatial and temporal diversity of begomoviral complexes in papayas with leaf curl disease. *Arch Virol* 157:1217–1232.

Sinha V, Kumar A, Sarin NB, Bhatnagar D (2016). Recombinant *croton yellow vein mosaic virus* associated with severe leaf curl disease of papaya in India. *Intl J Adv Res* 4(3):1598–1604.

Sivalingam PN, Malathi VG, Varma A (2010). Molecular diversity of the DNA-β satellites associated with tomato leaf curl disease in India. *Arch Virol* 155:757–764.

Snehi SK, Raj SK, Khan MS, Prasad V (2011a). Molecular identification of a new begomovirus associated with yellow mosaic disease of *Jatropha gossypifolia* in India. *Arch Virol* 156:2303–2307.

Snehi SK, Khan MS, Raj SK, Prasad V (2011b). Complete nucleotide sequence of *Croton yellow vein mosaic virus* and DNA-β associated with yellow vein mosaic disease of *Jatropha gossypifolia* in India. *Virus Genes* 43:93–101.

Snehi SK, Srivastava A and Raj SK (2012). Biological characterization and complete genome sequence of a possible strain of *Indian cassava mosaic virus* from *Jatropha curcas* in India. *J Phytopathol* 160:547–553.

Srivastava A, Raj SK, Kumar S, Snehi SK et al. (2013a). Molecular identification of *Ageratum enation virus*, betasatellite and alphasatellite molecules isolated from yellow vein diseased *Amaranthus cruentus* in India. *Virus Genes* 47(3):584–590.

Srivastava A, Raj SK, Kumar S, Snehi SK (2013b). New record of *Papaya leaf curl virus* and *Ageratum leaf curl betasatellite* associated with yellow vein disease of aster in India. *New Dis Rep* 28:6.

Srivastava A, Kumar S, Raj SK (2014). Association of *Pedilanthus leaf curl virus* with yellow mottling and leaf curl symptoms in two jasmine species grown in India. *J Gen Plant Pathol* 80:370–373.

Srivastava A, Jaidi M, Kumar S, Raj SK, Shukla S (2015a). Association of *Papaya leaf curl virus* with the leaf curl disease of grain amaranth (*Amaranthus cruentus* L.) in India. *Phytoparasitica* 43:97–101.

Srivastava A, Kumar S, Jaidi M, Raj SK (2015b). Characterization of a novel begomovirus associated with yellow mosaic disease of three ornamental species of Jatropha grown in India. *Virus Res* 201:41–49.

Swarnalatha P, Mamatha M, Manasa M, Singh RP, Reddy MK (2013). Molecular identification of *Ageratum enation virus* (AEV) associated with leaf curl disease of tomato (*Solanum lycopersicum*) in India. *Australas Plant Dis Notes* 8:67–71.

Tiwari N, Padmalatha KV, Singh VB, Haq QM, Malathi VG (2010). *Tomato leaf curl Bangalore virus* (ToLCBV): Infectivity and enhanced pathogenicity with diverse betasatellites. *Arch Virol* 155:1343–1347.

Usharani KS, Surendranath B, Paul Khurana SM, Garg ID, Malathi VG (2003). Potato leaf curl – a new disease of potato in northern India caused by a strain of *Tomato leaf curl New Delhi virus*. *New Dis Rep* 8:2.

Usharani KS, Surendranath B, Haq QM, Malathi VG (2004). Yellow mosaic virus infecting soybean in northern India is distinct from the species infecting soybean in southern and western India. *Curr Sci* 86:845–850.

Vanitha LS, Rangaswamy KT, Govindappa MR, Manjunatha L (2013). Survey, vector relationships and host range studies of *Tomato leaf curl Karnataka virus* causing sunflower leaf curl disease. *Trends Biosci* 6 (1):36–39.

Varma A, Malathi VG (2003). Emerging geminivirus problems: A serious threat to crop production. *Ann Appl Biol* 142:145–164.

Varsani A, Navas-Castillo J, Moriones E, Hernández-Zepeda C et al. (2014). Establishment of three new genera in the family *Geminiviridae: Becurtovirus, Eragrovirus* and *Turncurtovirus*. *Arch Virol* 159(8):2193–2203.

Vasudeva RS, Raj SM (1948). A leaf curl disease of tomato. *Phytopathology* 38:364–369.

Venkataravanappa V, Reddy CNL, Swaranalatha P, Jalali S et al. (2011). Diversity and phylogeography of begomovirus associated betasatellites of okra in India. *Virology J* 8:555.

Venkataravanappa V, Reddy CNL, Swaranalatha P, Devaraju et al. (2012a). Molecular evidence for association of *Cotton leaf curl Alabad virus* with yellow vein mosaic disease of okra in North India. *Arch Phytopathol Plant Prot* 45(17):2095–2113.

Venkataravanappa V, Reddy CNL, Jalali S, Reddy MK (2012b). Molecular characterization of distinct bipartite begomovirus infecting bhendi (*Abelmoschus esculentus* L.) in India. *Virus Genes* 44:522–535.

Venkataravanappa V, Reddy CNL, Jalali S, Reddy MK (2013). Molecular characterization of a new species of begomovirus associated with yellow vein mosaic of bhendi (Okra) in Bhubhaneswar, India. *Eur J Plant Pathol* 136:811–822.

Venkataravanappa V, Reddy CNL, Jalali S, Briddon RW, Reddy MK (2015). Molecular identification and biological characterisation of a begomovirus associated with okra enation leaf curl disease in India. *Eur J Plant Pathol* 141:217–235.

Venkataravanappa V, Swarnalatha P, Reddy CNL, Chauhan N, Reddy MK (2016). Association of recombinant *Chilli leaf curl virus* with enation leaf curl disease of tomato: A new host for chilli begomovirus in India. *Phytoparasitica* 44:213–223.

Venkatasalam EP, Singh S, Sivalingam PN, Malathi VG, Garg ID (2011). Polymerase chain reaction and nucleic acid spot hybridisation detection of begomovirus(es) associated with apical leaf curl disease of potato. *Arch Phytopathol Plant Prot* 44:987–992.

Zaffalon V, Mukherjee SK, Reddy VS, Thompson JR, Tepfer M (2012). A survey of geminiviruses and associated satellite DNAs in the cotton-growing areas of northwestern India. *Arch Virol* 157(3):483–495.

Zaim M, Kumar Y, Hallan V, Zaidi AA (2011). *Velvet bean severe mosaic virus*: A distinct begomovirus species causing severe mosaic in *Mucuna pruriens* (L.) DC. *Virus Genes* 43:138–146.

7 Sugarcane Bacilliform Viruses
Present Status

Govind Pratap Rao, Susheel Kumar Sharma, and P. Vignesh Kumar

CONTENTS

7.1 BADNAVIRUSES: GENETICALLY AND SEROLOGICALLY DIVERSE VIRUS SPECIES

Badnaviruses are a group of viruses having double-stranded DNA as genome, non-enveloped bacilliform-shaped virions in the family *Caulimoviridae* (King et al. 2012). The name Badna was derived from the bacilliform DNA viruses. *Commelina yellow mottle virus* (ComYMV) is the type species of genus *Badnavirus*. Genome size of badnaviruses varies from 7.2 to 9.2 kb (King et al. 2012). As per the International Committee on Taxonomy of Viruses (ICTV)-recognized virus species list, the genus *Badnavirus* comprises 37 recognized species (ICTV website, accessed February, 2017) and many tentative species (King et al. 2012). Geographically, badnaviruses are distributed across the tropical countries in Asia, Africa, and America, and in the Pacific Islands (Borah et al. 2013; Ishwara et al. 2016). They are highly heterogeneous at both the genetic and serological level (Geering et al. 2000; Harper et al. 2004, 2005; Jaufeerally-Fakim et al. 2005; Kenyon et al. 2008; Lockhart and Olszewski 1993), as they share very low nucleotide sequence identity, even between species of same genus.

7.2 SUGARCANE BACILLIFORM VIRUSES: HISTORY AND GEOGRAPHICAL DISTRIBUTION

Badnaviruses are mostly known to infect dicotyledonous plants except for two species groups, sugarcane bacilliform viruses (SCBV) and banana streak viruses (BSV), which infect two monocots—sugarcane and banana, respectively. A group of serologically and genetically heterogeneous badnaviruses associated with sugarcane leaf fleck disease are known as sugarcane bacilliform viruses. The first report of bacilliform viral particles in sugarcane came from Cuba (Rodriguez Lema et al. 1985). Later, Lockhart and Autrey (1988) reported this virus from Morocco and Hawaii. The viruses associated with leaf fleck disease were subsequently named sugarcane bacilliform viruses (SCBV), and have been shown to be closely related to BSV. Initially both SCBV and BSV were considered to belong to the same virus species (Lockhart and Olszewski 1993); however, later these two groups of viruses infecting two monocotyledonous plants were identified to be separate distinct complexes of *Badnavirus* species.

Infections of sugarcane bacilliform viruses have been reported from major sugarcane growing regions worldwide, including Australia, Madagascar, Madeira, Malawi, Mauritius, Morocco, Papua New Guinea, Reunion, South Africa, Taiwan, United States, The Dominican Republic, Thailand, China, Cuba, and India (Peralta et al. 1991; Autrey et al. 1995a,b; Viswanathan et al. 1996; Rott et al. 2000; Rao et al. 2002, 2014; Singh et al. 2003). SCBV infection has been intercepted in quarantine in sugarcane clones from Argentina, Barbados, Brazil, Puerto Rico, China, Laos, Mexico, and Vietnam (Autrey et al. 1995a). This suggests that the virus has spread throughout the world from an early time via the exchange of germplasm, mainly due to the lack of reliable symptoms and clonal propagation of sugarcane. It has been suggested that the virus originated in the center of origin of *Saccharum*, Papua New Guinea (Autrey et al. 1995b). To date, however, only commercial canes in Papua New Guinea were tested SCBV-positive, and were introduced from Australia. A more extensive survey of wild and commercial canes from Papua New Guinea and Indonesia as well as India and China, where other *Saccharum* species are thought to have been involved, shed some light on the origin and evolution of SCBV.

SCBV exhibit a significant sequence variability and serological heterogeneity (Braithwaite et al. 1995; Muller et al. 2011; Karuppaiah et al. 2013; Rao et al. 2014), thus suggesting the existence of complex SCBV population associated with the sugarcane. All the badnaviruses have a very narrow host range (King et al. 2012); however, SCBV has been shown to infect banana and rice through agroinfection (Bouhida et al. 1993) and to infect *Sorghum halepense* and *Brachiaria extensa* under experimental transmission (Frison and Putter 1993). Due to the genetic relatedness of SCBV and BSV as well as the ability of the infectious DNA of SCBV to infect banana, earlier workers considered both as strains of same virus (Lockhart and Autrey 1988). Sugarcane-infecting badnaviruses are genetically heterogeneous species complex. At present, four ICTV-recognized *Badnavirus* species are known to be associated with sugarcane leaf fleck disease worldwide: *Sugarcane bacilliform Guadeloupe A virus* (SCBGAV), *Sugarcane bacilliform Guadeloupe D virus* (SCBGDV), *Sugarcane bacilliform IM virus* (SCBIMV), and *Sugarcane bacilliform MO virus* (SCBMOV) (ICTV website). In addition, there are more than five recently proposed species from different parts of the world (Muller et al. 2011; King et al. 2012; Karuppaiah et al. 2013).

7.3 BIOLOGICAL PROPERTIES, HOST RANGE, AND ECONOMIC LOSSES

Symptoms of SCBV infection on sugarcane are highly variable and unreliable. On noble canes, symptoms may vary from asymptomatic expression to pronounced white, chlorotic flecks or streaks (Figure 7.1). SCBV infection leads to diverse symptoms including severe chlorotic speckles, chlorotic mottling, stunted growth, and pronounced freckles, depending on the sugarcane variety (Lockhart 1988; Viswanathan et al. 1996; Rao et al. 2002; Singh et al. 2009). Some sugarcane genotypes infected with SCBV displayed symptoms including chlorotic speckles starting at the leaf tip of newly formed leaves and moving downward, stunted growth in severely infected plants, poor or no tillering, reduction in internodal elongation, formation of a bunchy top, and premature death (Viswanathan et al. 1996). However, it is known that SCBV does not necessarily produce visible foliar symptoms in most of the clones tested, and no specific symptoms could be attributed to SCBV infection (Lockhart et al. 1992). SCBV are reported to infect a number of species of sugarcane (*Saccharum officinarum*, *S. barberi*, *S. sinense*, *S. robustum*, and *Saccharum* interspecific hybrids). Decreased juice, sucrose content, gravity, purity, and stalk weight were observed in SCBV-infected sugarcane in China (Li et al. 2010).

Sugarcane, a vegetatively propagated plant, is commonly infected with mixed viruses. Mixed infection of SCBV with other viruses has also been reported in India (Singh et al. 2009). Sugarcane

FIGURE. 7.1 Different types of symptoms caused by sugarcane bacilliform virus on different sugarcane cultivars: (a) Assam CoBlN 4172; (b) Bihar Bo 150; (c) UP 01235.

clones infected with *Sugarcane mild mosaic virus* also have been found infected with SCBV in the United States (Lockhart et al. 1992). These clones showed a variety of foliar symptoms ranging from no apparent symptoms to pronounced chlorotic striate mosaic. All infected clones showed evidence of retarded growth, narrowing of leaves, and occasional dieback of ratoon shoots. Mixed infection of SCBV with *Sugarcane mosaic virus* (SCMV) has also been reported, leading to more pronounced mosaic symptoms and higher SCBV particle numbers than when virus was present alone (Lockhart and Autrey 1988).

Systematic studies on determining the effects of SCBV infection on sugarcane yield have not been undertaken. Circumstantial evidence suggested that SCBV contributed to decreased productivity in the clone Mex 57-473 in Morocco (Lockhart and Autrey 1988). No reduction in growth has been observed in greenhouse-grown plants (Anonymous 1992). A field study designed to assess the effect of SCBV on sugarcane biomass production gave conflicting results (Comstock and Lockhart 1990). It is postulated that the virus originated at the center of sugarcane origin in Papua New Guinea and spread to other cultivating areas through the movement of germplasm, especially noble canes. Further dissemination has been caused by its mealy bug vector (*Sacharicoccus sachhari*). A Moroccan strain of BSV was serologically indistinguishable from some isolates of SCBV, indicating the relatedness of SCBV and BSV. Under experimental conditions, SCBV (and BSV) can also be transmitted mechanically to healthy sugarcane and bananas (Lockhart 1988). Based on polymerase chain reaction (PCR) assay, Braithwaite et al. (1995) reported the occurrence of SCBV in noble canes (*Saccharum officinarum*), commercial hybrids, and clones within the '*Saccharum* complex' including *S. robustum*, *S. spontaneum*, *S. barberi*, *Erianthus arundinaceus*, and *E. ravennae*. Singh et al. (2009) reported mixed infection of SCBV with other viruses in field sugarcane sampled from different sugarcane growing areas of India through PCR-based methods.

Badnaviruses in general have a narrow host range limited to different varieties of a host genus or species (Borah et al. 2013). SCBV is an exception, with a comparatively wide host range. SCBV is known to infect plant hosts of the families Poaceae (e.g., sugarcane, rice, Johnson grass, sorghum, *Panicum maximum*, *Rottboellia exaltata*, and *Brachiaraia* sp.), Musaceae (bananas), *Chenopodium quinoa*, or tobacco (Bouhida et al. 1993; Lockhart et al. 1996; Lockhart and Autrey 1998, 2000; Borah et al. 2013). SCBV infection was initially detected in *Saccharum officinarum* germplasm but its infection on commercial sugarcane hybrids (*Saccharum* interspecific hybrids) and other *Saccharum* species (*S. barberi*, *S. robustum*, *S. spontaneum*, and *S. sinensis*) was also observed (Lockhart et al. 1996).

7.4 TRANSMISSION

Under field conditions, SCBV is transmitted by infected vegetative propagative material in addition to the spread mediated by mealybug vectors (Lockhart et al. 1996; Viswanathan et al. 1996). Infected vegetative propagating materials, particularly cuttings, are the major means of SCBV spread to different geographical areas. Insect vectors such as pink sugarcane mealybug (*Saccharicoccus sacchari*) and the grey sugarcane mealybug (*Dysmicoccus boninsis)* are known to transmit SCBV in a semi-persistent manner (Lockhart and Autrey 1991; Lockhart et al. 1992; Bouhida et al. 1993). BSV is transmitted by the citrus mealybug (*Planococcus citri*), which does not colonize sugarcane naturally but can be used as an experimental vector for SCBV. Seed transmission has not yet been demonstrated for SCBV but does occur in several other badnaviruses. It has been reported that SCBV can be transmitted by *S. sacchari* from infected sugarcane to bananas cv. Grand Naine, with SCBV-infected bananas developing symptoms indistinguishable from those described for BSV under laboratory conditions; however, natural infection of SCBV in bananas in field conditions has not been reported. In addition, BSV was transmitted to sugarcane by mechanical inoculation (Lockhart and Autrey 1991). Although BSV and SCBV were earlier considered to belong to the same virus species, spread between the two hosts in the field may be limited due to vector host preference. A 1.1 genome length construct of SCBMOV was infectious to rice and banana following agro-inoculation (Bouhida et al. 1993). The agro-inoculated *Musa acuminata* plants showed typical streak symptoms similar to BSV infection 45 days post-inoculation, which conclusively proved that SCBV could infect banana upon agro-infection (Bouhida et al. 1993).

7.5 TAXONOMY AND VIRION PROPERTIES

SCBV is a plant pararetrovirus with non-enveloped bacilliform particles of 130 × 30 nm size (Figure 7.2) with circular, double-stranded DNA (dsDNA) genome of 7.5–8.0 kb (Bouhida et al. 1993; Geijskes et al. 2002; King et al. 2012). Badnaviruses have dsDNA as genome, which like other members of family *Caulimoviridae* replicates via reverse transcription (Geering and Hull 2012). Genome organization of SCBV is typical of badnaviruses, consisting of three open reading frames (ORFs) and a single intergenic region (King et al. 2012). The function of protein encoded by ORF1 is not known. ORF2 encodes a virion-associated protein (VAP) (Geering and Hull 2012). ORF3 encodes for a polyprotein which is cleaved post-translationally by aspartic protease to functional proteins with functions related to movement protein (MP), coat protein (CP), a reverse transcriptase (RT), and ribonuclease H (RNase H) (Bouhida et al. 1993; Geijskes et al. 2002; Muller et al. 2011; Karuppaiah et al. 2013). Complete genome sequences of 12 SCBV isolates have been reported (Bouhida et al. 1993; Geijskes et al. 2002; Muller et al. 2011; Karuppaiah et al. 2013; Sun et al. 2016)

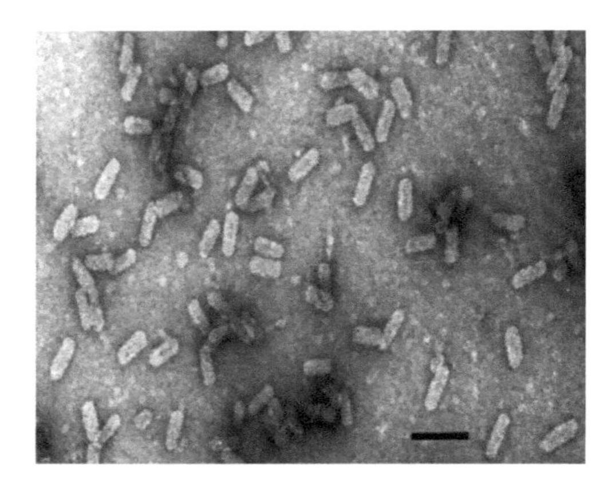

FIGURE 7.2 Non-enveloped bacilliform particles of sugarcane bacilliform virus (130 nm × 30 nm).

in addition to the demonstration of remarkable genetic diversity in SCBV isolates from different parts of the world (Gisjkes et al. 2002; Muller et al. 2011; Rao et al. 2014; Sun et al. 2016). Unlike other badnaviruses (banana streak viruses), SCBV are not known to have host-integrated endogenous counterparts.

The first complete genome sequence of SCBV isolate from Morocco (now known as SCBMOV) was reported in 1993 (Bouhida et al. 1993). SCBMOV was reported as having a genome of 7568 bp. Complete genome sequence of an Australian SCBV isolate, Ireng Maleng isolate (SCBIMV) was reported in 2002, having a genome length of 7687 bp (Geijskes et al. 2002). SCBMOV and SCBIMV were the only two badnaviruses species recognized by ICTV until 2012 (King et al. 2012). At genome level, SCBMOV and SCBIMV shared an identity of 72% and therefore are considered as two distinct species. After that, full genome sequences of three SCBV isolates from Guadeloupe (SCBGAV and SCBGDV) were reported (Muller et al. 2011). Guadaloupe SCBV isolates had genome length varying from 7317 to 7446 bp.

Karuppaiah et al. (2013) sequenced five full-length SCBV isolates from India. A full genome of SCBV-B091 (sampled from commercial cultivar BO91) and other four isolates, SCBV-BB, SCBV-BT, SCBV-BRU, and SCBV-Iscam were sampled from *S. officinarum* germplasm Boetatoe Bilatoe, Black Tannna, Black Reunion, and Iscambine, respectively. Genome length of these SCBV isolates was in the range of 7553–7884 bp (Karuppaiah et al. 2013). These Indian SCBV isolates shared identities of 69–85% for full genome sequences. Recently, two full genomes of SCBV isolates (SCBV-CHN1 and SCBV-CHN2) were reported from China (Sun et al. 2016). The full genome of SCBV-CHN1 and SCBV-CHN2 were 7764 and 7629 bp, respectively.

7.6 DIAGNOSIS

Several methods have been developed and are used for detection of SCBV infection. Symptom-based diagnosis is not considered reliable, as SCBV infection often remains asymptomatic and symptom development is dependent on sugarcane cultivars, prevailing environmental conditions, and viral genotypes. In addition, the appearances of visual symptoms are periodic and often confused with the symptoms caused by infection of other viruses. The virus concentration remains below detectable limits in many sugarcane genotypes, hence reliable detection is a challenge. Many cultivars, particularly commercial hybrids, do not show symptoms, and the severity of symptoms is not related to virus titer. SCBV is not easily transmissible mechanically to common indicator laboratory hosts. Examination for virions in leaf sap by negative-stain electron microscopy lacks sensitivity in most cases and is only suitable for cultivars with very high virus titer (Autrey et al. 1990).

In general, the symptoms caused by badnaviruses infections are variable. Due to narrow host range (Lockhart 1995), suitable indicator plants producing typical symptoms are not available; hence it is not possible to undertake biological detection. Serological detection systems have been developed for detection of SCBV (using purified virus preparations in general). Lockhart (1986) first reported the detection of BSV by serological methods. Badnaviruses pose a serious challenge in their detection due to high serological and genetic heterogeneity.

Lockhart and Autrey (1988) first reported serological detection of SCBV using antisera raised against SCBV and BSV. In immunosorbent electron microscopy (ISEM), SCBV antiserum and BSV antiserum trapped the virus in the partially purified suspension. Reliable detection of SCBV by enzyme immunoassay in sugarcane leaf tissue was obtained using an alkaline phosphatase–IgG conjugate dilution of 1:1000. Later, Autrey et al. (1990) found that the middle of the leaf-lamina generally contained more virus particles than the base or tip. Using the double-antibody sandwich (DAS)-enzyme-linked immunosorbent assay (ELISA) technique for routine diagnosis of SCBV, they readily detected the virus particles by ISEM with antisera to SCBV and BSV. They reported a relation between severity of symptoms and concentration of the virus particles in noble canes but not in interspecific hybrids. They also reported that some noble canes and interspecific hybrids with varying intensity of freckles, though negative by ELISA, were found to contain SCBV particles by

ISEM. Subsequently, Peralta et al. (1991) reported that SCBV virus particles reacted with SCBV antiserum in immunodiffusion and indirect ELISA tests. DAS-ELISA using BSV and SCBV antiserum is more useful for detection of the virus in crude plant extracts (Irey et al. 1992; Balamuralikrishnan and Vishwanathan 2005; Singh et al. 2003; Viswanathan and Premchandran 1998). In order to circumvent the problem of serological heterogeneity, antisera against a mixture of isolates was prepared initially, and cocktail antisera for different SCBV and BSV isolates were used. The isolates of SCBV were serologically related to BSV (Ndowora and Lockhart 2000). A mixture of purified virus preparation of 32 SCBV isolates, which were showing poor reaction with the previously prepared antisera (Lockhart 1986), representing the wide diversity of SCBV and a BSMYV isolate (as BSMYV did not react with SCBV antisera), were used for preparation of polyclonal antisera. Broad spectrum antisera were successfully employed for detection of serologically unrelated SCBV and BSV isolates in DAS-ELISA (Lockhart and Olszewski 1993).

In most of the suspected genotypes, the virus was found associated with clear foliar symptoms. However, certain symptom-free clones carried the virus as well. The virus was detected by immunoelectron microscopy (IEM) and ELISA in suspected clones. Virus titer was low in most of the genotypes. However, a close correlation between symptom expression and virus titer was reported to exist in some genotypes (Viswanathan et al. 1996). Detection can be improved to some extent by observing partially purified viral preparations (minipreps) under the electron microscope, but visualization can be improved greatly by trapping viral particles with an antiserum (ISEM; Autrey et al. 1990).

Occurrence of SCBV in the world sugarcane germplasm collections at the Sugarcane Breeding Institute, Coimbatore, India, was confirmed by Viswanathan in 1992–1993 through serological techniques such as ELISA and ISEM (Viswanathan et al. 1996). Direct antigen coating ELISA (DAC-ELISA) was standardized for the detection of the virus in crude extracts. An antigen dilution of 1:100 and antiserum dilution of 1:2000 gave the most reproducible results in DAC-ELISA for the detection of SCBV from certain genotypes (Viswanathan et al. 1996). With both ELISA and ISEM, similar results were obtained when BSV antiserum was used for the detection in place of SCBV antiserum (Balamuralikrishnan and Viswanathan 2005). There was a close correlation between symptom expression and virus titer in some genotypes. Reliable serological detection of SCBV is hindered by the existence of a wide degree of heterogeneity among naturally occurring isolates of the virus. These antisera raised against SCBV detect a wide range of SCBV isolates in DAS-ELISA. and a wider range of isolates can be detected by indirect ELISA using $F(ab')_2$ fragments of IgG [$F(ab')_2$-ELISA]. ISEM of the purified minipreps samples was the other commonly employed method of SCBV indexing during initial years (Lockhart et al. 1996). Although ELISA gave varying results in the detection of SCBV, it can still be used for indexing of the clones for infection by the virus. As discussed above, the current serological detection is now based on an antiserum raised against a mixture of SCBV and BSV isolates identified as being serologically different. The full extent of serological variation in the SCBV/BSV population is not known, and particles of some isolates may not be efficiently trapped.

Nucleic acid-based detection methods were widely employed for the detection of SCBV worldwide. The extensive genetic variation between SCBV isolates from different parts of the world has also limited the usefulness of DNA-based techniques for accurate detection. Simple techniques such as dot-blots are unsuitable for detection because cross-hybridization between DNA from genetically different isolates is very limited. This has been demonstrated by hybridizing labeled PCR-generated viral DNA-to-DNA amplified from different isolates. A number of distinct cross-hybridization groups can be identified within the SCBV/BSV population (Lockhart and Olszewski 1993; Smith 1996). PCR can be modified to account for genetic variation through the design of primers based on conserved sequences and use of degenerate primers. Both these approaches have been used for the detection of SCBV (Lockhart and Olszewski 1993; Braithwaite et al. 1995). Use of PCR probes for the detection of the SCBV was suggested as another method to overcome the barrier caused by the extremely heterogeneous nature of SCBV (Braithwaite et al. 1995). However, certain experimental results question whether the PCR product is always SCBV-specific, and the primers for routine indexing for SCBV should be used with caution until further experiments are completed (Lockhart et al. 1996). PCR-based detection has been standardized for screening of sugarcane varieties and germplasm (Singh et al. 2009; Rao et al. 2014). The RT and RNase H region of *Badnavirus* genome has been used to design the conserved primers for detection and phylogenetic analysis which was used for PCR-based detection and characterization of new SCBV isolates (Yang et al. 2003). Generally, primers to RT/RNase H region of genome or core RT/RNase H region were used in PCR-based detection (Singh et al. 2009; Karuppaiah et al. 2013; Rao et al. 2014; Sun et al. 2016). Since SCBV is not known to occur as integrated form in the host genome, PCR is considered the standard indexing procedure. Recently, Sharma et al. (2015) reported that RCA concentrated DNA can be used as a template in PCR for sensitive detection SCBV in field samples. In order to improve PCR-based sensitivity, the use of RCA concentrated DNA in PCR-based detection was recently demonstrated for SCBV detection (Sharma et al. 2015). RCA improved the concentration of virus nucleic acid, and the concentrated DNA can be used as template in PCR for sensitive detection SCBV in field samples. Utility of RCA in amplification of SCBV genome and detection was also demonstrated (Sharma et al. 2015).

7.7 GENETIC DIVERSITY AND RECOMBINATION

The identification and molecular characterization of SCBV from different parts of the world has demonstrated that the genetic structure of badnaviruses infecting sugarcane is highly complex and variable (Geijskes et al. 2002; Muller et al. 2011; Rao et al. 2014; Sun et al. 2016). As discussed earlier, ICTV recognizes four different species of SCBV: SCBIMV, SCBMOV, SCBGAV, and SCBGDV, based on the badnaviruses species demarcation threshold of 20% differences in the nucleotide sequences of polymerase gene (RT and RNase H) (King et al. 2012). Based on ICTV species demarcation criteria, new species have been reported from India (Karuppaiah et al. 2013). Further studies on analysis of population diversity of SCBV sampled from different parts of the world are therefore required.

Sequence analysis of RT/RNase H genomic region of SCBV isolates originating from different geographical areas of India showed a variation of up to 27% (Rao et al. 2014). Five isolates originating from Assam, Bihar, and Uttar Pradesh shared maximum sequence similarity (86–88%) with SCBIMV, whereas the three other isolates (Bihar, Uttar Pradesh, and Assam) shared maximum similarity with SCBMOV (75–79%) for the analyzed RT/RNase H sequences. This study indicated that genetically diverse SCBV isolates infect sugarcane in India. These eight SCBV isolates segregated in three diverse subclusters, out of which two were novel. Karuppaiah et al. (2013) reported full genome sequences of genetically diverse SCBV isolates. A full genome of SCBV-B091 (infecting commercial cultivar BO91, Coimbatore) and four other isolates, SCBV-BB, SCBV-BT, SCBV-BRU, and SCBV-Iscam, originated from *S. officinarum* germplasm Boetatoe Bilatoe, Black Tannna, Black Reunion, and Iscambine, respectively, maintained at the sugarcane germplasm collection, Kannur, Kerala. Genome length of these SCBV isolates ranged from 7553 to 7884 bp. These Indian SCBV isolates shared identities of 69–85% for full genome sequences. Phylogenetically, these isolates formed four new groups. Based on ICTV species demarcation criteria, SCBV-BT and SCBV-Iscam fell in SCBIMV and SCBMOV genotypes respectively; however, SCBV-BB, SCBV-BO, and SCBV-BRU represent novel *Badnavirus* species (Karuppaiah et al. 2013). Based on the sequence analysis for different genomic regions, isolate SCBV-BB was most distinct. SCBV-BRU has the largest genome size (7884 bp) among all sugarcane-infecting badnaviruses, and has probably evolved recently. Since earlier reported primers for SCBV were not able to detect these newly reported isolates from India, a new set of primers was designed that could detect all known SCBV species/variants prevalent in the Indian subcontinent (Karuppaiah et al. 2013). SCBV isolates from Guadeloupe were characterized based on the sequences of RT/RNase H region and full genome. Extensive genetic variability was found in the SCBV isolates (Muller et al. 2011). Two SCBV

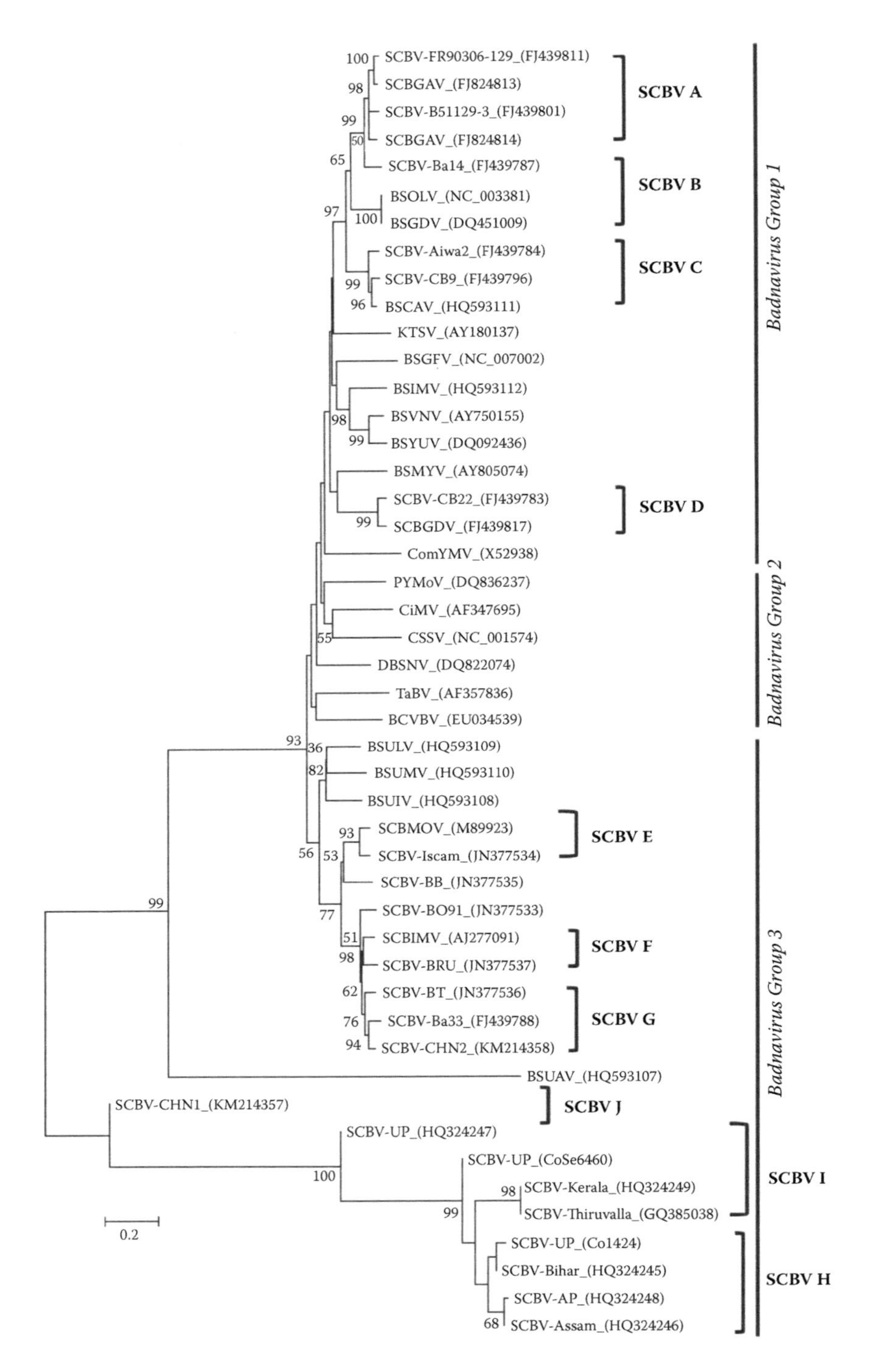

FIGURE 7.3 Neighbor joining (NJ) phylogenetic tree of RT/RNaseH sequences of sugarcane bacilliform viruses (SCBV) and other badnaviruses from elsewhere in world. The evolutionary history was inferred by the NJ method based on the Tamura-Nei model. The percentage of trees in which the associated taxa clustered together is shown next to the branches. Initial tree(s) for the heuristic search were obtained automatically as follows: 1000 bootstrap replicates were used in analysis. Evolutionary analyses were conducted in MEGA7. *Badnavirus* groups and SCBV subclades A–J have been presented.

isolates characterized from China (Sun et al. 2016) were analyzed for their genetic diversity and phylogenetic relationship. SCBV isolates formed nine phylogroups. SCBV-CHN1 was phylogenetically related to SCBV-BB, whereas SCBV-CHN2 was phylogenetically closer to SCBV-BT.

Neighbor-joining phylogenetic analysis indicated extensive genetic diversity among SCBV isolates. Out of three *Badnavirus* groups, SCBV were found distributed to two groups (1 and 3) (Figure 7.3). Ten overall SCBV phylogenetic groups were found based on total genetic diversity analyzed to date (Figure 7.2). SCBV groups A, B, C, and D belong to *Badnavirus* group 1. SCBV groups E to J belong to *Badnavirus* group 2. This indicated extensive genetic diversity among the badnaviruses infecting sugarcane.

Although genetically diverse, few studies have analyzed the role of recombination phenomenon in evolution of genetically diverse SCBV isolates. Rao et al. (2014) detected the recombinant origin of SCBV isolate [SCBV-UP (CoSe92423)] from two other Indian isolates, SCBV-Kerala (Co7219) as minor parent and SCBV-UP (Co1424) as major parent. This was the first report on the possible role of recombination in sugarcane bacilliform viruses. A systematic analysis of full genome sequences of SCBV detected 22 intra-SCBV recombination events (Sharma et al. 2015). Interestingly, this study also identified 32 inter BSV-SCBV recombination events, which may have arisen due to the exchange of sequence fragments among sugarcane-infecting badnaviruses and banana-infecting badnaviruses. Recombination hot spots were concentrated in IGR (3′ end after ORF3 and 5′ end after +1 start site) followed by ORF1, whereas the ORF3 genomic regions were recombination cold spots in *Badnavirus* genome (Sharma et al. 2015). Reports of recombination hot spots and cold spots in the genomic regions of sugarcane- and banana-infecting badnaviruses supported their relatedness and possible host shift. The recombination hot spots were concentered to the intergenic region (3′ end after ORF3 and 5′ end after +1 start site) and ORF1, whereas the recombination cold spots were concentrated to ORF3 of badnaviruses. Recently, two recombinant SCBV isolates were characterized from China (Sun et al. 2016). Isolate SCBV-CHN1 was a recombinant with SCBV-Iscam and SCBV-CHN2 as two parents for the recombination traces falling at N-terminal. For the recombination traces in the C-terminal of ORF3 in SCBV-CHN1, isolates SCBV-BRU and SCBMOV-MOR had contributed as putative parents. Isolate SCBV-CHN2 was also a recombinant with SCBV-CHN1 and SCBMOV-MOR/SCBV-Iscam as putative parents. Similar to the earlier findings, this study also reported IGR and C-terminal of ORF3 as recombination hot spot in SCBV genome (Sun et al. 2016).

7.8 USE OF THE SUGARCANE BACILLIFORM VIRUS ENHANCER IN TRANSGENIC MAIZE

Enhancers are DNA elements that are able to increase transcription by other promoters whether they are placed upstream or downstream of transcription start sites, and their promoter-enhancing activity is independent of orientation relative to the transcription start site (LaFleur et al. 1996; Jakowitsch et al. 1999). Enhancers that are effective in plants have been isolated from genes of plants as well as from genes of viruses and bacteria that infect plants. These include enhancers from the tobacco tCUP (Richert-Poggeler and Shepherd 1997; Kunii et al. 2004), the pea plastocyanin (Hansen and Heslop-Harrison 2004), the *Cauliflower mosaic virus* 35S (CaMV 35S) (Staginnus et al. 2007; Geering et al. 2014), the *Figwort mosaic virus* (Chabannes and Iskra-Caruana 2013), *Agrobacterium tumefacians* 780 (Borah et al. 2013) and *ocs* promoters (Medberry et al. 1990).

Plant virus-derived promoters have been shown to be a rich source of strong constitutive promoters for use in plant biology, and several have been shown to contain enhancer sequences (Chabannes and Iskra-Caruana 2013; Geering et al. 2014). The CaMV 35S promoter has been used extensively in driving transgenes in transgenic plants. Many other viral promoters have also been shown to effectively drive expression of transgenes; CaMV 19S, *Rice tungro bacilliform virus*

(RTBV), *Soybean chlorotic mottle virus* (Hohn and Richert-Poggeler 2006), *Mirabilis mosaic virus* (Staginnus et al. 2009; Côte et al. 2010), *Figwort mosaic virus* (FMV) (Dallot et al. 2001; Lheureux et al. 2003), *Peanut streak chlorotic virus* (Hearon and Locke 1984), Banana streak virus (Quainoo et al. 2008), *Cestrum yellow leaf curling virus* (CmYLCV) (Hareesh and Bhat 2010), and SCBV (Devitt et al. 2005; Quainoo et al. 2008; Deeshma and Bhat 2015). Among these, the CaMV 35S and the FMV promoters have been demonstrated to have enhancer sequences within the promoter (Staginnus and Richert-Poggeler 2006; Chabannes and Iskra-Caruana 2013).

A transcriptional enhancer from the Sugarcane bacilliform virus—Ireng Maleng isolate (SCBV-IM) that can cause increased transcription when integrated into the genome near maize genes has been identified (Davies et al. 2014). In transgenic maize, the SCBV-IM promoter was shown to be comparable in strength to the maize ubiquitin 1 promoter in young leaf and root tissues. The promoter was dissected to identify sequences that confer high activity in transient assays. Enhancer sequences were identified and shown to increase the activity of a heterologous truncated promoter. These enhancer sequences were shown to be more active when arrayed in four copy arrays than in one or two copy arrays. When the enhancer array was transformed into maize plants it caused an increase in accumulation of transcripts of genes near the site of integration in the genome. The SCBV-IM enhancer can activate transcription upstream or downstream of genes and in either orientation. It may be a useful tool to enhance from specific promoters, or in activation tagging.

7.9 MANAGEMENT

No serious attempts have been made for the management of badnaviruses. Conventional control measures (chemical, cultural, and physical) against the vectors are often the only methods used. However, they are not effective in those cases in which the transmission is not well understood and/or occurs by more than one means, as well as in the case of occurrence of integrated virus genomes. Standard viral eradication techniques such as tissue culture and heat treatment have failed to eradicate SCBV from sugarcane (Autrey et al. 1990; Egeskov et al. 1994). A major handicap in the development of control strategies against sugarcane bacilliform badnaviruses is the absence of comprehensive information on vectors or other modes of transmission and on the relationship between different strains as well as their virulence and distribution.

7.10 CONCLUSIONS

The extreme heterogeneity of sugarcane badna viral nucleotide sequences seems to suggest that this group of viruses, although ancient in origin, adapted so early and deeply to their respective hosts that most of the ancestral genes have been lost and new ones have been acquired for adaptation to lives restricted by the host. Probably another consequence of the above heterogeneity is the varying numbers of ORFs in badnaviruses, all within their rather narrow genomic size range. SCBV viral DNA cannot be mechanically inoculated onto the respective hosts, and hence agro-inoculation has been the only successful method for the reintroduction of cloned badna viral genomes back to the host plant. Although the reason for this lack of infectivity has not yet been addressed, it is probably because of the limitation and possibly adaptation of the virions to the vascular tissue of the host plant into which the vector delivers the viruses. Such possible adaptation, if any, should form the subject of active research in the future. The lack of infectious clones has also severely handicapped efforts to determine gene functions and their roles in pathogenesis by site-directed mutagenesis, an approach which can otherwise be very fruitful. Moreover, the high degree of variability makes the detection and taxonomic designation of SCBV difficult. Since the SCBV is vegetatively propagated, its importance as pathogens transmitted by infected sugarcane setts is an important consideration for quarantine purposes. Methods of detection of SCBV need to be made much more robust because of the atypical symptoms produced on different sugarcane cultivars in different sugarcane growing countries. More understanding of the molecular biology of the sugarcane infecting badnaviruses is therefore also required to achieve the ultimate goal of producing efficient and workable management strategies.

REFERENCES

Anonymous (1992). Mauritius Sug. Ind. Res. Annual Report 1992.

Autrey LJC, Saumtally S, Dookun A, Boolell S (1990). Occurrence of sugarcane bacilliform virus in Mauritius. *Proc S Afri Sugar Technol Assoc* 64: 34–39.

Autrey LJC, Boolell S, Jones P (1995a). Distribution of sugarcane bacilliform virus in various geographical regions. *Proc Intl Soc Sugar Cane Technol* 21: 657.

Autrey LJC, Boolell S, Lockhart BEL, Jones P, Nadif A (1995b). The distribution of sugarcane bacilliform virus in various geographical regions. *Proc Intl Soc Sugarcane Technol* 21: 527–541.

Balamuralikrishnan M, Vishwanathan R (2005). Comparison of PCR and DAC-ELISA for the diagnosis of Sugarcane bacilliform virus in Sugarcane. *Sugar Tech* 7(4): 119–122.

Borah BK, Sharma S, Kant R, Johnson AMA, Sai Gopal, DVR, Dasgupta I (2013). Bacilliform DNA-containing plant viruses in the tropics: Commonalities within a genetically diverse group. *Mol Plant Pathol* 14: 759–771.

Bouhida M, Lockhart BEL, Olszewski NE (1993). An analysis of the complete sequence of a sugarcane bacilliform virus genome infectious to banana and rice. *J Gen Virol* 74: 15–22.

Braithwaite KS, Egeskov NM, Smith GR (1995). Detection of sugarcane bacilliform virus using the polymerase chain reaction. *Plant Disease* 79: 792–796.

Chabannes M, Iskra-Caruana ML (2013). Endogenous pararetroviruses - a reservoir of virus infection in plants. *Curr Opin Virol* 3: 615–620.

Comstock JC, Lockhart BEL (1990). Widespread occurrence of sugarcane bacilliform virus in U.S. sugarcane germplasm collection. *Plant Disease* 74: 530.

Côte FX, Galzi S, Folliot M, Lamagnere Y, Teycheney P-Y, Iskra-Caruana M-L (2010). Micropropagation by tissue culture triggers differential expression of infectious endogenous banana streak virus sequences (eBSV) present in the B genome of natural and synthetic interspecific banana plantains. *Mol Plant Pathol* 11: 137–144.

Dallot S, Acuna P, Rivera C, Ramirez P, Côte F, Lockhart BEL, Iskra-Caruana ML (2001). Evidence that the proliferation stage of micropropagation procedure is determinant in the expression of banana streak virus integrated into the genome of the FHIA 21 hybrid (*Musa* AAAB). *Arch Virol* 146: 2179–2190.

Davies JP, Reddy V, Liu XL, Reddy AS, Ainley WM, Thompson M, Sastry-Dent L, Cao Z, Connell J, Gonzalez DO, Wagner DR (2014). Identification and use of the sugarcane bacilliform virus enhancer in transgenic maize. *BMC Plant Biology* 14: 359.

Deeshma KP, Bhat AI (2015). Complete genome sequencing of *Piper yellow mottle virus* infecting black pepper, betelvine, and Indian long pepper. *Virus Genes* 50: 172–175.

Devitt L, Ebenebe A, Gregory H, Harding R, Hunter D, Macanawai A (2005). Investigations into the seed and mealybug transmission of Taro bacilliform virus. *Aust Plant Pathol* 34: 73–76.

Egeskov N, Braithwaite KS, Smith GR (1994). Development of technique for the eradication of sugarcane bacilliform badanavirus from infected cane. *Proc Intl Soc Sugarcane Pathology Workshop* 4: 39.

Frison EA, Putter CAJ (1993). *FAO/IBPGR Technical Guidelines for the Safe Movement of Cocoa Germplasm.* Rome: Food and Agriculture Organization of the United Nations/International Board for Plant Genetic Resources.

Geering ADW, Hull R (2012). Family *caulimoviridae*. In: AMQ King, MJ Adams, EB Carestens, EJ Lefkowitz (Eds.) *Virus Taxonomy Classification and Nomenclature of Viruses. Ninth Report of the International Committee on Taxonomy of Viruses* (pp. 424–443). San Diego: Elsevier.

Geering ADW, McMichael LA, Dietzgen RG, Thomas JE (2000). Genetic diversity among banana streak virus isolates from Australia. *Phytopathology* 90: 921–927.

Geering ADW, Maumus F, Copetti D, Choisne N, Zwick DJ, Zytnicki M, McTaggart AR, Scalabrin S, Vezzulli S, Wing RA, Quesneville H, Teycheney P-Y (2014). Endogenous florendoviruses are major components of plant genomes and hallmarks of virus evolution. *Nature Communications* 5: 5269.

Geijskes RJ, Braithwaite KS, Dale JL, Harding RM and Smith GR (2002). Sequence analysis of an Australian isolate of sugarcane bacilliform badnavirus. *Arch Virol* 147: 2393–2404.

Hansen CN, Heslop-Harrison JS (2004). Sequences and phylogenies of plant pararetroviruses, viruses and transposable elements. *Advances Bot Res* 41: 165–193.

Hareesh PS, Bhat AL (2010). Seed transmission of *Piper yellow mottle virus* in black pepper (*Piper nigrum* L). *J Plant Crops* 38: 62–65.

Harper G, Hart D, Moult S, Hull R (2004). Banana streak virus is very diverse in Uganda. *Virus Res* 100: 51–56.

Harper G, Hart D, Moult S, Hull R, Geering A, Thomas J (2005). The diversity of banana streak virus isolates in Uganda. *Arch Virol* 150: 2407–2420.

Hearon SS, Locke JC (1984). Graft, pollen and seed transmission of an agent associated with top spotting in *Kalanchoë blossfeldiana. Plant Dis* 68, 347–350.

Hohn T, Richert-Poeggeler KR (2006). Replication of plant pararetroviruses. In: *Recent Advances in DNA Virus Replication.* KL Hefferson (Ed.) (pp. 289–319). Research Signpost 37/661, Kerala, India.

Irey MS, Baucum LE, Lockhart BEL (1992). Occurrence of sugarcane bacilliform virus in Florida collection. *J Amer Soc Sugarcane Technol* 12: 16–21.

Ishwara BA, Hohn T, Selvarajan R (2016). Badnaviruses: The current global scenario. *Viruses* 8: 177.

Jakowitsch J, Mette MF, van der Winden J Matzke MA, Matzke AJM (1999). Integrated pararetroviral sequences define a unique class of dispersed repetitive DNA in plants. *Proc Natl Acad Sci U S A* 13241–13246.

Jaufeerally-Fakim Y, Khorugdharry A, Harper G (2005). Genetic variants of banana streak virus in Mauritius. *Virus Research* 115: 91–98.

Karuppaiah R, Viswanathan R, Ganesh Kumar V (2013). Genetic diversity of sugarcane bacilliform virus isolates infecting *Saccharum* spp. in India. *Virus Genes* 46: 505–516.

Kenyon L, Lebas BSM, Seal SE (2008). Yams (*Dioscorea* spp.) from the South Pacific Islands contain many novel badna viruses: Implications for international movement of yam germplasm. *Arch Virol* 153: 877–889.

King AMQ, Adams MJ, Carstens EB, Lefkowitz EJ (2012). *Virus Taxonomy: Classification and Nomenclature of Viruses. Ninth Report of International Committee on Taxonomy of Viruses.* Elsevier Academic Press, San Diego, CA.

Kunii M, Kanda M, Nagano H, Uyeda I, Kishima Y, Sano Y (2004). Reconstruction of putative DNA virus from endogenous *Rice tungro bacilliform virus*-like sequences in the rice genome: Implications for integration and evolution. *BMC Genomics* 5: 80.

LaFleur DA, Lockhart BEL, Olszewski NE (1996). Portions of the banana streak badnavirus genome are integrated in the genome of its host *Musa* sp. (Abstr.) *Phytopathology* 86: S100.

Lheureux F, Carreel F, Jenny C, Lockhart BEL, Iskra-Caruana ML (2003). Identification of genetic markers linked to banana streak disease expression in inter-specific *Musa* hybrids. *Theor Applied Genet* 106: 594–598.

Li W-F, Huang Y-K, Jiang D-M, Zhang Z-X, Zhang B-L, Li S-F (2010). Detection of Sugarcane bacilliform virus isolate and its influence on yield and quality of cane in Yunnan. *Acta Phytopathol Sin* 6: 651–654. [In Chinese]

Lockhart BEL (1986). Purification and serology of a bacilliform virus associated with banana streak disease. *Phytopathology* 76: 995–999.

Lockhart BEL (1988). Occurrence in sugarcane of a bacilliform virus related serologically to banana streak virus. *Plant Disease* 72: 230–233.

Lockhart BEL (1995). *Banana Streak Badnavirus Infection in Musa: Epidemiology, Diagnosis, and Control.* ASPAC Food & Fertilizer Technology Center, Taipei, Taiwan, Volume 143.

Lockhart BEL, Autrey LJC (1988). Occurrence in sugarcane of a bacilliform virus related serologically to banana streak virus. *Plant Disease* 72: 230–233.

Lockhart BEL, Autrey LJC (1991). Mealy bug transmission of sugarcane bacilliform and clostero-like viruses. In: *Third ISSCT Sugarcane Pathological Workshop*, Chiang Mai, Thailand.

Lockhart BEL, Autrey LJC (2000). Sugarcane bacilliform virus. In: *A Guide to Sugarcane Diseases*, P Rott, RA Bailey, JC Comstock, BJ Croft, AS Saumtally (Eds.) CIRAD and ISSCT, Montepellier France, pp. 268–272.

Lockhart BEL, Olszewski NE (1993). Serological and genomic heterogeneity of banana streak badnavirus: Implications for virus detection in *Musa* germplasm. In: *Breeding Banana and Plantain for Resistance to Diseases and Pests*, Ganry J (Ed.) CIRAD/INIBAP; Publication Press, France, pp. 105–113.

Lockhart BEL, Autrey LJC, Comstock JC (1992). Partial purification and serology of sugarcane mild mosaic virus, a mealy bug transmitted closterolike virus. *Phytopathology* 82: 691–695.

Lockhart BEL, Irey MJ, Comstock JC (1996). Sugarcane bacilliform virus, sugarcane mild mosaic virus and sugarcane yellow leaf syndrome. In: *Sugarcane Germplasm Conservation and Exchange*, BJ Croft, CM Piggin, ES Wallis, DM Hogarth (Eds.) ACIAR Proc. No. 67, Canberra, Australia, pp. 108–112.

Medberry SL, Lockhart B, Olszewski NE (1990). Properties of *Commelina yellow mottle virus*'s complete DNA sequence, genomic discontinuities and transcript suggest that it is a pararetrovirus. *Nucleic Acids Res* 18: 5505–5513.

Muller E, Dupuy V, Blondin L, Bauffe F, Daugrois J, Nathalie L, Iskra-Caruana M-L (2011). High molecular variability of sugarcane bacilliform viruses in Guadeloupe implying the existence of atleast three new species. *Virus Research* 160: 414–419.

Ndowora TCR, Lockhart BEL (2000). Development of a serological assay for detecting serologically diverse banana streak virus isolates. In: *Proceedings of the First International Conference on Banana and Plantain for Africa*, K Creanen, EB Karamura, D Vulsteke, D (Eds.) International Society for Horticultural Science, Leuven. *Acta Horticulturae* 540: 377–388.

Peralta EL, Anoheta O, Carajal O, Martinez Y (1991). Studies on *Sugarcane bacilliform virus*. *Proc. ISSCT Pathology Workshop* 3: 14.

Quainoo AK, Wetten AC, Allainguillaume J (2008). The effectiveness of somatic embryogenesis in eliminating the cocoa swollen shoot virus from infected cocoa trees. *J Virol Methods* 149: 91–96.

Rao GP, Singh M, Rishi N, Bhargava KS (2002). Century status of sugarcane virus diseases research in India. In: *Sugarcane Crop Management*. SB Singh, GP Rao, S Easwaramoorthy (Eds.) Sci Tech Publishing, Houston, TX, pp. 223–254.

Rao GP, Sharma SK, Singh D, Arya M, Singh P, Baranwal VK (2014). Genetically diverse variants of sugarcane bacilliform virus infecting sugarcane in India and evidence of a novel recombinant Badnavirus variant. *J Phytopathol* 162: 779–787.

Richert-Poggeler KR, Shepherd RJ (1997). Petunia vein clearing virus: A plant pararetrovirus with the core sequence of an integrase function. *Virology* 236: 137–146.

Rodriguez Lema E, Rodriguez D, Fernandez E, Acevador R, Lopez D (1985). Report of a new sugarcane virus. *Ciencias Agriculture* 23: 130.

Rott P, Bailey RA, Comstock JC, Croft BJ, Saumtally AS (2000). *A Guide to Sugarcane Disease*. ISSCT, CIRAD, Montepellier, France.

Sharma SK, Vignesh Kumar P, Geetanjali AS, Pun KB, Baranwal VK (2015). Subpopulation level variation of banana streak viruses in India and common evolution of banana and sugarcane badnaviruses. *Virus Genes* 50: 450–465.

Singh D, Tewari AK, Rao GP, Karuppaiah R, Viswanathan R, Arya M, Baranwal VK (2009). RT-PCR/PCR analysis detected mixed infection of DNA and RNA viruses infecting sugarcane crops in different states of India. *Sugar Tech* 11: 373–380.

Singh M, Gaur RK, Rao GP (2003). Distribution of sugarcane bacilliform badna virus in India. *J Mycol Pl Pathol* 33(3): 406–410.

Smith GR (1996). Sugarcane mosaic and Fiji disease. In: *Sugarcane Germplasm Conservation and Exchange*. BJ Croft, CM Piggin, ES Wallis, DM Hogarth (Eds.). ACIAR Proceedings No. 67, Canberra, pp. 120–122.

Staginnus C, Gregor W, Mette MF, Teo CH, Borroto-Fernandez EG, Machado ML, Matzke M, Schwarzacher T (2007). Endogenous pararetroviral sequences in tomato (*Solanum lycopersicum*) and related species. *BMC Plant Biology* 7: 24.

Staginnus C, Richert-Poggeler KR (2008). Endogenous pararetroviruses: Two-faced travelers in the plant genome. *Trends Plant Sci* 11: 485–491.

Staginnus C, Iskra-Caruana M-L, Lockhart BEL, Hohn T, Richert-Poggeler KR (2009). Suggestions for nomenclature of endogenous pararetroviral (EPRV) sequences in plants. *Arch Virol* 154: 1189–1193.

Sun S-R, Damaj MB, Alabi OJ, Wu X-B, Mirkov TE, Fu H-Y, Chen R-K, Rao S-J (2016). Molecular characterization of two divergent variants of sugarcane bacilliform viruses infecting sugarcane in China. *Eur J Plant Pathol* 145: 375.

Viswanathan R, Alexander KC, Garg ID (1996). Detection of sugarcane bacilliform virus in sugarcane germplasm. *Acta Virology* 40: 5–8.

Viswanathan R, Premchandran MN (1998). Occurrence and distribution of sugarcane bacilliform virus in sugarcane germplasm collection in India. *Sugar Cane* 6: 9–18.

Yang IC, Hafner GJ, Revill PA, Dale JL, Harding RM (2003). Sequence diversity of South Pacific isolates of *Taro bacilliform virus* and the development of a PCR-based diagnostic test. *Arch Virol* 148: 1957–1968.

8 Wheat Streak Mosaic Virus
Cereal Pathogen with Growing Importance

Khushwant Singh and Jiban Kumar Kundu

CONTENTS

8.1 INTRODUCTION

Wheat streak mosaic virus (WSMV) is the type species of the genus *Tritimovirus* in the family *Potyviridae*. WSMV was first recognized nearly 100 years ago in the central Great Plains of the United States (McKinney, 1937). WSMV was initially described as "*yellow mosaic*" in 1922 in Nebraska (Hunger, 2010). The virus is widely distributed in most wheat growing regions of the world including the United States, Canada, Mexico, Brazil, Argentina, Europe, Turkey, Iran, Australia, and New Zealand (Hadi et al., 2011; Navia et al., 2013). WSMV was formerly placed in the genus *Rymovirus* along with mite-transmitted viruses of the family *Potyviridae*. Later, the complete genome sequence and phylogenetic analysis of WSMV demonstrated that it shares ancestry with the whitefly-transmitted *Sweet potato mild mottle virus* and not with *Ryegrass mosaic virus*, the type member of genus *Rymovirus* (Stenger et al., 1998). The finding proposed a new genus known as "*Tritimovirus*" within the family *Potyviridae*, of which *Wheat streak mosaic virus* is the type member (Rabenstein et al., 2002).

WSMV infects many plant species of the family *Poaceae*, including wheat (*Triticum aestivum* L.), oat (*Avena sativa* L.), barley (*Hordeum vulgare* L.), maize (*Zea mays* L.), millet (*Panicum*, *Setaria*, and *Echinochloa* spp.), and several grasses (French and Stenger, 2002; Dráb et al., 2014;

Chalupníková et al., 2017). Given the potentially devastating impact of WSMV on affected cereal crops, the appearance of this disease on wheat production has been a cause for concern, as it can range from minimal to complete crop failure (French and Stenger, 2003; Hunger, 2004). Improvement in disease resistance is an important aspect of wheat production, and the development of existing disease resistance cultivars has helped to increase production (Price et al., 2010a).

WSMV imposes a devastating impact on wheat growing countries around the globe and is of great significance. This chapter describes various aspects of WSMV, including history of the virus, genome architecture, disease cycle and symptoms, mechanism of transmission, host range, genetic diversity of WSMV, resistance against WSMV, development of diagnostic tests for WSMV, and crop management.

8.2 CAUSAL AGENT OF WHEAT STREAK MOSAIC DISEASE

The causal agent of wheat streak mosaic disease is WSMV. WSMV is a non-enveloped, flexible, filamentous rod-shaped virus composed of a positive-sense single-strand RNA genome (ssRNA+), and can be either mono- or bipartite. The genome size of WSMV is approximately 9.3–9.4 kb and has a single open reading frame, which is transcribed into a large polyprotein. This polyprotein is composed of ten proteins: P1, HC-Pro, P3, 6K1, Cl, 6K2, NIa (VPg and Pro), Nib, and CP (Stenger et al., 1998; Choi et al., 2002) (Figure 8.1). 5′ terminus has a genome-linked protein (VPg) and 3′ terminus has a poly (A) tail. The RNA is infectious and serves as both the genome and viral messenger.

Every individual protein in the virus genome has its own specific function (Oana et al., 2009), as follows. P1 is development of symptoms of disease, genome amplification, and binding of RNA. HC-Pro is transmission of the virus vector, the motion of virion from the cell into the cell genome amplification, determination of pathogenicity, development of symptoms of the disease, and the binding of RNA. P3 is responsible for pathogenicity and replication of the viral genome. 6K1, together with P3, is involved in pathogenicity and replication of the virus genome. C1 is responsible for the movement of virions from the cell into the cell genome amplification and RNA binding. 6K2 is involved with placement of viral replication complexes in the cell walls and amplification of the virus genome. VPg is responsible for amplification of the viral genome and binding of RNA. NIa is responsible for binding RNA amplification genome, protein-protein interactions. NIb is responsible for replication of the viral genome, RNA-dependent RNA polymerase (RdRp) and the binding of RNA. CP is transmission of the virus, virion movement from cell to cell and long-distance movement, assembly of virions, RNA binding, and amplification of the virus genome.

8.3 DISEASE CYCLE

WSMV spreads primarily by the vector wheat curl mite (WCM), *Aceria tosichella*. The disease cycle begins in autumn when WCMs migrate from nearby infected volunteer wheat, barley, corn, millet, or grasses on newly emerged crop hosts (Figure 8.2). Some grasses are hosts for the mites but not for the

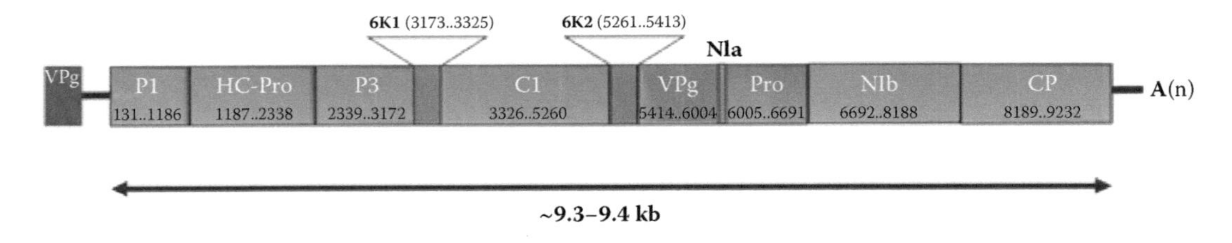

FIGURE 8.1 Genome architecture of *Wheat streak mosaic virus*. The genome size of WSMV is approximately 9.3–9.4 kb and possesses a single open reading frame, which is transcribed into a large polyprotein. This polyprotein is composed of ten proteins: P1, HC-Pro, P3, 6K1, Cl, 6K2, NIa (VPg and Pro), Nib, and CP.

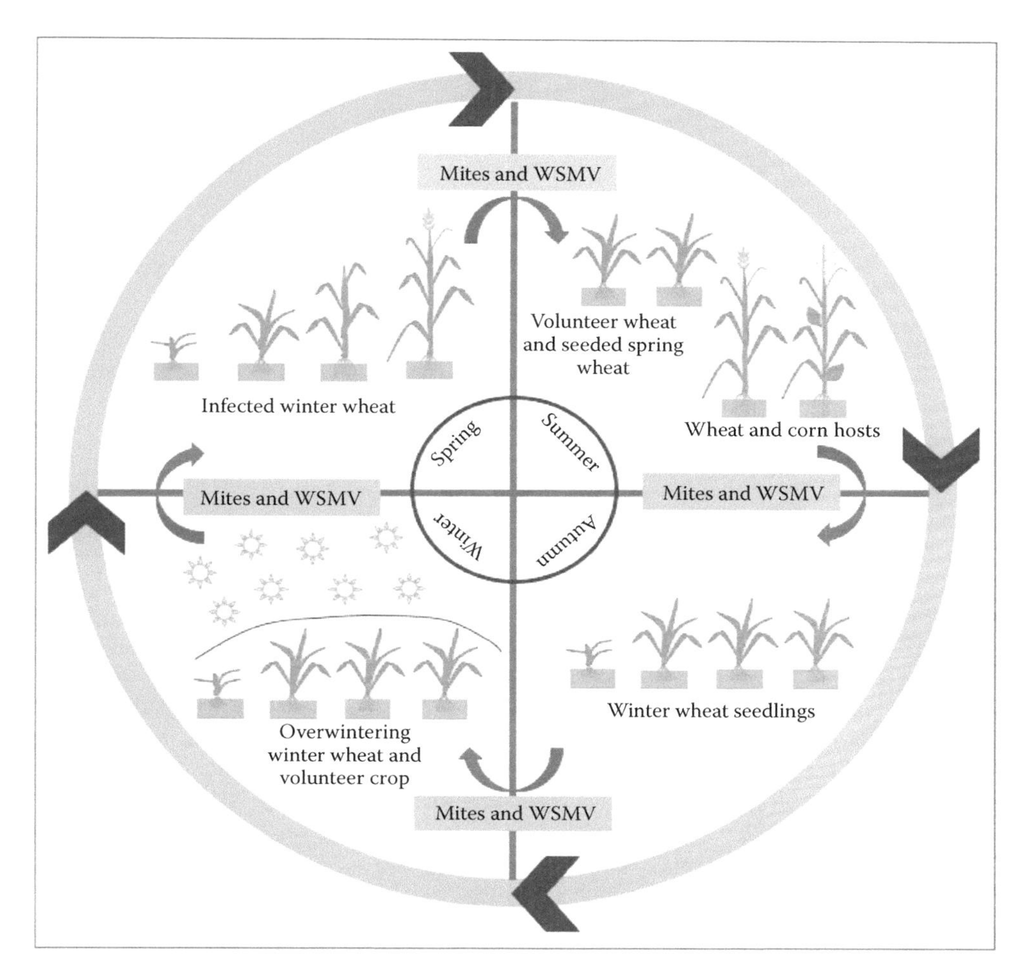

FIGURE 8.2 Disease cycle of *Wheat streak mosaic virus*. (Adaption of the diagram according to Watkins JE, French R, Campbell JB et al. (1989). *Wheat Streak Mosaic Virus*. Nebraska Cooperative Extension EC89-1871. http://digitalcommons.unl.edu/cgi/viewcontent.cgi?article=2232&context=extensionhist (accessed February 3, 2017.)

WSMV. Some accommodate both the mite and virus. Some are susceptible to the virus but are not good hosts for mites, and others are immune to both (Murray et al., 2005). Early infection in young growing plants results in a greater yield loss of winter wheat (Slykhuis et al., 1957; Hunger et al., 1992). Winter wheat can also be infected in the spring by virus, but typically spring infections have no significant impact on yield losses (Somsen and Sill, 1970). During the spring, virus-infected winter wheat and perennial grasses serve as sources of virus and mites for spring wheat infection. Spring wheat planted near infected winter wheat fields during this time is at greater risk of WSMV infection (Langham and Glover, 2005). Humid weather during summer that facilitates lush growth of volunteer wheat also aids in high population of WCM and WSMV for successive winter wheat infection (Christian and Willis, 1993; Thomas and Hein, 2003).

8.4 DISEASE SYMPTOMS

WSMV-infected plants normally are stunted and have mottled and streaked leaves. Initial symptoms include small chlorotic lines on individual leaves which elongate to form discontinuous yellow to

(a) (b) (c)

FIGURE 8.3 *Wheat streak mosaic virus* disease symptoms on hosts. (a) Wheat cv. Hymack showing chlorotic mosaic pattern developed due to WSMV. (b) Wheat cv. Turondot showing advanced symptoms with linear streaks coalescing into almost solid fellow areas. (c) Wheat cv. Ludwing showing severe symptoms progressing into leaf tissue necrosis. (Courtesy of Jiban K. Kundu.)

pale green streaks, forming a mosaic pattern in the leaves, usually in spring (Vacke et al., 1986). As the weather warms, the symptoms become more severe and the stripes may merge, forming large chlorotic areas, commonly resulting in symptoms progressing into leaf tissue necrosis and plant death. Symptoms can vary with the cultivar, strain of the virus, time of infection, and environmental conditions. Wheat fields infected with WSMV exhibit yellowing and stunting in irregular areas, often at field margins or volunteer hosts. As the season progresses, infection may spread and symptoms may appear within the field. The infected plant often appears on the field margins as the WCMs emigrate from bordering crops and weeds (Hunger, 2010). Leaf mite-infested leaves tend to remain erect, with their lateral margins rolled toward the upper midrib. Symptoms caused by WSMV infection in the wheat crops are shown in Figure 8.3a–c.

8.5 TRANSMISSION

Under natural conditions, WSMV is transmitted primarily by WCM *A. tosichella*, and to a lesser extent in seeds of infected plants. In experimental conditions the virus can also be transmitted by mechanical inoculation by infected plant. Different modes of transmission are described below.

8.5.1 Transmission by Wheat Curl Mite *Aceriatosichella*

WSMV is transmitted primarily by the mite *A. tosichella*, which belongs to the family *Eriophyidae* (Skoracka et al., 2014). Vector WCM is white and cigar-shaped, possesses four legs near the head, and is 0.3 mm long (Figure 8.4). WCM lacks wings, so it is usually carried by wind from plant to plant and field to field. The life stages of the WCM are egg, two larval stages, and adult. Females lay between 3 and 25 eggs, usually about 1 per day, during their life. The minute eggs get fastened to the leaves in straight lines, parallel to the leaf veins. Under ideal conditions, one adult could produce 3 million offspring in 60 days. Development is very slow at lower temperatures of 8–10°C and essentially stops at 0°. All stages can survive at least 3 months at near-freezing temperatures and can survive for several days at 16–20°C.

At temperatures of around 25–28°C WCM complete their life cycle within 8–10 days. The WCM live and feed on the host's green leaves, which are critical for mite survival. High humidity is another key factor in mite survival (Slykhuis, 1955). Without food, WCM can survive for less than 8 hours at 24°C, and nearly 2–3 days at 15°C (Slykhuis, 1955). Eggs survived longer under sub-zero temperatures; over 25% of eggs exposed to 15°C for 8 days were able to hatch when moved to room temperature (Slykhuis, 1955). Orlob (1966) suggested that slower mite development under near-freezing temperature may play a role in maintaining longer virus retention periods. A single female

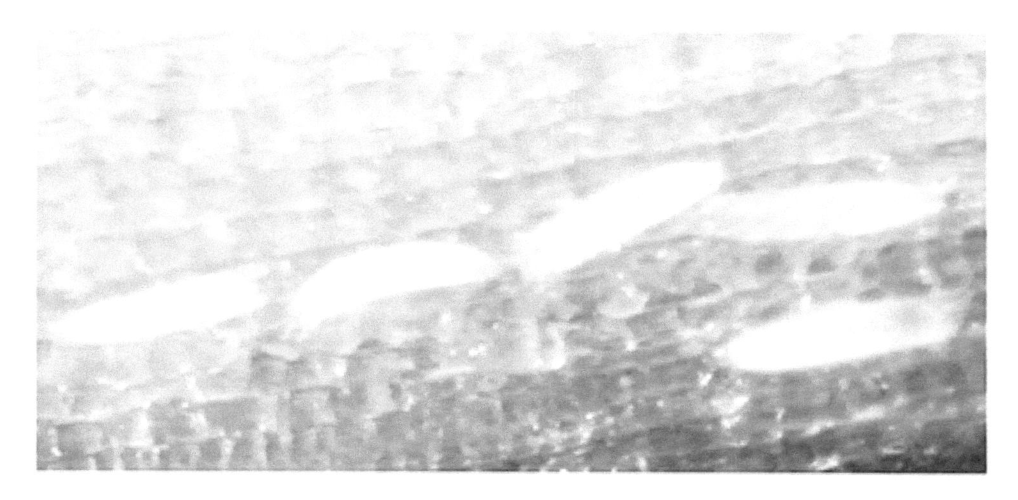

FIGURE 8.4 Causal agent of *Wheat streak mosaic disease*. Wheat curl mite, *Aceria tosichella*, 0.02 mm long. (Photo by Dr. Anna Skoracka, Adam Mickiewicz University, Poznań, Poland.)

WCM is capable of laying 12–20 eggs and can produce 3 million more descendants within 60 days. Population of WCM increase dramatically at warm summer temperatures around 20°C, but are limited by temperatures above 30°C. Conversely, at temperatures lower than 15°C their reproductive ability declines very rapidly.

The virus can be picked up during a 10- to 30-minute feeding period and remains active in the mite for 7–9 days. The virus remains in the mite as it molts from stage to stage but cannot be passed to the next mite generation through the egg stage. Mites may directly damage the young plants or influence their growth, causing the leaves to remain curled. Mites in these curled leaves survive for several months and can even survive well in the husk and leaf sheaths. In colder climates, mites can survive up to 2 weeks without any food. Warmer temperatures and adequate course of rainfall lead to rapid growth of grass host, which result in accelerating population of mites and accumulation of virus in host plants.

8.5.2 Transmission by Seeds

Transmission of the WSMV by seed was first described in the early 1970s from a number of maize inbred lines (*Zea mays* L.) plants in seed production fields in Iowa (Hill et al., 1974). A very low percentage of seed transmission (0.1%) of the virus was found. Jones et al. (2005) evaluated the seed transmission of WSMV in different wheat genotypes. Seed transmission of virus depends on the wheat genotypes, i.e., the rate of transmission is lower across the wheat breeding collection tested (0.2–0.5%) and higher in individual genotypes (up to 1.5%) (Jones et al., 2005). The seed in the infected seed samples from individual genotypes is likely to have come from healthy plants; the inherent transmission rate in seed only from infected plants is probably higher (Jones, 2000).

8.6 HOST RANGE OF WSMV

WSMV has sufficient host range including cereals and grasses. Wheat is an excellent host for virus and is a favored food for the WCM. Other cereal hosts susceptible to WSMV are oats (*Avena sativa*), barley (*Hordeum vulgare*), rye (*Secalecereale*), corn (*Zea mays*), foxtail millet (*Setariaitalica*), and broom-corn millet, or proso (*Panicummili aceum*) (Table 8.1). The mite also feeds and reproduces on these cereals. However, some cereals are susceptible to the virus but are not good hosts for mites; for example, barley (*Hordeum vulgare*) and rye (*Secalecereale*). Several annual and perennial grasses may also serve as a host of WSMV. A complete host range of WSMV is given in Table 8.1.

TABLE 8.1
Host Range of Wheat Streak Mosaic Virus

Host	Common Name	References
	Cereals	
Avenabarbata	Bearded oat	Coutts et al., 2014
Avenasativa	Oat	Brakke, 1971
Hordeumvulgare	Barley	Brakke, 1971
Panicummillaceum	Broomcorn millet	Sill and Agusiobo, 1955; Vacke et al., 1986; Ellis et al., 2004
Pennisetumglaucum	Pearl millet	Seifers et al., 1996
Secalecereale	Cereal rye	Vacke et al., 1986; Ito et al., 2012
Setariaitalica	Foxtail millet	Truol et al., 2010
Sorghum bicolor	Sorgum	Seifers et al., 1996
Triticumaestivum	Wheat	Brakke, 1971
Zea mays	Maize	Brakke, 1971
	Grasses	
Aegilopscylindrica	Jointed goatgrass	Sill and Connin, 1953
Agropyronrepens	Couch grass	Dráb et al., 2014
Agrostiscapillaris	Common Bent	Chalupníková et al., 2017
Alopecuruspratensis	Meadow foxtail	Dráb et al., 2014
Anthoxanthumodoratum	Sweet vernal-grass	Chalupníková et al., 2017
Arrhenatherumelatius	False oat-grass	Dráb et al., 2014
Austrostipacompressa	Speargrass	Vincent et al., 2014
Avenafatua	Wild oat	Vacke et al., 1986
Avenastrigesa	Wild oats	Vacke et al., 1986
Avenasterilis	Wild oats	Murray et al., 2005
Briza maxima	Blowfly grass	Coutts et al., 2014
Bromusarvenis	Field brone	Sill and Connin, 1953
Bromusdiandrus	Great brome	Murray et al., 2005
Bromusjaponicus	Japanese brome	Wegulo et al., 2008
Bromusrigidus	Brome grass	Coutts et al., 2014
Bromussecalinus	Cheat grass	Sill and Connin, 1953
Bromustectorum	Downy brome	Sill and Connin, 1953
Cenchruslongispinus	Mat sandbur	Connin, 1956
Cenchruspauciflours	Sandbur	Wegulo et al., 2008
Cynodondactylon	Couch grass	Ellis et al., 2004
Digitariasanguinalis	Hairy crab grass	Vacke et al., 1986; Somsen and Sill, 1970
Echinochloa crus-galli	Barnyardgrass	Sill and Connin, 1953
Echinochloacolonum	Junglerice	Khadivar and Nasrolahnejad, 2009
Elymusrepens	Quackgrass	Ito et al., 2012
Eragrostiscilianensis	Stink grass	Connin, 1956
Eragrostiscurvula	African lovegrass	Ellis et al., 2004
Eriochloaacuminata	Tapertipcupgrass	Seifers et al., 2010
Eriochloacontracta	Prairie cupgrass	Christian and Willis, 1993

(Continued)

TABLE 8.1 (CONTINUED)
Host Range of Wheat Streak Mosaic Virus

Host	Common Name	References
Eleusineindica	Crowsfoot	Murray et al., 2005
Eleusinetristachya	Spike goosegrass	Ellis et al., 2004
Elymuscanadensis	Canada wild rye	Ito et al., 2012
Holcuslanatus	Soft-grass	Chalupníková et al., 2017
Holcusmollis	Creeping soft grass	Chalupníková et al., 2017
Hordeumleporinum	Barley grass	Coutts et al., 2014
Lagurusovatus	Hare's-tail	Vacke et al., 1986
Loliummitiflorum	Annual ryegrass	Vacke et al., 1986; Ellis et al., 2004
Loliumrigidum	Ryegrass	Murray et al., 2005; Coutts et al., 2014
Panicumdichotomiflorum	Fall panicgrass	Sill and Connin, 1953
Panicumcapillare	Witch grass	Coutts et al., 2008a, b
Phalarisaquatica	Phalaris	Ellis et al., 2004
Phleumpratense	Timothy-grass	Dráb et al., 2014
Poapratensis	Bluegrass	Ito et al., 2012; Dráb et al., 2015
Setariaverticellata	Whorled pigeon grass	Murray et al., 2005
Setariaviridis	Green bristlegrass	Sill and Connin, 1953
Tragus australianus	Small burr grass	Coutts et al., 2008a, b

8.7 WORLDWIDE OCCURRENCE OF WSMV

Since, the first report of WSMV in the central Great Plains in North America in 1922 (McKinney, 1937), the incidence of the virus has been reported in North and South America, Eastern Europe, the Middle East, Western Asia, Australia, and New Zealand (Hadi et al., 2011; Navia et al., 2013) (Figure 8.5). Production losses in wheat caused by WSMV infection are significant. In Kansas, yield losses ranged from 7 to 13% (Hansing et al., 1950; Christian and Wills, 1993); about 18% reported yield losses in Canada (Atkinson and Grant, 1967). Significant losses due to virus have been reported in Colorado and Oklahoma (Shahwan and Hill, 1984; Hunger, 2004). In Australia, damage to production caused by WSMV is significantly higher—up to 83% of production was destroyed (Lanoiselet et al., 2008). WSMV is reported to be a major limiting factor of wheat production in Texas panhandle (Velandia et al., 2010). The economic impact of these losses can be devastating to farmers and the wheat industry. Velandia et al., (2010) showed a marginal loss up to $464.5/hectares due to WSMV. This loss was higher in irrigated wheat than in dryland wheat due to the reduced water use efficiency that results when wheat is infected by WSMV (Price et al., 2010a).

8.8 DETECTION OF WSMV

WSMV infestation has historically been detected by means of symptoms on leaves. Alternative diagnostic approaches of WSMV have included laboratory testing using techniques such as serological methods enzyme-linked immunosorbent assay (ELISA) (DAS- and TAS-ELISA). More recently, various diagnostic tests based on the polymerase chain reaction (PCR), PCR, RT-PCR, multiplex PCR, and RT-qPCR, have been developed for the detection of WSMV (Gadiou et al., 2009; Tatineni et al., 2010; Dráb et al., 2014; Schubert et al., 2015).

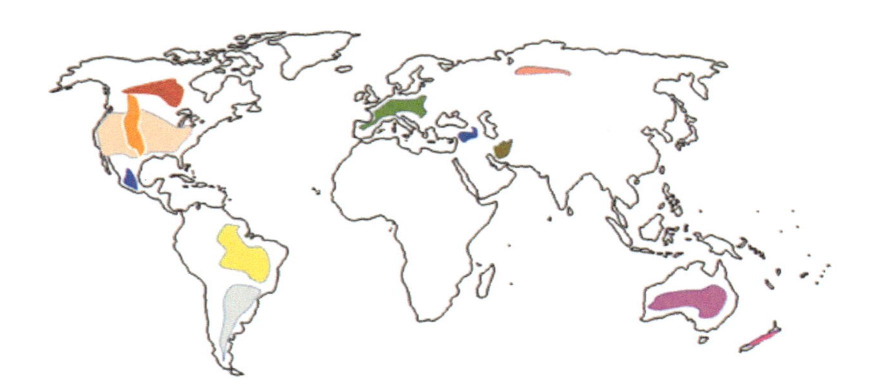

FIGURE 8.5 *Wheat streak mosaic virus* distribution around the globe. The virus is present in the United States (all states, highlighted in orange), Canada (highlighted in red), Mexico (highlighted in blue), Brazil (highlighted in yellow), Argentina (highlighted in grey), Europe (highlighted in green), Turkey (highlighted in light blue), Iran (highlighted in gold), Russia (highlighted in rose), Australia (highlighted in lavender), and New Zealand (highlighted in pink). (The scheme of WSMV distribution retrieved from French R, Stenger DC (2003). Evolution of *wheat streak mosaic virus*: Dynamics of population growth within plants may explain limited variation. *Annual Review of Phytopathology* 41:199–214; Gadiou S, Kúdela O, Ripl J et al. (2009). An amino acid deletion in *Wheat streak mosaic virus* capsid protein distinguishes a homogeneous group of European isolates and facilitates their specific detection. *Plant Disease* 93:1209–1213, and Schubert J, Ziegler A, Rabenstein F (2015). First detection of *wheat streak mosaic virus* in Germany: Molecular and biological characteristics. *Archives of Virology* 160:1761–1766.)

8.8.1 Serological Detection

For a long time, serological testing has been the standard for plant virus disease diagnostics. Two similar serological methods are available for the detection of WSMV, DAS-ELISA (double antipodes sandwich enzyme linked immuno-sorbent assay) and TAS-ELISA (triple antipodes sandwich enzyme linked immuno-sorbent assay). ELISA combines the specificity of antibodies with the sensitivity of simple enzyme assay. It can be run in a qualitative and quantitative format. ELISA is an effective method when large number of samples must be assayed, when results are needed rapidly. Unlike DAS-ELISA, TAS-ELISA uses two conjugates: A (Conjugate A), which is a monoclonal antibody (mouse) and B (Conjugate B), which is a specific murine antibody (anti-mouse Ig species). However, both are linked together with alkaline phosphatase.

ELISA is the most established method for monitoring viruses, but is less effective due to (i) low sensitivity compared with methods based on cDNA amplification (PCR) (Izzo et al., 2012); (ii) its inability to recognize all related viral strains (Coutts et al., 2011); and (iii) it is inefficient to interpret the viral accumulations (Schubert et al., 2015). WSMV ELISA detection kits are available through several commercial companies including SEDIAGSAS, France; Agdia Inc, USA; Thermofisher, USA; and others.

8.8.2 PCR-Based Detection

Application of molecular approaches has provided much greater sensitivity in detecting plant viral pathogens, including cereal viral pathogens, when compared to the serological testing (Ay et al., 2008). WSMV has been detected by molecular methods such as RT-PCR and RT-qPCR (Gadiou et al., 2009; Dráb et al., 2014; Schubert et al., 2015). Most PCR-based detection protocols have targeted the viral coat protein gene sequence (Gadiou et al., 2009). WSMV was detected using multiplex RT-PCR reactions along with meta-community of wheat viral pathogens such as *Barley yellow dwarf* (BYDV), *Cereal yellow dwarf virus* (CYDV), *Wheat spindle streak virus*, and

Soil-borne wheat mosaic virus (Deb and Anderson, 2008; Price et al., 2010b). Multiplex RT-PCR is being used not only for detection of viral pathogens, but for strain identification of viral pathogens. RT-PCR and multiplex RT-PCR provide an indication only of the presence or absence of the WSMV, rather than the virus titer analysis in a particular sample using RT-qPCR (Chalupníková et al., 2017).

8.8.3 WSMV Quantification

Serological and PCR-based tests serve as important tools for detection of WSMV. However, these methods cannot interpret the abundance of virus in given samples. In contrast, the advent technology known as quantitative reverse transcription polymerase chain reaction (RT-qPCR) has enabled the quantification of the virus concentration of several plant RNA viruses including WSMV (Dráb et al., 2014; Chalupníková et al., 2017). The method is preferred for absolute quantification of virus/viruses to study virus biology, virus gene expression, and virus–host and virus–vector interactions. Tatineni et al. (2010) quantified WSMV concentrations in single and double infection (WSMV and *Triticum mosaic virus* TriMV), in wheat cultivars Araphose, Tomahawk, and Mace by RT-qPCR and revealed that WSMV and TriMV induced cultivar-specific disease synergism in wheat. Schubert et al. (2015) revealed the higher accumulation of RNA in USA PV57 strain compared to European isolates using FAM and Atto-labeled sequence-specific probes in RT-qPCR (Table 8.2).

8.9 GENETIC DIVERSITY OF WSMV

WSMV is reported on all continents, and the extent of genetic diversity has been evaluated between various isolates from different origins. Based on sequence and phylogenetic analysis of coat protein gene, WSMV isolates have been divided into four major clades/groups, A to D (French and Stenger, 2002; Stenger and French, 2009). However, recent analysis of complete genome sequences from WSMV revealed the existence of three types: A, B, and D (Schubert et al., 2015) (Table 8.3). Type C represents high diversity between America and Australia isolates viz. FJ348358_WA98, AF511643_WA99, AF511630_MON96, AF511618_ID96, AF285169_USA (PV57), and AF957533_Sidney81, and merged with type D isolates (Schubert et al., 2015). The phylogenetic analysis of complete genome sequences of WSMV is shown in Figure 8.6. Clade/type A represents isolates from Mexico known as El-Batán (accession no. AF285170); Clade/type B contains isolates from Europe, Russia, and Asia (Saadat-Shahr from Iran, accession no. EU914918). Type B isolates are known as *WSMV-ΔE* and are characterized by a deletion of three nucleotides GCA coding for glycine at amino acid position 2761 in the sequence of the coat protein (Gadiou et al., 2009). Schubert et al., (2015) showed whole-genome comparisons of type B isolates and revealed the differences in putative protein P1/helper-component proteinase (HC-Pro) protease cleavage site in addition to the CPgene (Choi et al., 2002). Type D comprises isolates from the United States, Argentina, Brazil, Australia, Turkey, and Canada (Dwyer et al., 2007; Stenger and French, 2009). The deletion of 3-nucleotide was later identified in isolates originating from the United States (Robinson and Murray, 2013). Some of the isolates collected from Turkey and clustered in clade B, resulting in the hypothesis of two distinct genotypes coexisting in Turkey (Gadiou et al., 2009). Recent isolates collected from Iran also clustered into distinctive types, i.e., Saadat-Shahr is type B and Naghadeh is type D (Schubert et al., 2015). Type A isolates share ~79% similarity with type B and type D isolates. Type B share ~88% with type D (Figure 8.7).

Recombination event analysis of complete WSMV sequences from various types revealed that type B isolates recombined only with isolates from within this cluster, whereas isolates from type A and type D also contain sequences of type B isolates (Schubert et al., 2015). More polymorphic sites and parsimony informative sites, as well as increased diversity, were observed for the CP sequences of type D isolates than those of type B isolates, suggesting a more recent establishment of the virus in the latter (Robinson and Murray, 2013).

TABLE 8.2
PCR Primer Used for Detection of WSMV

Name	5′–3′	Amplification (bp)	RT-PCR/ RT-qPCR	Target Gene	Reference
WSMVF	TCGAGTAGTGGAAGCACTCA	~948	PCR	CP	Mar et al., 2013
WSMVR	CCTCACATCATCTGCATCAT				
WSMVL2	CGACAATCAGCAAGAGACCA	~198	RT-PCR	VPg-NIa	Deb and Anderson, 2008
WSMVR2	TGAGGATCGCTGTGTTTCAG				
WSMV-CP1	CGGACGGATTTAGGAGAAGAG	~634	RT-PCR	CP	Dwyer et al., 2007
WSMV-CP3	ATGCAGTGGTAGACCCATC				
WSMV-CP2	AAGGGCTTGACGTGACAGAGG	~581			
WSMV-CP4	CGCTCAAATCCTGGTACTC				
WSM1	CAGGCACGACTGAGTGCGG	~270	PCR	CP	French and Robertson, 1994
RCF1	AGCTGGATCCTTTTTTTTTTTTTTT				
WS-8166F	GAGAGCAATACTGCGTGTACG	~750	RT-PCR	CP	Kudela et al., 2008
WS-8909R	GCATAATGGCTCGAAGTGATG				
WsmF7288	CAAAGCTGTGGTTGATGAGTTCA	~55	RT-PCR	NIb	Price et al., 2010b
WsmR7343	TTGATTCCGACAGTCCATGGT				
WsmP7312	CAAATTCTTCTACACAAAGCATTTGCGCG	probe			
WSMVF	GTTGGGAGGCTTAATTGAAGTG	~720	PCR	CP	Byamukama et al., 2016
WSMVR	CAGCCATTACTCGTGTTATCC A				
SB104	GGGCTTGATGTRACAGAGG	~493	RT-PCR	CP	Jones et al., 2005
SB105	TCACATCATCTGCATCATGACGTG				
WSMV-F	AAGTGCAGAACAGCGTTG	~104	RT-qPCR	CP	Tatineni et al., 2010
WSMV-R	AAACTGTGCGTGTTCTCC				
WSMV-probe	ACTGAGTGCGGGTACTAATGAGGAC	probe			
XV1	GATCCGTTGAGGATTTGTACTT	~1267	PCR	CP	Hall et al., 2001
XC1	AACCCACACATAGCTACCAAG				
WSMspeFw	GCCTCGACACGGGAGCTA	~354	PCR	CP	Gadiou et al., 2009
WSMspeRv	ACCCATCCAGGAAGCAAGG				
WS5-7750	CTTATCAATGCCGACACAAAGGA	~98	RT-PCR/ RT-qPCR	CP and NIb	Schubert et al., 2015
WS3-7895	GCTTCATGAATGTGTGTGACATGTA				
WS3-8910	GCATAATGGCTCGAAGTGATG				
as7855WSEUFam	TCGATGCTACGATACGCTTCATCAAATGAA		probe		
as7850WSUSAtto	CTACGGTAAGCCTCGTCAAATGAGTAA		probe		

8.10 RESISTANCE OF CEREALS TO WSMV

Efforts have been made to find the sources of resistance in cultivated wheat, which is the most effective way to prevent further transmission and reduce yield losses (Thomas et al., 2004; Richardson et al., 2014). WSMV resistance was first reported in perennial Triticeae relatives such as *Thinopyrum intermedium* (2n = 6x = 42, JJsS) and *Th. ponticum* (2n = 10x = 70, JJJJsJs) (Friebe et al., 1993; Harvey et al., 1999; Chen et al., 2003). So far, three resistance genes have been identified: *Wsm1*, *Wsm2*, and *Wsm3* (Friebe et al., 2009; Haley et al., 2002; Lu et al., 2012; Fahim

TABLE 8.3
Diversity of *Wheat Streak Mosaic Virus* Isolates

Isolate	Country of Origin	NCBI Accession No.	Type	Reference
El-Batán	Mexico	AF285170	A	Gadiou et al., 2009
CZ		AF454454	B	Gadiou et al., 2009
Margmagne	France	HG810953	B	Schubert et al., 2015
	Hungary	AF454456	B	Gadiouet al., 2009
Hoym	Germany	HG810954	B	Schubert et al., 2015
	Austria	LN624217	B	
	Russia	AF454459	B	Gadiou et al., 2009
	Poland	KP261825	B	
Naghadeh	Iran	EU914917	B	Schubert et al., 2015
Saadat_shahr	Iran	EU914918	D	
ATCC-PV57	USA	AF285169	D	Schubert et al., 2015
ID99		AF511619	D	French and Stenger, 2002
Sidney81		AF957533	D	Gadiou et al., 2009
WA94		FJ348358	D	Schubert et al., 2015
WA99		AF511643	D	
TX96		AF511642	D	French and Stenger, 2002
Burdett	Australia	DQ888801	D	Dwyer et al., 2007
Gibson		DQ888803	D	
Tamworth		AY327866	D	
Bordertown		AY327870	D	
Arg1	Argentina	FJ348356	D	French and Stenger, 2002
Arg2		FJ348359	D	
Arg3		FJ348357	D	

et al., 2012). These resistant genes were introduced into cultivated wheat lines. The resistance gene *Wsm1* is associated with chromosome 4D and has led to the release of the cultivar (cv.) such as Mace (Graybosch et al., 2009). The resistance gene *Wsm2* is associated with chromosome arm 3BS and is identified in germplasm line CO960293. The most widely used resistance gene, *Wsm2*, was successfully implemented in several varieties of wheat such as Snowmass (Haley et al., 2011), Ron L (Seifers et al., 2006), Clara CL (Martin et al., 2014) and Oakley CL (Zhang et al., 2015). However, both *Wsm1* and *Wsm2* are ineffective at higher growth temperature (Seifers et al., 2013). The third true resistance, C2652, named *Wsm3*, has recently been identified and is proven to be effective at higher growth temperatures as compared to *Wsm1* and *Wsm2* (Fahim et al., 2012). However, *Wsm3* is not yet available in any commercial wheat varieties (Richardson et al., 2014). All these varieties come from America. In Europe, there are no reports about the wheat varieties or any other cereal species which are potentially resistant to WSMV.

WCM resistance genes have been identified in grass, *Aegilo pstauschii* (Coss), *Thinopyrum ponticum*, and *Th. Intermedium* (Qi et al., 1979; Fedak and Han, 2005; Fahim et al., 2011). The grass genes intercrossed to hexaploid wheat, but only a few wheat varieties possess effective resistance against WCM, due to virulent WCM populations (Martin et al., 1976; Hakizimana et al., 2004; Murugan et al., 2011). However, WCM resistance remains a compelling approach to reduce losses due to WSMV. Distant hybrids between spring wheat and *Agropyrong laucum*, Zhong1, and Zhong2 showed effective resistance to both WSMV and its vector WCM (Qi et al., 1979; Han et al., 2003; Chen et al., 2003).

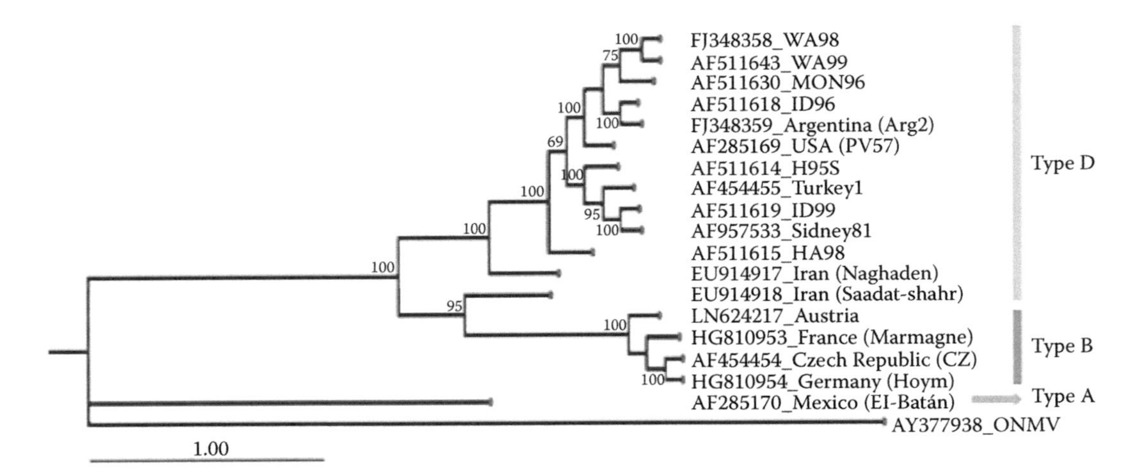

FIGURE 8.6 Various types of WSMV. WSMV isolate from Mexico (El-Batán) represents type A. The European isolates represent type B and include Asian isolate from Iran (Saadat-shahr). Isolates from the United States, Argentina, Turkey, and Australia represent type D. *Oat necrotic mottle virus* ONMV (AY377938_ONMV) is used as an outgroup. For generation of the tree, the whole genome nucleotide sequences were aligned using ClustalX2 (Larkin MA, Blackshields G, Brown NP et al. (2007). Clustal W and Clustal X version 2.0. Bioinformatics 23:2947–2948) and the tree was constructed using MEGA 7 (Kumar S, Stecher G, Tamura K (2016). MEGA7: Molecular Evolutionary Genetics Analysis version 7.0 for bigger datasets. *Molecular Biology and Evolution* 33:1870–1874) as described earlier (Singh K, Zouhar M, Mazakova J et al. (2014). Genome wide identification of the immunophilin gene family in *Leptosphaeria maculans*: A casual agent of blackleg disease in oilseed rape (*Brassica napus*). *Omics—A Journal of Integrative Biology* 18:645–657; Singh K, Winter M, Zouhar M, Rysanek P (2017). Cyclophilins: Less studied proteins with critical roles in pathogenesis. *Phytopathology* doi: 10.1094/PHYTO-05-17-0167-RVW). Tree viewed using GLC genomic workbench (QIAGEN, USA). Neighbor-joining method was used for construction of the tree and the reliability of the branches was inferred from a bootstrap analysis of 1000 replicates. Bootstrap values above 60% have been shown as numbers.

8.11 WSMV DISEASE MANAGEMENT

WSMV disease management should focus on the regulation of virus sources and reservoirs and control of abundancy of WCM vector in the agro-ecosystem. The risk factors of WSMV endemic includes character of the habitat, weather conditions, and incorrect agriculture practices which support the development of the population of vectors and reservoir of the virus in areas of intensive cultivation of cereals (mainly wheat) (Murray et al., 2005).

The factors influencing the occurrence of virus are summarized as follows (Wegulo et al., 2008):

- Stubble is an important source of WSMV infection and the WCM migrate to the volunteer cereals and grasses "green bridge" from where they spread the in newly emerging crop.
- Presence of grass weeds in cereal crop (mainly wheat) increases the risk of vector abundance; therefore adequate protection against weeds is crucial to regulate WSMV infection.
- Cultivation of spring cereals in the vicinity of the WSMV-infected winter cereal crops increases disease viability.
- The proximity of forests or wooded stripes creates better conditions for WCM.
- Warm weather during the summer provides better living conditions for WCM. Warm and dry weather in autumn creates ideal conditions for the reproduction and spreading of mites and helps in disease development in newly emerge cereals (wheat).
- Mild temperatures in early spring increase the activity of the mites, and along with windy weather increase the spread of the WCM even in long distances; hence an increase in WSMV transmission in winter and spring cereals.

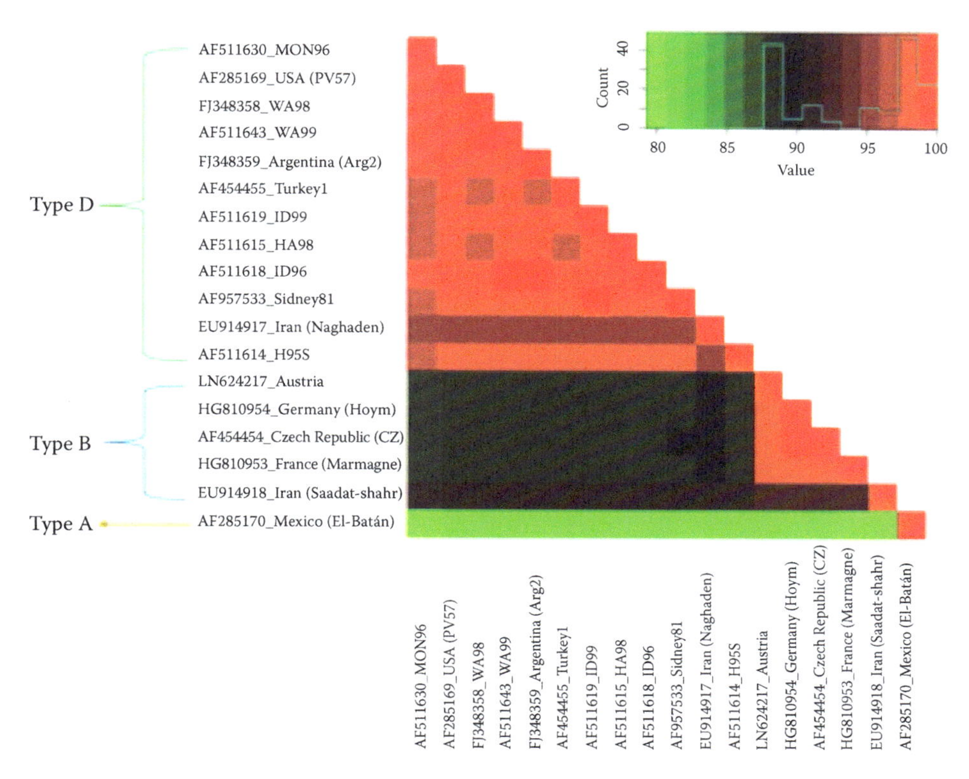

FIGURE 8.7 Identity matrix between various type pf WSMV. WSMV isolate from Mexico (El-Batán) represents type A. The European isolates represent type B and include Asian isolate from Iran (Saadat-shahr). Isolates from the United States, Argentina, Turkey, and Australia represent type D. Heatmap was drawn using ggplots of R software using R script (Singh K, Zouhar M, Mazakova J et al. (2014). Genome wide identification of the immunophilin gene family in *Leptosphaeria maculans*: A casual agent of blackleg disease in oilseed rape (*Brassica napus*). *Omics—A Journal of Integrative Biology* 18:645–657). Color key represents the scale distribution of the matrix.

Protection of crops against WSMV depends on good agronomical practices that minimize the activity of vectors that transmit WSMV and destroying infected host plants that serve as the source of the virus. Seed and foliar insecticide (acaracide) application are not effective for controlling the WSMV vector, WCM, and prevention of virus spread. The control strategies for managing the virus are described below (Watkins et al., 1989; Wegulo et al., 2008).

- Destroy volunteer wheat and grasses in adjoining fields 10 days prior to planting, and 2–4 weeks before sowing in the field to be seeded. Such destruction of volunteer wheat (cereals) and grasses through tillage or herbicide spray has been shown to minimize the size of migrating mite population from those plants to crops.
- Avoid early planting of winter wheat in areas with high infection pressure of viruses to escape migrations of the mite from volunteer wheat or grasses to crops.
- Control grass weeds in cereals, which are hosts for the virus and mites.
- Plant a variety with resistance to the virus or vector.
- Limit minimization technologies in corn growing in areas with a greater incidence of viruses.
- Control grasses and grassy weeds around the borders of recently sown crops.
- Break the disease cycle by controlling summer cereal volunteers (the "green bridge") to reduce the number of mites that can invade the following autumn sown crop.

ACKNOWLEDGMENT

This work was supported by the grants RO0415 and QJ1230159 from the Ministry of Agriculture, Czech Republic.

REFERENCES

Atkinson TG, Grant MN (1967). An evaluation of streak mosaic losses in winter wheat. *Phytopathology* 57:188–192.

Ay Z, Kerenyi Z, Takacs A et al. (2008). Detection of cereal viruses in wheat (*Triticum aestivum* L.) by serological and molecular methods. *Cereal Research Communications* 36:215–224.

Brakke MK (1971). *Wheat streak mosaic virus. CMI/AAB Descriptions of Plant Viruses*. No. 48. Association of Applied Biologists, Wellesbourne, UK.

Byamukama E, Tatineni S, Hein G et al. (2016). Incidence of *Wheat streak mosaic virus*, *Triticum mosaic virus*, and *Wheat mosaic virus* in wheat curl mites recovered from maturing winter wheat spikes. *Plant Disease* 100:318–323.

Chalupníková J, Kundu JK, Singh K et al. (2017). *Wheat streak mosaic virus*: Incidence in field crops, potential reservoir within grass species and uptake in winter wheat cultivars. *Journal of Integrative Agriculture* 16:60345–60357.

Chen Q, Conner RL, Li HJ et al. (2003). Molecular cytogenetic discrimination and reaction to *Wheat streak mosaic virus* and the wheat curl mite in Zhong series of wheat–*Thinopyrum intermedium* partial amphiploids. *Genome* 46:135–145.

Choi IR, Horken KM, Stenger DC et al. (2002). Mapping of the P1 proteinase cleavage site in the polyprotein of *Wheat streak mosaic virus* (genus *Tritimovirus*). *Journal of General Virology* 83:443–450.

Christian ML, Willis WG (1993). Survival of *wheat streak mosaic virus* in grass hosts in Kansas from what harvest to fall wheat emergence. *Plant Disease* 77:239–242.

Connin RV (1956). The host range of the wheat curl mite, vector of *wheat streak mosaic*. *Journal of Economic Entomology* 48:1–4.

Coutts BA, Hammond NEB, Kehoe MA et al. (2008a). Finding *Wheat streak mosaic virus* in southwest Australia. *Australian Journal of Agricultural Research* 59:836–843.

Coutts BA, Strickland GR, Kehoe MA et al. (2008b). The epidemiology of *Wheat streak mosaic virus* in Australia: Case histories, gradients, mite vectors, and alternative hosts. *Australian Journal of Agricultural Research* 59:844–853.

Coutts BA, Kehoe MA, Webster CG et al. (2011). *Zucchini yellow mosaic virus*: Biological properties, detection procedures and comparison of coat protein gene sequences. *Archives of Virology* 156:2119–2131.

Coutts BA, Banovic M, Kehoe MA et al. (2014). Epidemiology of *Wheat streak mosaic virus* in wheat in a Mediterranean-type environment. *European Journal of Plant Pathology* 140:797–813.

Deb M, Anderson JM (2008). Development of a multiplexed PCR detection method for Barley and *Cereal yellow dwarf viruses*, *Wheat spindle streak virus*, *Wheat streak mosaic virus* and Soil-borne wheat mosaic virus. *Journal of Virological Methods* 148:17–24.

Dráb T, Svobodová E, Ripl J et al. (2014). SYBR Green I based RT-qPCR assays for the detection of RNA viruses of cereals and grasses. *Crop and Pasture Science* 65:1323–1328.

Dwyer GI, Gibbs MJ, Gibbs AJ et al. (2007). *Wheat streak mosaic virus* in Australia: Relationship to isolates from the Pacific Northwest of the USA and its dispersion via seed transmission. *Plant Disease* 91:164–170.

Ellis MH, Rebetzke GJ, Kelman WM et al. (2004). Detection of *Wheat streak mosaic virus* in four pasture grass species in Australia. *Plant Pathology* 53:239.

Fahim M, Mechanicos A, Ayala-Navarrete L et al. (2011). Resistance to *Wheat streak mosaic virus*–A survey of resources and development of molecular markers. *Plant Pathology* 61:425–440.

Fahim M, Larkin PJ, Haber S et al. (2012). Effectiveness of three potential sources of resistance in wheat against *Wheat streak mosaic virus* under field conditions. *Australas Plant Pathology* 41:301–309.

Fedak G, Han F (2005). Characterization of derivatives from wheat-Thinopyrum wide crosses. *Cytogenetic and Genome Research* 109:360–367.

French R, Robertson NL (1994). Simplified sample preparation for detection of *Wheat streak mosaic virus* and *Barley yellow dwarf virus* by PCR. *Journal of Virological Methods* 49:93–99.

French R, Stenger DC (2002). *Wheat streak mosaic virus. CMI/AAB Descriptions of Plant Viruses*. 398. Assoc. Appl. Biol., Wellesbourne, UK. http://www.dpvweb.net/dpv/showdpv.php?dpvno=393 (accessed February 2, 2017).

French R, Stenger DC (2003). Evolution of *wheat streak mosaic virus*: Dynamics of population growth within plants may explain limited variation. *Annual Review of Phytopathology* 41:199–214.

Friebe B, Jiang J, Gill BS et al. (1993). Radiation-induced nonhomologous wheat *Agropyron intermedium* chromosomal translocations conferring resistance to leaf rust. *Theoretical and Applied Genetics* 86:141–149.

Friebe B, Qi LL, Wilson DL et al. (2009). Wheat-*Thinopyrum intermedium* recombinants resistant to *Wheat streak mosaic virus* and *Triticum mosaic virus*. *Crop Science* 49:1221–1226.

Gadiou S, Kúdela O, Ripl J et al. (2009). An amino acid deletion in *Wheat streak mosaic virus* capsid protein distinguishes a homogeneous group of European isolates and facilitates their specific detection. *Plant Disease* 93:1209–1213.

Graybosch RA, Peterson CJ, Baenziger PS et al. (2009). Registration of "Mace" hard red winter wheat. *Journal of Plant Registrations* 3:51–56.

Hadi BAR, Langham MAC, Osborne L et al. (2011). *Wheat streak mosaic virus* on wheat: Biology and management. *Journal of Integrated Pest Management* 2:1–5.

Hakizimana F, Ibrahim AMH, Langham MAC et al. (2004). Generation means analysis of *wheat streak mosaic virus* resistance in winter wheat. *Euphytica* 139:133–139.

Haley SD, Martin TJ, Quick JS et al. (2002). Registration of CO960293-2 wheat germplasm resistant to *wheat streak mosaic virus* and Russian wheat aphid. *Crop Science* 42:1381–1382.

Haley SD, Johnson JJ, Peairs FB et al. (2011). Registration of "Snowmass" wheat. *Journal of Plant Registrations* 5:87–90.

Hall JS, French R, Hein GL et al. (2001). Three distinct mechanisms facilitate genetic isolation of sympatric *wheat streak mosaic virus* lineages. *Virology* 282:230–236.

Han FP, Fedak G, Benabdelmouna A et al. (2003). Characterization of six wheat × *Thinopyrum intermedium* derivatives by GISH, RFLP, and multicolor GISH. *Genome* 46:490–495.

Hansing ED, Melcher LE, Fellows H et al. (1950). Kansas Phytopathological notes: 1949. *Transaction of Kansas Academy of Science* 53:344–354.

Harvey TL, Seifers DL, Martin TJ et al. (1999). Survival of wheat curl mites on different sources of resistance in wheat. *Crop Science* 39:1887–1889.

Hill JH, Martinson CA, Russell WA (1974). Seed transmission of maize dwarf mosaic and *Wheat streak mosaic viruses* in maize and response of inbred lines. *Crop Science* 14:232–235.

Hunger RM (2004). *Wheat streak mosaic virus* prevalent in Western Oklahoma and the Panhandle. *Plant Disease and Insect Advisory* 3(7). http://entoplp.okstate.edu/pddl/2004/PDIA3-7.pdf/view (accessed October 20, 2017).

Hunger RM (2010). *Wheat streak mosaic virus*. In WW Bockus, RL Bowden, RM Hunger, WL Morrill, TD Murray, RW Smiley (Eds.) *Compendium of Wheat Diseases and Pests*, 3rd edition. American Phytopathological Society Press, St. Paul, MN.

Hunger RM, Sherwood JL, Evans CK et al. (1992). Effects of planting date and inoculation date on severity of *wheat streak mosaic* in hard red winter wheat cultivars. *Plant Disease* 76:1056–1060.

Ito D, Miller Z, Menalled F et al. (2012). Relative susceptibility among alternative host species prevalent in the Great Plains to *Wheat streak mosaic virus*. *Plant Disease* 96:1185–1192.

Izzo MM, Kirkland PD, Gu X et al. (2012). Comparison of three diagnostic techniques for detection of *Rotavirus* and *Coronavirus* in calf faeces in Australia. *Australian Veterinary Journal* 90:122–129.

Jones RAC (2000). Determining "threshold" levels for seed-borne virus infection in seed stocks. *Virus Research* 71:171–183.

Jones RAC, Coutts BA, Mackie AE et al. (2005). Seed transmission of *Wheat streak mosaic virus* shown unequivocally in wheat. *Plant Disease* 89:1048–1050.

Khadivar RS, Nasrolahnejad S (2009). Serological and molecular detection of *Wheat streak mosaic virus* (WSMV) in cereal fields of Golestan province, Northern Iran. *Journal of Plant Production* 16:4.

Kudela O, Kudelova M, Novakova S et al. (2008). First report of *Wheat streak mosaic virus* in Slovakia. *Plant Disease* 92:1365.

Kumar S, Stecher G, Tamura K (2016). MEGA7: Molecular Evolutionary Genetics Analysis version 7.0 for bigger datasets. *Molecular Biology and Evolution* 33:1870–1874.

Langham M, Glover K (2005). Effects of *Wheat streak mosaic virus* (genus: *Tritimovirus*; family: *Potyviridae*) on spring wheat. *Phytopathology* 95:S56.

Lanoiselet VM, Hind-Lanoiselet T, Land Murray GM (2008). Studies on the seed transmission of *Wheat streak mosaic virus*. *Australasian Plant Pathology* 37:584–588.

Larkin MA, Blackshields G, Brown NP et al. (2007). Clustal W and Clustal X version 2.0. *Bioinformatics* 23:2947–2948.

Lu H, Kottke R, Devkota R et al. (2012). Consensus mapping and identification of markers for marker-assisted selection of Wsm2 in wheat. *Crop Science* 52:720–728.

Mar TB, Lau D, Schons J et al. (2013). Identification and characterization of *Wheat streak mosaic virus* isolates in wheat-growing areas in Brazil. *International Journal of Agronomy* (Article ID 983414).

Martin TJ, Harvey TL, Livers RW (1976). Resistance to *Wheat streak mosaic virus* and its vector *Aceriatulipae*. *Phytopathology* 66:346–349.

Martin TJ, Zhang G, Fritz AK et al. (2014). Registration of "Clara CL" Wheat. *Journal of Plant Registrations* 8:38–42.

McKinney HH (1937). *Mosaic diseases of wheat and related cereals*. US Department of Agriculture Circular No 442:1–23. https://archive.org/details/mosaicdiseasesof442mcki (accessed January 15, 2017).

Murray GM, Knihinicki DK, Wratten K et al. (2005). *Wheat Streak Mosaic and the Wheat Curl Mite*. NSW Department of Primary Industries, Orange NSW Australia. Primefact 99. https://www.ag.ndsu.edu/pubs/plantsci/smgrains/pp646.pdf (accessed December 12, 2016).

Murugan M, Sotelo Cardona P, Duraimurugan P et al. (2011). Wheat curl mite resistance: Interactions of mite feeding with *Wheat streak mosaic virus* infection. *Journal of Economical Entomology* 104(4):1406–1414.

Navia D, de Mendonca RS, Skoracka A et al. (2013). Wheat curl mite, *Aceriatosichella*, and transmitted viruses: An expanding pest complex affecting cereal crops. *Experimental and Applied Acarology* 59:95–143.

Oana D, Ziegler A, Torrance L et al. (2009). *Potyviridae Family-short review*. http://www.usab-tm.ro/Journal-HFB/romana/Lucrari_2009_paginate/94.pdf (accessed January 5, 2017).

Orlob GB (1966). Feeding and transmission characteristics of *Aceria tulipae* Keifer as vector of wheat streak mosaic virus. *Journal of Phytopathology* 55:218–238.

Price J, Workneh F, Evett S et al. (2010a). Effects of *Wheat streak mosaic virus* on root development and water-use efficiency of hard red winter wheat. *Plant Disease* 94:766–770.

Price JA, Smith J, Simmons A et al. (2010b). Multiplex real-time RT-PCR for detection of *Wheat streak mosaic virus* and *Triticum mosaic virus*. *Journal of Virological Methods* 165:198–201.

Qi SY, Yu S, Zhang XYH et al. (1979). Studies on distant hybridization between spring wheat and *Agropyron glaucum*. *Scientia Agricultura Sinica* 2:1–11.

Rabenstein F, Seifers DL, Schubert J et al. (2002). Phylogenetic relationships, strain diversity and biogeography of *Tritimoviruses*. *Journal of General Virology* 83:895–906.

Richardson K, Miller AD, Hoffmann AA et al. (2014). Potential new sources of wheat curl mite resistence in wheat to prevent the spread of yield-reducing pathogens. *Experimental and Applied Acarology* 64:1–19.

Robinson MD, Murray TD (2013). Genetic variation of *Wheat streak mosaic virus* in the United States Pacific Northwest. *Phytopathology* 103:98–104.

Schubert J, Ziegler A, Rabenstein F (2015). First detection of *wheat streak mosaic virus* in Germany: Molecular and biological characteristics. *Archives of Virology* 160:1761–1766.

Seifers DL, Harvey TL, Kofoid KD et al. (1996). Natural infection of pearl millet and sorghum by *Wheat streak mosaic virus* in Kansas. *Plant Disease* 80:179–185.

Seifers DL, Martin TJ, Harvey TL et al. (2006). Temperature sensitive and efficacy of *Wheat streak mosaic virus* resistance derived from CO960293 wheat. *Plant Disease* 90:623–628.

Seifers DL, Martin TJ, Fellers JP (2010). An experimental host range for *Triticum mosaic virus*. *Plant Disease* 94:1125–1131.

Seifers DL, Haber S, Martin TJ et al. (2013). New sources of temperature sensitive resistance to *Wheat streak mosaic virus* in wheat. *Plant Disease* 97:1051–1056.

Shahwan IM, Hill JP (1984). Identification and occurrence of Wheat streak mosaic virus in winter wheat in Colorado and its effects on several wheat cultivars. *Plant Disease* 68:579–581.

Sill Jr. WH, Connin RV (1953). Summary of the known host range of Wheat streak mosaic virus. *Transactions of the Kansas Academy of Science* 56:411–417.

Sill Jr. WH, Agusiobo PC (1955). Host range studies of the *Wheat streak mosaic virus*. *Plant Disease Reporter* 39:633–642.

Singh K, Zouhar M, Mazakova J et al. (2014). Genome wide identification of the immunophilin gene family in *Leptosphaeria maculans*: A casual agent of blackleg disease in oilseed rape (*Brassica napus*). *Omics—A Journal of Integrative Biology* 18:645–657.

Singh K, Winter M, Zouhar M, Rysanek P (2017). Cyclophilins: Less studied proteins with critical roles in pathogenesis. *Phytopathology* doi: 10.1094/PHYTO-05-17-0167-RVW.

Skoracka A, Rector B, Kuczyński L et al. (2014). Global spread of wheat curl mite by its most polyphagous and pestiferous lineages. *Annals of Applied Biology* 165:222–235.

Slykhuis JT (1955). *Aceria tulipae* Keifer (Acarina: Eriophyidae) in relation to the spread of *Wheat streak mosaic*. *Phytopathology* 45:116–128.

Slykhuis JT, Andrews JE, Pittman UJ (1957). Relation of date of seeding winter wheat in southern Alberta to losses from *wheat streak mosaic*, root rot, and rust. *Canadian Journal of Plant Science* 37:113–127.

Somsen HW, Sill WH (1970). *The wheat curl mite, AceriatulipaeKeifer, in relation to epidemiology and control of wheat streak mosaic*. https://www.ksre.k-state.edu/historicpublications/pubs/STB162.pdf (accessed January 20, 2017).

Stenger DC, French R (2009). *Wheat streak mosaic virus* genotypes introduced to Argentina are closely related to isolates from the American Pacific Northwest and Australia. *Archives of Virology* 154:331–336.

Stenger DC, Hall JS, Choi I et al. (1998). Phylogenetic relationships within the family *Potyviridae: Wheat streak mosaic virus* and *Brome streak mosaic virus* are not members of the genus Rymovirus. *Phytopathology* 88:782–787.

Tatineni S, Graybosch RA, Hein GL et al. (2010). Wheat cultivar-specific disease synergism and alteration of virus accumulation during co-infection with *Wheat streak mosaic virus* and *Triticum mosaic virus*. *Phytopathology* 100:230–238.

Thomas JA, Hein GL (2003). Influence of volunteer wheat plant condition on movement of the wheat curl mite, *Aceria tosichella*, in winter wheat. *Experimental and Applied Acarology* 31:253–268.

Thomas JB, Conner RL, Graf RJ (2004). Comparison of different sources of vector resistance for controlling *Wheat streak mosaic* in winter wheat. *Crop Science* 44:125–130.

Truol G, Sagadin M, Rodriguez M (2010). Fox tail millet (*Setaria italica* L.): A new reservoir species of the *Wheat streak mosaic virus* (WSMV) in the province of Buenos Aires. *Biocell* 34:A135.

Vacke J, Zacha V, Jokeš M (1986). Identification of virus in wheat new to Czechoslovakia. In: *Proceedings X Czechoslovak Plant Protection*. Conference, September 2–5, pp. 209–210. Brno, Czechoslovakia.

Velandia M, Rejesus RM, Jones DC et al. (2010). Economic impact of *Wheat streak mosaic virus* in the Texas High Plains. *Crop Protection* 29:699–703.

Vincent SJ, Coutts BA, Jones RAC (2014). Effects of introduced and indigenous viruses on native plants: Exploring their disease-causing potential at the agro-ecological interface. *PLoS ONE* 9:e91224.

Watkins JE, French R, Campbell JB et al. (1989). *Wheat Streak Mosaic Virus*. Nebraska Cooperative Extension EC89–1871. http://digitalcommons.unl.edu/cgi/viewcontent.cgi?article=2232&context=extensionhist (accessed February 3, 2017).

Wegulo SN, Hein GL, Klein RN et al. (2008). *Managing Wheat Streak Mosaic*. University of Nebraska, Lincoln (Extension EC1871). http://extensionpublications.unl.edu/assets/pdf/ec1871.pdf (accessed January 20, 2017).

Zhang G, Martin TJ, Fritz AK et al. (2015). Registration of "Oakley CL" Wheat. *Journal of Plant Registrations* 9:190–195.

9 Population Structure and Diversity of *Banana Bunchy Top Virus* and *Banana Bract Mosaic Virus*

R. Selvarajan, V. Balasubramanian, and C. Anuradha

CONTENTS

9.1 INTRODUCTION

Bananas and plantains are ranked fourth, after wheat, rice, and maize, in importance as a food crop in the world (Kumar et al. 2011; Perrier et al. 2011) and are grown in over 130 countries, producing 150 million tonnes from an area of 6 million ha. The fruit is consumed both as a staple food and as fruit, and serves as a source of livelihood for more than 100 million in Africa. Bananas are produced by parthenocarpic, mainly triploid hybrids derived from *Musa acuminata* alone and sometimes combined with *M. balbisiana*. Present-day cultivated bananas originated from New Guinea, Indonesia, the Philippines, or the Southeast Asia Peninsula (Perrier et al. 2011) between 7000 and 10,000 years ago (Denham et al. 2003). It is assumed that different banana varieties might have been introduced and reintroduced to Africa and the southwest Indian Ocean Islands many times during the period between 1500 and 700 years ago (Lejju et al. 2006; Randrianja and Ellis 2009). The viruses, the obligate pathogens infecting bananas, are believed to have spread with the domestication process of edible bananas for human consumption. About 20 plant virus species have been reported in banana; the major ones inflicting serious crop damage are *Banana bunchy top virus* (BBTV), various *Banana streak virus* species (BSVs), *Banana bract mosaic virus* (BBrMV), and *Cucumber mosaic virus* (CMV). In tissue culture banana, BBTV is a major threat and caused a loss of \$50 million in India in 2009–2011 (Kumar et al. 2015). BBTV has been identified as one of the 100 worst alien invasive species by the International Union for Conservation of Nature. Globally, this disease is a major production constraint, and fortunately it is not found in Central and Latin American countries. Another emerging virus is BBrMV, belonging to the genus *Potyvirus* and the family *Potyviridae*.

This virus causes a yield loss of 40% in the Philippines (Thomas and Magnaye 1996) and 40–70% loss in India (Cherian et al. 2002; Selvarajan and Jeyabaskaran 2006).

It is well known that in all living entities, reproduction may result in the generation of individuals that differ genetically from their parents. Such variants are called mutants. A rapid multiplication rate of viruses in living infected cells leads to genetic changes in their genome, so the plant virus populations are always genetically heterogeneous in nature. However, various selection pressures act upon the virus population which may lead to positive (useful changes will be retained by the virus), negative (deleterious mutations will be often removed), and neutral selection in the virus. The frequency distribution of genetic variants in the population of an organism may change with time, and this process is called evolution; population of a variant is called genetic structure of the pathogen. Most plant viruses have RNA as their genome, and some are DNA viruses which cause major diseases in crop plants, leading to enormous losses. Plant viruses replicate in a short time in a cell and tissues, where they live parasitically and are known to evolve very quickly owing to their short generation times and high replication rates. RNA viruses often get mutated while they replicate in the host cell due to error-prone RNA polymerases. Mutations, reassortment, and recombination are the main mechanisms that cause rapid genetic changes and evolution in plant viruses (Garcia-Arenal et al. 2001). There is evidence that the multipartite single-stranded DNA (ssDNA) viruses are more prone to genetic changes in the population, and the mechanisms of such changes are reported to be reassortment and recombination rather than mutations. There might be additional reasons for the evolution of segmented genomes, but the consequence is a potential gain which leads to increased genetic diversity in virus population. One well-recognized outcome of this potential genomic reassortment mechanism among different isolates of a virus is *Influenza A virus* (Nelson et al. 2008), which generates new "combinations" of segments in a genome. Genome segmentation also sets up a stage where different genomic segments may potentially undergo intermolecular recombination leading to increased genetic diversity by the phenomenon of intragenomic recombination.

It is necessary to study the population structure of a pathogen to develop strategies for managing the diseases caused by those pathogens. Host resistance to pathogens is often lost during evolution of new variants that emerge frequently due to various factors including climate, host, and environmental influence on the pathogens. With the advent of fast sequencing of complete genomes and next-generation sequencing systems and the data analysis of population of any pathogen, it has become easier to understand the population structure of pathogens at any point of time.

9.2 *BANANA BUNCHY TOP VIRUS*

Banana bunchy top disease (BBTD), caused by banana bunchy top virus, is considered as one of the most serious viral diseases of banana. BBTV is one of the members of the genus *Babuvirus* of the newly described plant virus family, the *Nanoviridae* (Vetten et al. 2005). The virus infects only banana, plantains, ensete, and abaca. It is known to be phloem-limited, and is transmitted persistently by the banana black aphid *Pentalonia nigronervosa* Coquerel (Bell et al. 2002). Occurrence of BBTD in tissue culture plantations in 2007–2011 in Jalgaon, Maharashtra, India and Kodur, Andhra Pradesh caused an annual loss of production worth US$50 million (Selvarajan and Balasubramanian 2014). Cook et al. (2012) simulated the benefit of the exclusion of BBTV and reported that an amount of AUS$ 15.9–27 million is being saved every year in Australia by excluding the virus through a surveillance and eradication program. Substantial production losses ranging from 5% to 70% due to BBTV in the regions across Sub-Saharan Africa (SSA) have been reported (Kumar et al. 2011; Niyongere et al. 2013; Boloy et al. 2014; Mukwa et al. 2014). The genome of BBTV is multisegmented and consists of six ssDNA components, each packed individually in an icosahedral particle (Figure 9.1).

The components are DNA-R, -U, -S, -M, -C, and -N, and they encode for the master replication initiation protein (R), unknown protein with no function assigned (U), capsid protein (S), movement protein (M), cell cycle link protein (Clink, C), and nuclear-shuttle protein (N), respectively (Hafner et al. 1997a; Aronson et al. 2000; Wanitchakorn et al. 2000). Distinct from DNA-R, some additional

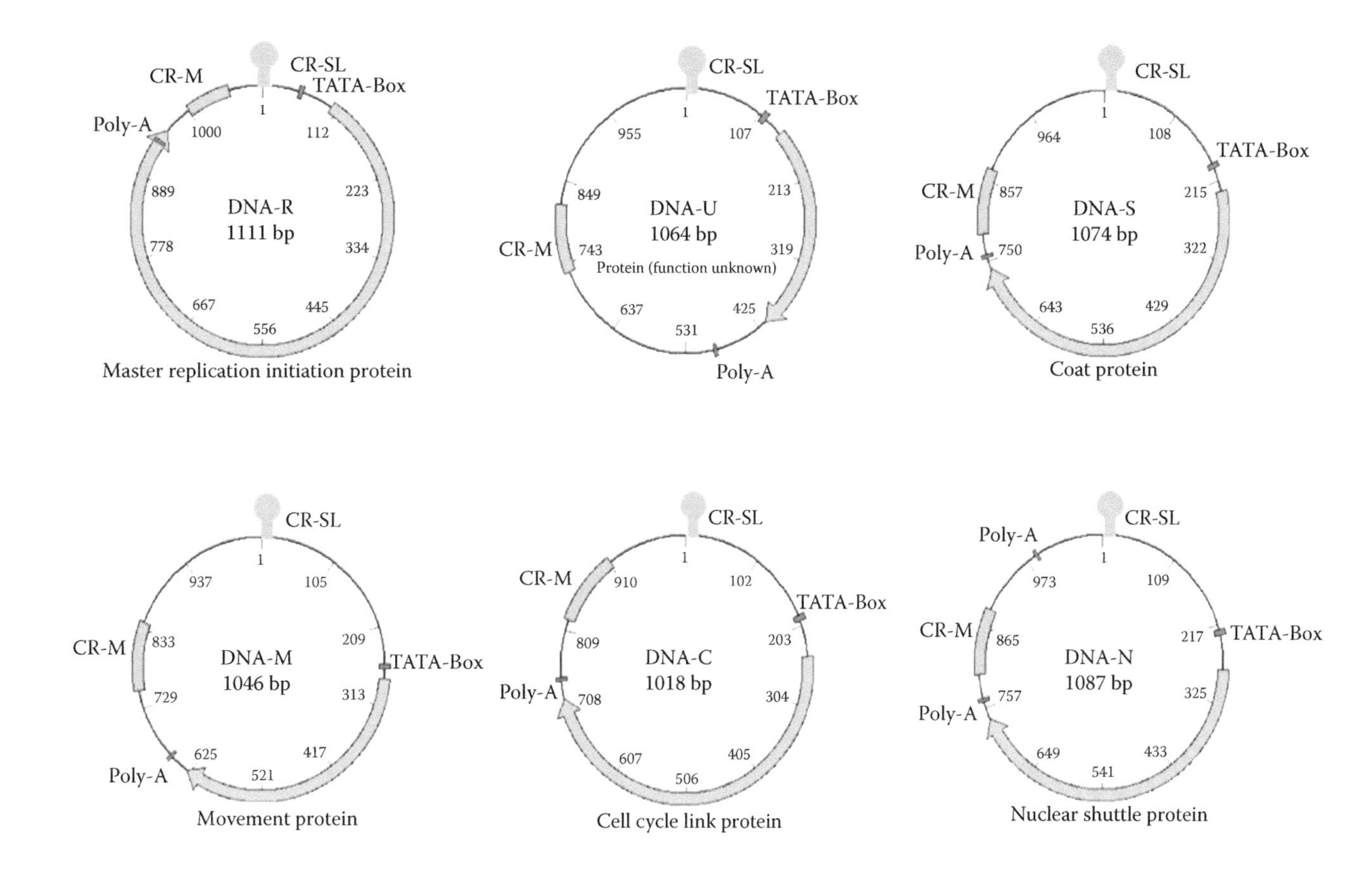

FIGURE 9.1 Genome organization of BBTV. Six ss circular DNA components of BBTV; DNA-R, codes for master replication initiation protein; DNA-U, function of this protein coded by this component is not known; DNA-S, codes for coat protein; DNA-M, codes for movement protein; DNA-C, codes for cell cycle linking protein; DNA-N, codes for nuclear shuttle protein; CR-M, major common region; CR-SL, common region-stem–loop; Poly-A, Polyadynylation signal sequence; TATA-Box, transcription initiating site.

replication-associated protein (Rep) encoding components are also associated with some isolates of babuvirus, but these Rep proteins only replicate their own cognate DNAs (Timchenko et al. 1999). All six BBTV DNA components have two highly conserved regions that contain regulatory elements. These are the stem-loop common region (CR-SL), containing the origin of replication for rolling-circle replication, and the major common region (CR-M), to which small ssDNA primers are bound to initiate complementary-strand DNA synthesis after entering a host cell (Hafner et al. 1997b). The genes, or at least some parts of genetic components of BBTV, are evolving through diverse evolutionary processes. Their Rep encoding components constitute a multigene family believed to follow birth-and-death type evolution, where new components have evolved by the process of duplication and some have been lost during evolution (Hughes 2004). Since this virus is multicomponent and its genome and has common regions in all the components, frequent recombination and reassortment of components have recently been reported. In this chapter, the diversity and population structures of BBTV is reviewed in detail.

9.3 GENETIC DIVERSITY OF BBTV

Though BBTV is a ssDNA virus, the basic molecular mechanisms of evolution including mutation, recombination, and reassortment are qualitatively similar to those of other plant viruses with positive-sense RNA (Hu et al. 2007; Garcia-Arenal et al. 2001). The initial phylogenetic analysis of BBTV performed by Karan et al. (1994) based on the sequence of DNA-R component has grouped BBTV isolates into two distinct lineages, namely the South Pacific group (SPG) and the Asian group (AG). The SPG consists of isolates from Australia, India, Pakistan, Egypt, Hawaii, Sri Lanka, and all SSA countries, where recent reports were recorded, and the AG consists of isolates from Taiwan, the Philippines, China, Indonesia, and Vietnam. Two whole-genome sequencing and analysis of BBTV in India has confirmed that they belong to the SPG (Vishnoi et al. 2009; Selvarajan et al. 2010). The monophyletic population of BBTV is proposed to have originated from a recent single introduction in Pakistan with expected low level of sequence divergence (Amin et al. 2008; Yasmin et al. 2005). Hyder et al. (2011) reported that among the BBTV components, the DNA-U3 appeared to be the most diverse component. A detailed inspection of stem-loop sequences of all the genomic components reported for BBTV in GenBank indicated the most frequent sequence type (73%) which is present in DNA-R, -S, -C, and -M. However, none of DNA-U3 share most frequent sequence type except for the Pakistan isolate KHI. Sequence analysis indicates that deletions and point mutations are also present in some stem-loop sequences and most of the mutations occur in the stem region and are non-complementary, resulting in imperfect stem structures (Hyder et al. 2011).

Phylogenetic network analysis of coding sequences of DNA-R and DNA-S reconfirmed two major groups of BBTV but showed web-like phylogenies for some genes (Hu et al. 2007). Kumar et al. (2011) reported that after the first record of BBTV in the Democratic Republic of Congo (DRC) it has spread to 11 countries in SSA, and based on the sequence analysis of DNA-S and DNA-R, all these isolates belong to the South Pacific phylogenetic group, which is comprised of isolates of Australia, India, South Asia, and South Pacific countries. Banerjee et al. (2014) reported a distinct BBTV isolate from Umiam, Meghalaya, India; however, this isolate also belonged to the Pacific Indian Ocean (PIO) group. Twenty nucleotide intergenic regions in DNA-R and ORF in DNA-U3 were missing in this isolate, which was the highlight for its distinction. The DNA-R and -S of BBTV isolates of the Great Lakes region of Africa spanning Burundi, the DRC, and Rwanda have been sequenced and analyzed by Niyongere et al. (2015), and these isolates were shown to belong to the SPG along with Indian isolates. Chiaki et al. (2015) reported that BBTV population of Sumatra Island in Indonesia belonged to AG rather than SPG, and analysis of DNA-R and DNA-S of 61 isolates and DNA-U3 of 37 isolates revealed there were no recombination and very low diversity compared to other areas. Complete genome of BBTV of the Kandy region of Sri Lanka has been reported and it belongs to the PIO group (previously called SPG) (Wikaramaarachchi et al. 2016), and it had closest similarity to a Tamil Nadu isolate than North and Northeast Indian isolates, which supports the hypothesis of introduction of BBTV to India from Sri Lanka in the 1940s. Mukwa et al. (2016) surveyed in the DRC and reported very low genetic variations among the 52 isolates of BBTV collected from different parts of the DRC. Based on the coding region of DNA-R, they have estimated the haplotype diversity to be 0.944 ± 0013 with 30 haplotypes from 68 isolates in DRC, which shows the possibility of a single BBTV introduction followed by its spread to different regions.

9.4 RECOMBINATION IN BBTV

A review of the literature on the recombination events reported shows that genetic recombinations between different BBTV isolates (intergenomic) as well as among the genomic components of same isolate (intragenomic) are playing a major role in the evolution of BBTV. Intragenomic recombination is unique and rare among multicomponent ssDNA viruses and is a novel way to generate genetic diversity. Different methods used to find the recombinants were RDP, GENCONV, BOOTSCAN, MAXCHI, CHIMERA, SISCAN, and 3SEQ (Martin et al. 2015). The number of intragenomic recombinant sequences for each component of BBTV is furnished in Table 9.1. The highest number of recombinations is observed in DNA-U3 (Hyder et al. 2011; Stainton et al. 2015), followed by DNA-N, -R, and -S. Most of the recombination occurred in the conserved region, CR-SL. All these events were detected by at least three methods with significant p-value. The number of major parents contributing to this intragenomic recombination is higher than the minor parents, but in the case of intergenomic recombination the number of minor parents contributing is quite high (Table 9.2).

The case of intercomponent recombination in BBTV has been reported by Stainton et al. (2012). They have shown more frequent recombinations in the components M (53), S (50), and U3 (37), followed by C (12). The breakpoints of recombination fell mostly in the conserved CR-M and CR-SL regions. We have detected recombinations within Indian isolates by at least three methods with high confidence. The list of recombinant isolates is furnished in Table 9.3. Most recombinants were shown to be from isolates from north and northeastern hill regions of India, which showed that the selection pressure might be due to the wild varieties and species of banana. Recombinants detected in the DNA-S and -N components of a Tamil Nadu isolate were contributed by the same major parent from Bihar isolate and a minor parent from Umiam isolate, and the breakpoints were from the CR-SL region.

Recombination in different strains and among different species is common among geminiviruses (Padidam et al. 1999; Garcia-Arenal et al. 2001; Lefeuvre et al. 2009) but its level in nanoviruses is significantly low (Lefeuvre et al. 2009). Several studies have identified potential recombination events in BBTV (Fu et al. 2009; Islam et al. 2010; Hyder et al. 2011; Stainton et al. 2012; Banerjee et al. 2014; Stainton et al. 2015). BBTV isolates of Pakistan were subjected to sequencing, and sequence analysis revealed intergenomic recombination in DNA-U3 among the isolates of two subgroups and a very rare intragenomic recombination in the BBTV population of Pakistan (Hyder et al. 2011). Wang et al. (2013) have developed a motif-based tool for studying the evolution of multipartite ssDNA viruses including BBTV. They have shown several unusual recombination events to the evolution of BBTV and concluded translocation of short conserved sequences in all of the BBTV genomic components. Stainton et al. (2012) used alignment-based methods, which easily identify recombination events in the conserved regions, whereas Wang et al. (2013) detected that recombination outside the conserved region, for example a missing motif in CR-SL of DNA-U3, which has been relocated to a different position may not be detectable by alignment-based methods. In addition, it is stated that the evolutionary processes including intergenomic and intragenomic recombination among the genomic components of the same isolate may also have a significant contribution in the evolution of BBTV genome. Intragenomic recombination therefore appears to be a unique way to generate genetic diversity in multicomponent ssDNA viruses like BBTV (Hyder 2009; Hyder et al. 2011). Evidence of recombination was found in five isolates of BBTV component

TABLE 9.1
Intracomponent Recombination Sequences Detected in DNA Components of BBTV Isolates

DNA Component	Number of Recombinant Sequences	Recombination Breakpoints	Number of Major Parents	Number of Minor Parents	*p* Value (Method of Detection)	Number of Methods Detected the Recombinants	Reference
DNA-R	03	325–732	37	Unknown	1.70×10^{-05}(M)	3	Stainton et al. 2015
	01	1102–46	45	Unknown	8.04×10^{-05} (R)	4	
	02	1088–378	03	02	3.68×10^{-04}(M)	3	
	03	169–312	Unknown	02	2.95×10^{-03}(R)	3	
	05	97–579	01	01	3.70×10^{-03}(S)	3	
	06	150–373	01	01	3.080×10^{-02}(R)	4	Banerjee et al. 2014
	01	162–251	01	01	3.080×10^{-02}(R)	4	
	03	151–374	01	01	3.080×10^{-02}(R)	4	
	01	150–311	01	01	3.080×10^{-02}(R)	4	
	01	162–251	01	01	3.080×10^{-02}(R)	4	
	01	1050–18	01	30	2.61×10^{-04}(R)	3	Stainton et al. 2012
	35	718–1057	25	Unknown	8.27×10^{-08}(S)	2	
	01	1002–228	08	Unknown	1.96×10^{-03}(M)	2	
	02	206–203	01	02	1.18×10^{-02}(T)	2	
DNA-U	05	247–1002	01	01	2.69×10^{-27} (T)	7	Mukwa et al. 2016
	48	1558–1778	82	Unknown	1.17×10^{-17}(G)	6	Stainton et al. 2015
	02	1674–17	24	Unknown	7.77×10^{-12}(G)	5	
	07	1706–214	91	06	4.07×10^{-08}(G)	3	
	01	1321–1789	04	Unknown	1.22×10^{-04}(M)	3	
	28	1585–353	36	Unknown	5.93×10^{-19}(M)	4	
	01	1608–38	07	Unknown	7.39×10^{-11} (R)	6	
	02	356–567	01	16	7.21×10^{-05}(R)	5	
	07	374–534	Unknown	03	1.86×10^{-03} (G)	3	
	01	460–1011	02	Unknown	6.68×10^{-05}(M)	4	
	02	357–563	01	01	1.07×10^{-04}(S)	4	
	01	1010–1558	33	01	1.85×10^{-05}(S)	3	
	01	371–504	01	Unknown	2.47×10^{-03}(G)	3	
	01	126–192	01	Unknown	2.93×10^{-02}(C)	5	Banerjee et al. 2014
	01	917–31	01	01	1.385×10^{-11}(C)	6	
	01	419–557	01	01	5.510×10^{-01}(R)	3	
	01	451–562	01	01	5.510×10^{-01}(R)	3	
	01	373–529	01	01	5.510×10^{-01}(R)	3	
	01	417–491	01	01	5.510×10^{-01}(R)	3	
	01	943–31	01	01	2.341×10^{-11}(M)	6	

(Continued)

TABLE 9.1 (CONTINUED)
Intracomponent Recombination Sequences Detected in DNA Components of BBTV Isolates

DNA Component	Number of Recombinant Sequences	Recombination Breakpoints	Number of Major Parents	Number of Minor Parents	*p* Value (Method of Detection)	Number of Methods Detected the Recombinants	Reference
	01	419–527	01	01	5.510×10^{-01}(R)	3	
	01	901–56	01	Unknown	2.249×10^{-08}(S)	7	
	01	914–75	01	Unknown	2.250×10^{-08}(S)	7	
	01	901–56	01	Unknown	2.250×10^{-08}(S)	7	
	01	940–4	04	16	4.07×10^{-16}(G)	7	Stainton et al. 2012
	02	1012–4	16	03	6.87×10^{-15}(B)	6	
	01	949–1026	02	19	3.64×10^{-08}(M)	3	
	03	911–11	16	Unknown	6.53×10^{-07}(M)	4	
	03	944–14	10	Unknown	2.29×10^{-17}(S)	6	
	01	692–751	Unknown	02	1.95×10^{-05}(G)	3	
	01	998–568	01	Unknown	1.27×10^{-04}(S)	3	
	01	301–340	01	Unknown	8.60×10^{-03}(M)	3	
	01	612–760	01	Unknown	7.60×10^{-03}(R)	4	
	03	571–867	01	02	1.52×10^{-09}(G)	5	
	02	930–49	01	15	3.27×10^{-51} (L)	7	Hyder et al. 2011
	09	863–39	05	04	8.31×10^{-23}(L)	8	
	09	874–27	02	04	8.797×10^{-13}(L)	6	
DNA-S	04	130–1019	01	01	1.41×10^{-20} (S)	3	Mukwa et al. 2016
	03	1193–68	37	Unknown	3.96×10^{-09}(T)	3	Stainton et al. 2015
	36	362–730	Unknown	03	3.72×10^{-06}(T)	6	
	01	421–478	12	Unknown	6.90×10^{-04}(B)	3	
	02	392–514	01	Unknown	2.90×10^{-03}(B)	3	
	03	1193–64	01	Unknown	7.27×10^{-03}(G)	3	
	01	1042–129	01	Unknown	2.057×10^{-02}(C)	5	Banerjee et al. 2014
	05	294–420	09	12	7.74×10^{-03}(M)	3	
DNA-M	01	406–759	47	07	9.97×10^{-05}(S)	4	Stainton et al. 2015
	01	1285–15	01	70	2.19×10^{-04}(S)	3	
	01	987–42	01	Unknown	1.297×10^{-02}(T)	4	Banerjee et al. 2014
	01	278–373	18	Unknown	4.59×10^{-09}(R)	5	Stainton et al. 2012
DNA-C	01	277–780	139	01	1.03×10^{-07}(T)	5	Stainton et al. 2015
	06	1068–28	08	Unknown	1.41×10^{-06}(R)	3	
	02	1108–500	01	Unknown	4.42×10^{-06}(S)	3	

(*Continued*)

TABLE 9.1 (CONTINUED)
Intracomponent Recombination Sequences Detected in DNA Components of BBTV Isolates

DNA Component	Number of Recombinant Sequences	Recombination Breakpoints	Number of Major Parents	Number of Minor Parents	*p* Value (Method of Detection)	Number of Methods Detected the Recombinants	Reference
	01	430–482	01	Unknown	6.65×10^{-03}(T)	3	
	01	231–707	01	01	2.175×10^{-03}(G)	7	Banerjee et al. 2014
DNA-N	01	300–318	25	Unknown	2.95×10^{-10}(B)	3	Stainton et al. 2015
	12	732–1091	47	44	2.26×10^{-08}(T)	6	
	01	804–1229	01	Unknown	5.26×10^{-22}(S)	4	
	01	1167–1198	80	01	7.42×10^{-06}(G)	3	
	02	1172–16	21	Unknown	1.03×10^{-03}(G)	3	
	05	1179–1202	01	01	7.28×10^{-03}(G)	3	
	02	519–978	01	Unknown	2.248×10^{-02}(B)	7	Banerjee et al. 2014
	03	659–980	01	Unknown	2.248×10^{-02}(B)	7	
	01	658–978	01	Unknown	2.248×10^{-02}(B)	7	
	01	237–764	Unknown	01	6.060×10^{-04}(M)	6	
	01	621–924	05	Unknown	8.26×10^{-10}(T)	6	Stainton et al. 2012
	06	657–969	06	05	2.56×10^{-07}(T)	6	

Note: RDP (R), GENCONV (G), BOOTSCAN (B), MAXCHI (M), CHIMERA (C), SISCAN (S), and 3SEQ (T). Minor Parent = Parent contributing the smaller fraction of sequence. Major Parent = Parent contributing the larger fraction of sequence. Unknown = Only one parent and a recombinant need be in the alignment for a recombination event to be detectable. The sequence listed as unknown was used to infer the existence of a missing parental sequence.

DNA-U3 and four isolates in DNA-S, whereas none of the DNA-R component of all the isolates exhibited recombinations (Mukwa et al. 2016), and interestingly, the potential parents of recombination for DNA-S were from India and Sri Lanka, and DNA-U3 was from Taiwan. Banerjee et al. (2014) reported two intracomponent and five intercomponent recombination events in BBTV-Umiam, India isolate but none of them were unique. It has been shown that CR-SL region of DNA-U3 from a Taiwanese isolate has been replaced due to recombination with an additional Rep encoding component DNA-Y (Fu et al. 2009). Similarly, evidence for recombination of Indian BBTV isolates has recently been presented by Islam et al. (2010). Hyder et al. (2009) reported that DNA-U3 is the most diverse component of the BBTV genome of Pakistan isolate. In this study the sequence of DNA-U3 component of Pakistan isolates was believed to have resulted from a single introduction (Amin et al. 2008). Hu et al. (2007) used 102 major common regions (CR-Ms) from all six components and showed possible concerted evolution within SPG isolates of BBTV which is likely due to recombination in this region, and they reported web-like phylogenies for some genes of BBTV. These analyses revealed that all components displayed at least some evidence of recombination, with the greatest number of recombination events being detected in DNA-U3 (12 events) and the fewest in DNA-M (two events). All components carried evidence of recombinant regions involving the CR-SL region (with breakpoints falling within and/or on either side of this region), but only DNA-U3 and -N have recombination regions involving the CR-M, all of which had breakpoints

TABLE 9.2
Intercomponent Recombinant Sequences Detected in DNA Components of BBTV Isolates

Break Point	Number of Recombinant Sequences	Number of Major Parents	Number of Minor Parents	*p* Value (Method of Detection)	No of Methods Which Detected the Recombinants	Reference
1069–49	DNA-R (01)	DNA-R (42)	DNA-M (11)	7.31×10^{-10}(B)	3	Stainton et al. 2012
1001–1063	DNA-R (01)	DNA-R (1)	DNA-N (01)	1.44×10^{-09}(B)	4	
926–75	DNA-U (03)	DNA-U	DNA-C (16)	4.22×10^{-29}(G)	7	
883–216	DNA-U (01)	DNA-U (19)	DNA-N (24)	1.62×10^{-42}(S)	7	
772–813	DNA-U (04)	Unknown	DNA-R (36)	1.36×10^{-10}(G)	4	
1021–67	DNA-U (06)	DNA-U (17)	DNA-S (26) DNA-R (02) DNA-M (10)	6.50×10^{-15}(B)	3	
750–1037	DNA-U (06)	Unknown	DNA-R (37)	2.17×10^{-13}(R)	7	
774–815	DNA-U (12)	Unknown	DNA-R	3.46×10^{-07}(G)	3	
748–806	DNA-U (05)	DNA-U (03)	DNA-S (23) DNA-M (01) DNA-N (14)	3.59×10^{-05}(G)	3	
1045–157	DNA-N (01)	DNA-N (18)	DNA-S (01)	5.29×10^{-27}(B)	7	
20–161	DNA-S (01)	DNA-S (31)	DNA-N (01)	9.88×10^{-24}(G)	7	
802–1038	DNA-S (25)	Unknown	DNA-M (180)	1.60×10^{-30}(G)	7	
750–893	DNA-S (21) DNA-M (01)	DNA-S (25)	DNA-N (060)	4.60×10^{-16}(R)	7	
758–1047	DNA-M (01)	Unknown	DNA-S (25)	2.41×10^{-41}(G)	7	
790–1047	DNA-M (01)	Unknown	DNA-S (06)	8.16×10^{-36}(G)	7	
974–1033	DNA-M (01)	Unknown	DNA-R (01)	4.06×10^{-12}(G)	7	
953–8	DNA-M (01)	Unknown	DNA-R (01)	5.12×10^{-09}(R)	3	
877–1040	DNA-M (23)	Unknown	DNA-S (06)	2.15×10^{-25}(S)	6	
798–846	DNA-M (18)	Unknown	DNA-R (36)	7.63×10^{-13}(G)	4	
794–868	DNA-M (05)	Unknown	DNA-R (03)	6.90×10^{-09}(R)	4	
1031–106	DNA-M (01)	DNA-M (03)	Unknown	2.31×10^{-10}(T)	4	
62–391	DNA-M (01)	DNA-R (75)	DNA-M (05)	7.79×10^{-12}(S)	3	
24–86	DNA-C (06)	Unknown	DNA-M (04)	6.44×10^{-11}(G)	3	
852–984	DNA-C (06)	Unknown	DNA-R (36)	2.48×10^{-08}(R)	3	
811–922	DNA-N (01)	DNA-N (05)	Unknown	1.52×10^{-07}(S)	6	
828–913	DNA-N (06)	DNA-N (18)	DNA-M (05)	2.17×10^{-11}(G)	5	

Note: RDP (R), GENCONV (G), BOOTSCAN (B), MAXCHI (M), CHIMERA (C), SISCAN (S), and 3SEQ (T). Minor Parent = Parent contributing the smaller fraction of sequence. Major Parent = Parent contributing the larger fraction of sequence. Unknown = Only one parent and a recombinant need be in the alignment for a recombination event to be detectable. The sequence listed as unknown was used to infer the existence of a missing parental sequence.

falling on either side of this region. Of the 18 recombination events that were identified within multiple isolates, nine are seen within isolates from multiple countries. As with the reassortment events that are observed in multiple different genomes, these recombination events apparently occurred within genomes that were ancestral to two or more of the sequences analyzed and indicated that at least some BBTV recombinants are epidemiologically relevant. Twenty-two recombination

TABLE 9.3
Recombination Sequences Detected in DNA Components of BBTV Indian Isolates

DNA Component	Recombinant Sequences	Recombination Breakpoints	Number of Major Parents	Number of Minor Parents	*p* Value Shown Is for the Method	Number of Detection Methods
DNA-R	TripuraSH4	432–986	Manipur9-gn	Unknown	$7.827.851\times10^{-05}$ (T)	7
	736-4-1997	72–1018	Unknown	Assam1-GC	2.652×10^{-06} (M)	4
	523-6B-1991	167–1036	Unknown	523-6A-1991	3.321×10^{-22} (S)	6
	Arunachal Pradesh-P1	1110–1018	Unknown	Assam1-GC	3.068×10^{-11} (S)	6
	736-4-1997	62–1018	Unknown	Tamilnadu	8.292×10^{-11} (M)	7
	Arunachal Pradesh-P1	1109–1022	Assam1-GC	Tripura-N6	1.751×10^{-11} (S)	6
	736-4-1997	432–986	Unknown	Manipur-9	7.287×10^{-05} (T)	7
DNA-U	Bangalore- BT Bihar	584–772	Lucknow	Bihar	6.851×10^{-14} (R)	7
	Bihar	1060–370	Umiam	Lucknow	2.778×10^{-10} (B)	7
	Bihar	1061–382	Bangalore-BT1	Unknown	5.845×10^{-11} (G)	7
	Bangalore-GKVK	729–732	Meghalaya	736-4-1997	4.88×10^{-04} (G)	3
	Bangalore-GKVK	376–675	Unknown	Bangalore-BT1	4.266×10^{-05} (S)	5
	Q524-2	983–226	Unknown	Q524-2	6.812×10^{-13} (M)	6
DNA-S	Tamilnadu	1042–157	Umiam	Bihar	7.594×10^{-03} (M)	5
	Tripura-DH4	485–683	Tripura -KH4	Unknown	1.971×10^{-02} (S)	3
	Umiam	523–1050	Bangalore-GKVK	Unknown	6.728×10^{-29} (S)	7
	Umiam	186–552	Lucknow	Unknown	1.957×10^{-06} (T)	7
	Umiam	1055–315	Unknown	Bangalore-GKVK	1.509×10^{-13} (T)	7
	Tripura-N-6	485–683	Unknown	Tripura-KH4	2.618×10^{-02} (T)	4
	Tripura-N-6	996–724	Unknown	Q524-2	5.350×10^{-09} (5)	4
	Tripura-N-6	986–724	Unknown	Tripura-G08	2.157×10^{-08} (S)	3
	Tripura-N6	684–905	Tripura-GO8	Unknown	5.939×10^{-08} (S)	5
	Tripura-N6	724–977	Tripura-G08	Unknown	6.559×10^{-04} (T)	5
DNA-M	Bihar	1046–126	Unknown	Bangalore BT1	2.649×10^{-07} (R)	3
	Bihar	57–113	Unknown	Bangalore BT1	5.295×10^{-28} (R)	3
DNA-C	Umiam	242–731	Bihar Unknown	Lucknow Umiam	1.001×10^{-11}(S)	6
DNA-N	Umiam	1086–161	Bangalore-BT	Unknown	7.764×10^{-14} (B)	6
	Q524-3	996–1002	Q524-1	523-6B-1991	6.841×10^{-13} (R)	3
	Umiam	16–161	Bihar	Unknown	3.789×10^{-11} (T)	5
	Tamilnadu	1083–161	Umiam	Bihar	7.115×10^{-15} (B)	7

(Continued)

TABLE 9.3 (CONTINUED)
Recombination Sequences Detected in DNA Components of BBTV Indian Isolates

DNA Component	Recombinant Sequences	Recombination Breakpoints	Number of Major Parents	Number of Minor Parents	*p* Value Shown Is for the Method	Number of Detection Methods
	Bihar	992–380	Unknown	Bangalore-GKVK	7.292×10^{-14} (M)	3
	Umiam	16–161	Bihar	Unknown	2.8052×10^{-09} (G)	5

Note: RDP (R), GENCONV (G), BOOTSCAN (B), MAXCHI (M), CHIMERA (C), SISCAN (S), and 3SEQ (T). Minor Parent = Parent contributing the smaller fraction of sequence. Major Parent = Parent contributing the larger fraction of sequence. Unknown = Only one parent and a recombinant need be in the alignment for a recombination event to be detectable. The sequence listed as unknown was used to infer the existence of a missing parental sequence.

events were detected within the components encoding genes of known function as has been found in previous nanovirus recombination studies (Grigoras et al. 2014). Savory and Ramakrishnan (2014) detected similar numbers of recombination breakpoints within the noncoding and coding regions in cardamom bushy dwarf virus (CBDV) (24 and 20 breakpoints, respectively). In total, 13 events resulted in recombinant genes that could express chimeric proteins. However, all 13 of these events involved recombination between closely related BBTV variants, meaning that these recombination events would have only a minimal impact on encoded protein amino acid sequences (Lefeuvre et al. 2009). Another possible sign of protein-coding sequences having an impact on recombination patterns in BBTV is that the DNA-U3 component, which has no confirmed protein coding function, has a higher concentration of detectable recombination breakpoints than those of the known protein-coding genes of other components. Interestingly, DNA-U3 is also the component that appears to be most frequently exchanged by reassortment in BBTV. High frequencies of recombination in this component might reflect the fact that it is mostly evolving neutrally with no risk that recombinants might express defective chimeric proteins (Lefeuvre et al. 2009) and that there is little conservation of co-evolved epistatic interactions within this component.

9.5 REASSORTMENT IN BBTV

Reassortment is an evolutionary mechanism of segmented/multipartite RNA or DNA viruses that plays an important role in virus evolution. Reassortment, or pseudo-recombination, occurs only in viruses with segmented genomes and involves the exchange of discrete genome components between different species or genetically distinct strains which co-replicate within the same host cell. This process generates hybrid progeny with novel combinations of genome components inherited from different parental viruses and may lead to the emergence of highly virulent strains (Pita et al. 2001; Gu et al. 2007; Chakraborty et al. 2008; Nelson et al. 2008; Chen et al. 2009). Identifying geographic locations where reassortant lineages are most likely to emerge could be a valuable strategy for making decisions for disease management and surveillance efforts. Reassortment is most commonly detected through incongruencies in phylogenetic relationships among the different segments of a viral genome (Lindstrom et al. 1998; Holmes et al. 2005; Nelson et al. 2008; Vijaykrishna et al. 2011; Westgeest et al. 2014), as gene segments from the same virus isolate occupy conflicting phylogenetic positions due to differences in their evolutionary histories. This has led to the development of several automated reassortment detection methodologies

(Suzuki 2010; Nagarajan and Kingsford 2011; Svinti et al. 2013; Yurovsky and Moret 2011), but phylogeny-based methods have remained the most robust and popular methods for detecting reassortment (Savory et al. 2014). Phylogenetic inference of BBTV movement dynamics might, however, be confounded by two other evolutionary processes that occur in BBTV, namely, genome component reassortment and homologous recombination (Stainton et al. 2015). The evolutionary studies reported on BBTV have shown genetic reassortment of components among its isolates and revealed a concerted type evolution in noncoding regions of its genome.

BBTV isolates that have genome components derived from two or more different parental viruses have been inferred using a variety of phylogenetic (Hu et al. 2007; Yu et al. 2012) and statistical recombination detection methods (Martin et al. 2010; Stainton et al. 2012). Number of reassorted sequences of BBTV observed by different studies is shown in Table 9.4. Totally, 136 reassortant sequences were detected in different DNA components of BBTV. The greatest number of reassortment events occurred with component M (55) and U3 (47), and much fewer in all other components. The *p*-value is highly significant with reassortment events detected by Stainton et al. (2012, 2015), who have used nearly 1191 full-length components of isolates collected at different periods and across the world where BBTV is spread. Hu et al. (2007) demonstrated existence of chimeric BBTV isolate, TW4, as having a mixture of the Asian and the Pacific components in a single isolate; TW4 likely contain DNA-U3, -S, and -C of the Pacific group and DNA-M of the AG, possibly by genome reassortment, which is supported by CR-M and ORF phylogenies. Interestingly, the two DNA-R components of Asian and the Pacific groups were present in TW4. Since BBTV genome components are individually encapsulated, it is expected that mixed infections will often result in reassortment of genome components (Hu et al. 2007; Stainton et al. 2012). For the greatest accuracy in phylogeny, it is important to remove sequences derived from recombination and reassortment events. Being multicomponent, genetic reassortment is possible in nanoviruses such as *Faba bean necrotic yellow virus* (FBNYV) (Timchenko et al. 2000) and has been shown to occur in BBTV isolates (Hu et al. 2007) belonging to the South Pacific and the Asian subgroups. Reassortment occurs in the species of *Nanoviridae*, and the evidence shows that BBTV and FBNYV have the following similarities: (i) reassortment seems to generally involve exchanges of only one genome component; (ii) DNA-M component seems particularly prone to reassortment, accounting for 37.5% and 25% of observed reassortment in the genomes of samples of babu- and nanoviruses, respectively; (iii) DNA-S and DNA-C components of viruses in both genera display the lowest frequencies of reassortment (Grigoras et al. 2014). Savory and Ramakrishnan (2014) observed that DNA-U3, which is present only in babuviruses (BBTV, *Abaca bunchy top virus* [ABTV], and CBDV), was not found to be among the most frequently transferred genome components, and in contrast to this observation, Stainton et al. (2015) have shown 47 reassorted components involving DNA-U3 of BBTV. They have identified a total of 75 isolates as reassortant which have been contributed from 40 different reassortment events by employing a huge set of full-length genome sequences, whereas previously only 10 reassortment events were detected with a limited set of data (Hu et al. 2007; Stainton et al. 2012). Out of 40 reassortment events detected, 11 were of DNA-U3, eight of DNA-M, seven of both DNA-S and DNA-N, five from DNA-C, and the last two events were from DNA-R. Savory and Ramakrishnan (2014) suggested that the patterns of component transfer are not absolutely conserved between different species of babuviruses, as they did not see similar reassortment pattern in CBDV, a distinct species of *Babuvirus* genus.

Stainton et al. (2015) analyzed 855 full-length sequences of BBTV genome components from 171 banana leaf samples collected from 14 countries, and they have inferred that recombination and genome component reassortment contributed to global geographic structuring of BBTV populations. They suggested that human-mediated intercontinental transfers of epidemiologically important BBTV genotypes have occurred through relatively infrequent movement events over the past 300 years primarily from its diversity hot-spots in India and SEA.

TABLE 9.4
Reassortment Sequences Detected in DNA Components of BBTV Isolates

DNA Component	Number of Reassortment Sequences	Number of Major Parents	Number of Minor Parents	*p* Value	No of Methods Which Detected the Reassortment	Reference
DNA-R	01	21	01	1.59×10^{-15} (S)	6	Stainton et al. 2015
	03	Unknown	29	2.13×10^{-21} (S)	6	
	01	Unknown	05	2.048×10^{-50} (G)	7	Stainton et al. 2012
	01	04	06	1.61×10^{-16} (S)	7	
DNA-U	01	23	65	3.51×10^{-135} (T)	6	Stainton et al. 2015
	01	01	35	1.17×10^{-86} (T)	6	
	01	24	124	7.04×10^{-78} (T)	6	
	01	21	73	2.46×10^{-81} (T)	6	
	05	25	37	7.39×10^{-14} (S)	5	
	03	01	25	2.76×10^{-22} (S)	6	
	02	17	23	4.78×10^{-05} (M)	3	
	27	17	27	3.03×10^{-14} (S)	6	
	01	15	01	2.47×10^{-21} (T)	7	
	01	Unknown	01	3.48×10^{-16} (S)	3	
	01	Unknown	01	9.02×10^{-8} (T)	6	
	03	08	01	1.68×10^{-05} (T)	6	Stainton et al. 2012
DNA-S	01	18	120	8.44×10^{-96} (T)	5	Stainton et al. 2015
	01	64	Unknown	2.83×10^{-24} (G)	6	
	01	20	80	1.34×10^{-86} (M)	6	
	01	44	49	1.10×10^{-11} (T)	4	
	01	23	111	2.15×10^{-64} (T)	7	
	01	29	103	8.39×10^{-66} (T)	6	
	01	01	01	8.15×10^{-04} (T)	3	
	01	01	01	2.11×10^{-08} (S)	3	Stainton et al. 2012
DNA-M	01	38	38	3.38×10^{-25} (T)	6	Stainton et al. 2015
	01	33	40	9.34×10^{-23} (T)	6	
	01	29	03	3.33×10^{-22} (S)	3	
	01	43	45	3.37×10^{-19} (T)	7	
	26	32	38	6.87×10^{-20} (S)	7	
	05	01	01	3.84×10^{-12} (T)	4	
	01	35	44	8.43×10^{-20} (S)	7	
	10	39	31	6.63×10^{-18} (T)	7	
	07	05	Unknown	5.68×10^{-16} (T)	7	Stainton et al. 2012
	01	03	Unknown	9.15×10^{-15}(B)	7	
	01	17	04	2.52×10^{-52}(G)	7	

(*Continued*)

TABLE 9.4 (CONTINUED)
Reassortment Sequences Detected in DNA Components of BBTV Isolates

DNA Component	Number of Reassortment Sequences	Number of Major Parents	Number of Minor Parents	*p* Value	No of Methods Which Detected the Reassortment	Reference
DNA-C	01	08	115	1.49×10^{-69} (T)	6	Stainton et al. 2015
	04	23	26	8.82×10^{-35} (T)	5	
	01	Unknown	03	3.95×10^{-06} (G)	4	
	01	01	08	9.33×10^{-03} (M)	4	
	02	01	Unknown	1.51×10^{-08} (S)	2	
DNA-N	01	19	01	3.85×10^{-27} (T)	4	Stainton et al. 2015
	01	Unknown	29	1.77×10^{-17} (T)	6	
	04	35	39	8.84×10^{-09} (T)	6	
	01	Unknown	01	5.89×10^{-05} (S)	3	
	01	24	44	1.53×10^{-55} (T)	6	
	01	39	09	1.55×10^{-25} (T)	6	
	01	27	109	6.07×10^{-70} (G)	6	
	01	04	06	1.61×10^{-16}(S)	7	Stainton et al. 2012
	01	12	02	2.13×10^{-16}(S)	7	

Note: RDP (R), GENCONV (G), BOOTSCAN (B), MAXCHI (M), CHIMERA (C), SISCAN (S), and 3SEQ (T). Minor Parent = Parent contributing the smaller fraction of sequence. Major Parent = Parent contributing the larger fraction of sequence. Unknown = Only one parent and a recombinant need be in the alignment for a recombination event to be detectable. The sequence listed as unknown was used to infer the existence of a missing parental sequence.

9.6 POPULATION STRUCTURE OF BBTV AT THE GLOBAL LEVEL

Removal of recombinants and reassortant sequences from the population is necessary to draw the molecular phylogeny of any organism. Stainton et al. (2015) have elaborately analyzed the BBTV spread across the world over the years. They have employed Bayesian Monte Carlo Markov Chain analysis using 224 BBTV sequences to draw a maximum clade credibility (MCC) tree. They also deduced a nucleotide substitution rate of 2.916×10^{-4} substitutions/site/year, whereas Almeida et al. (2009) reported just half of what Stainton and his co-workers reported. The time since the most common ancestor of BBTV sequences, as per MCC tree, was 1086 years. They have shown that out of 14 movement events, the possible first of these movements of BBTV could be from South East Asia (SEA) to the Indian subcontinent approximately 1000 years ago, and the latest being 30 years ago from the same region to Egypt (Stainton et al. 2015). The Indian subcontinent has been shown to be the major hub for long-distance movements in both outward and inward directions. They have brought out interesting facts on the possible movements of BBTV from Samoa to Hawaii estimated to be between 1961 and 1978, and from SEA to Africa between 1982 and 2010. Stainton et al. (2015) suggested that the movement of BBTV was consistent with neither the movement of banana across the globe nor the human-mediated transcontinental spread, but it was more gradual, natural, and human-facilitated movements of infected virus or viruliferous aphids over the past 300 years from its centers of diversity in India and SEA/Far East across the banana-growing regions of the world.

9.7 MUTATION

Mutations occur at a frequency of 10^{-4} to 10^{-5} and the plant viruses can replicate billions of copies in a short time, so it is obvious that variants can often emerge in a population; stability of such variants depends upon the advantages offered to the survival of the viral pathogen, and such positive selection is more rare than neutral/negative selection. As mutations accumulate in future generations of the virus, the virus "drifts" away from its ancestor strain.

9.8 *BANANA BRACT MOSAIC VIRUS*

Bract mosaic disease is one of the recently reported diseases of banana. It is widely distributed in the Philippines, Sri Lanka, and India (mostly in southern states). This disease was first reported in the Philippines in 1979 (Magnaye and Espino 1990), and subsequently from India, Sri Lanka, Western Samoa, Thailand, Vietnam, and Ecuador (Bateson and Dale 1995; Rodoni et al. 1997; Quito-Avila et al. 2013). In India, a loss of Rs.38.7 crores per annum in a French plantain cultivar Nendran has been extrapolated based on the survey and yield loss data (Selvarajan and Balasubramanian 2008). This virus causes characteristic symptoms of spindle-shaped pinkish to reddish mosaic on bracts and pseudostem and also on midribs. The fingers in the bunches of affected plants fail to develop, causing severe yield losses. This virus is transmitted by banana black aphids in a non-persistent manner, and other aphids such as *Aphis gossypii* and *Rhopalosiphum maidis* are also known to transmit the disease in the same way. BBrMV has also been reported to infect small cardamom, *Elettaria cardamomum*, in India and flowering ginger, *Alpinia purpurata*, in Hawaii (Siljo et al. 2012; Wang et al. 2010; Zhang et al. 2016).

To date four complete genomes of the virus have been sequenced and characterized. In addition to whole genomes, the coat protein coding gene and helper component protease (HC-Pro) genes of many isolates have been sequenced, and diversity analyses have been done. The details of genetic diversity and the possible means of evolution has been furnished elsewhere in this chapter.

9.9 GENETIC DIVERSITY IN BBrMV

BBrMV is a member of the genus *Potyvirus* of the family *Potyviridae*, and has flexuous filamentous particles of ca. 750×11 nm which encapsidates a single-stranded positive-sense RNA as its genome (Bateson and Dale 1995; Thomas et al. 1997). The RNA genome is 9711 nucleotides long, excluding the 3′ terminal poly A tail. The virus contains a typical large open reading frame (ORF) coding for a polyprotein of 3125 amino acids (Figure 9.2).

BBrMV contains nine protease cleavage sites, yielding ten matured functional proteins that have motifs conserved among homologous protein of other potyviruses (Balasubramanian and Selvarajan 2012; Ha et al. 2008). In addition, an overlapping coding sequence for pipo (pretty interesting

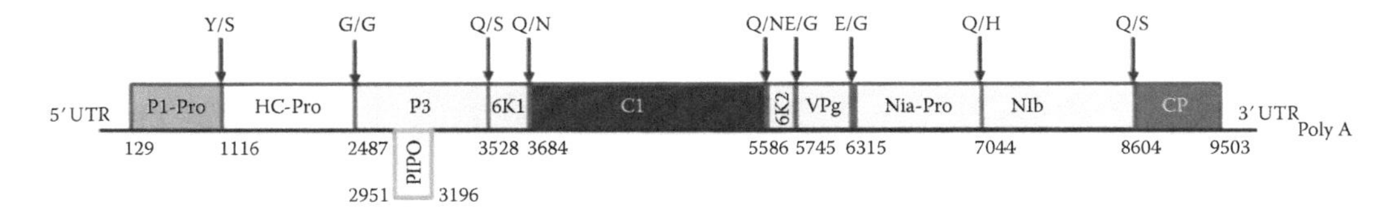

FIGURE 9.2 Genome organization of BBrMV. The position and dipeptide motif of the cleavage sites are indicated.

potyviridae; ORF) encoding a 7 kDa protein (Chung et al. 2008) that exists in the +2-reading frame within the protein 3 (P3)-encoding region has also been found in BBrMV isolates. The whole genome of BBrMV-TRY (India) and BBrMV-PHI (the Philippines) had 94% nucleotide (nt) sequence identity and 88–98% amino acid (aa) sequence identities (Balasubramanian and Selvarajan 2012). Percentage pairwise nt identities of coat protein gene of BBrMV were determined using sequence demarcation tool (SDT) v1.2 (Muhire et al. 2014) with the MUSCLE-based alignment option. The coat protein-encoding gene of BBrMV isolates shared an identity of 79–100% at nt (Figure 9.3) and 80–100% at aa level (Figure 9.4) (Balasubramanian and Selvarajan 2014).

Two distinct isolates, TN14 and TN16, had a divergence of 21 and 18% at nt and 20 and 12% at aa level, respectively, with other isolates. Among the isolates of non-Indian origin, 96–100% similarity at nt and aa levels was recorded. Except for isolates TN14 and TN16, all the Indian isolates shared 94–100% and 92–100% identity at nt and aa level, whereas the partial gene sequences from Hawaii and Ecuador are 99% identical, compared with the corresponding fragment of a BBrMV isolate from the Philippines (Quito-Avila et al. 2013; Wang et al. 2010). Sequence identity of BBrMV isolates encoding HC-Pro was 92–100% at both the nt and aa levels (Balasubramanian et al. 2014).

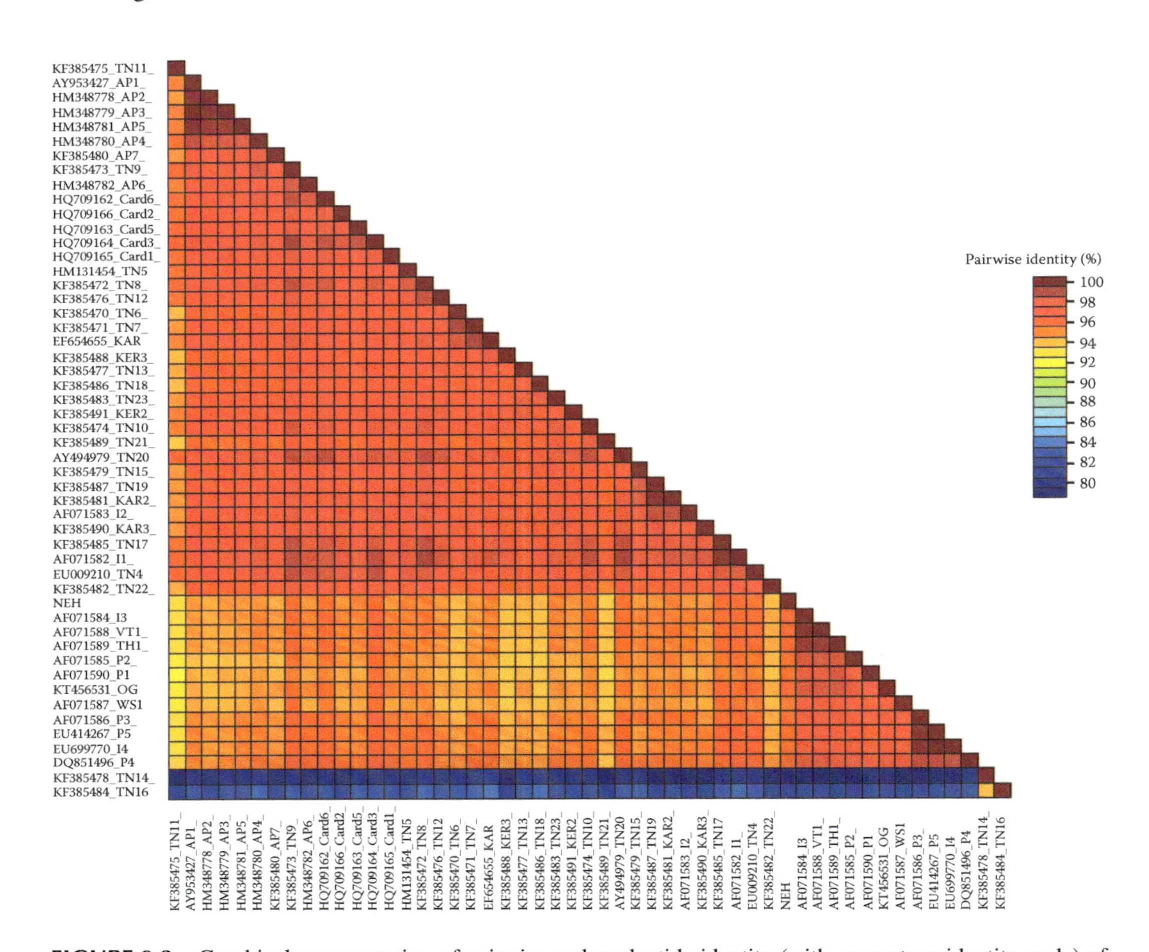

FIGURE 9.3 Graphical representation of pairwise and nucleotide identity (with percentage identity scale) of 51 BBrMV isolates. Isolates used in this study: TN, Tamil Nadu; AP, Andhra Pradesh; KER Kerala; KAR, Karnataka; NEH, Northeastern hill states. TN, AP, KER, KAR, I, NEH are of Indian origin. P, Philippines; WS, Western Samoa; VT, Vietnam; THI, Thailand. Card-sequences of coat protein gene of BBrMV infecting cardamom; OG, ornamental ginger.

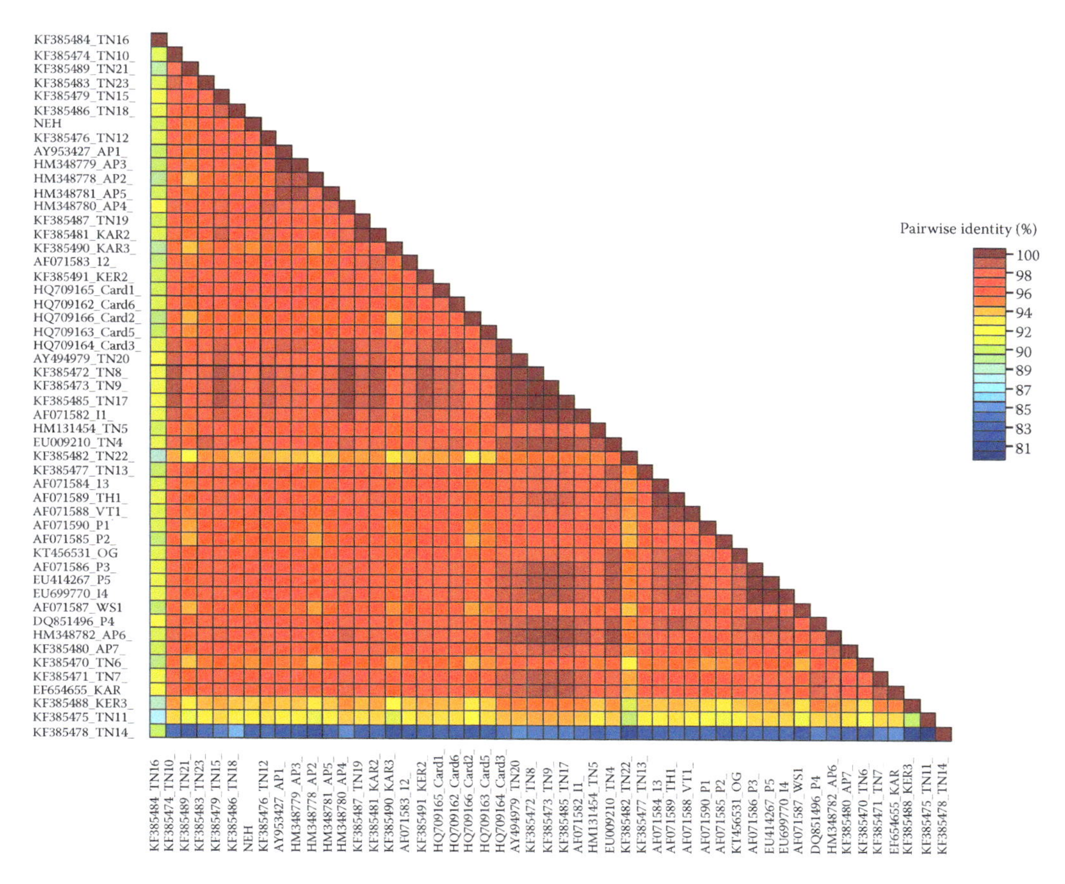

FIGURE 9.4 Graphical representation of pairwise and amino acid identity (with percentage identity scale) of 51 BBrMV isolates. (Details on the name of the isolates are furnished in Figure 9.3.)

9.10 RECOMBINATION

To gain insight into potential recombination in CP gene of BBrMV, all the available sequences (51 isolates) were examined with various recombination detection algorithms to identify putative recombinants and recombination breakpoints. A total of 10 potential recombinants were detected, shown in Table 9.5. Out of 51 BBrMV isolates analyzed, isolate TN22 showed evidence of recombination by seven recombination detection algorithms. Recombinant KER3was detected by five, and TN11 was detected by four recombination detection methods. The remaining recombinants were detected by three recombination detection methods. The relationship of putative recombinants to their contributing major and minor parental isolates could not be attributed to either geographical locations or the host cultivar. However, these recombinants would have occurred due to mixed infection of isolates in a banana variety and later it would have been transmitted through vectors to other banana varieties. Three isolates, AP6, KAR3, TN22, and TN6, were involved as major or minor parental isolates in a majority of the recombinants. Interestingly, a Philippine isolate P1 was a major parent for the Indian isolate TN11. This suggests a possible derivation of Southeast Asian isolate by genetic exchange involving Indian isolates.

Based on HC-Pro gene, four potential recombinants were detected. Six recombination breakpoints were detected in the isolate TN32, followed by five in TN16, three in TN28, and one in the PHI isolate. All six recombination breakpoints of recombinant TN32 were detected from N-terminal

TABLE 9.5
Recombination Sequences Detected in the Coat Protein Gene of BBrMV Isolates

Recombination Breakpoints	Recombinant Sequence(s)	Major Parents	Minor Parents	Detection Methods	*P* Value
324–522	TN11	TN22	Unknown	RMST	1.834×10^{-8}
136–498	TN11	TN4	Unknown	BMST	1.201×10^{-7}
250–480	TN11	I2	Unknown	MCST	4.357×10^{-7}
890–522	TN11	TN22	Unknown	MST	7.136×10^{-8}
134–479	TN11	P1	Unknown	MCS	3.069×10^{-4}
130–294	KER3	KER2, TN18	Unknown	RBMCT	4.196×10^{-4}
106–332	TN8, TN13,15 TN18, TN21, TN22, TN23	KER3, Unknown	KAR3, KAR3, TN12	MST	3.954×10^{-4}
658–892	TN22	TN1	Unknown	RGBMCST	1.684×10^{-6}
658–106	TN22	AP6	TN6	MCS	7.620×10^{-3}
476–895	TN22	AP6	TN11	MCT	5.019×10^{-6}
476–136	TN11, TN13, TN15	AP6, KER3	CARD1,KAR3	MST	3.954×10^{-4}
302–890	TN13	TN22	Unknown	MCS	5.228×10^{-4}
52–658	TN22	Unknown	TN4	MST	9.554×10^{-4}
556–895	TN15, TN23	TN6, TN22	TN22, Unknown	MCS	1.822×10^{-4}
536–3	TN21	TN6	TN22	MCS	1.822×10^{-4}
567–103	TN8	TN6	TN22	MCS	1.822×10^{-4}
549–07	TN10	TN6	TN22	MCS	1.822×10^{-4}
448–11	TN18	TN22	Unknown	MCS	1.822×10^{-4}

Note: RDP (R), GENCONV (G), BOOTSCAN (B), MAXCHI (M), CHIMERA (C), SISCAN (S), and 3SEQ (T). Minor Parent = Parent contributing the smaller fraction of sequence. Major Parent = Parent contributing the larger fraction of sequence. Unknown = Only one parent and a recombinant need be in the alignment for a recombination event to be detectable. The sequence listed as unknown was used to infer the existence of a missing parental sequence.

region of HC-Pro gene, whereas five recombination breakpoints detected in the recombinant isolate TN16 were located in the C-terminal region. Of the four recombinants identified, two (TN32 and TN16) were detected by five to eight programs with significant *p*-values always lower than 10^{-5}, adding to the validity of the results. Indian isolate TN11 appeared as parental isolate for a recombinant PHI isolate and PHI isolate as parent for recombinant TN16 (Balasubramanian et al. 2014).

9.11 CONCLUSIONS

BBTV is a cause of a very serious disease in banana which was initially reported in 1889 in Fiji and has spread to most of the banana growing continents except Central and Latin America. The association of ssDNA in bunchy top disease was proved only in 1991. The existence of this invasive virus along with the introduction of banana and plantain from Southeast Asia and the Pacific islands to different countries of all the continents has led to the diversity and evolution of BBTV. The initial phylogenetic analysis of BBTV performed based on the sequence of component DNA-R have grouped BBTV isolates into two distinct lineages, the SPG and the AG. However, during its spread, mutation, recombination, and reassortment have led to the variations in the genome, especially in the places of its origin, and subsequently in other places of introduction. The review has shown that genetic recombination between different BBTV isolates as well as among the genomic components of same isolate played a major role in the evolution, and intragenomic recombination is unique and rare among multicomponent ssDNA viruses and is a novel way to generate genetic diversity. Because of genome components each being packaged individually into separate virions, new infections that are propagated from mixed BBTV infections will frequently contain an assortment of different genome components. In total, 136 reassortant sequences were detected in different DNA components of BBTV. The greatest number of reassortment events occurred with component M (55) and U3 (47), and much less in all other components revealed the significance of recombinations and reassortment which have contributed to the global geographic structuring of BBTV populations. Recent studies suggested that human-mediated intercontinental transfers of epidemiologically important BBTV genotypes have occurred through relatively infrequent movement events over the past 300 years, primarily from its diversity hot-spots in India and Southeast Asia. Most of the facts on recombination, reassortment, and population structure of BBTV came out of the outstanding work of Stainton and his co-workers in 2012 and 2015. They have predicted that multiple movements of BBTV would have occurred in the past 100 years between Taiwan, China, the Indian subcontinents, and Indonesia. It is necessary to have similar studies in India as most of the wild bananas are naturally present in North Eastern Region of India where BBTV is omnipresent. In contrast to BBTV, the BBrMV is of very recent origin and though it was first reported as Kokkan disease in India of unknown etiology, the association of the virus was authentically reported from the Philippines. Based on our study using more than 50 sequences of coat protein gene, many recombinant events were detected. That this virus has now spread to small cardamom in India and to ornamental ginger, *Alpinia purpurata*, in Hawaii suggests that many surveys and intensive surveillance need to be taken up to collect and sequence a wide range of geographic isolates to find the perfect genetic structure of this virus so that management strategies can be developed for these viral diseases in the future.

REFERENCES

Almeida RPP, Bennett GM, Anhalt MD, Tsai CW, O'Grady P (2009). Spread of an introduced vector-borne banana virus in Hawaii. *Mol Ecol* 18: 136–146.

Amin I, Qazi J, Mansoor S, Ilyas M, Briddon RW (2008). Molecular characterization of Banana bunchy top virus (BBTV) from Pakistan. *Virus Genes* 36: 191–198.

Aronson MN, Meyer AD, Györgyey J, Katul L, Vetten HJ, Gronenborn B, Timchenko T (2000). Clink, a nanovirus-encoded protein, binds Both pRB and SKP1. *J Virol* 74: 2967–2972.

Balasubramanian V, Selvarajan R (2012). Complete genome sequence of a Banana bract mosaic virus (BBrMV) isolate infecting the French plantain cv Nendran in India. *Arch Virol* 157: 397–400.

Balasubramanian V, Selvarajan R (2014). Genetic diversity and recombination analysis in the coat protein gene of Banana bract mosaic virus. *Virus Genes* 8(3): 9–17.

Balasubramanian V, Sukanya RS, Anuradha C, Selvarajan R (2014). Population structure of Banana bract mosaic virus reveals recombination and negative selection in the helper component protease (HC-Pro) gene. *Virus Disease* 25(4): 460–466.

Banerjee A, Roy S, Behere GT, Roy SS, Dutta SK, Ngachan SV (2014). Identification and characterization of a distinct Banana bunchy top virus isolate of Pacific-Indian Oceans group from North-East India. *Virus Res* 183: 41–49.

Bateson MF, Dale JL (1995). Banana bract mosaic virus: Characterization using potyvirus specific degenerate PCR primers. *Arch Virol* 140: 515–527.

Bell KE, Dale JL, Ha CV, Vu MT, Revill PA (2002). Characterisation of Rep-encoding components associated with banana bunchy top nanovirus in Vietnam. *Arch Virol* 147: 695–707.

Boloy FN, Nkosi BI, Losimba JK, Bungamuzi CL, Siwako HM, Balowe FW, Lohaka JW, Dhed'a Djailo B, Lepoint P, Sivirihauma C, Blomme G (2014). Assessing incidence, development and distribution of banana bunchy top disease across the main plantain and banana growing regions of the Democratic Republic of Congo. *African J Agri Res* 9: 2611–2623.

Chakraborty S, Vanitharani R, Chattopadhyay B, Fauquet CM (2008). Super virulent pseudo recombination and asymmetric synergism between genomic components of two distinct species of begomovirus associated with severe tomato leaf curl disease in India. *J Gen Virol* 89: 818–828.

Chen LF, Rojas M, Kon T, Gamby K, Xoconostle-Cazares B, Gilbertson RL (2009). A severe symptom phenotype in tomato in Mali is caused by a reassortant between a novel recombinant begomovirus (Tomato yellow leaf curl Mali virus) and a betasatellite. *Mol Plant Pathol* 10: 415–430.

Cherian AK, Menon R, Suma A, Nair S, Sudheesh MV (2002). Impact of banana bract mosaic diseases on the yield of commercial banana varieties of Kerala. In: *Global Conference on Banana and Plantain*, Bangalore, 28–31 October 2002, 155.

Chiaki Y, Nasirl N, Herwina H, Jumjunidang, Sonoda A, Fukumoto T, Nakamura M, Iwai H (2015). Genetic structure and diversity of the Banana bunchy top virus population on Sumatra Island, Indonesia. *Eur J Plant Pathol* 143: 113–122.

Chung BYW, Miller WA, Atkins JF, Firth AE (2008). An overlapping essential gene in the Potyviridae. *Proc Natl Acad Sci U S A* 105: 5897–5902.

Cook DC, Liu S, Edwards J, Villalta ON, Aurambout JP, Kriticos DJ, Drenth A, De Barro PJ (2012). Predicting the benefits of banana bunchy top virus exclusion from commercial plantations in Australia. *PLoS One* 7: e42391.

Denham TP, Haberle SG, Lentfer C, Fullagar R, Field J, Therin M, Porch N, Winsborough B (2003). Origins of agriculture at Kuk Swamp in the Highlands of New Guinea. *Science* 301: 189–193.

Fu HC, Hu JM, Hung TH, Su HJ, Yeh HH (2009). Unusual events involved in banana bunchy top virus strain evolution. *Phytopathology* 99: 812–822.

Garcia-Arenal F, Fraile A, Malpica JM (2001). Variability and genetic structure of plant virus populations. *Ann Rev Phytopathol* 39: 157–186.

Grigoras I, Ginzo AI, Martin DP, Varsani A, Romero J, Mammadov ACh, Huseynova IM, Aliyev JA, Kheyr-Pour A, Huss H, Ziebell H, Timchenko T, Vetten HJ, Gronenborn B (2014). Genome diversity and evidence of recombination and reassortment in nano viruses from Europe. *J Gen Virol* 95: 1178–1191.

Gu H, Zhang C, Ghabrial SA (2007). Novel naturally occurring Bean pod mottle virus reassortants with mixed heterologous RNA1 genomes. *Phytopathology* 97: 79–86.

Ha C, Coombs S, Revil PA, Harding RM, Vu M, Dale JL (2008). Design and application of two novel degenerate primer pairs for the detection and complete genomic characterization of poty viruses. *Arch Virol* 153: 25–36.

Hafner GJ, Stafford MR, Wolter LC, Harding RM, Dale JL (1997a). Nicking and joining activity of Banana bunchy top virus replication protein in vitro. *J Gen Virol* 78: 1795–1799.

Hafner GJ, Harding RM, Dale JL (1997b). A DNA primer associated with Banana bunchy top virus. *J Gen Virol* 78: 479–486.

Holmes EC, Ghedin E, Miller N, Taylor J, Bao Y, George KS, Grenfell BT, Salzberg SL, Fraser CM, Lipman, DJ, Taubenberger JK (2005). Whole-genome analysis of human influenza A virus reveals multiple persistent lineages and reassortment among recent H3N2 viruses. *PLoS Biology* 3: 1579–1589.

Hu JM, Fu HC, Lin CH, Su HJ, Yeh HH (2007). Re-assortment and concerted evolution in Banana bunchy top virus genomes. *J Virol* 81: 1746–1761.

Hughes AL (2004). Birth-and-death evolution of protein-coding regions and concerted evolution of non-coding regions in the multicomponent genomes of nano viruses. *Mol Phylogenet Evol* 30: 287–294.

Hyder MZ (2009). Sequencing and genetic characterization of major DNA components of Banana bunchy top virus. PhD Thesis, PirMehr Ali Shah, Arid Agriculture University Rawalpindi, Pakistan.

Hyder MZ, Shah SH, Hameed S, Naqvi SMS (2011). Evidence of recombination in the Banana bunchy top virus genome. *Infect Genet Evol* 11: 1293–1300.

Islam MN, Naqvi AR, Jan AT, Haq QMR (2010). Genetic diversity and possible evidence of recombination among Banana bunchy top virus (BBTV) isolate. *Int Res J Microbiol* 1: 1–12.

Karan M, Harding RM, Dale JL (1994). Evidence for two groups of Banana bunchy top virus isolates. *J Gen Virol* 75: 3541–3546.

Kumar PL, Hanna R, Alabi OJ, Soko MM, Oben TT, Vangu GH, Naidu RA (2011). Banana bunchy top virus in sub-Saharan Africa: Investigations on virus distribution and diversity. *Virus Res* 159: 171–182.

Kumar PL, Selvarajan R, Iskra-Caruana ML, Chabannes M, Hanna R (2015). Biology, etiology, and control of virus diseases of banana and plantain. *Adv Virus Res* 91: 229–269.

Lefeuvre P, Lett JM, Varsani A, Martin DP (2009). Widely conserved recombination patterns among single-stranded DNA viruses. *J Virol* 83: 2697–2707.

Lejju BJ, Robertshaw P, Taylor D (2006). Africa's earliest bananas. *J Arch Science* 33: 102–113.

Lindstrom SE, Hiromoto Y, Nerome R, Omoe K, Sugita S, Yamazaki Y, Takahashi T, Nerome K (1998). Phylogenetic analysis of the entire genome of influenza A (H3N2) viruses from Japan: Evidence for genetic reassortment of the six internal genes. *J Virol* 72: 8021–8031.

Magnaye LV, Espino RRC (1990). Note: Banana bract mosaic, a new disease of banana I. Symptomatology. *The Philippine Agriculturist* 73: 55–59.

Martin DP, Murrell B, Golden M, Khoosal A, Muhire B (2015). RDP4: Detection and analysis of recombination patterns in virus genomes. *Virus Evol* 1(1): vev003.

Muhire BM, Varsani A, Martin DP (2014). A virus classification tool based on pairwise sequence alignment and identity calculation. *PLoS One* 9: e108277.

Mukwa LFT, Muengula M, Zinga I, Kalonji A, Iskra-Caruana ML, Bragard C (2014). Occurrence and distribution of banana bunchy top virus related agro-ecosystem in south western Democratic Republic of Congo. *Am J Plant Sci* 5: 647–658.

Mukwa LF, Gillis A, Vanhese V, Romay G, Galzi S, Laboureau N, Kalonji-Mbuyi A, Iskra-Caruana ML et al. (2016). Low genetic diversity of Banana bunchy top virus, with a sub-regional pattern of variation, in Democratic Republic of Congo. *Virus Genes* 52: 900–905.

Nagarajan N, Kingsford C (2011). GiRaF: Robust, computational identification of influenza reassortments via graph mining. *Nucleic Acids Res* 39: e34.

Nelson MI, Viboud C, Simonsen L, Bennett RT, Griesemer SB, George KS, Taylor J, Spiro DJ, Sengamalay NA, Ghedin E, Taubenberger JK, Holmes EC (2008). Multiple reassortment events in the evolutionary history of H1N1 influenza A virus since 1918. *PLoS Pathogens* 4(2): e1000012.

Niyongere C, Losenge T, Ateka ME, Ntukamazina N, Ndayiragije P, Simbare A, Cimpaye P, Nintije P, Lepoint P, Blomme G (2013). Understanding banana bunchy top disease epidemiology in Burundi for an enhanced and integrated management approach. *Plant Pathol* 62: 562–570.

Niyongere C, Lepoint P, Losenge T, Blomme G, Ateka EM (2015). Towards understanding the diversity of banana bunchy top virus in the Great Lakes region of Africa. *African J Agri Res* 10(7): 702–709.

Padidam M, Sawyer S, Fauquet CM (1999). Possible emergence of new Gemini viruses by frequent recombination. *Virology* 265: 218–225.

Perrier X, De Langhe E, Donohue M, Lentfer C, Vrydaghs L, Bakry F, Carreel F, Hippolyte I, Horry J-P, Jenny C, Lebot V, Risterucci A-M, Tomekpe K, Doutrelepont H, Ball TB, Manwaring J, De Maret P, Denham T (2011). Multidisciplinary perspectives on banana (Musa spp.) domestication. *Proc Natl Acad Sci U S A* 108(28): 11311–11318.

Pita JS, Fondong VN, Sangare A, Otim-Nape GW, Ogwal S, Fauquet CM (2001). Recombination, pseudo recombination and synergism of Gemini viruses are determinant keys to the epidemic of severe cassava mosaic disease in Uganda. *J Gen Virol* 82: 655–665.

Quito-Avila DF, Ibarra MA, Alvarez RA, Ratti MF, Espinoza L, Cevallos Cevallos JM, Peralta EL (2013). First report of banana bract mosaic virus in "Cavendish" banana in Ecuador. *Plant Disease* 97: 1003.

Randrianja S, Ellis S (2009). *Madagascar: A Short History*. Chicago, IL: University of Chicago Press.

Rodoni BC, Ahlawat YS, Varma A, Dale JL, Harding RM (1997). Identification and characterization of banana bract mosaic virus in India. *Plant Disease* 81: 669–672.

Savory FR, Ramakrishnan U (2014). Asymmetric patterns of reassortment and concerted evolution in cardamom bushy dwarf virus. *Infect Genet Evol* 24: 15–24.

Savory FR, Varun V, Ramakrishnan U (2014). Identifying geographic hot spots of reassortment in a multipartite plant virus. *Evol App* 7(5): 569–579.

Selvarajan R, Balasubramanian V (2008). Banana viruses. In: *Characterization, Diagnosis and Management of Plant Viruses*. 2. GP Rao, A Myrta, K-S Ling (eds), 109–124. Houston, TX: Studium Press.

Selvarajan R, Balasubramanian V (2014). Host–virus interactions in banana-infecting viruses. In: *Plant Virus–Host, Interaction Molecular Approaches and Viral Evolution*. RK Gaur, T Hohn, P Sharma (eds), 57–78. Boston, MA: Elsevier Academic Press.

Selvarajan R, Jeyabaskaran KJ (2006). Effect of banana bract mosaic virus (BBrMV) on growth and yield of cultivar Nendran (plantain, AAB). *Indian Phytopathol* 59: 496–500.

Selvarajan R, Sheeba M, Balasubramanian V, Rajmohan R, Lakshmi Dhevi N, Sasireka T (2010). Molecular characterization of geographically different banana bunchy top virus (BBTV) isolates in India. *Indian J Virol* 21(2): 110–116.

Siljo A, Bhat AI, Biju CN, Venugopal MN (2012). Occurrence of Banana bract mosaic virus on cardamom. *Phytoparasitica* 40: 77–85.

Stainton D, Kraberger S, Walters M, Wiltshire EJ, Rosario K, Halafihi M, Lolohea S, Katoa I, Faitua TH, Aholelei W, Taufa L, Thomas JE, Collings DA, Martin DP, Varsani A. (2012). Evidence of intercomponent recombination, intracomponent recombination and reassortment in banana bunchy top virus. *J Gen Virol* 93: 1103–1119.

Stainton D, Martin DP, Muhire BM, Lolohea S, Halafihi M, Lepoint P, Blomme G, Crew KS, Sharman M, Kraberger S, Dayaram A, Walters M, Collings DA, Mabvakure B, Lemey P, Harkins GW, Thomas JE, Varsani A (2015). The global distribution of Banana bunchy top virus reveals little evidence for frequent recent, human-mediated long-distance dispersal events. *Virus Evol* 1(1): vev009.

Suzuki Y (2010). A phylogenetic approach to detecting reassortments in viruses with segmented genomes. *Gene* 464: 11–16.

Svinti V, Cotton JA, McInerney JO (2013). New approaches for unravelling reassortment pathways. *BMC Evol Biol* 13: 1.

Thomas JE, Magnaye LV (1996). Banana bract mosaic disease. Musa Disease Fact Sheet 1. INIBAP, Montpellier, France.

Thomas JE, Geering ADW, Gambley CF, Kessling AF, White M (1997). Purification, properties and diagnosis of banana bract mosaic potyvirus and its distinction from abaca mosaic potyvirus. *Phytopathology* 87: 698–705.

Timchenko T, de Kouchkovsky F, Katul L, David C, Vetten HJ, Gronenborn B (1999). A single Rep protein initiates replication of multiple genome components of Faba bean necrotic yellows virus, a single-stranded DNA virus of plants. *J Virol* 73: 10173–10182.

Timchenko T, Katul L, Sano Y, de Kouchkovsky F, Vetten HJ, Gronenborn B (2000). The master rep concept in nanovirus replication: Identification of missing genome components and potential for natural genetic reassortment. *Virology* 274: 189–195.

Vetten HJ, Chu PWG, Dale JL, Harding R, Hu J, Katul L, Kojima M, Randles JW, Sano Y, Thomas JE (2005). Nanoviridae, virus taxonomy. *Eighth Report of the International Committee on Taxonomy of Viruses*. CM Fauquet, MA Mayo, J Maniloff, U Desselberger, LA Ball (eds), 343–352. San Diego, CA: Elsevier Academic Press.

Vijaykrishna D, Smith GJ, Pybus OG, Zhu H, Bhatt S, Poon LL, Riley S, Bahl J, Ma SK, Cheung CL, Perera RA, Chen H, Shortridge KF, Webby RJ, Webster RG, Guan Y, Peiris JS (2011). Long-term evolution and transmission dynamics of swine influenza A virus. *Nature* 473: 519–522.

Vishnoi R, Raj SK, Prasad V (2009). Molecular characterization of an Indian isolate of banana bunchy top virus based on six genomic DNA components. *Virus Genes* 38: 334–344.

Wang HI, Chang CH, Lin PH, Fu HC, Tang C, Yeh HH (2013). Application of motif-based tools on evolutionary analysis of multipartite single-stranded DNA viruses. *PLoS ONE* 8(8): e71565.

Wang IC, Sether DM, Melzer MJ, Borth WB, Hu JS (2010). First report of Banana bract mosaic virus in flowering ginger in Hawaii. *Plant Disease* 94(7): 921.

Wanitchakorn R, Harding RM, Dale JL (2000). Sequence variability in the coat protein gene of two groups of banana bunchy top isolates. *Arch Virol* 145: 593–602.

Westgeest KB, Russell CA, Lin X, Spronken MJI, Bestebroer TM, Bahl J, van Beek R, Skepner E, Halpine RA, de Jong JC, Rimmelzwaan GF, Osterhaus ADME, Smith DJ, Wentworth DE, Fouchier RAM, de Graaf M (2014). Genome wide analysis of reassortment and evolution of human influenza A(H3N2) viruses circulating between 1968 and 2011. *J Virol* 88: 2844–2857.

Wickramaarachchi WA, Shankarappa KS, Rangaswamy KT, Maruthi MN, Rajapakse RG, Ghosh S (2016). Molecular characterization of banana bunchy top virus isolate from Sri Lanka and its genetic relationship with other isolates. *Virus Disease* 27(2): 154–160.

Yasmin T, Naqvi SMS, Shah H, Khalid S (2005). RFLP-based relationship of Pakistani isolate of Banana bunchy top virus with South Pacific virus group. *Pakistan J Botany* 37: 399–406.

Yurovsky A, Moret BME (2011). FluReF, an automated flu virus reassortment finder based on phylogenetic trees. *BMC Genomics* 12(Suppl 2): S3.

Zhang J, Borth WB, Lin B, Dey KK, Melzer MJ, Shen H, Pu X, Sun D, Hu JS (2016). Deep sequencing of banana bract mosaic virus from flowering ginger (Alpiniapurpurata) and development of an immunocapture RT-LAMP detection assay. *Arch Virol* 161(7): 1783–1795.

Section II

Host Interaction

10 Molecular Dynamics of Geminivirus-Host Interactome

V.G. Malathi, P. Renukadevi, and S. Rageshwari

CONTENTS

10.1 INTRODUCTION

Among the plant-infecting viruses, only a few have the DNA genome, in contrast to other prokaryotes, invertebrates, and vertebrates, which are infected by nearly a thousand DNA viruses. This small number of plant-infecting DNA viruses suggests the fundamental differences in the molecular challenges DNA viruses meet within the plant. Among the plant-infecting DNA viruses, members of the family *Caulimoviridae* have double-stranded circular DNA genome, replicate via reverse transcription, and replication occurs in cytoplasm. Contrastingly, members of the other two families of plant-infecting DNA viruses, *Geminiviridae* and *Nanoviridae*, have single-stranded (ss) circular DNA genome and are confined to the nucleus. They directly involve with DNA metabolism of the host, DNA replication, repair, recombination, and methylation. They interfere with cell cycle events to convert the DNA synthesis machinery to their advantage and emerge as successful molecular parasites. The recent upsurge in genomics-related tools has helped research workers to delineate the molecular events of pathogenesis of geminiviruses which are briefly discussed here. The reader is referred to excellent reviews such as Stanley (1985); Harrison and Robinson (1999); Briddon et al. (2003); Hanley Bowdoin et al. (1999, 2013); Gutierrez (2000); Rojas et al. (2005); Yadava et al. (2010); Fondong (2013); and Rizvi et al. (2015).

10.2 *GEMINIVIRIDAE*

Geminiviruses have ss circular DNA genome (2.5–5.2 kb) encapsidated within characteristic geminate virion particles (38 × 22 nm) (Figure 10.1) consisting of two incomplete icosahedra (T=1) containing 110 coat protein subunits organized as 22 pentameric capsomers. The symptoms of the disease caused by geminiviruses are vascular streak, yellow mosaic, veinal yellowing, severe curling, and stunting (Figure 10.2). In recent years the incidence and intensity of diseases caused by geminiviruses have increased. The devastating diseases caused by geminiviruses in economically important crops are challenging ones such as maize streak, beet curly top, cotton leaf curl, cassava mosaic, and legumes yellow mosaic.

Virus members of the family *Geminiviridae* are categorized into nine genera (Table 10.1) on the basis of the vector, hosts, and genome organization (Zerbini et al. 2017). The genera *Becurtovirus*, *Curtovirus*, *Capulavirus*, *Eragrovirus*, *Grablovirus*, *Mastrevirus*, *Topocuvirus*, and *Turncurtovirus* have monopartite genome, i.e., the genetic information required for viral life cycle is contained in one DNA molecule. The virus species belonging to the genus *Begomovirus* have either monopartite or bipartite genome designated as DNA A and DNA B. All the virus members share the basic features of genome organization that the transcription is bidirectional and coding regions are present in both virion and complementary strands which diverge from a non-coding intergenic region. The viral strand has coat protein gene (VI/AVI/CP) and the complementary strand has the gene-encoding replication initiation or associated proteins (CI/ACI/Rep). In the genera mastreviruses, eragroviruses, becurtoviruses, capulaviruses, and grabloviruses, there are two intergenic regions (IRs)—one large intergenic region (LIR) at 5′ start of the coding region, and one small intergenic region (SIR) at 3′ termini of the coding region. In the other four genera, the intergenic region between 3′ termini of viral and complementary sense transcripts is only few base-pair length.

The IR/LIR contains the origin of replication of viral sense strand and the SIR the transcription termination signals of viral and complementary strand. In addition to coat protein gene, one more open reading frame (ORF) V2/AV2/pre-coat protein gene is present downstream of IR in all geminiviruses except New World bipartite begomoviruses. The genera differ in number of genes encoded in the complementary strand. The complementary strands of mastre and becurtoviruses have only replication-associated protein (Rep), which is translated from ORFs C1:C2 by transcript splicing in the

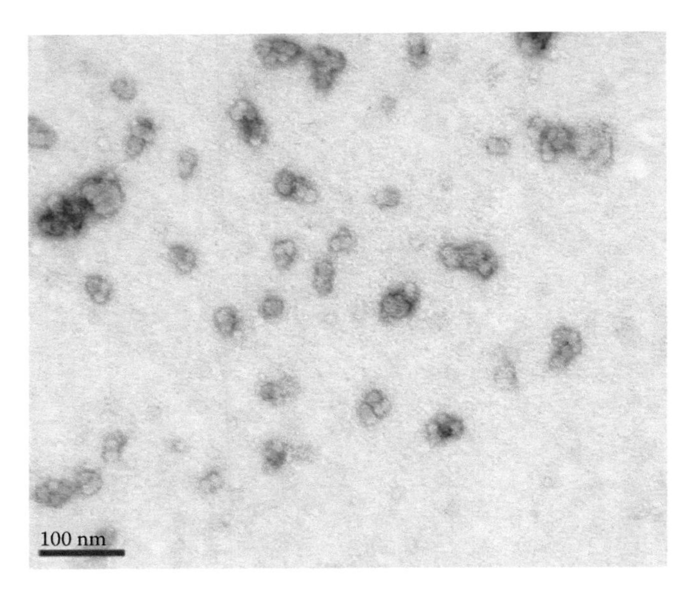

FIGURE 10.1 Purified particles of MYMV stained with uranyl acetate showing typical twinned quasi-isometric subunits. The bar represents 100 nm.

FIGURE 10.2 Symtoms caused by geminiviruses in different hosts. (a) Maize streak virus causing streak symptoms in maize. (b) Curly top symptom in beet by beet curly top virus. (c) Yellow mosaic symptom in grain legumes by yellow mosaic viruses. (d), (e), (f) Leaf curl symptoms caused by begomoviruses in tomato, cotton, and potato. (g) Yellow vein mosaic symptom in okra by bhendi yellow vein mosaic virus. (h) Leaf curl in papaya by papaya leaf curl virus.

TABLE 10.1
Genera of the Family *Geminiviridae*

Genus	Genome	Vector	Hosts	No. of Species
Becurtovirus	Monopartite	Leafhopper	dicot	2
Capulovirus	Monopartite	One species transmitted by aphid	dicot	4
Curtovirus	Monopartite	Leafhopper	dicot	3
Eragrovirus	Monopartite	?	monocot	1
Grablovirus	Monopartite	?	dicot	1
Mastrevirus	Monopartite	Leafhopper	monocot and dicot	>30
Topocuvirus	Monopartite	Treehopper	dicot	1
Turncurtovirus	Monopartite	Leafhopper	dicot	1
Begomoirus	Monopartite and bipartite	Whitefly	dicot	>320

Source: Zerbini FM, Briddon RW, Idris A, Martin DP, Moriones E, Nacas-Castillo J, Bustamante R, Roumagnac P, Varsani A et al. (2017). Geminiviridae. *Journal of General Virology* 98: 131–133.

complementary strand. However, three ORFs, transcriptional activator protein gene which also functions as silencing suppressor gene (C2/AC2/TrAP/SS), replication enhancer gene (C3/AC3/REn), and symptom determinant gene (C4/AC4/sd) are present in the complementary strand in curto, topocu, turncurto, and begomoviruses. The genus *Eragrovirus* encodes a gene for homologue of C2 and the genera *Capulovirus* and *Grablovirus* encode for another gene, whose function is not resolved.

In the case of begomoviruses, almost all the begomoviruses of the New World and some viruses of the Old World are bipartite; DNA A and DNA B are encapsidated separately in geminate particles. Both DNA A and DNA B are required for infectivity. In DNA B, the viral sense strand has one ORF, BVI/NSP coding for nuclear shuttle protein, and one ORF in the complementary strand-encoding movement protein gene (BCI/MP).

Details of viral proteins encoded by the geminiviruses and the functions attributed to them are provided in Table 10.2. Within the IR there is a stretch of 200nt extremely conserved between DNA A and DNA B of the virus, which is referred to as common region (CR). The IR/CR of all geminiviruses has repetitive sequence elements called iterons, which represent the Rep binding sites. The sequence and arrangement of iterons are specific for a particular lineage of the virus.

In IR/CR an extremely conserved stem-loop structure is present which has the invariant nonanucleotide sequence TAA TAT TAC in the loop, and the nicking between the 7th and 8th nucleotide by Rep initiates replication of positive-sense strand.

The monopartite begomoviruses and few of the bipartite begomoviruses of Old World are associated with additional single stranded circular DNA components considered as satellites. There are three types of satellites. The alphasatellites (1:1 kb length), earlier known as DNA 1, which resemble DNA R component of nanoviruses, encode only Rep gene on the viral strand, a stem-loop structure in the noncoding region, and the nonanucleotide sequence TAG TAT TAC. The Rep protein encoded by this satellite shares some features with that of nanoviruses. The most important satellites are betasatellites; they have 1.3 kb circular ss DNA, which shares the origin of replication sequences of begomoviruses and encodes one ORF (beta C1) which is the pathogenicity determinant and also functions as silencing suppressor. All betasatellites have an extremely conserved region upstream of origin of replication, referred to as satellite conserved region (SCR), which is required for replication. In some of the monopartite begomoviruses, though DNA A alone can infect the host plants, severe symptoms are expressed only when DNA A of helper virus is accompanied with betasatellite.

TABLE 10.2
Geminiviral Proteins Encoded and Their Predicted Function

Proteins of the Genes[a]	Abbreviated Name	Molecular Weight (kDa)	Predicted Function
V2/AV2	MP	13	Movement (silencing suppressor) in the case of monopartite viruses
U1/AV1	CP	29	Encapsidation on vector transmission Co-ordination of movement in monopartite
C1/AC1	Rep	40	Replication initiation protein; reprogramming cell cycle
C2/AC2	TrAP/ss	15–17	Transcription activator of the virion sense gense silencing suppressor
In viral sense, C3/AC3	REn	15	Replication enhancer
C4/AC4	Sd	11	Symptom determinant
BV1	NSP	29	Nuclear shuttle protein, movement of viral DNA from nucleus to cytosol
BC1	MP	32	Movement of viral DNA from cell to cell
βC1	Beta C1 or βC1	12–15	Silencing suppressor, virulence factor

[a] For bipartite viruses, prefix A or B is added.

The betasatellite is dependent on DNA A for replication, encapsidation, and vector transmission. The third group of satellites, recently designated as delta satellites (Lozano et al. 2016) represents the subgenomic betasatellites identified earlier with tomato leaf curl virus (ToLCV) from Australia. They have stem-loop, nonanucleotides, SCR, and an "A" rich region similar to betasatellites, but lack the beta C1 ORF. The role of alpha and delta satellites in the viral pathogenicity is not yet fully understood.

10.3 LIFE CYCLE

The deep probing mouth parts of the vector place the virus in the phloem parenchyma and companion cells. Inside the plant cell, the particles are directed to the nucleus, presumably guided by the nuclear localization signals of the coat protein; the ss DNA is released from the virion in the nucleus and the complementary strand is synthesized by host DNA polymerase. In the case of begomoviruses, ribo-priming has been confirmed (Saunders et al. 1992); in the mastreviruses, complementary DNA strands 78–88 nucleotides in length were found associated with virion DNA, which may prime the complementary strand synthesis (Andersen et al. 1988). The ds DNA is the template both for transcription and replication. The host RNA polymerase II transcribes the Rep gene (early gene), the Rep initiates replication. The replication is either by rolling circle mode or by the recombination-dependent mode. The newly synthesized strands will enter replication cycles or may get encapsulated. From the nucleus the viral genomic DNA (ss and ds—it is still disputed) is moved to the periphery by the coordinated activities of V2/CP/C4 proteins in monopartite viruses, by NSP in bipartite viruses from where they are docked on to plasmodesmata to facilitate cell to cell movement. The ss DNA gets encapsulated, and the assembled virion in the young leaves are acquired by the vector, which transmits the virions to healthy plants while ingesting the plant sap.

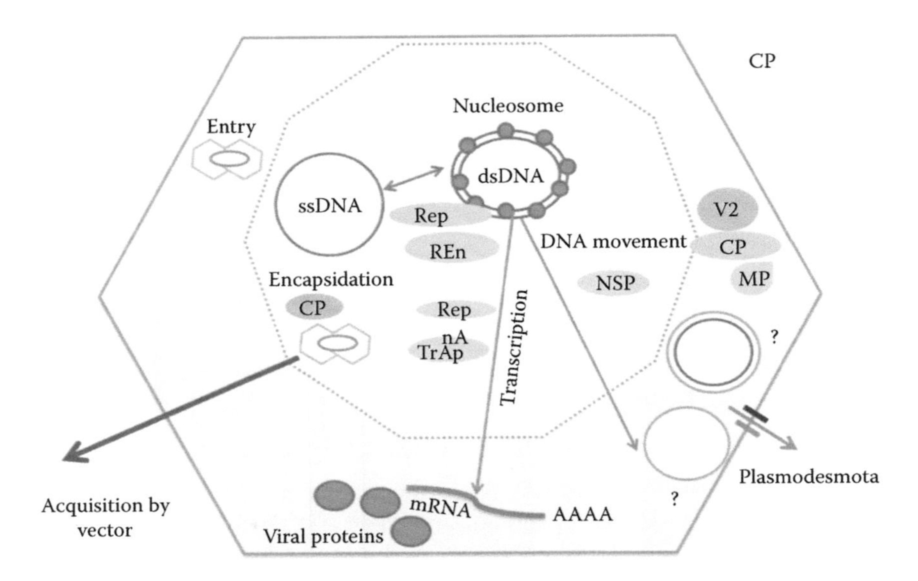

FIGURE 10.3 The geminivirus life cycle. ssDNA of the introduced virion particle is released into the nucleus. Host DNA polymerase copies it; the ds DNA formed is the template for replication and transcription. Rep protein initiates replication and movement is facilitated by NSP and MP in bipartite viruses, by CP/V2 in monopartite viruses.

From the initial event of infecting the cells to its final maturity stage (Figure 10.3), the encounter between the host and the geminiviruses can be analyzed under five headings:

- Reprogramming of host cell cycle and replication
- Voyage of viral DNA by hijacking transport proteins
- Interference with host defense
- Disruption of signaling pathway
- Deregulation of host gene expression

Reprogramming Cell Cycle and Viral Replication

Geminiviruses are transmitted by arthropod vectors, which place the virion particles into fully differentiated proto phloem cells; in the plants, once the cells are fully differentiated, they have exited from cell cycles and are not rich in proteins involved in DNA synthesis like host DNA polymerase and many accessory factors. The cell cycle of the normal plant is shown in Figure 10.4. It is separated into two major phases that alternate with each other the interphase during which cells grow, prepare for mitosis, and duplicate into DNA, and the mitotic M phase in which the cell divides into two daughter cells. The interphase consists of three phases. G1 phase: cell prepares for division; S phase: when DNA synthesis takes place, genetic material is replicated; G2 phase: metabolic changes will lead to assembly of cytoplasmic material for mitosis and cell division; and M phases: when nucleo division followed by cytokinesis takes place. During geminivirus infection, alteration of transcription of all genes related to cell cycles occurs. Nagar et al. (1995) showed that in TGMV-infected *Nicotiana benthamiana* under virus infection, proliferating cell nuclear antigen (PCNA) accumulates. PCNA is a processivity factor for host polymerase and interacts with many proteins involved in cell cycle regulation, DNA replication, and DNA repair. Further, by transcription profiling, Ascencio-Ibanez et al. (2008) revealed enhanced expression of cell cycle-associated genes in virus-infected Arabidopsis. The genes associated with early G1 phase and late G2 phases are upregulated, thereby facilitating re-entry of the cells into S phase to ensure availability of DNA synthesis proteins.

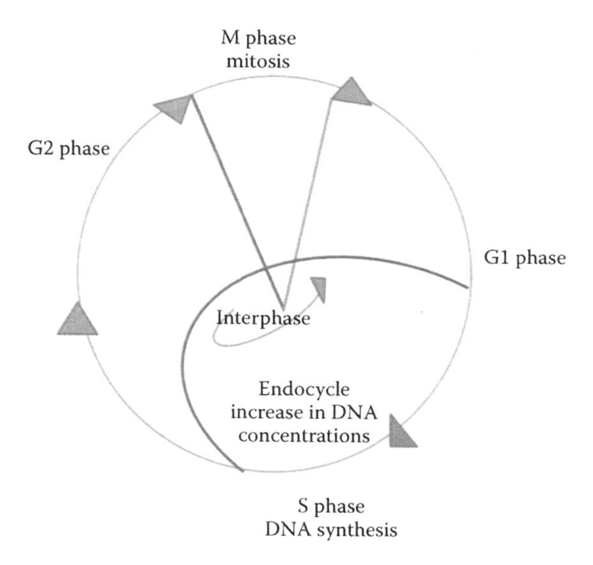

FIGURE 10.4 The cell cycle in normal cell. Two major phases alternate with each other: interphase, during which the cell grows, prepares for mitosis and duplicates its DNA, and the mitotic (M) phase are shown. Interphase is divided into three phases: G_1, S, and G_2.

Once the environment conducive for DNA synthesis is created, geminivirus replication is initiated. The replication is initiated by rolling circle amplification mode, which is as follows (Figure 10.5). Three phases of replication are recognized: initiation, elongation, and termination. The geminivirus-encoded replication initiation protein Rep performs catalytic action on the phosphodiester bond between the 7th and 8th nucleotide of the nonanucleotide loop, which is highly conserved across all geminiviruses. The Rep executes ATP-dependent isomerase1, ATPase, and Helicase activities. Rep, the 40 kDa protein having 362 amino acids, has the following domains. Domains 1–120 amino acid residues govern DNA nicking, and 1–130 residues govern DNA-binding activities. The Walker A and

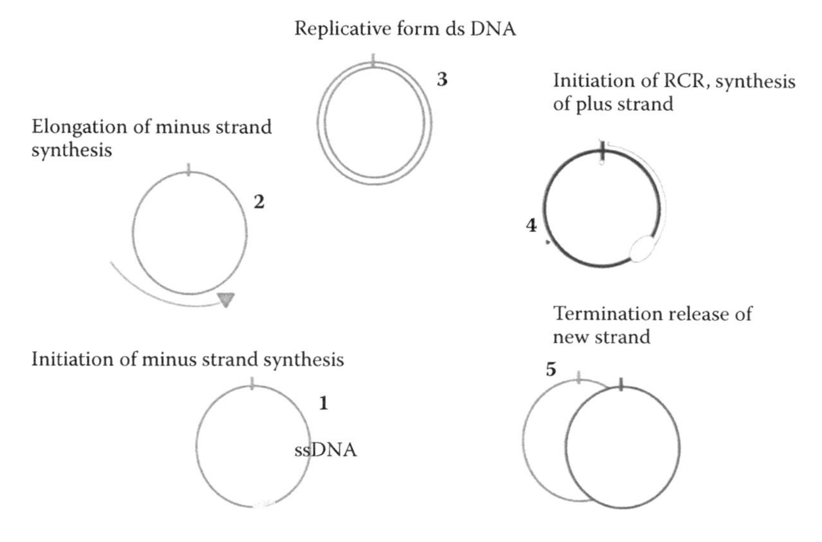

FIGURE 10.5 Rolling circle replication. Minus strand synthesis is by host DNA polymerase; initiation of RCR by nicking at nonanucleotide sequences, elongation followed by termination and release of sscircular DNA.

Walker B motif are characterized in rolling circle replication. The initiation proteins span from amino acids 221–228 and 261–264 residues, respectively; 3′ to 5′ helicase activity lies between amino acids 120–361 (Hanley Bowdoin et al. 1999, 2013; Pant et al. 2001; Choudhury et al. 2006; Yadava et al. 2010; Clerot and Bernardi 2006). The Rep protein is vital in recruiting other accessory factors to initiate replication. A complex assembly called replisome, consisting of viral proteins, Rep and Ren, and host factors like PCNA—factors involved in DNA replication and repair—is formed to initiate replication. Rep binds to the iteron sequences in a cooperative manner, leading to conformational changes in DNA such as formation of cruciform structure. More than one molecule of Rep is required to perform nicking and ligation activity, hence Rep oligomerizes, and about eight molecules of Rep are involved in the events of replication. Hetero-oligomers are also formed between Rep and REn.

The initiation phase is marked by binding of Rep and nicking, creating 3′OH end, which is followed by elongation phase. During elongation, Rep recruits many host factors, such as a large subunit of replication factor C complex (Luque et al. 2002) which is required to facilitate interaction between PCNA and DNA, and a 32 kDa subunit of replication protein A (Singh et al. 2007) which binds to ss DNA. Thus many host factors and Rep assemble at the 5′–3′ region, and elongation of the strand is executed by host DNA polymerase. The Rep of geminiviruses was also shown to have Helicase or strand displacement activity (Choudhury et al. 2006; Clerot and Bernardi 2006; George et al. 2014). At the end, the nascent concatenated DNA is subjected to Rep-mediated nicking and ligation, releasing the full length ss circular DNA which enters into another round of replication or is encapsidated.

During the process of replication, Rep and REn interact with several host proteins: PCNA, replication factor C complex, and replication protein A (see Table 10.3).

Rep also binds to a protein having vital role in recombination, RAD54 (Kaliappan et al. 2012). Interestingly, RAD 54 enhances ATPase and Helicase activities of Mungbean yellow mosaic India virus (MYMIV) in yeast system contrasting to PCNA, which downregulates nicking and ligating activities. Both PCNA and RAD54 may modulate rolling circle replication and recombination mediated replication.

The most interesting interaction of Rep with the host proteins is its interaction with retinoblastoma-related proteins (RBR). The plant RBR controls the genes governing cell cycle, cell specification, and differentiation. RBR interacts with the transcriptional factors E2F, which govern the expression of genes governing host replication proteins. The interaction between RBR with E2F is controlled by phosphorylation. The phosphorylation of RBR disturbs its binding with E2F, and as a result E2F factor is released. The released E2F leads to transcription of genes in late G1 phase, allowing entry to S phase. The Rep protein of geminivirus interferes with RBR-E2F complex, which has been shown for Rep A protein of mastrevirus (Liu et al. 1999). The binding of Rep A to RBR is through its LXCXE motif, similar to RBR binding domain of proteins of mammalian viruses. The begomoviruses bind to RBR through a different motif. By binding to RBR, releasing E2F factor, enhanced gene expression of late G1 phase occurs, resulting in its re-entry into S phase.

The binding of REn with RBR may help in overcoming RBR-mediated inhibition of virus replication in mature leaves. The interaction involving G1 cyclins and their cyclin-dependent kinase (CDK) downstream of RBR action may vary depending on the cell type, host, and viruses involved. In many geminivirus-infected cells, plant cells do not progress to mitosis (increased cell division) but re-enter into the endocycle, leading to enhanced replication of both viral and plant chromosomal DNA (Ascencio-Ibanez et al. 2008; Nagar et al. 2002). The occurrences of enations and thickening of veins in geminivirus infection do suggest active cell division and mitosis cell cycle, as reported by Briddon et al. (2003). For example, Lai et al. (2009) showed that C4 proteins of beet severe curly top virus induce the synthesis of E3 ligase, which inactivates cyclin kinase inhibitors required for degradation, leading to plant cell proliferation. A similar type of enhanced cell division has also been demonstrated for beta C1 proteins (Yang et al. 2008).

The Rep protein also interacts with two closely related protein kinases, Geminivirus Rep interacting kinase 1 (GRIK 1) and GRIK 2 (Kang et al. 2005; Shen et al. 2006, 2009). The GRIK kinase are involved in plants response to abiotic and biotic stress and are upstream regulators of SNFI-related

TABLE 10.3
Summary of Host Proteins Interacting with Geminiviral Proteins

Viral Protein	Host Interacting Proteins	Role in Viral Pathogenesis
V1/AV1/CP	α karyopherin α1 α importin	Facilitating viral DNA movement (Kunik et al. 1999; Guerra-Peraza et al. 2005)
V2/AV2/MP	Suppressor of gene silencing 3 (SGS 3) protein, binds to 21nt siRNA and miRNA	Interference with host gene silencing activity (Zhang et al. 2012)
C1/AC1/Rep	proliferating cell nuclear antigen (PCNA), retinoblastoma-related protein (RBR), RAD 54, Geminivirus Rep interacting kinase 1 (GRIK1), kinase 2 (GRIK2). >150 proteins identified E2 enzyme SUMO-conjugating enzyme-1 (SCE-1)	Reprogramming of cell, proteasome pathway (Hanley-Bowdoin et al. 2013) Sumoylation (Castillo et al. 2004)
C2/AC2/ TrAP	SNF1-related protein kinase (SNRK1), Adenosine kinase (ADK) S-adenosyl methionine decarboxylase (SAMDC1) CSN complex	Interference with defense (Hao et al. 2003) Methylation pathway (Wang et al. 2005; Zhang et al. 2011) Ubiquitylation (Lozano-Duran et al. 2011)
C3/AC3/REn	RBR	To counter RBR initiation in mature leaves (Settlage et al. 2001)
C4/AC4/sd	Shaggy-related kinases (specifically, Brassinosteroid- insensitive 2 (BIN 2) Ring fungus protein (RKP)	Brassinosteroid signaling pathway (Piroux et al. 2007; Dogra et al. 2009) Proteosomal degradation (Lai et al. 2009)
BV1/NSP	Nuclear shuttle protein interacting kinases NIK1, NIK2 NIK 3 Histone Arabidopsis thaliana acetyltransferase (AtNS1) NSP interacting GTPase (NIG)	Translational repression (Carvalho et al. 2008) Nucleosome binding to viral DNA along with NSP for movement (Zhou et al. 2011) Inhibition of histone acetylation (Carvalho et al. 2006) Exit from nucleus into cytosol (Carvalho et al. 2008)
BC1/MP	Histone Chloroplast heat shock protein (HSP 70) Synaptotagnim (SYTA)	Movement across plamodesmata (Zhou et al. 2011) Transport of viral DNA (Krenz et al. 2010) Transport of macromolecule (Lewis et al. 2010)
βC1	SNRK1 UBC 3 S-Adenosyl homosysteine hydrolase (SAHH)	Phosphorylation of βC1 interferes with infection (Shen et al. 2011) Accumulation of ubiquitylated proteins, symptom severity (Eini et al. 2009) Viral genome methylation prevented (Yang et al. 2011b)

protein kinases (SNRPK 1). The SNRPK 1 influences development of symptoms and accumulation of viral DNA and binds to C2/TrAP and beta C1 protein (Hao et al. 2003; Shen et al. 2011). SNRPK 1 is involved in defense of the plants, and it is hypothesized that its activity may be neutralized by TrAP/C2 protein. The involvement of GRIK and SNRPK1 in geminivirus infection is not yet fully understood, but its influences on viral DNA accumulation suggest its action is through replication.

The Rep protein also binds with histones (Kong et al. 2002). Binding with histone assumes significance in the context of formation of mini-chromosomes by geminivirus ds DNA with 11–12 nucleosomes (Pilartz and Jeske 1992). Binding with histone will lead to displacement of nucleosomes from DNA, which will allow replication machinery to have access to DNA. Indirectly, the binding may prevent methylation of histones (at H3 lysine 9), which is known to impair viral replication. Interestingly, Rep is demonstrated to bind to mitotic kinensin (Kong et al. 2002) and with mini-chromosome maintenance protein 2 (MCM2) (Suyal et al. 2013). Yadava et al. (2010) cloned the MYMIV replicon in yeast vector YrP34 and examined the eukaryotic factors interacting with Rep, isolated more than 150 factors, and showed that about two dozen factors are required for MYMIV replication.

Rep, being a multifunctional protein, interferes with various host pathways. The events prior to replication and post replication can be summarized as below. As soon as geminivirus infection occurs and Rep is translated, Rep binds to RBR, thereby reprogramming the cell cycle to enter active DNA synthesis phase (G1–S phase), which is conducive to rolling circle DNA replication. Besides RCR, geminiviruses can also replicate by the recombination-dependent replication (RDR) mechanism (Jeśke 2009; Jeske et al. 2001; Erdmann et al. 2010). Accumulation of viral DNA, nicked open circular DNA, and ss DNA trigger overexpression of genes encoding DNA repair and recombination proteins (Ascencio-Ibanez et al. 2008). Excessive expression of DNA recombination and repair protein may alter the viral DNA replication from rolling circle mode to RDR mode. In this mode of replication, a small ss DNA fragment may anneal to an analogous region in the circular ds DNA facilitated by host recombination enzymes. The DNA polymerase extends the linear ss DNA fragment using a circular template. After one or two rounds of replication, linear ss DNA is converted to linear ds DNA by host DNA polymerase complex primed by a short fragment of viral DNA. Thus a high population of heterogenous ds DNA is produced, which becomes substrate for methylation; while RDR priming does not require Rep, Rep is needed for nicking and circularizing ss DNA, which may enter into RCA mode of replication. The RDR type of replication explains why recombination is so frequent among geminiviruses.

10.4 VOYAGE OF VIRAL DNA BY HIJACKING TRANSPORT PROTEINS

The coat protein of the monopartite viruses, tomato yellow leaf curl virus, was demonstrated to have functional nuclear localizing signals in the amino acid residues 3–20 (arginine rich). It was found to be bipartite classes of NLS, as other additional residues were observed between amino acid residues 36–61 (Kunik et al. 1998) the TYLCV CP interacted with karyopherin α (Kunik et al. 1999). The interaction mediates its nuclear import by a karyopherin α-dependent mechanism. Its coordinated action of CP/V2/C4 which brings about movement of viral DNA outside the nucleus in the cell periphery and onto plasmodesmata. In the case of bipartite begomovirus Mungbean yellow mosaic virus (MYMV), Guerra-Peraza et al. (2005) located two nuclear localizing signals (residue 3, KRR, and 41 KRRR) and demonstrated that they interact with nuclear import factor importing α; thus CP may get transported inside the nucleus through an importing α-dependent pathway. In the case of bipartite begomoviruses, the nuclear shuttle protein (BV1/NSP1) and movement proteins (MP) execute the movement of viral DNA (ss or ds is not clear). BV 1 performs the function of CP/V2 of monopartite virus and is responsible for the exit of the DNA from the nucleus into the cytoplasm periphery. BC1 facilitates docking of viral DNA into plasmodesmata and relocalization of NSP into the nucleus.

NSP is a dynamic multifaceted protein, i.e., it interacts with histones H3 (Zhou et al. 2011) and with *A. thaliana* acetyl transferase (AtNS1) (Carvalho et al. 2004). Carvalho et al. (2006) hypothesized that viral ds DNA is packaged by histones into nucleosome, which is further compacted by NSP. Since NSP interacts with AtSN1, histone acetylation is suppressed. Interaction of MP was also shown with histone protein, suggesting that viral DNA moves from cell to cell as nucleosome. It is possible that mini-chromosome may be involved through ER medicated tubule. The NSP interacts with GTPase (nuclear shuttle protein interacting GTPase) available on the nucleus envelope which may allow the NSP exit into the cytosol. The transfer of DNA from NSP-DNA complex may occur when NIG-catalyzed GTP hydrolysis occurs. The MP also interacts with HSP 70 in the chloroplast and with synaptotagamin proteins (Krenz et al. 2010; Lewis et al. 2010). Thus the evidence has been gathered to show that geminiviral proteins hijack the host transport proteins to gain intracellular and intercellular movement.

10.5 INTERFERENCE WITH HOST DEFENSE PATHWAYS

Antiviral immunity against geminiviruses can be discussed under three topics: effector-triggered immunity, the acquired immunity triggered by dsRNA, and the third novel type of translational repression, a unique mode of viral resistance in geminiviruses.

Effector-Triggered Immunity

The host defense against pests and diseases is increasingly understood as an inherent or innate evolutionarily conserved mechanism to recognize the microbe-associated molecular pattern (MAMP) or pathogen-associated molecular pattern (PAMP) by the transmembrane recognition receptors (PRKn), leading to PAMP-triggered immunity, or PTI. PTI involves cascade of mitogen-activated kinase actions leading to activation of transcriptional factors and enhanced expression of defense-related genes. This innate immunity in response to PAMP molecules is more common against fungal and bacterial pathogens. To counter this defense, the pathogen releases effector molecule, which interferes with PTI, and the susceptible response is soon challenged by the R gene factor which recognizes effector, neutralizes it, and switches on the effector-triggered immunity (ETI). In the case of plant viruses, PAMP molecules have not been characterized, and whether any epitope of coat protein could function as PAMP molecules and be recognized by pattern receptor-like kinases is not yet known. This kind of effector/immunoreceptor resistance has been reported in only one geminivirus, MYMIV (Maiti et al. 2012). The dominant R gene locus characterized in *Vigna mungo* is termed CYR1, which the characteristic C-NBS-LRR features. It was suggested to interact with MYMIV-CP (Maiti et al. 2012) to activate the cascade of defense.

Translation Suppression

A novel type of host resistance to geminivirus was revealed by Fontes et al. (2004), Santos et al. (2009), and Carvalho et al. (2008). They identified three leucine-rich receptor (LRR)-like kinases (NIK1, NIK2, NIK3) in tomato and soybean interacting with nuclear shuttle protein of TYLCV and tomato crinkle leaf yellow virus (TCrLYV). NIK1 proteins (NSP interacting kinases) are typical kinases in that they do get autophosphorylated and get involved in phosphorylation of different substrates. NSP binds with NIK at kinase domain, inhibits autophosphorylation, and as a result they cannot phosphorylate downstream effector ribosomal protein RPL10. The NIK family belongs to the same subfamily of LRR-RLK family as BRI1-associated receptor kinase (BAK1). Zorzatto et al. (2015) demonstrated that the translocation of RPL10 to the nucleus, where it interacts with MYB-like protein (L10-interacting MYB domain-containing proteins [LIM7B]) and downregulates translational machinery. When LIMYB is overexpressed, the ribosomal protein genes are repressed at transcriptional level, protein synthesis is inhibited, and viral messenger RNA association with polysome is reduced, and Arabidopsis plants show tolerance phenotype to cabbage leaf curl virus. Conversely, the loss of LIMYB function results in release of repression of translation-related genes and increased susceptibility. This is the first instance in plants that activation of global translation suppression is employed as a mechanism to fight viruses.

10.6 RNAi DEFENSE PATHWAY

The most effective and versatile defense pathway of the host against viruses is the RNA-silencing pathway that employs small interfering RNA (siRNAs) to target viruses, transposan, or any invasive nucleic acid. This is considered an acquired immunity, and ds DNA is considered a PAMP molecule that primes the RNAi pathway. This phenomenon in RNA viruses is only a post-transcriptional gene silencing (PTGS) occurring in cytoplasm, but in the case of geminiviruses it is both transcriptional (TGS) and post-transcriptional silencing. Normally the silencing pathway involves recognition and cleavage of ds RNA by DICER-like (DCL) ds RNA-binding proteins. It is diced into 21–24 nt small RNA, RNA in duplexes with 2nt overhangs. The 5′ of the sRNA is phosphorylated, and the second nucleotide at the 3′ end is methylated. Of these duplex RNAs, one of the strands associates with Argonaute (AGO) and other proteins to form RNA-induced silencing complexes (RISC) accompanied by degradation of the other strand. Depending upon the effect of protein argonautes (AGO), either siRNAs are engaged in chromatin modification (RNA-induced transcriptional silencing [RTS]) or in translational inhibition and cognate RNA degradation (slicing). The cleavage products, "aberrent RNA," serve as a template for host RNA-dependent RNA polymerase producing more ds RNA and amplifying the siRNA (Ding and Voinnet 2007; Hohn and Vazquez 2011; Garcia-Ruiz et al. 2010; Wang et al. 2011).

There are many deviations observed in geminivirus infection compared to RNA silencing in RNA virus infection. While silencing of RNA viruses occurs predominantly in the cytoplasm by PTGS, in the case of DNA viruses it occurs in both the cytoplasm and nucleus. The production of ds DNA to prime RNAi has always been an interesting question. It was clearly shown in the case of Cabbage leaf curl virus (CaLCuV) that the bidirectional read-through transcription at the 3′ end of rightward and leftward transcripts provides dsRNA to prime RNA silencing. An overlap of 12–16 nt length generates a ds RNA priming RNAi activities. Contrasting to RNA viruses, production of viral small RNA (vs RNA) of 24 nt length by DCL 3 were shown in the case of geminiviruses.

The region in the viral genome targeted for siRNA production also varies. In CaLCuV/Arabidopsis and Squash leaf curl China virus (SLCNNV)/cassava, siRNA were produced for CP and NSP transcripts, and more from DNA B genome than DNA A. In the case of Pepper golden mosaic virus (PepGMV)/pepper, mainly 24 nt siRNAs are produced for the non-coding region. Generally, 24 nt vsRNA targeting intergenic promoter-containing regions is more predominant.

There is enough evidence to suggest that the main pathway by which hosts encounter the infection of geminiviruses is by methylation of viral DNA. It occurs along the entire genome, but depending on the virus-host combination, the regions may differ (Akbergenov et al. 2006; Rodriguez-Negrete et al. 2009; Yang et al. 2011a; Aregger et al. 2012). The enzyme complexes involved in methylation process are many. To start with, the 24 nt ds RNA is methylated at 2 OH group by HEN1 methyl transferase, then the guide strand binds to AGO4, or AGO6, or AGO 9, and these along with scaffold transcripts made by Pol IV chromoproteins, histones H3K9 methylase, and DNA-rearranged methyl (DRM) chromomethylase (DRM 1), form RNA-triggered silencing, resulting in DNA methylation (Pooggin 2013). In geminiviruses, methylation can cover the entire viral genome, and the region targeted varies from host/virus combination. For example, in pepper golden mosaic virus–infected pepper plants, methylation of intergenic region is higher than coat protein region (Rodriguez-Negrete et al. 2009). However, geminiviruses are known to escape methylation. Interestingly, Brough et al. (1992) observed that TGMV DNA, when methylated in vitro and introduced into plant cells though replication, is reduced, non-methylated progeny DNAs are freshly produced. Paprotka et al. (2011) suggested that only linear heterogenous viral DNA, which does not take part in active replication, is methylated. Thus, geminiviruses are not much affected by methylation as they can depend on unmethylated DNA for replication or by inhibition of methylation pathway by viral suppressors.

The viral suppressors encoded by virus interfere with the RNAi pathway at different steps to gain control over the defense of hosts. In the case of geminiviruses, there are multiple viral suppressors of RNA silencing (VSR): C2/AC2, C4/AC4, and V2/AV2 encoded by the monopartite DNA A component of the begomoviruses. NSP encoded by DNA B and the beta C1 protein encoded by betasatellites are also demonstrated to have suppressor activity (Table 10.4). These proteins mainly target the methylation pathway of the host. For example, Wang et al. (2012) showed that the C2/AC2/TrAP proteins interact with and inactivate adenosine kinase (ADK), which is required for synthesis of the methyl donor S-adenosyl methionine (SAM) (Wang et al. 2005). The C2 homologue in curtovirus

TABLE 10.4
Suppressors Encoded by Geminiviruses

Virus	Viral Genes Having Silencing Suppressor Activity
Beet curly top virus (BCTV)	C2
African cassava mosaic virus (ACMV)	AC2, AC4
East African cassava mosaic virus (EACMV)	AC2, AC4
Mungbean yellow mosaic virus (MYMV)	AC2
Mungbean yellow mosaic India virus (MYMIV)	AC2, AC4
Sri Lankan cassava mosaic virus (SLCMV)	AC2, AC4
Bhendi yellow vein mosaic virus (BYVMV)	C2, C4, βC1
Cotton leaf curl Multan virus (CLCuMV)	C2, C4, V2, βC1, α satellite Rep
Tomato leaf curl China virus (TLCCV)	C2, βC1
Tomato yellow leaf curl virus (TYLCV)	C2, V2
Tomato leaf curl Java virus (ToLCJV)	βC1
Tomato leaf curl Philippines virus (ToLCPV)	C2, C4, βC1

interacts with SAM decarboxylase, which enhances SAM decarboxylation (Zhang et al. 2011). The beta C1 (Yang et al. 2011b) showed interaction with S-adenosyl homocysteine hydrolase (SAHH), an enzyme which is required for methylation. Interestingly, Rep and C4 downregulates DNA methyl transferase (METI) chromomethylase 3 (CMT3) enzymes involved in maintenance of methylation (Rodriguez-Negrete et al. 2013).

This conclusion is also corroborated by studies on symptom recovery phenomenon commonly observed in begomovirus infection. In cases of viruses showing recovery of symptoms (Raja et al. 2008; Yang et al. 2011a; Hagen et al. 2008), increased viral DNA methylation has been reported in all those systems showing recovery phenotypes. As stated earlier, the methylation and vsRNA are more targeted to the intergenic region that contains the promoters and origin of replication. Yadava et al. (2010) found that methylation of the intergenic region was greater in resistant cultivars of soybean than the susceptible ones.

The role of secondary siRNA, produced by RNA-dependent RNA polymerase (RDR), was examined. Aregger et al. (2012) found that rdr mutants of Arabidopsis did not show recovery phenotype, suggesting that RDR role and secondary siRNA production are equally important in geminiviruses. In this context, it is interesting to note that the only geminivirus resistance gene cloned and sequenced, *TY1*, is an RNA RDRγ tomato with homology to RDR 3, RDR 4, and RDR 5 of *Arabidopsis thaliana*. It is proposed that RDR activity is required for DNA methylation (Verlaan et al. 2013).

In addition to methylation, VSR such as V2, which binds to siRNA, may also bind to host-silencing protein SGS3 and interfere with the normal developmental process; V2 may bind to miRNA-governing auxin-related genes, which may cause deregulation of auxin-related activites and contribute to symptom production (Zhang et al. 2012).

10.7 INTERACTION WITH KINASES, HORMONE SIGNALING PATHWAY GENES

The shaggy-related kinases, homologues to glycogen synthase kinases (GSKn), are being characterized in plants. Analysis of Arabidopsis genome revealed ten GSK genes categorized into four distinct subfamilies. These are involved in flower development, brassinosteroid signaling, and sodium chloride and wound stress. Interestingly, the C4 and AC4 proteins of geminiviruses of interact with one of the shaggy kinases of Arabidopsis, Brassinosteroid insensitive 2 (BIN 2), and although both

C4 and AC4 interact with BIN2, their modes of action are different. In bipartite begomoviruses, AC4 protein binds to BIN2 and upregulates brassinosteroid pathway genes (Dogra et al. 2009). In contrast, C4 protein of monopartite virus stops B1N2 inactivation and maintains inhibition of the brassinosteroid pathway (Piroux et al. 2007). This is a clear indication of how viruses may differ in their modes of action with the same host protein.

Geminiviruses also interfere with salicylic acid and ethylene pathways. Increased salicylic acid or more expression of gene in the pathway confer resistance to infection (Ascencio-Ibanez et al. 2008; Pierce and Rey 2013; Chen et al. 2010; Garcia-Neria and Rivera-Bustamante 2011). Conversely, the jasmonic acid pathway is suppressed by viral protein-like beta C1 (Yang et al. 2008; Lozano-Duran et al. 2011; Soitamo et al. 2012). Symptoms of vein thickening, enations often observed in geminivirus infection, are suggestive of altered cytokinin and auxin pathways. The cytokinin-responsive genes are activated (Baliji et al. 2010; Park et al. 2004). Wang et al. (2003) suggested that adenosine kinase, the enzyme required for phosphorylation of cytokinin, is converted to inactive form by C2/AC2/TrAP.

It is perplexing that as such, cell-death or necrosis phenotype is still not observed in any geminivirus infection. Rep, V2, and NSP cause cell death in transient agroinfiltration analyses and many host genes associated with cell death are upregulated.

10.8 PROTEOSOME/UBIQUITYLATION PATHWAY

Modification of proteins by ubiquitin and ubiquitin-like proteins is one important post-translational event that modulates protein function and thereby influences many developmental processes and responses to biotic and abiotic stress. The ubiquitin molecule is covalently linked to the lysine residue in the target protein by conservative action of three enzymes: El ubiquitin-activating enzymes, E2 ubiquitin-conjugating enzymes, and an E3 ubiquitin ligase. The polyubiquitylation of the protein is recognized by the 26S proteasome, where it is degraded. In addition, there are small ubiquitin-like modifier proteins (SUMO) associated with another set of E1, E2, and E3 enzymes. Monoubiquitylation and sumoylation alter the protein activities, its interaction, and sub-cellular localization. Geminiviral proteins have been demonstrated to interact with ubiquitin and ubiquitin-like pathways. For example, βC 1 protein of the tomato yellow leaf curl china satellite TYLCCB binds to E2 ubiquitin-conjugating enzymes (UBC3) and reduces the level of polyubiquitylated protein, resulting in severe symptom production (Eini et al. 2009; Bachmair et al. 1990). The C2 protein interacts with many key regulators such as E3SKPl, CUL-1, and F-box-containing SCF ligases. Overexpression of C2 alters several plant hormone responses regulated by CUL1-based SCF ubiquition E3 ligases. The C4 protein of curtoviruses and monopartite begomoviruses induces cell proliferation by interacting and activating host RING finger protein (RKP) (Lai et al. 2009). It is important to note here that RING finger proteins target cyclin kinase inhibitors for proteosomal degradation. Rep also interacts with the E2 enzyme Sumoconjugation enzymel (SCE1), and this interaction is necessary for viral replication. In transient assays, Rep alters the sumoylation status of host proteins that might have roles in replication (Lozano-Duran et al. 2011; Castillo et al. 2004).

10.9 DEREGULATION OF HOST GENE EXPRESSION

Upon geminivirus infection, the plant remodels its cellular components involved in diverse processes, such as transcription, hormone signaling, metabolic pathway, and defense-related processes. Transcript profiling of host under geminivirus infection has been accomplished for many virus/host combinations, which are reviewed here.

A study by Sahu et al. (2010) highlighted the gene expression changes during incompatible interaction between tolerant tomato plant and *Tomato leaf curl New Delhi virus* (ToLCNDV). A suppression-subtractive hybridization library (SSH) was prepared in a naturally tolerant cultivar of tomato, namely H-88-78-1. This study revealed that tolerant tomato plants have an enhanced level of transcript related to cell cycle and DNA/RNA processing, signaling molecules, transporters, and transcription factors along with the proteins of unknown functions. An interesting observation was that apart from these genes, classes of host ubiquitin proteasome pathway genes were also highly expressed in tolerant cultivar in comparison to a susceptible cultivar followed by ToLCNDV.

Sahu et al. (2016) subsequently attempted to characterize 26S proteosomal subunit RPT4a (SIRPT4) gene under ToLCNDV pathogenesis in the tolerant cultivar H-88-78-1. They showed specific binding of SIRPT4 at the stem-loop region of IR in both DNA A and DNA B. They noted that this binding is secondary structure specific, and binding at IR inhibited RNA pol II activity, thereby reducing bidirectional transcription. When they silenced the SIRPPT4 gene, the tolerant phenotype of H 88-78-1 was converted to susceptible phenotype. Overexpression of SIRPT4 gene resulted in programmed cell death and hypersensitive reactions. They suggest that SIRPT4 interference of viral pathogenicity is due more to specific binding, and not due to any proteolytic function.

Mandal et al. (2015) studied the transcript level of the TORNADOI (SITRNI) gene, which is important for cell expansion and vein formation. Though SITRN1 has two start sites, under viral pathogenesis there is a preferential use of one start site only. They found that the promoter sequences of SITRN1 have multiple W boxes, which mediate induction of SITRN1 under ToLCNDV infection. They postulate that during stress, the SA pathway gets activated, which induces WRKY16, leading to transcription of the SITRN1 gene.

Kundu et al. (2011) identified levels of about 29 proteins involved in stress response, photosynthesis, transport, and signal transduction to be enhanced upon SA treatment, which conferred protection against MYMIV infection. Castillo et al. (2012) analyzed transcriptome of PepGMV-infected pepper plants from initial stage of infection to recovery of symptoms. They identified novel genes, pepper-RRP1, and histone proteins under pathogenesis. Kaur et al. (2014) investigated differential gene expression in tomato inoculated ToLCNDV. They concluded that 652 genes belonging to 77 known pathways are altered. These genes are related to increased rate of respiration, decreased rate of photosynthesis, and accumulation of sugars and higher level of amino acid synthesis. Miozzi et al. (2013) made extensive analysis of tomato yellow leaf curl Sardinia virus-infected tomato and found upregulation of genes involved in hormoneand nucleic acid metabolism, ubiquitin-proteosome, and autophagy. They also recorded induction of GA-ABA-responsive genes and activation of autophagic process.

A study on soybean plant susceptible to MYMIV was used to study host gene response. A high-throughput microarray analysis of MYMIV-infected soybean probed with 17000 genes (Yadav and Chattopadhyay 2014) revealed programmed cell death and disease resistance response in the enhanced expression of various genes linked with SAR.

An attempt to identify the Chilli leaf curl disease resistance genes was carried out by Kushwaha et al. (2015). A comprehensive transcriptome analysis was performed to identify the host defense-responsive genes exploiting a ChiLCV-resistant variety of chilli, Punjab Lal. This study reveals that during incompatible interaction between ChiLCV and Punjab Lal, the host plant activates numerous genes involved in defense, transcription, DNA replication, transport, signaling, and translation process. A comparative expression analysis between contrasting cultivars of chilli differences in ChiLCV tolerance traits reveals the differential expression of diverse genes such as NBS-LRR, polyphenol oxidase, lipid transfer protein, thionin, ATP/ADP transporter, and histone H1.

A proteomic approach by Carmo et al. (2015) to investigate proteins associated with Tyking-derived recessive resistance to tomato chlorotic mottle virus (ToCMoV) revealed that proteins associated to chromatin structure, cytoskeleton structure cuticle biosynthesis, and ubiquitin pathway to play a vital role in viral pathogenesis. A comparative proteomic study in tomato following infection by TYLCV by Huang et al. (2015) revealed 86 differentially expressed proteins. Proteins such as CDC48, CH1, and HSC70 were vital in viral infection.

The molecular mechanism behind resistance to pepper leaf curl virus in pepper cultivars was probed by Rai et al. (2016), who discovered upregulation of about 234 unigenes. Liu et al. (2014) studied the transcriptional analysis of cassava affected by ACMV, and explained the chlorotic symptoms to be due to upregulation of chlorophyll degradation genes such as chlorophyllase, pheophytinase, and pheophorbide oxygenase and downregulation of genes governing apoproteins in light harvesting complex II.

Jeevalatha et al. (2017) studied the transcript profile of potato inoculated by ToLCNDV, and observed changes in defense-related genes. Czosnek et al. (2013) attempted to identify the genes involved in resistance phenotype under infection by TYLCV. Extensive analysis showed that Permease-1 like gene, Hexose transporter LeHT1, and Lipocalin protein are required for maintaining resistance.

A similar analysis by Navqui et al. (2011) revealed that the genes related to innate immunity, metabolism, and ethylene signaling are implicated in the systemic infection during ToLCNDV infection in tomato.

10.10 DEREGULATION OF miRNA

MicroRNAs (miRNAs) are the category of small non-coding RNAs that function as a crucial regulatory molecule of post-transcriptional gene silencing (PTGS) in plants. Briefly, biogenesis of miRNAs begins with the action of RNA polymerase II on target loci for transcription and processing to ~100 nucleotide-long primary miRNA (pri-miRNA) transcripts. This pri-miRNA is processed to precursor miRNA (pre-miRNA) by Dicer-like 1 (DCL1) protein, which further cleaves it to release a miRNA/miRNA* duplex within the nucleus (Ha and Kim 2014). The nuclear HEN1 protein assists in the 2′-O-methylation process at the 3′-end of miRNA/miRNA* duplex to avert non-templated 3′-polymerization, thus speeding up miRNA yield (Ha and Kim 2014). These miRNA/miRNA* are transported to cytoplasm by a plant homologue of Exportin-5 known as HASTY protein. Loading of this duplex onto AGO1 (a protein with the slicing characteristics) directs the endonucleolytic cleavage of target RNAs (Ha and Kim 2014).

To date, a total of 28645 hairpin precursor miRNAs, expressing around 7399 mature miRNAs, are reported from plant species (miRBase Release 21, June 2015). Report on antiviral characteristics of host miRNAs are well documented in animal–virus interactions; however, not many examples of miRNAs regulating geminivirus–plant interaction have been illustrated. Geminivirus proteins are known to modify the host PTGS pathways, which in turn deform normal cellular activities, leading to disease development/resistance (Sahu et al. 2014). Efforts have also been made to identify the ToLCNDV-responsive miRNAs responsible either for disease development or in providing tolerance (Naqvi et al. 2010; Pratap et al. 2015). An NGS platform was used to identify the ToLCNDV-responsive miRNAs, which further resulted in detection of 53 novel miRNAs (Pradhan et al. 2015). These novel miRNAs were not only involved in targeting leaf architecture and plant development-related host genes, but were also implicated in plant defense response. For example, novel miRNAs such as Tom 14 and Tom 43 have been shown to target transcription factors such as AP2/ERF and teosinte branched1/cycloidea/PCF (TCP) transcription factor, respectively, and might lead to leaf curl phenotype in plant. In addition, a disease resistance gene such as CC-NBS-LRR type protein was also shown to be targeted by a novel miRNA Tom 17. The role of conserved miRNAs has also been examined by Naqvi et al. (2010), revealing that the differential accumulation of miR159/319 and miR172 have a correlation with leaf curl symptom development in tomato. Interestingly, the authors have also postulated the role of miR168 and miR162 in the alteration of global miRNA flux by targeting DCL1 and AGO1.

Numerous reports on computational prediction of virus genome-derived miRNAs are available, but few of them are reported to be involved in disease resistance (Naqvi et al. 2011; Shweta and Khan 2014). Overall, these studies clearly suggest the involvement of miRNA(s) as prospective markers in terms of either disease development or disease tolerance during plant–geminivirus interaction.

Naqvi et al. (2010) found that miR159/319 and miR172 were associated with symptoms, the levels of which increased during ToLCNDV infection. The accumulation of miRNA in *N. benthamiana* following transient expression of four begomoviruses genes was examined by Amin et al. (2011a,b). In the case of agroinfiltration of βC 1 of cotton leaf curl multan betasatellite, there was significant reduction in the level of miR164, miR165, miR166, miR169, and miR170. However, there was a significant increase in accumulation of miR159 and miR160. The mRNA transcript levels are predicted to be regulated by these miRNAs, which may explain many of the abnormal developments of levels in begomovirus infection. To a great extent, disruption in the host gene expression may be the one mechanism of many specific symptoms under geminiviral pathogenesis.

10.11 CONCLUSION

Analysis of molecular events following geminivirus infection reveals that

- They reprogram the cell cycle-to-DNA synthesis phase and take control of genes related to cell cycle, DNA synthesis.
- They interact with kinases related to various pathways and interfere with hormone signaling.
- The RNAi pathway of the host generates mainly 24 nt vsiRNA targeting the intergenic region, and it is both TGS and PTGS.
- The methylation-mediated defense of the host is countered by the geminivirus suppressors effectively, which interact with various methylating enzyme complexes.
- The siRNA-generated and viral suppressors binding with host miRNA might function as determinants of symptoms.

10.12 FUTURE AREAS OF RESEARCH

Despite the explosion of information in the post-genomic era, there are still vital gaps that need to be understood if these diseases are to be managed. In the initial step after introduction of the virus in the cell, release of the ss DNA, the second-strand synthesis is facilitated by host DNA polymerase. It will be relevant to ask whether the DNA polymerase decides the host specificity. It is worthwhile to identify the differences between DNA polymerase of host/non-host-resistant/susceptible genotypes. Among the molecular events, it is essential to resolve any pattern or commonalities. It is possible that the mechanism of interaction may be common between one family of hosts or one host of one particular habitat. It is essential to figure out the common vital steps, so that a strategy is conceived to neutralize those steps. The sequences in both coat protein and genomic DNA that decide the origin of assembly of virion particles are not resolved. The intriguing geminate morphology common between the leaf hopper, plant hopper, tree hopper whitefly, and aphid-transmitted viruses offers a challenge for recognition mechanism between the vector and virus. The genomic information generated for various host plants should help to evolve a sustainable management strategy.

REFERENCES

Akbergenov R, Si-Ammour A, Blevins T, Amin I, Kutter C, Vanderschuren H, Zhang P, Gruissem W, Meins F Jr, Hohn T, Pooggin MM (2006). Molecular characterization of geminivirus-derived small RNAs in different plant species. *Nucleic Acids Research* 34: 462–471.

Amin I, Hussain K, Akbergenov R, Yadav JS, Qazi J, Mansoor S et al. (2011a). Suppressors of RNA silencing encoded by the components of the cotton leaf curl begomovirus-betasatellite complex. *Molecular Plant Microbe Interaction* 24: 973–983.

Amin I, Patil B, Briddon R, Mansoor S, Fauquet C (2011b). Comparison of phenotypes produced in response to transient expression of genes encoded by four distinct begomoviruses in Nicotiana benthamiana and their correlation with the levels of developmental miRNAs. *Virol Journal* 8: 238.

Andersen MT, Richardson KA, Harbison SA, Morris BA (1988). Nucleotide sequence of the geminivirus chloris striate mosaic virus. *Virology* 164: 443–449.

Aregger M, Borah BK, Seguin J, Rajeswaran R, Gubaeva EG, Zvereva AS et al. (2012). Primary and secondary siRNAs in geminivirus-induced gene silencing. *PLoS Pathogens* 8: e1002941.

Ascencio-Ibanez JT, Sozzani R, Lee TJ, Chu TM, Wolfinger RD, Cella R, Hanley-Bowdoin L (2008). Global analysis of Arabidopsis gene expression uncovers a complex array of changes impacting pathogen response and cell cycle during geminivirus infection. *Plant Physiology* 148: 436–454.

Bachmair A, Becker F, Masterson RV, Schell J (1990). Perturbation of the ubiquitin system causes leaf curling, vascular tissue alterations and necrotic lesions in a higher plant. *EMBO Journal* 9(13): 4543–4549.

Baliji S, Lacatus G, Sunter G (2010). The interaction between geminivirus pathogenicity proteins and adenosine kinase leads to increased expression of primary cytokinin-responsive genes. *Virology* 402: 238–247.

Briddon RW, Bull SE, Amin I, Idris AM, Mansoor S, Bedford ID, Dhawan P, Rishi N, Siwatch SS, Abdel-Salam AM, Brown JK, Zafar Y, Markham P (2003). Diversity of DNA b, a satellite molecule associated with some monopartite begomoviruses. *Virology* 312: 106–121.

Brough CL, Gardiner WE, Inamdar N, Zhang XY, Ehrlich M, Bisaro DM (1992). DNA methylation inhibits propagation of tomato golden mosaic virus DNA in transfected protoplasts. *Plant Moleular Biology* 18: 703–712.

Carmo LST, Murad AM, Resende RO, Boiteux LS, Ribeiro SG, Jorrin-Novo JV, Mehta A (2015). Plant responses to tomato chlorotic mottle virus: Proteomic view of the resistance mechanisms to a bipartite begomovirus in tomato. *Journal of Proteomics* 151: 284–292.

Carvalho MF, Lazarowitz SG (2004). Interaction of the movement protein NSP and the Arabidopsis acetyl-transferase AtNSI is necessary for Cabbage leaf curl geminivirus infection and pathogenicity. *Journal of Virology* 78(20): 11161–11171.

Carvalho MF, Turgeon R, Lazarowitz SG (2006). The geminivirus nuclear shuttle protein NSP inhibits the activity of AtNSI, a vascular-expressed Arabidopsis acetyl transferase regulated with the sink-to-source transition. *Plant Physiology* 140: 1317–1330.

Carvalho CM, Santos AA, Pires SR, Rocha CS, Saraiva DI, Machado PB, Mattos EC, Fietto LG, Fontes EPB (2008). Regulated nuclear trafficking of rpL10A mediated by NIK1 represents a defense strategy of plant cells against virus. *PLoS Pathogens* 4: e1000247.

Castillo AG, Kong LJ, Hanley-Bowdoin L, Bejarano ER (2004). Interaction between a geminivirus replication protein and the plant sumoylation system. *Journal of Virology* 78: 2758–2769.

Castillo GE, Laclette IE, Trejo-Saavedra DL, Rivera-Bustamante RF (2012). Transcriptome analysis of symptomatic and recovered leaves of geminivirus-infected pepper (Capsicum annuum). *Virology Journal* 9: 295.

Chen H, Zhang Z, Teng K, Lai J, Zhang Y, Huang Y, Li Y, Liang L, Wang Y, Chu C et al. (2010). Up-regulation of LSB1/GDU3 affects geminivirus infection by activating the salicylic acid pathway. *The Plant Journal* 62: 12–23.

Choudhury NR, Malik PS, Singh DK, Islam MN, Kaliappan K, Mukherjee SK (2006). The oligomeric Rep protein of Mungbean yellow mosaic India virus (MYMIV) is a likely replicative helicase. *Nucleic Acids Research* 34: 6362–6377.

Clerot D, Bernardi F (2006). DNA helicase activity is associated with the replication initiator protein rep of tomato yellow leaf curl geminivirus. *Journal of Virology* 80: 11322–11330.

Czosnek H, Eybishtz A, Sade D, Gorovits R, Sobol I, Bejarano E, Rosas-Díaz T, Rosa Duran L (2013). Discovering host genes involved in the infection by the tomato yellow leaf curl virus complex and in the establishment of resistance to the virus using tobacco rattle virus-based post transcriptional gene silencing. *Viruses* 5: 998–1022.

Ding SW, Voinnet O (2007). Antiviral immunity directed by small RNAs. *Cell* 130: 413–426.

Dogra SC, Eini O, Rezaian MA, Randles JW (2009). A novel shaggy-like kinase interacts with the Tomato leaf curl virus pathogenicity determinant C4 protein. *Plant Molecular Biology* 71: 25–38.

Eini O, Dogra S, Selth LA, Dry IB, Randles JW et al. (2009). Interaction with a host ubiquitin-conjugating enzyme is required for the pathogenicity of a geminiviral DNA beta satellite. *Molecular Plant Microbe Interaction* 22: 737–746.

Erdmann JB, Shepherd DN, Martin DP, Varsani A, Rybicki EP, Jeske H (2010). Replicative intermediates of maize streak virus found during leaf development. *Journal of General Virology* 91: 1077–1081.

Fondong VN (2013). Geminivirus protein structure and function. *Molecular Plant Pathology* 14(6): 635–649.

Fontes EP, Santos AA, Luz DF, Waclawovsky AJ and Chory J (2004). The geminivirus nuclear shuttle protein is a virulence factor that suppresses transmembrane receptor kinase activity. *Genes and Development* 18: 2545–2556.

Garcia-Neria MA, Rivera-Bustamante RF (2011). Characterization of geminivirus resistance in an accession of Capsicum chinense Jacq. *Molecular Plant Microbe Interaction* 24: 172–182.

Garcia-Ruiz H, Takeda A, Chapman EJ, Sullivan CM, Fahlgren N, Brempelis KJ et al. (2010). Arabidopsis RNA-dependent RNA polymerases and dicer-like proteins in antiviral defense and small interfering RNA biogenesis during turnip mosaic virus infection. *Plant Cell* 22: 481–496.

George B, Ruhel B, Mazumder R, Sharma M, Jain VK, Gourinath SK, Chakraborty S (2014). Mutational analysis of the helicase domain of a replication initiator protein reveals critical roles of Lys 272 of the B′ motif and Lys 289 of the β-hairpin loop in geminivirus replication. *Journal of General Virology* 95(7): 1591–1602.

Guerra-Peraza O, Kirk D, Seltzer V, Veluthambi K, Schmit AC, Hohn T, Herzog E (2005). Coat proteins of Rice tungro bacilliform virus and Mungbean yellow mosaic virus contain multiple nuclear-localization signals and interact with importin alpha. *Journal of General Virology* 86: 1815–1826.

Gutierrez C (2000). Geminiviruses and the plant cell cycle. *Plant Molecular Biology* 43: 763–772.

Ha M, Kim VN (2014). Regulation of microRNA biogenesis. *Nature Reviews Molecular Cell Biology* 15: 509–524.

Hagen C, Rojas MR, Kon T, Gilbertson RL (2008). Recovery from cucurbit leaf crumple virus (family Geminiviridae, genus Begomovirus) infection is an adaptive antiviral response associated with changes in viral small rnas. *Phytopathology* 98: 1029–1037.

Hanley-Bowdoin L, Settlage SB, Orozco BM, Nagar S, Robertson D (1999). Geminiviruses: Models for plant DNA replication, transcription, and cell cycle regulation. *Critical Reviews in Plant Sciences* 18: 71–106.

Hanley-Bowdoin L, Bejarano ER, Robertson D, Mansoor S (2013). Geminiviruses: Masters at redirecting and reprogramming plant processes. *Nature Reviews Microbiology* 11(11): 777–788.

Hao L, Wang H, Sunter G, Bisaro DM (2003). Geminivirus AL2 and L2 proteins interact with and inactivate SNF1 kinase. *Plant Cell* 15: 1034–1048.

Harrison BD, Robinson DJ (1999). Natural genomic and antigenic variation in whitefly-transmitted Geminiviruses (Begomoviruses). *Annual Reviews of Phytopathology* 37: 369–398.

Hohn T, Vazquez F (2011). RNA silencing pathways of plants: Silencing and its suppression by plant DNA viruses. *Biochimica et Biophysica Acta* 1809(11–12): 588–600.

Huang Y, Zhang BL, Sun S, Xing GM, Wang F, Li MY, Tian YS, Xiong AS (2006). AP2/ERF transcription factors involved in response to *Tomato yellow leaf curl virus* in tomato. *The Plant Genome* 9, doi: 0.3835 /plantgenome2015.09.0082.

Jeevalatha A, Siddappa S, Kumar A, Kaundal P, Guleria A, Sharma S, Singh BP (2017). An insight into differentially regulated genes in resistant and susceptible genotypes of potato in response to Tomato leaf curl New Delhi virus-[potato] infection. *Virus Research* 232: 22–33.

Jeske H (2009). Geminiviruses. *Current Topics in Microbiology and Immunology* 331: 185–226.

Jeske H, Lutgemeier M, Preiss W (2001). DNA forms indicate rolling circle and recombination-dependent replication of Abutilon mosaic virus. *EMBO Journal* 20: 6158–6167.

Kaliappan K, Choudhury NR, Suyal G, Mukherjee SK (2012). A novel role for RAD 54: This host protein modulates geminiviral DNA replication. *FASEB Journal* 26: 1142–1160.

Kang BC, Yeam I, Jahn MM (2005). Genetics of plant virus resistance. *Annual Reviews of Phytopathology* 43: 581–621.

Kaur H, Yadav CB, Alatar AA, Faisal M, Jyothsna P, Malathi VG, Praveen S (2014). Gene expression changes in tomato during symptom development in response to leaf curl virus infection. *Journal of Plant Biochemistry and Biotechnology* 24: 347–354.

Kong LJ, Hanley-Bowdoin L (2002). A geminivirus replication protein interacts with a protein kinase and a motor protein that display different expression patterns during plant development and infection. *Plant Cell* 14: 1817–1832.

Krenz B, Wege C, Jeske H (2010). Cell-free construction of disarmed Abutilon mosaic virus-based gene silencing vectors. *Journal of Virological Methods* 169: 129–137.

Kundu S, Chakraborty D, Pal A (2011). Proteomic analysis of salicylic acid induced resistance to Mungbean Yellow Mosaic India Virus in Vigna mungo. *Journal of Proteomics* 74(3): 337–349.

Kunik T, Palanichelvam K, Czosnek H, Citovsky V, Gafni Y (1998). Nuclear import of the capsid protein of Tomato yellow leaf curl virus (TYLCV) in plant and insect cells. *Plant Journal* 13: 393–399.

Kunik T, Mizrachy L, Citovsky V, Gafni Y (1999). Characterization of a tomato karyopherin alpha that interacts with the Tomato yellow leaf curl virus (TYLCV) capsid protein. *Journal of Exerimental Botany* 50: 731–732.

Kushwaha N, Singh A, Basu S, Chakraborty S (2015). Differential response of diverse solanaceous hosts to Tomato leaf curl New Delhi virus infection indicates coordinated action of NBS-LRR and RNAi-mediated host defense. *Archives of Virology* 160: 1499–1509.

Lai J, Chen H, Teng K, Zhao Q, Zhang Z, Li Y, Liang L, Xia R, Wu Y, Guo H, Xie Q (2009). RKP, a RING finger E3 ligase induced by BSCTV C4 protein, affects geminivirus infection by regulation of the plant cell cycle. *Plant Journal* 57: 905–917.

Lewis MW, Leslie ME, Fulcher EH, Darnielle L, Healy P, Youn JY, Liljegren SJ (2010). The SERK1 receptor-like kinase regulates organ separation in Arabidopsis flowers. *Plant Journal* 62: 817–828.

Liu L, Saunders K, Thomas CL, Davies JW, Stanley J (1999). Bean yellow dwarf virus RepA, but not Rep, binds to maize retinoblastoma protein, and the virus tolerates mutations in the consensus binding motif. *Virology* 256: 270–279.

Liu J, Yang J, Bi H, Zhang P (2014). Why mosaic? Gene expression profiling of African cassava mosaic virus-infected cassava reveals the effect of chlorophyll degradation on symptom development. *Journal of Integrated Plant Biology* 56(2): 122–132.

Lozano G, Trenado HP, Fiallo-Olivé E, Chirinos D, Geraud-Pouey F, Briddon RW, Navas-Castillo J (2016). Characterization of non-coding DNA satellites associated with sweepoviruses (genus Begomovirus, Geminiviridae)—Definition of a distinct class of begomovirus-associated satellites. *Frontiers in Microbiology* 7: 162.

Lozano-Duran R, Rosas-Diaz T, Gusmaroli G, Luna AP, Taconnat L, Deng XW, Bejarano ER (2011). Geminiviruses subvert ubiquitination by altering CSN-mediated derubylation of SCF E3 ligase complexes and inhibit jasmonate signaling in Arabidopsis thaliana. *Plant Cell* 23: 1014–1032.

Luque A, Sanz-Burgos AP, Ramirez-Parra E, Castellano MM, Gutierrez C (2002). Interaction of geminivirus Rep protein with replication factor C and its potential role during geminivirus DNA replication. *Virology* 302: 83–94.

Maiti S, Paul S, Pal A (2012). Isolation, characterization, and structure analysis of a non-TIR-NBS- LRR encoding candidate gene from MYMIV-resistant Vigna mungo. *Molecular Biotechnology* 52: 217–233.

Mandal A, Sarkar D, Kundu S, Kundu P (2005). Mechanism of regulation of tomato TRN1 gene expression in late infection with tomato leaf curl New Delhi virus (ToLCNDV). *Plant Science* 241: 221.

Miozzi L, Gambino G, Burgyan J, Pantaleo V (2013). Genome-wide identification of viral and host transcripts targeted by viral siRNAs in Vitis vinifera. *Molecular Plant Pathology* 14: 30–43.

Nagar S, Pedersen TJ, Carrick KM, Hanley-Bowdoin L, Robertson D (1995). A geminivirus induces expression of a host DNA synthesis protein in terminally differentiated plant cells. *Plant Cell* 7: 705–719.

Nagar S, Hanley-Bowdoin L, Robertson D (2002). Host DNA replication is induced by geminivirus infection of differentiated plant cells. *Plant Cell* 14: 2995–3007.

Naqvi AR, Haq QMR, Mukherjee SK (2010). MicroRNA profiling of Tomato leaf curl New Delhi virus (ToLCNDV) infected tomato leaves indicates that deregulation of mir159/319 and mir172 might be linked with leaf curl disease. *Virology Journal* 7: 281.

Naqvi AR, Choudhury NR, Mukherjee SK, Haq QM (2011). In silico analysis reveals that several tomato microRNA/microRNA sequences exhibit propensity to bind to Tomato leaf curl virus (ToLCV) associated genomes and most of their encoded open reading frames (ORFs). *Plant Physiology and Biochemistry* 49: 13–17.

Pant V, Gupta D, Choudhury NR, Malathi VG, Varma A, Mukherjee SK (2001). Molecular characterization of the Rep protein of the blackgram isolate of Indian mungbean yellow mosaic virus. *Journal of General Virology* 82: 2559–2567.

Paprotka T, Deuschle K, Metzler V, Jeske H (2011). Conformation-selective methylation of geminivirus DNA. *Journal of Virology* 85: 12001–12012.

Park J, Hwang H, Shim H, Im K, Auh CK, Lee S, Davis KR (2004). Altered cell shapes, hyperplasia, and secondary growth in Arabidopsis caused by beet curly top geminivirus infection. *Molecules and Cells* 17: 117–124.

Pierce EJ and Rey MEC (2013). Assessing global transcriptome changes in response to South African cassava mosaic virus [za-99] infection in susceptible Arabidopsis thaliana. *PLoS ONE* 8: e67534.

Pilartz M, Jeske H (1992). Abutilon mosaic geminivirus double-stranded DNA is packed into minichromosomes. *Virology* 189: 800–802.

Piroux N, Saunders K, Page A, Stanley J (2007). Geminivirus pathogenicity protein C4 interacts with Arabidopsis thaliana shaggy-related protein kinase AtSKeta, a component of the brassinosteroid signaling pathway. *Virology* 362: 428–440.

Pooggin M, Shivaprasad PV, Veluthambi K, Hohn T (2003). RNAi targeting of DNA virus in plants. *Nature Biotechnology* 21: 131–132.

Pratap D, Kashikar AR, Mukherjee SK (2011). Molecular characterization and infectivity of a *Tomato leaf curl New Delhi virus* variant associated with newly emerging yellow mosaic disease of eggplant in India. *Virology Journal* 8: 305.

Rai VP, Rai A, Kumar R, Kumar S, Kumar S, Singh M, Singh SP (2016). Microarray analyses for identifying genes conferring resistance to pepper leaf curl virus in chilli pepper (Capsicum spp.). *Genomics Data* 9: 140–142.

Raja P, Sanville BC, Buchmann RC, Bisaro DM (2008). Viral genome methylation as an epigenetic defense against geminiviruses. *Journal of Virology* 82: 8997–9007.

Rizvi I, Choudhury NR, Tuteja N (2015). Insights into the functional characteristics of geminivirus rolling-circle replication initiator protein and its interaction with host factors affecting viral DNA replication. *Archives of Virology* 160: 375–387.

Rodriguez-Negrete EA, Carrillo-Tripp J, Rivera-Bustamante RF (2009). RNA silencing against geminivirus: Complementary action of posttranscriptional gene silencing and transcriptional gene silencing in host recovery. *Journal of Virology* 83(3): 1332–1340.

Rodriguez-Negrete E, Lozano-Duran R, Piedra-Aguilera A, Cruzado L, Bejarano ER, Castillo AG (2013). Geminivirus Rep protein interferes with the plant DNA methylation machinery and suppresses transcriptional gene silencing. *New Phytologist* 199: 464–475.

Rojas MR, Hagen C, Lucas WJ, Gilbertson RL (2005). Exploiting chinks in the plant's armor: Evolution and emergence of geminiviruses. *Annual Review of Phytopathololgy* 43: 361–394.

Sahu PP, Rai NK, Chakraborty S, Singh M, Chandrappa PH, Ramesh B, Chattopadhyay D, Prasad M (2010). Tomato cultivar tolerant to Tomato leaf curl New Delhi virus infection induces virus-specific short interfering RNA accumulation and defence-associated host gene expression. *Molecular Plant Pathology* 11(4): 531–544.

Sahu PP, Sharma N, Puranik S, Muthamilarasan M, Prasad M (2014). Involvement of host regulatory pathways during geminivirus infection: A novel platform for generating durable resistance. *Functional and Integrative Genomics* 14: 47–58.

Sahu PP, Sharma N, Puranik S, Chakraborty S, Prasad M (2016). Tomato 26S Proteasome subunit RPT4a regulates ToLCNDV transcription and activates hypersensitive response in tomato. *Scientific Reports* 6: 27078.

Santos AA, Carvalho CM, Florentino LH, Ramos HJ, Fontes EP (2009). Conserved threonine residues within the A-loop of the receptor NIK differentially regulate the kinase function required for antiviral signaling. *PLoS ONE* 4: e5781.

Saunders K, Lucy A, Stanley J (1992). RNA-primed complementary-sense DNA synthesis of the geminivirus African cassava mosaic virus. *Nucleic Acids Research* 20: 6311–6315.

Settlage SB, Miller AB, Gruissem W, Hanley-Bowdoin L (2001). Dual interaction of a geminivirus replication accessory factor with a viral replication protein and a plant cell cycle regulator. *Virology* 279: 570–576.

Shen W, Hanley-Bowdoin L (2006). Geminivirus infection up regulates the expression of two Arabidopsis protein kinases related to yeast SNF1- and mammalian AMPK-activating kinases. *Plant Physiology* 142: 1642–1655.

Shen W, Reyes MI, Hanley-Bowdoin L (2009). Arabidopsis protein kinases GRIK1 and GRIK2 specifically activate SnRK1 by phosphorylating its activation loop. *Plant Physiology* 150: 996–1005.

Shen Q, Liu Z, Song F, Xie Q, Hanley-Bowdoin L, Zhou X (2011). Tomato SlSnRK1 protein interacts with and phosphorylates βC1, a pathogenesis protein encoded by a Geminivirus β-satellite. *Plant Physiology* 157: 1394–1406.

Shweta, Khan J (2014). In silico prediction of cotton (Gossypium hirsutum) encoded microRNAs targets in the genome of Cotton leaf curl Allahabad virus. *Bioinformation* 10: 251–255.

Singh DK, Islam MN, Choudhury NR, Karjee S, Mukherjee SK (2007). The 32 kDa subunit of replication protein A (RPA) participates in the DNA replication of Mung bean yellow mosaic India virus (MYMIV) by interacting with the viral Rep protein. *Nucleic Acids Research* 35: 755–770.

Soitamo AJ, Jada B, Lehto K (2012). Expression of geminiviral AC2 RNA silencing suppressor changes sugar and jasmonate responsive gene expression in transgenic tobacco plants. *BMC Plant Biology* 12: 204.

Stanley J (1985). The molecular biology of geminiviruses. *Advances in Virus Research* 30: 139–177.

Suyal G, Mukherjee SK, Srivastava PS, Choudhury NR (2013). *Arabidopsis thaliana* MCM2 plays role(s) in mungbean yellow mosaic India virus (MYMIV) DNA replication. *Archives of Virology* 158: 981–992.

Verlaan MG, Hutton SF, Ibrahem RM, Kormelink R, Visser RG, Scott JW, Edwards JD, Bai Y (2013). The Tomato Yellow Leaf Curl Virus resistance genes Ty-1 and Ty-3 are allelic and code for DFDGD-class RNA-dependent RNA polymerases. *PLoS Genetics* 9(3): e1003399.

Wang H, Hao L, Shung CY, Sunter G, Bisaro DM (2003). Adenosine kinase is inactivated by geminivirus AL2 and L2 proteins. *Plant Cell* 15: 3020–3032.

Wang H, Buckley KJ, Yang X, Buchmann RC, Bisaro DM (2005). Adenosine kinase inhibition and suppression of RNA silencing by geminivirus AL2 and L2 proteins. *Journal of Virology* 79: 7410–7418.

Wang MB, Masuta C, Smith NA, Shimura H (2012). RNA silencing and plant viral diseases. *Molecular Plant–Microbe Interaction* 25: 1275–1285.

Wang XB, Jovel J, Udomporn P, Wang Y, Wu Q, Li WX, Gasciolli V, Vaucheret H, Ding SW (2011). The 21 nucleotide, but not 22 nucleotide, viral secondary small interfering RNAs direct potent antiviral defense by two cooperative Argonautes in *Arabidopsis thaliana*. *Plant Cell* 23: 1625–1638.

Yadav RK, Chattopadhyay D (2014). Differential soybean gene expression during early phase of infection with Mungbean yellow mosaic India virus. *Molecular Biology Reports* 41: 5123–5134.

Yadava P, Suyal G, Mukherjee SK (2010). Begomovirus DNA replication and pathogenicity. *Current 360 Science* 98: 360–368.

Yang J-Y, Iwasaki M, Machida C, Machida Y, Zhou X, and Chua N-H (2008). βC1, the pathogenicity factor of TYLCCNV, interacts with AS1 to alter leaf development and suppress selective jasmonic acid responses. *Genes Dev.* 22: 2564–2577.

Yang X, Wang Y, Guo W, Xie Y, Xie Q, Fan L et al. (2011a). Characterization of small interfering RNAs derived from the geminivirus/betasatellite complex using deep sequencing. *PLoS ONE* 6(2): e16928.

Yang X, Xie Y, Raja P, Li S, Wolf JN, Shen Q, Bisaro DM, Zhou X (2011b). Suppression of methylation-mediated transcriptional gene silencing by betaC1-SAHH protein interaction during geminivirus-betasatellite infection. *PLoS Pathogens* 7: e1002329.

Zerbini FM, Briddon RW, Idris A, Martin DP, Moriones E, Nacas-Castillo J, Bustamante R, Roumagnac P, Varsani A et al. (2017). Geminiviridae. *Journal of General Virology* 98: 131–133.

Zhang Z, Chen H, Huang X, Xia R, Zhao Q, Lai J, Teng K, Li Y, Liang L, Du Q et al. (2011). Bsctv c2 attenuates the degradation of samdc1 to suppress DNA methylation-mediated gene silencing in arabidopsis. *Plant Cell* 23: 273–288.

Zhang X, Zhang X, Singh J, Li D, Qua, F (2012). Temperature-dependent survival of Turnip crinkle virus-infected Arabidopsis plants relies on an RNA silencing-based defense that requires DCL2, AGO2, and HEN1. *Journal of Virology* 12: 6847–6854.

Zhou Y, Rojas MR, Park MR, Seo YS, Lucas WJ, Gilbertson RL (2011). Histone H3 interacts and colocalizes with the nuclear shuttle protein and the movement protein of a geminivirus. *Journal of Virology* 85: 11821–11832.

Zorzatto C, Machado JPB, Lopes KVG, Nascimento KJT, Pereira WA et al. (2015). NIK1-mediated translation suppression functions as a plant antiviral immunity mechanism. *Nature* 520: 679–682.

11 Host–Virus Interactions from Potyvirus Replication to Translation

Swarnalok De, Andres Lõhmus, Maija Pollari, Shreya Saha, and Kristiina Mäkinen

CONTENTS

11.1 INTRODUCTION

Potyviruses are a large and agriculturally important group of plant positive-stranded RNA ((+)RNA) viruses. The translational strategy exploited by potyviruses is the production of two polyproteins which are subsequently cleaved into 11 mature functional proteins (reviewed in Ivanov et al. 2014). Many potyviral proteins are multifunctional and bind to multiple host proteins, which ensures that all functions required for infection can be carried out with this relatively low number of viral proteins. Severe losses in crop quality and yield are attributable to potyvirus infections as exemplified by potato virus Y (PVY) in potato, plum pox virus (PPV) in stony fruits, soybean mosaic virus (SMV) in soybean and zucchini yellow mosaic virus (ZYMV) in cucurbits. Gibbs et al. (2008) dated the emergence of potyviruses to circa 6600 years ago and suggested that human agricultural activity has contributed to the high species richness of potyviruses. Metagenome sequencing approaches have revealed that potyviruses are abundantly present also in wild plants (Roossinck 2012). The broad geographical distribution of potyviruses and their capability to infect a wide range of host plants shows their outstanding evolutionary success.

Because of the heavy impact of potyviruses on food and plant production, it is important to tackle them. Compatible interactions between virus and host proteins essential for viral multiplication present potential opportunities for discovering mechanisms for virus resistance. Natural variants or intentionally mutated host proteins, which can carry out their cellular functions but not the functions associated with virus infection, can be used for breeding for virus resistance (reviewed in Hashimoto et al. 2016). The molecular nature of natural recessive resistance against potyviruses in a wide variety of crop plants has revealed one such host factor, the eukaryotic translation initiation factor eIF4E or eIF(iso)4E (Robaglia and Caranta 2006). In addition to the natural variants, potyvirus resistance was also conferred by genome edited eIF4E/(iso)4E genes (Chandrasekaran et al. 2016). In order to achieve broad and durable resistance to potyviruses it is important to explore and expand the genetic resources available for resistance breeding.

Although viral translation and replication are considered physically separated phenomena, they were suggested to occur in close association in potyviruses (Cotton et al. 2009; Hafrén et al. 2010).

In a recent composition study, ribosomal proteins were shown to be abundantly associated with the 6K2-induced viral replication complexes (VRCs) (Lõhmus et al. 2016), highlighting the close relationship between replication and translation. The mechanism of how the replicated viral RNA molecules exit potyviral VRCs has not been studied in depth. As the infection continues with virus replication, progeny (+)RNA strands are utilized in the further production of viral proteins. Multiple viral RNA replication-translation cycles ensure robust virus titers during infection. During viral replication in the VRCs the double-stranded RNA (dsRNA) replication intermediates are protected from cellular inhibitors of infection. After release from the VRC, newly formed viral RNAs are exposed to antiviral defense mechanisms. Without viral countermeasures, the host's RNA silencing-mediated antiviral defenses have the potential to substantially reduce the accumulation of potyvirus proteins and RNA. To counteract antiviral silencing, most plant viruses encode one or more silencing suppressor proteins (reviewed in Csorba et al. 2015). Potyviral proteins acting as silencing suppressors are the helper component-proteinase (HCpro) and viral protein genome-linked (VPg).

11.2 RNA SILENCING AGAINST POTYVIRAL RNA

RNA silencing (RNAi) is a sequence-specific RNA degradation and translation repression mechanism triggered by dsRNA. It is highly conserved in eukaryotes and serves as an important antiviral mechanism in plants (reviewed in Szittya and Burgyan 2013). The silencing of viral RNA is triggered by self-complementary secondary structures or other exposed viral dsRNA molecules. The main components of the silencing machinery are dicer-like (DCL) endonucleases and dsRNA-binding proteins that recognize dsRNA and digest it to 21–24 nt fragments; argonaute (AGO) proteins that cleave or translationally repress their target RNA guided by the complementarity of the small RNA molecule they carry; and RNA-dependent RNA polymerases (RDRs) involved in the synthesis of dsRNA. This helps in the generation of secondary siRNAs contributing to the spread of the silencing signal throughout the plant. Figure 11.1 summarizes the main components and processes involved in the silencing of potyviruses and highlights potyviral silencing suppression mechanisms.

During an antiviral response against potyviruses in *Arabidopsis* the majority of viral small-interfering RNAs (vsiRNAs) are 21 bp long, suggesting that DCL4 generates most siRNAs during potyvirus infection. In addition, DCL2, generating 22 bp siRNA, has been shown to be functional against potyviruses in inflorescence tissue (Garcia-Ruiz et al. 2010). The *Arabidopsis* genome encodes six RDR homologs (RDR1-6) out of which RDR1 and to a lesser extent RDR6 are associated with siRNA production in turnip mosaic virus (TuMV). Both *RDR1* and *RDR6* are required to achieve full antiviral defense in *Arabidopsis* (Garcia-Ruiz et al. 2010).

After the generation of ds-siRNA molecules, they are stabilized by 2′-O-methylation at their 3′ ends by HUA enhancer 1 (HEN1) (Yu et al. 2005). HEN1 is involved in silencing against potyvirus infection, as its knockdown partially rescues gene expression of HCpro-deficient potato virus A (PVA; potyvirus; Ivanov et al. 2016), suggesting that RNA-silencing suppression by HCpro interferes with HEN1 methylation. In addition, PVA HCpro inhibits the activity of S-adenosyl-L-methionine synthetase required for production of S-adenosyl-L-methionine (SAM; Ivanov et al. 2016). SAM serves as methyl donor for HEN1, and its absence was suggested to inhibit HEN1 and debilitate siRNA stabilization (Ivanov et al. 2016). Additionally, S-adenosyl-L-homocysteine hydrolase (SAHH) interacts with the HCPro of PVY, and down-regulation of SAHH decreases sRNA accumulation and locally suppresses silencing (Canizares et al. 2013). These findings support the idea that potyviruses interfere with the methionine cycle in order to suppress RNAi.

AGO2 has been shown to have a major role against potyvirus infection both locally and systemically in *Arabidopsis* (Garcia-Ruiz et al. 2015; Carbonell et al. 2012). In addition, AGO1 and AGO10 have smaller overlapping antiviral effects in the inflorescences (Garcia-Ruiz et al. 2015). In the absence of HCpro AGO1, 2 and 10 proteins bind TuMV siRNAs, whereas in the presence of

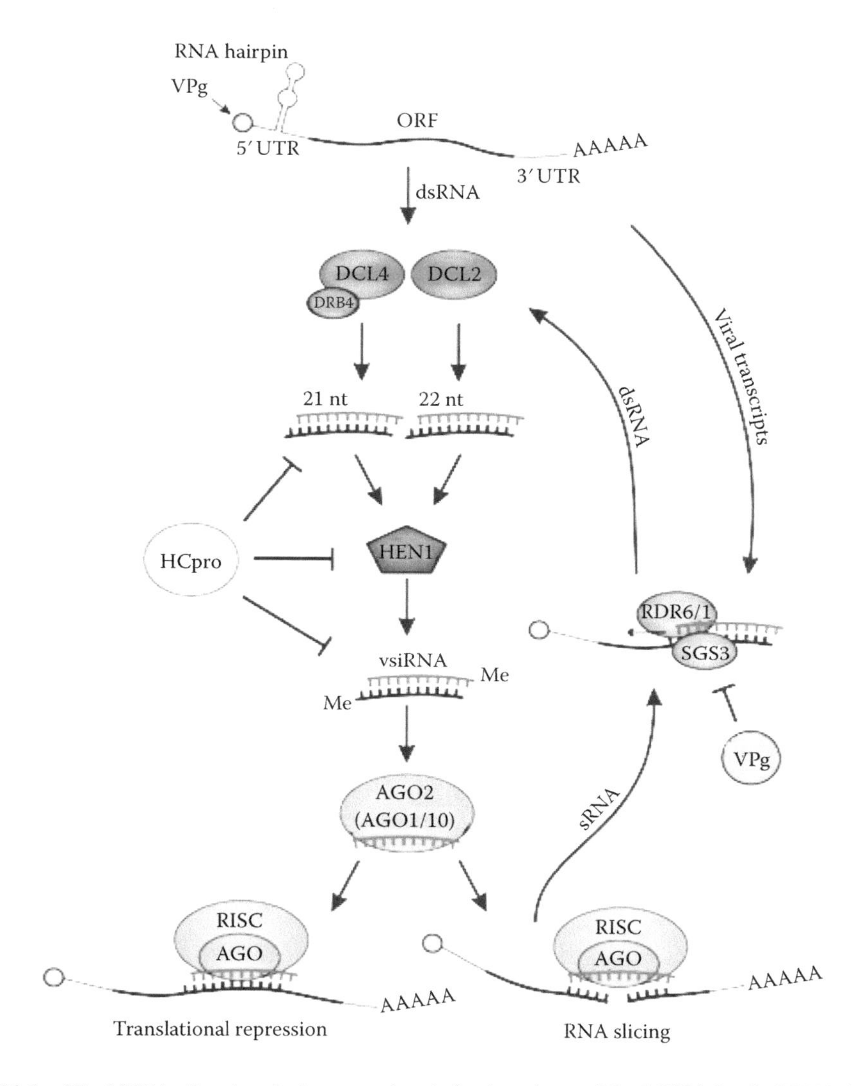

FIGURE 11.1 Viral RNA silencing during potyvirus infection. Accessible dsRNA in viral RNA hairpins and replicative intermediates serve as templates for DCLs that produce primary siRNA. During potyvirus infection mainly 21 nt DCL4-dependent and 22 nt DCL2-dependent siRNAs are produced with the help of dsRNA binding protein 4 (DRB4). HEN1-mediated methylation of the 3′ ends of siRNA serves to stabilize the molecules. siRNAs are loaded to AGO proteins, forming RNA-induced silencing complex (RISC). siRNA guides the RISC to target RNA, which is then translationally repressed or cleaved by the complex. Single-stranded cleavage products can be converted to double-stranded form by the action of RDR6 and SGS3 and again serve as substrates to DCLs. The potyviral silencing suppressors HCpro and VPg interfere with several steps of the silencing pathway. HCpro binds siRNAs, preventing loading of RISC. In addition, HCpro interferes with methylation of siRNAs, possibly leading to their suboptimal stability. VPg interacts with SGS3 and promotes its degradation together with RDR6.

HCpro the programming of AGOs with vsiRNAs is inhibited (Garcia-Ruiz et al. 2015). The HCpro of PVA co-localizes with AGO1 to potyvirus-induced RNA granules (PGs; Hafrén et al. 2015) and also ribosome-associated multiprotein complexes (Ivanov et al. 2016). Both of these complexes have been hypothesized to be the functional sites for silencing suppression by the HCpro of PVA.

Suppressor of gene silencing 3 (SGS3) protein aggregates together with RDR6 to cytoplasmic bodies which are referred to as siRNA bodies (Kumakura et al. 2009). They contribute to the conversion of ssRNA transcripts and viral RNAs to dsRNA molecules and to the production of various types of siRNA molecules (Fukunaga and Doudna 2009). VPg, the second potyviral silencing suppressor (Rajamäki and Valkonen 2009), was reported to co-localize to these SGS3/RDR6 bodies, and its interaction with SGS3 benefited PVA infection (Rajamäki et al. 2014). The interaction of TuMV VPg and SGS3 leads to the degradation of SGS3 via the 20S proteasome and autophagy (Cheng and Wang 2017). In addition, it was shown that in the presence of VPg the SGS3 binding partner RDR6 became degraded (Cheng and Wang 2017).

11.3 RNA GRANULES ASSOCIATED WITH POTYVIRUS INFECTION

RNA metabolism in the cytoplasm has an important role in the control of gene expression. In eukaryotic cells, storage of translationally inactive mRNA takes place in ribonucleoprotein complexes termed as stress granules (SGs), while mRNA decay occurs in processing bodies (PBs) (reviewed by Kedersha et al. 2005). Although PBs have housekeeping functions and are constitutively present, SGs are formed around stalled translation initiation complexes in response to stresses and are linked to host defense responses including antiviral mechanisms (Weber et al. 2008; Rozelle et al. 2014). Proteins in SGs and PBs participate in viral infections either by repressing the functions of viral RNA and obstructing the viral life cycle or by promoting infections (Beckham and Parker 2008).

Invading viruses have developed a myriad of strategies for interacting with host proteins within SGs and PBs in order to avoid the host's lines of defense and to ensure a successful infection (reviewed by Poblete-Duran et al. 2016). For example, (+)RNA viruses, such as members of the *Picornaviridae*, have evolved ways to escape degradation by disrupting SGs, while others, for example hepatitis C virus (HCV), belonging to the *Flaviviridae*, manipulate granule functions to enhance virus accumulation (Pager et al. 2013; Poblete-Duran et al. 2016). Although still less studied than animal viruses, plant viruses have similarly been found to interact with SG and PB components. For example, brome mosaic virus RNA has been shown to accumulate in PBs where it specifically interacts with mRNA decapping-associated Lsm proteins to promote viral replication and translation (Galao et al. 2010).

Some alpha viruses can effectively block SG formation by an interaction between FGDF-amino acid motifs on the viral nuclear shuttle protein and host G3BP, a Ras-GAP SH3 domain-binding protein, which regulates the assembly of SGs (Panas et al. 2015). Krapp et al. (2017) have recently identified a G3BP homolog in *Arabidopsis thaliana* and showed its interaction with similar motifs in the nuclear shuttle proteins from abutilon mosaic virus (family *Geminiviridae*) and pea necrotic yellow dwarf virus (family *Nanoviridae*).

PVA can induce the formation of cytoplasmic RNA granules that contain both viral and host proteins (Hafrén et al. 2015). However, based on host factor composition, the PVA-induced granules do not fit canonical plant SG or PB categories. Instead, they combine a mixture of hallmark proteins from both SGs (e.g., UBP-1, eIF4E, PABP) and PBs (VCS, AGO1) (Hafrén et al. 2015, Figures 11.2 and 11.3a). The importance of these factors in PVA infection was underlined by impaired virus accumulation in experiments where granule-associated host proteins UBP1, VCS, P0, and eIF4Es were knocked down using hairpin constructs (Hafrén et al. 2015). A corresponding example from the animal virus field shows that HCV accumulation diminished when genes expressing the homologs of UBP1 and VCS (TIA1 and Ge-1, respectively) were similarly silenced (Pager et al. 2013).

The multifunctional potyviral HCpro has emerged as an important player in the relationship between viral RNA, cytoplasmic granules, and RNA-silencing suppression. PVA, TuMV, and PVY HCpro proteins all have the capacity to induce the formation of granules even in the absence of the virus (Hafrén et al. 2015; del Toro et al. 2014). As HCpro is a key viral silencing suppressor, the biological significance of RNA granules in potyvirus infections could be related to the protection of viral RNA from the host's silencing machinery. This link is supported by the finding that an HCpro

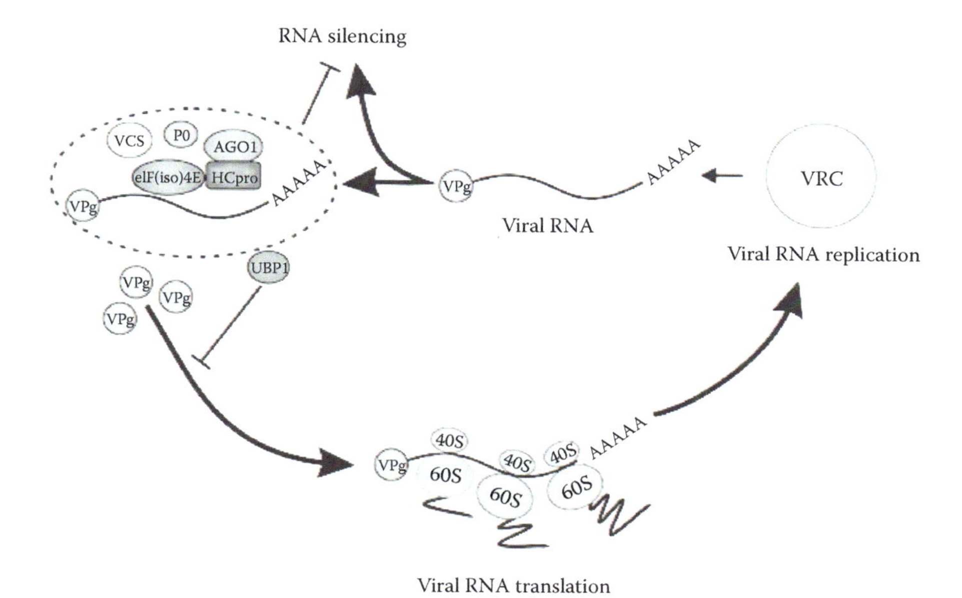

FIGURE 11.2 A model of a potyvirus-induced RNA granule and its convergence with potyviral translation. Many of the host factors having a role in regulation of active potyviral RNA translation together with VPg have been identified to associate with HCpro and viral RNA into the granules when the amount of VPg is limiting viral translation.

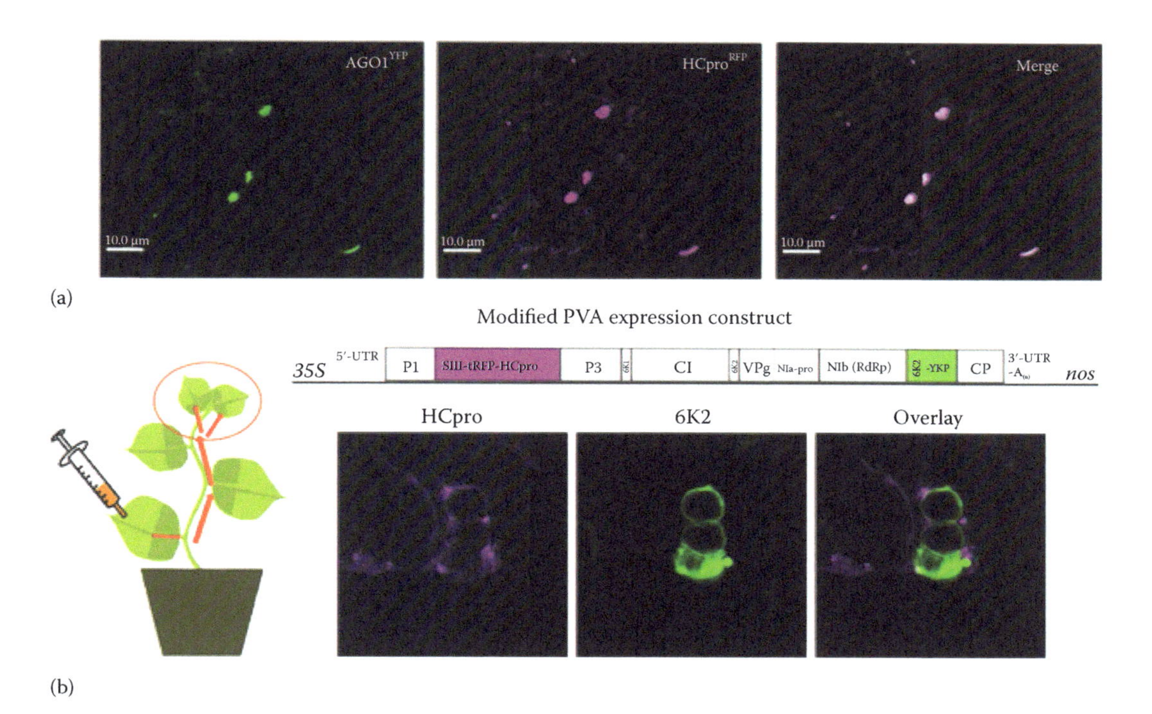

FIGURE 11.3 (a) Confocal fluorescence microscope image showing YFP-tagged AGO1 and RFP-tagged HCpro colocalization in potyvirus-induced granules. (b) Confocal fluorescence microscope image of PVA-induced granules (white arrows, labeled with RFP-tagged HCpro) near viral replication complexes (labeled with YFP-tagged 6K2, a VRC marker) surrounding chloroplasts.

mutant deficient in silencing suppression also failed to form PVA-induced granules (Hafrén et al. 2015). PVA-induced granules are candidate sites for HCpro-mediated suppression of RNA silencing, as both AGO1 and HCpro localize to them together with viral RNA (Figure 11.3a; Hafrén et al. 2015).

PVA-induced granules often associate closely with virus replication complexes (Figure 11.3b). Their physical proximity could be an advantage in the competition for progeny RNA against the host's silencing mechanisms.

11.4 HOST–VIRUS INTERACTIONS IN POTYVIRUS TRANSLATION

As viruses lack their own translational apparatus, they must hijack the host's protein synthesis machinery for the translation of their own proteins. Since the same machinery is also used by the host, competition between the host and the virus becomes inevitable. Viruses have evolved distinct strategies to gain advantage over their hosts. An efficient infection is generally associated with enhanced channeling of ribosomes and translation factors toward the viral genome with a simultaneous repression of host protein synthesis. This requires complex and dynamic interactions between several host and viral components (Au and Jan 2014; Walsh et al. 2013).

Genome-wide expression profiling has on multiple occasions shown the potyvirus infection to be associated with the transcriptional upregulation of genes encoding ribosomal proteins (Dardick 2007; Yang et al. 2007). However, the number of ribosomes did not increase in the infected cells and not all ribosomal proteins were found to be equally upregulated (Yang et al. 2007, 2009). Hence it has been speculated that either viral infection might alter ribosomal composition to favor viral RNA translation or these differentially upregulated proteins are involved in some other processes yet to be investigated (Ivanov et al. 2014).

Most eukaryotic mRNAs have a 7-methyl guanosine (m7GpppG) cap at the 5′ end and a poly(A)-tail at the 3′ end (for an advanced review, see Topisirovic et al. 2011). The 5′ cap region interacts with cap-binding protein eIF4E/(iso)4E and the 3′poly(A)-tail region with poly(A)-binding protein (PABP). A large adapter protein, eIF4G/(iso)4G, interacts simultaneously with eIF4E/(iso)4E and PABP and recruits other proteins, including the eIF4A helicase. The formation of this eIF4F/(iso)4F-complex is required to recruit ribosomes to mRNA and to initiate the scanning of mRNA to find the first start codon. Many plant and animal viruses, including potyviruses (Niepel and Gallie 1999), have uncapped 5′ UTR, and viral translation depends on internal ribosomal entry site (IRES) in the 5′ UTR region. Potyviruses have a poly(A)-tail at the 3′ end of the genome but instead of a cap structure they carry VPg covalently attached to the 5′ end.

VPg enhances the translation of potyviral RNA (Figure 11.3) and its accumulation *in planta* (Eskelin et al. 2011). This occurs both for wild-type PVA RNA carrying VPg at its 5′ end and non-replicating capped PVA transcripts. PVA 5′ UTR contributes to this boost in viral RNA translation, whereas the 3′ UTR does not. Members of the eIF4E family play a role in VPg-mediated enhancement of PVA gene expression. Simultaneous silencing of eIF4E and eIF(iso)4E decreased (Eskelin et al. 2011) and co-expression eIF(iso)4E with VPg increased (Hafrén et al. 2013) VPg-mediated translational enhancement. One interpretation is that free VPg may recruit host factors including eIF4E for efficient translation through VPg-VPg dimerization at the 5′ UTR of PVA RNA or VPg-eIF4E-m7GpppG cap interaction in the case of non-replicating viral transcripts.

The relevance of the interaction between the 5′-terminal VPg and eIF4E/(iso)4E in the initiation of translation is not immediately obvious, as IRES-mediated translation does not require eIF4E for initiation (Gallie 2001, Iwakawa and Tomari 2013). When eukaryotic initiation factors (eIFs) become limiting in an in vitro translation mix tobacco etch virus (TEV) 5′ UTR has a translational advantage over capped transcripts. TEV IRES recruits the limiting eIFs, especially eIF4G, which binds a pseudoknot structure on the TEV 5′ UTR more efficiently than capped transcripts (Gallie 2001).

In addition to eIF4E/(iso)4E, several other host proteins are involved in the regulation of VPg-mediated translation (Figure 11.2). Ribosomal stalk protein P0 and the WD40-domain protein VCS activate VPg-mediated translation, while UBP1 represses it (Hafrén et al. 2013, 2015). Viral HCpro contributes to the upregulation of PVA gene expression but only in the presence of VPg, HCpro, P0, eIF(iso)4E, and VCS, which all have roles in potyviral translation, are also components of PGs forming an interesting link between PGs and translation (Hafrén et al. 2015). The number of PGs decreases significantly when VPg is abundant, suggesting an inverse correlation between the presence of PGs and the efficiency of potyviral translation.

VPg has been shown to suppress the expression of a capped reporter mRNA *in planta* (Eskelin et al. 2011). Even though VPg may have a role in the downregulation of capped nonviral transcripts, there is no evidence for universal translational inhibition, since polysome profiling has shown upregulation of polysome-bound transcripts during infection, providing evidence for rather efficient translation (Moeller et al. 2012).

In addition to antiviral RNA silencing by slicing, translational repression is another obstacle that needs to be overcome during the course of infection. Although the phenomenon has only recently been identified in plants, RISC-mediated translational repression appears to be as prevalent in them as it is in animals (Iwakawa and Tomari 2013).

Iwakawa and Tomari (2013) demonstrated that cap-independent translation of heterologous RNA fused to the 5′ IRES of TEV can be strongly repressed by a small RNA with perfect complementarity to a target sequence in the 5′ UTR or the open reading frame (ORF). In addition, the discovery that potyvirus translation and translational repression both occur on ER membranes strengthens the idea that RISC-mediated translational repression could be a host defense response against potyviruses (Wei et al. 2010; Ivanov et al. 2016).

A proposed model for RISC-mediated translational repression by Iwakawa and Tomari (2013) describes two possible mechanisms. In both cases RISC, i.e., AGO1 carrying sRNAs highly complementary to their targets in the 5′ UTR or ORF, acts as the effector complex. Downregulation of translation does not take place via viral RNA slicing. Instead, the blocking of translation initiation or elongation could be responsible for translation inhibition. According to the first model, AGO1 carrying a guide RNA binds to its complementary target sequence on the 5′ UTR and causes the detachment of the DExD/H-box helicase eIF4A from viral mRNA. This could result in the disruption of subsequent steps in translation initiation. In the second model, AGO1 carrying a guide RNA sequence complementary to the ORF has been proposed to bind its target and prevent the forward movement of ribosomes via steric hindrance.

Both models would require close associations between the RISC complex, viral RNA and polysomes. Ivanov et al. (2016) provided evidence that polysomes purified from PVA-infected *N. benthamiana* plants contain AGO1, which advocates the possible role of RISC-mediated translational repression in countering PVA infection. In addition to AGO1, ribosome-associated high-molecular-weight complexes in PVA-infected cells contain the viral proteins HCpro, CI, and VPg. Currently it is not known if all of them are part of the same complex or if there are several distinct complexes performing different functions. Based on these findings, it has been hypothesized that the ribosome-associated complexes might help the virus to overcome RISC-mediated translational repression and also to promote the translation of its own RNA.

Figure 11.4 shows hypothetical mechanisms for relieving the two forms of translational repression as proposed for potyviruses by Ivanov et al. (2016). Mechanism A proposes that virus-specific protein complexes are formed during potyviral translation initiation. These complexes might prevent RISC-mediated dissociation of eIF4A and allow the formation of the eIF4F complex followed by translation initiation. According to mechanism B, HCpro along with other potyviral and host proteins could cause the displacement of the RISC complex from the ORF on viral RNA to secure the uninterrupted elongation of the viral polypeptide. CI, the potyviral SF2 helicase, could participate in the remodeling of ribonucleoprotein complexes for the benefit of viral translation.

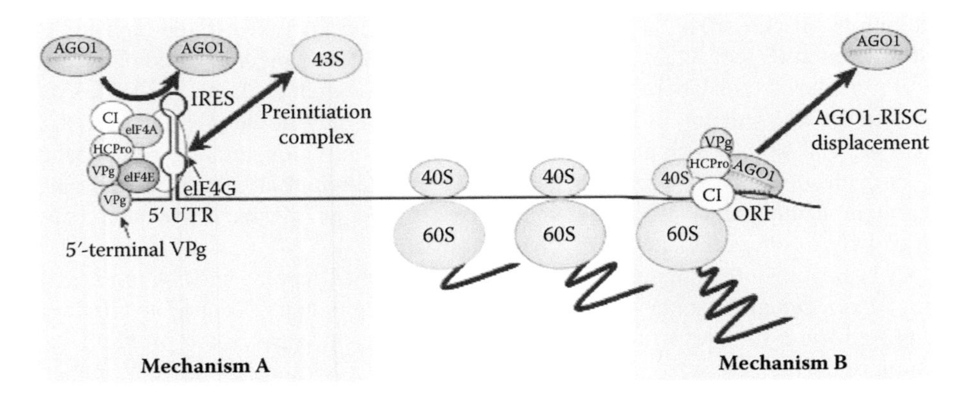

FIGURE 11.4 A model of the role of HCpro in relieving RISC-mediated translational repression (Modified from Ivanov K I et al., 2016, *Plant J* 85 (1):30–45.) Overall mechanism has been divided into two parts. Mechanism A describes relieving translational repression during initiation phase, while mechanism B depicts the probable role of HCpro along with other viral proteins in rescuing translational blockage during elongation phase.

11.5 CONCLUSION

The complexity of potyviral infection has become very clearly understood after the identification of novel viral RNA-binding factors and other players regulating viral gene expression. We now have ample knowledge about viral and host factors contributing to RNA silencing and its suppression during potyvirus infection. The validity of novel hypotheses on the translational repression of viral RNA by host and its relief by HCpro-containing ribosome-associated complexes as a part of the never-ending molecular "arms race" between the host and the virus as well as that on the convergence of potyviral RNA silencing suppression and translation are yet to be confirmed. The presence of a mechanism directing selective translation of viral RNA can't be denied, but many mechanisms, including the physiological relevance of the interaction between the 5′-terminal VPg and eIF4E/(iso) 4E in potyviral translation, are yet to be elucidated. A better description of the molecular processes underpinning efficient viral gene expression and virus accumulation should be developed to adopt new ways of controlling economically devastating potyviruses.

REFERENCES

Au H H and Jan E (2014). Novel viral translation strategies. *Wiley Interdiscip Rev RNA* 5 (6):779–801. doi: 10.1002/wrna.1246.

Beckham C J and Parker R (2008). P bodies, stress granules, and viral life cycles. *Cell Host Microbe* 3 (4):206–212.

Canizares M C, Lozano-Duran R, Canto T, Bejarano E R, Bisaro D M, Navas-Castillo J and Moriones E (2013). Effects of the crinivirus coat protein-interacting plant protein SAHH on post-transcriptional RNA silencing and its suppression. *Mol Plant Microbe Interact* 26 (9):1004–1015. doi: 10.1094/MPMI-02-13-0037-R.

Carbonell A, Fahlgren N, Garcia-Ruiz H, Gilbert K B, Montgomery T A, Nguyen T, Cuperus J T and Carrington J C (2012). Functional analysis of three Arabidopsis ARGONAUTES using slicer-defective mutants. *Plant Cell* 24 (9):3613–3629. doi: 10.1105/tpc.112.099945.

Chandrasekaran J, Brumin M, Wolf D, Leibman D, Klap C, Pearlsman M, Sherman A, Arazi T and Gal-On A (2016). Development of broad virus resistance in non-transgenic cucumber using CRISPR/Cas9 technology. *Mol Plant Pathol* 17 (7):1140–1153. doi: 10.1111/mpp.12375.

Cheng X and Wang A (2017). The potyvirus silencing suppressor protein VPg mediates degradation of SGS3 via ubiquitination and autophagy pathways. *J Virol* 91 (1). doi: 10.1128/JVI.01478-16.

Cotton S, Grangeon R, Thivierge K, Mathieu I, Ide C, Wei T, Wang A and Laliberte J F (2009). Turnip mosaic virus RNA replication complex vesicles are mobile, align with microfilaments, and are each derived from a single viral genome. *J Virol* 83 (20):10460–10471. doi: 10.1128/JVI.00819-09.

Csorba T, Kontra L and Burgyan J (2015). Viral silencing suppressors: Tools forged to fine-tune host-pathogen coexistence. *Virology* 479–480:85–103. doi: 10.1016/j.virol.2015.02.028.

Dardick C (2007). Comparative expression profiling of Nicotiana benthamiana leaves systemically infected with three fruit tree viruses. *Mol Plant Microbe Interact* 20 (8):1004–1017. doi: 10.1094/MPMI-20-8-1004.

del Toro F, Fernandez F T, Tilsner J, Wright K M, Tenllado F, Chung B N, Praveen S and Canto T (2014). Potato virus Y HCPro localization at distinct, dynamically related and environment-influenced structures in the cell cytoplasm. *Mol Plant Microbe Interact* 27 (12):1331–1343. doi: 10.1094/MPMI-05-14-0155-R.

Eskelin K, Hafrén A, Rantalainen K I and Mäkinen K (2011). Potyviral VPg enhances viral RNA translation and inhibits reporter mRNA translation in planta. *J Virol* 85 (17):9210–9221. doi: 10.1128/JVI.00052-11.

Fukunaga R and Doudna J A (2009). dsRNA with 5′ overhangs contributes to endogenous and antiviral RNA silencing pathways in plants. *EMBO J* 28 (5):545–555. doi: 10.1038/emboj.2009.2.

Galao R P, Chari A, Alves-Rodrigues I, Lobao D, Mas A, Kambach C, Fischer U and Diez J (2010). LSm1-7 complexes bind to specific sites in viral RNA genomes and regulate their translation and replication. *RNA* 16 (4):817–827. doi: 10.1261/rna.1712910.

Gallie D R (2001). Cap-independent translation conferred by the 5′ leader of tobacco etch virus is eukaryotic initiation factor 4G dependent. *J Virol* 75 (24):12141–12152.

Garcia-Ruiz H, Carbonell A, Hoyer J S, Fahlgren N, Gilbert K B, Takeda A, Giampetruzzi A, Garcia Ruiz M T, McGinn M G, Lowery N, Martinez Baladejo M T and Carrington J C (2015). Roles and programming of Arabidopsis ARGONAUTE proteins during Turnip mosaic virus infection. *PLoS Pathog* 11 (3):e1004755. doi: 10.1371/journal.ppat.1004755.

Garcia-Ruiz H, Takeda A, Chapman E J, Sullivan C M, Fahlgren N, Brempelis K J and Carrington J C (2010). Arabidopsis RNA-dependent RNA polymerases and dicer-like proteins in antiviral defense and small interfering RNA biogenesis during Turnip Mosaic Virus infection. *Plant Cell* 22 (2):481–496.

Gibbs A J, Ohshima K, Phillips M J and Gibbs M J (2008). The prehistory of potyviruses: Their initial radiation was during the dawn of agriculture. *PLoS One* 3 (6):e2523. doi: 10.1371/journal.pone.0002523.

Hafrén A, Lõhmus A and Mäkinen K (2015). Formation of Potato virus A-induced RNA granules and viral translation are interrelated processes required for optimal virus accumulation. *PLoS Pathog* 11 (12): e1005314. doi: 10.1371/journal.ppat.1005314.

Hafrén A, Eskelin K and Mäkinen K (2013). Ribosomal protein P0 promotes Potato virus A infection and functions in viral translation together with VPg and eIF(iso)4E. *J Virol* 87 (8):4302–4312. doi: 10.1128/JVI.03198-12.

Hafrén A, Hofius D, Rönnholm G, Sonnewald U and Mäkinen K (2010). HSP70 and its cochaperone CPIP promote potyvirus infection in Nicotiana benthamiana by regulating viral coat protein functions. *Plant Cell Online* 22 (2):523–535. doi: 10.1105/tpc.109.072413.

Hashimoto M, Neriya Y, Yamaji Y and Namba S (2016). Recessive resistance to plant viruses: Potential Resistance genes beyond translation initiation factors. *Front Microbiol* 7:1695. doi: 10.3389/fmicb.2016.01695.

Ivanov K I, Eskelin K, Basic M, De S, Lõhmus A, Varjosalo M and Mäkinen K (2016). Molecular insights into the function of the viral RNA silencing suppressor HCPro. *Plant J* 85 (1):30–45. doi: 10.1111/tpj.13088.

Ivanov K I, Eskelin K, Lõhmus A and Mäkinen K (2014). Molecular and cellular mechanisms underlying potyvirus infection. *J Gen Virol* 95:1415–1429. doi: 10.1099/vir.0.064220-0.

Iwakawa H O and Tomari Y (2013). Molecular insights into microRNA-mediated translational repression in plants. *Mol Cell* 52 (4):591–601. doi: 10.1016/j.molcel.2013.10.033.

Kedersha N, Stoecklin G, Ayodele M, Yacono P, Lykke-Andersen J, Fritzler Marvin J, Scheuner D, Kaufman R J, Golan D E and Anderson P (2005). Stress granules and processing bodies are dynamically linked sites of mRNP remodeling. *J Cell BioL* 169 (6):871–884. doi: 10.1083/jcb.200502088.

Krapp S, Greiner E, Amin B, Sonnewald U and Krenz B (2017). The stress granule component G3BP is a novel interaction partner for the nuclear shuttle proteins of the nanovirus pea necrotic yellow dwarf virus and geminivirus abutilon mosaic virus. *Virus Res* 227:6–14. doi: 10.1016/j.virusres.2016.09.021.

Kumakura N, Takeda A, Fujioka Y, Motose H, Takano R and Watanabe Y (2009). SGS3 and RDR6 interact and colocalize in cytoplasmic SGS3/RDR6-bodies. *FEBS Lett* 583 (8):1261–1266. doi: 10.1016/j.febslet.2009.03.055.

Łõhmus A, Varjosalo M and Mäkinen K (2016). Protein composition of 6K2-induced membrane structures formed during Potato virus A infection. *Mol Plant Pathol* 17 (6):943–958. doi: 10.1111/mpp.12341.

Moeller J R, Moscou M J, Bancroft T, Skadsen R W, Wise R P and Whitham S A (2012). Differential accumulation of host mRNAs on polyribosomes during obligate pathogen-plant interactions. *Mol Biosyst* 8 (8):2153–2165. doi: 10.1039/c2mb25014d.

Niepel M and Gallie D R (1999). Identification and characterization of the functional elements within the tobacco etch virus 5′ leader required for cap-independent translation. *J Virol* 73 (11):9080–9088.

Pager C T, Schutz S, Abraham T M, Luo G and Sarnow P (2013). Modulation of hepatitis C virus RNA abundance and virus release by dispersion of processing bodies and enrichment of stress granules. *Virology* 435 (2):472–484. doi: 10.1016/j.virol.2012.10.027.

Panas M D, Schulte T, Thaa B, Sandalova T, Kedersha N, Achour A and McInerney G M (2015). Viral and cellular proteins containing FGDF motifs bind G3BP to block stress granule formation. *PLoS Pathog* 11 (2):e1004659. doi: 10.1371/journal.ppat.1004659.

Poblete-Duran N, Prades-Perez Y, Vera-Otarola J, Soto-Rifo R and Valiente-Echeverria F (2016). Who regulates whom? An overview of RNA granules and viral infections. *Viruses* 8 (7). doi: 10.3390/v8070180.

Rajamäki M L and Valkonen J P (2009). Control of nuclear and nucleolar localization of nuclear inclusion protein a of picorna-like Potato virus A in Nicotiana species. *Plant Cell* 21 (8):2485–2502. doi: 10.1105/tpc.108.064147.

Rajamäki M-L, Streng J and Valkonen J P (2014). Silencing suppressor protein VPg of a potyvirus interacts with the plant silencing-related protein SGS3. *Molec Plant-Microbe Interact MPMI* 27 (11):1199–1210. doi: 10.1094/MPMI-04-14-0109-R.

Robaglia C and Caranta C (2006). Translation initiation factors: A weak link in plant RNA virus infection. *Trends Plant Science* 11 (1):40–45. doi: 10.1016/j.tplants.2005.11.004.

Roossinck M J (2012). Plant virus metagenomics: Biodiversity and ecology. *Annu Rev Genet* 46:359–369. doi: 10.1146/annurev-genet-110711-155600.

Rozelle D K, Filone C M, Kedersha N and Connor J H (2014). Activation of stress response pathways promotes formation of antiviral granules and restricts virus replication. *Mol Cell Biol* 34 (11):2003–2016. doi: 10.1128/MCB.01630-13.

Szittya G and Burgyan J (2013). RNA interference-mediated intrinsic antiviral immunity in plants. *Curr Top Microbiol Immunol* 371:153–181. doi: 10.1007/978-3-642-37765-5_6.

Topisirovic I, Svitkin Y V, Sonenberg N and Shatkin A J (2011). Cap and cap-binding proteins in the control of gene expression. *Wiley Interdiscip Rev RNA* 2 (2):277–298. doi: 10.1002/wrna.52.

Walsh D, Mathews M B and Mohr I (2013). Tinkering with translation: protein synthesis in virus-infected cells. *Cold Spring Harb Perspect Biol* 5 (1):a012351. doi: 10.1101/cshperspect.a012351.

Weber C, Nover L and Fauth M (2008). Plant stress granules and mRNA processing bodies are distinct from heat stress granules. *Plant J Cell Molec Biol* 56 (4):517–530. doi: 10.1111/j.1365-313X.2008.03623.x.

Wei T, Huang T-S, McNeil J, Laliberté J-F, Hong J, Nelson R S and Wang A (2010). Sequential recruitment of the endoplasmic reticulum and chloroplasts for plant potyvirus replication. *J Virol* 84 (2):799–809. doi: 10.1128/JVI.01824-09.

Yang C, Guo R, Jie F, Nettleton D, Peng J, Carr T, Yeakley J M, Fan J B and Whitham S A (2007). Spatial analysis of arabidopsis thaliana gene expression in response to Turnip mosaic virus infection. *Mol Plant Microbe Interact* 20 (4):358–370. doi: 10.1094/MPMI-20-4-0358.

Yang C, Zhang C, Dittman J D and Whitham S A (2009). Differential requirement of ribosomal protein S6 by plant RNA viruses with different translation initiation strategies. *Virology* 390 (2):163–173. doi: 10.1016/j.virol.2009.05.018.

Yu B, Yang Z, Li J, Minakhina S, Yang M, Padgett R W, Steward R and Chen X (2005). Methylation as a crucial step in plant microRNA biogenesis. *Science* 307 (5711):932–935. doi: 10.1126/science.1107130.

12 Molecular Interactions between Plant Viruses and Their Biological Vectors

Avinash Marwal, Rakesh Kumar Verma, Khurana SMP, and R.K. Gaur

CONTENTS

12.1 INTRODUCTION

For the successful viability of Geminiviruses to endure and maintain itself in order to complete their life cycle, their transmission from one host to another is a necessity (van Den Heuvel et al. 1994). All the Geminivirus species exploit homopterans as a carrier/vector for infectivity and survival in plants, which is regarded by a particular level of specificity (Fereres and Moreno 2009). Geminiviruses are one of the largest group of plant viruses containing single-stranded circular DNA encapsulated in geminate particles and prevalent in the tropical and subtropical regions of the world (Gaur et al. 2011; Marwal et al. 2013a,b). Geminiviruses have been found associated with other genomic components such as betasatellite and alphasatellite, and cause a noteworthy loss to agriculture and horticulture worldwide (Marwal et al. 2012, 2014; Prajapat et al. 2013, 2012). Based on genome-wide pairwise sequence identity, genome organization, host range, and insect vector, family *Geminiviridae* has recently been classified into nine genera: *Mastrevirus*, *Curtovirus*, *Begomovirus*, *Turncurtovirus*, *Topocuvirus*, *Eragrovirus*, *Becurtovirus*, *Capulavirus*, and *Grablovirus* (Adams et al. 2013; Varsani et al. 2014, 2017).

In nature, Geminiviruses are transmitted by phloem-feeding insects, including various species of leafhoppers (family *Cicadellidae*) (Nault and Ammar 1989), treehoppers (family *Membracidae*) (Dietrich 2009), and whiteflies (family *Aleyrodidae*) (Cicero and Brown 2011). Transmission of Geminiviruses has all three modes: persistent, semipersistent, and nonpersistent (Moreno et al. 2012), which can be synergistic, neutral, or antagonistic as characterized by complex direct and indirect interactions (Kluth et al. 2002; Hogenhout et al. 2008), whereas it is a matter of a few seconds to minutes, hours to days, or days to weeks for the arthropod vector to acquire the virus and broadcast it to a new host (Dietzgen et al. 2016). Geminiviruses with persistent nature are internalized by the insect vector, with virus residing in the salivary glands of the insect, and are characterized further into (a) circulative, (b) propagative and circulative, and (c) semi persistent (Drucker and Then 2015; Whitfield et al. 2005). Viruses having a tendency to flee from the insect gut and harbor in surrounding tissues, finally reaching to the salivary glands for transmission, are known as circulative (Pan et al 2011). Semi-persistent Geminiviruses are reserved in the gut lining of the insect vector by binding them to chitin, but do not appear to enter tissues, whereas the non-persistent viruses remain engaged to the insect's stylet (Ng and Zhou 2015; Ng and Falk 2006).

Direct interactions amid virus and the insect carrier comprises transmission and spread of the virus by the arthropod for replicating inside the host plant, both deriving food from the same host plant (Blanc and Michalakis 2016). On the other hand, one is responsible for any changes in the plant, making it unfit as food source for the other (Stout et al. 2006). Both direct and indirect effects of the pathogen on the vector, or vice versa, can be beneficial or harmful, depending on the species (Blanc et al. 2011). With the molecular approaches or biochemical means, momentous advancement has been made over the past 30 years in understanding of the Geminiviruses and their diverse insect vectors (Mauck et al. 2010). Unscrambling the horizon on the molecular biology of vector and virus and understanding the relationship/behavior/interaction between the two is imperative to recognize aspects for developing novel control measures. This forms the main objective of this chapter, highlighting the molecular aspects of macromolecules responsible for virus transmission through interaction with the insect vectors and also other recent advances in the field.

12.2 PATHWAYS FOR TRANSMISSION/INTERACTION

Proteins represent the vertebrae of cellular function by carrying out the tasks programmed in the genes articulated by a given cell type. Moreover, it remains exigent to competently categorize the equipped role of the individual protein entities identified in such procedures (Bairoch 2000). Practical properties of a protein domain, such as its enzymatic action or the capability to work together with other proteins, can often result from the estimated spatial arrangement of its amino acid chain in the folded position (Hannum et al. 2009). The journey of viruses is thought to involve viral proteins, plant proteins, and vector proteins. Vector specificity is the major way in which the family *Geminiviridae* virus genera are transmitted by the vectors. For carrying and spreading infection, the Begomoviruses start their journey with whiteflies feeding on the phloem sap of infected plants, whereby virions are ingested, reaching the alimentary canal (Moreno-Delafuente et al. 2013). The virions negotiate the gut at the midgut district and the mainstream of them gets accumulated in the filter chamber section. Mostly, Begomoviruses replicate in the host plants and make way for the virions to move through the whitefly gut, traverse the whitefly hemocele, embark in the primary salivary glands for virus transmission, and finally encounter in different salivary glands for transmission to occur (Figure 12.1). The progression of the genus *Mastrevirus* in the vector leaf hopper (*Psammotettix alienus*) takes around five minutes for acquisition of virus for transmission (Martin et al. 1997). A latent period of 6–8 hours is required for Mastreviruses, while eight hours in the case of Begomoviruses. Geminiviruses' coat protein (CP) is the foremost protein responsible for transmission via arthropod vector, as suggested by most evidence (Chen et al. 2011; Andret-Link and Fuchs 2005). Universal twinned icosahedral symmetry is only revealed by CP of Geminiviruses. An attempt to develop a chimeric virus by switching the CP open reading frams (ORFs) among *Beet curly top virus* (transmitted by vector leaf hopper) and *African cassava mosaic virus* (transmitted by vector whitefly) resulted in vector "exchange," suggesting that CP is mainly responsible in determining vector specificity (Briddon et al. 1990). In another study, the initial step in virus acquisition highlights the true proof of CP serving as the attachment protein of Geminiviruses to insect vector guts (Table 12.1). Leafhopper (*P. alienus*), when fed on infected plants harboring a recombinant CP of *Wheat dwarf virus* (WDV, *Mastrevirus*), resulted in a decreased amount of virus in vector leafhoppers in comparison to the wild-type WDV-fed leafhoppers. A similar experiment was performed with Begomoviruses (*Tomato yellow leaf curl virus*), where recombinant CP got localized in the midgut of feeding whitefly vectors, thus resulting in reduction of virus transmission (Ohnesorge et al. 2009).

By analyzing the complete transcriptome, around 1606 genes were identified in whiteflies which were considerably regulated in response to Geminivirus acquisition, indicating 157 biochemical

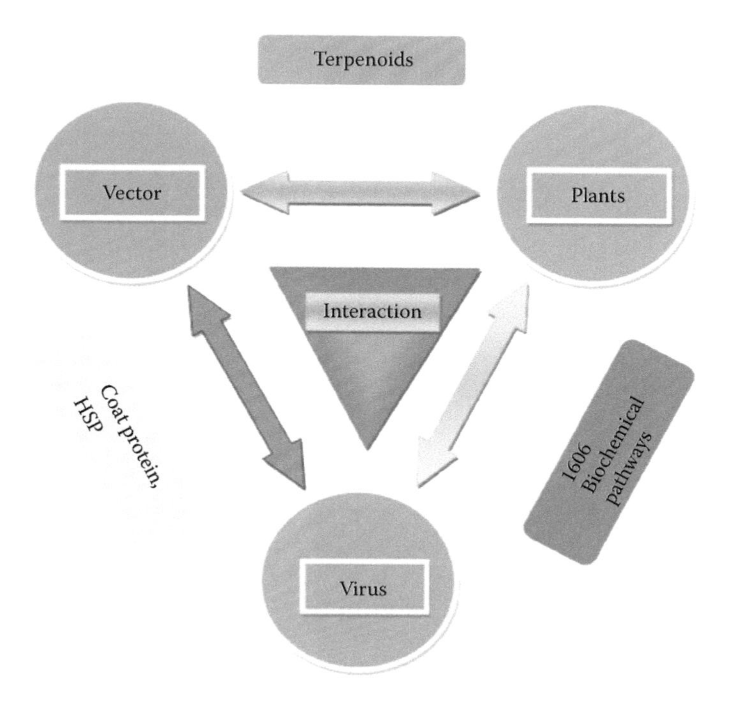

FIGURE 12.1 Various pathways/macromolecules for virus–insect–vector interactions for virus transmission.

pathways. This study was done on *B. tabaci* MEAM1 (full form, first time) in response to TYLCV reported from China. Gene or protein synthesis, cell cycle, autophagy genes, lipid metabolism, primary metabolism, lysosome function, immune response, and so forth were upregulated in the whiteflies (Luan et al. 2011) and the genes implicated for downregulation by TYLCV were that for apoptosis, toll-like and mitogen-activated protein kinase (MAPK) pathways, signal transduction of the immune response, and so forth. The experiments conducted were steady with whiteflies escalating a defense against Geminivirus incursion and the virus neutralizing this commencement of the immune response (Figure 12.2). The equilibrium between *B. tabaci* MEAM1 and TYLCV might be at variance for different Geminiviruses and whitefly biotypes, suggesting a rationalization for the disparity observed in transmission efficiency and vector specificity (Gottlieb et al. 2010).

For tackling arthropods, plants have developed a defense mechanism known as jasmonic acid (JA) pathway. Any hindrance of this pathway paves the way for insects to feed. Zhang et al. (2012) confirmed that *Tomato yellow leaf curl China betasatellite* (TYLCCNB) βC1 protein operates for a potent downregulation of the JA pathway and aids in the gathering and augmentation of *B. tabaci* MEAM1 on plants (Pieterse and Dicke 2007). A similar approach was performed by Luan et al. (2013), confirming that terpenoid synthesis and its release were not seen/revealed by the Geminivirus-infected plants, being favored by whiteflies to encounter them in comparison to the uninfected plants. Gene pathways such as detoxification of enzyme, oxidative phosphorylation pathway, and so forth were downregulated in *B. tabaci* MEAM1 feeding on virus-infected plants, finally attracting most of the whiteflies (Table 12.1).

Like many homopterans, whiteflies also nurture mycetome endosymbionts (bacteria) which produce a 63-kDa GroEL (belonging to the *Chaperonin* family of molecular chaperones) (De Barro et al. 2011; Dinsdale et al. 2010). In a study, an Israeli strain of *Tomato yellow leafcurl virus* (TYLCV; KC106635) intermingles with an endosymbiotic GroEL homologue (van Den Heuvel et al. 1994, 1997). Later the whiteflies were given a diet of anti-Buchnera GroEL antiserum before TYLCV acquisition. Remarkably, an 80% decrease in virus transmission was observed in

TABLE 12.1
Comparative Approach of Interacting Molecules among Different Plant Viruses

Interacting Molecules	Source	Activity Profile	Reference
CP (Coat Protein)	Geminivirus	Determining vector specificity, attachment protein to vectors, binds to the GroEL and HSPs	Briddon et al. 1990; Ohnesorge et al. 2009
GroEL	Mycetome endosymbionts (bacteria)	Binds to heat shock protein 16, Geminiviruses get stabilized upon interacting with GroEL	Morin et al. 2000; De Barro et al. 2011
HSPs (heat shock proteins)	*Bemisia tabaci*	HSP16 routing of Geminivirus across the gut epithelia into the haemolymph	Ohnesorge and Bejarano 2009; Rubinstein and Czosnek 1997
Cyclophilins	*Bemisia tabaci*	Involved in protein-protein interactions for Geminivirus	Kanakala and Ghanim 2016
Terpenoids	Host plants	More volatile terpens get released by healthy plants than the TYLCV-infected plants and *Bemisia tabaci* prefer infected plants	Fang et al. 2013
JA (jasmonic acid)	Host plants	βC1 protein of Geminivirus operate as a potent down regulation of the JA	Pieterse and Dicke 2007
G Proteins (glycoprotein)	Rhabdovirus	Interaction in the midgut of plant hopper	Ammar et al. 2009; Jackson et al. 2005; Mann and Dietzgen 2014
C-RTP Proteins (C-terminal of read-through protein)	Luteovirus	The virus protein interact with the aphids in their hindgut	Gray and Gildow 2003
Symbionin	Endosymbiotic bacteria of the genus *Buchnera*	Symbionin protect the Luteovirus from aphid immune system	van Den Heuvel et al. 1997
CP, P2, P3 (CaMV proteins)	Caulimovirus	Acquisition through the aphid's stylet, which is the point of interaction	Hoh et al. 2010; Leh et al. 1999
CP, HC-Pro (helper component proteinase)	Potyvirus	Acquisition through the aphid's stylet, which is the point of interaction	Wang et al. 1996; Blanc et al. 1998
CPm proteins (capsid protein)	Closterovirus	Viruses interact with the whiteflies in the foregut	Satyanarayana et al. 2004
P2 (capsid protein)	Reovirus	Filter chamber and midgut of leafhopper are the places of virus interaction	Zhou et al. 2007; Omura et al. 1998
CP (Coat Proteins)	Bromovirus	At the entry of the stylet CP interacts with Aphids	Osman et al. 1997; Brumfield et al. 2004

comparison to the control insects feeding on normal serum. Hemolymph of the whiteflies were occupied by active antibodies, clearly suggesting that the antibody interferes with the interaction between TYLCV and the GroEL homologue (Table 12.1). Riased Anti-GroEL antibodies coated with gold suggested that GroEL is produced by a secondary symbiotic bacteriosome *Hamiltonella* in *B. tabaci* MEAM1, hence making whiteflies a much more proficient vector for TYLCV than *B. tabaci* Mediterranean (MED), which does not nourish *Hamiltonella* (Morin et al. 1999, 2000).

Geminiviruses encode for CP, which easily binds to the GroEL in the gut of arthropods and also binds to heat shock protein 16 (HSP16) encoded by *B. tabaci*. Virions might get stabilized while interacting with HSP16 and GroEL chaperone proteins which have been implicated in transport to and stabilization in the hemolymph, respectively. A further study was conducted using CP of *Tomato yellow leaf curl Sardinia virus* in a yeast two-hybrid and screening it against a *B. tabaci* MED (Q biotype) cDNA library (Bosco et al. 2004; Ohnesorge and Bejarano 2009). The CP unites with a 16 kDa small heat shock protein (coined BtHSP16, belonging to the HSP20/alpha crystallin family). The CP-BtHSP16-interacting domain was positioned within the conserved region of the N-terminal part of the TYLCSV CP (amino acids 47–66), overlapping almost entirely with the nuclear localization signal (Figure 12.1). This study identified a good candidate for involvement in virus translocation (Rubinstein and Czosnek 1997; Hanley-Bowdoin et al. 2013). Another protein, cyclophilin B (CypB), was assessed in relation to the transmission of TYLCV by *B. tabaci*. Cyclophilins belong to the *Prolylisomerase* family involved in protein-protein interactions within the cell. Upon acquisition and retention of TYLCV Geminivirus, CypB expression in *B. tabaci* increased, hindering virus transmission (Kanakala and Ghanim 2016). With the help of co-immunolocalization and immunocapture-PCR, an interaction between TYLCV and CypB was observed in eggs, salivary glands, and midgut. Whiteflies were subjected to TYLCV-infected plants and on anti-CypB antibodies (membrane feeding), which clearly revealed reduction in the transmission of TYLCV, implying a major role of CypB in TYLCV transmission (Kanakala and Ghanim 2016).

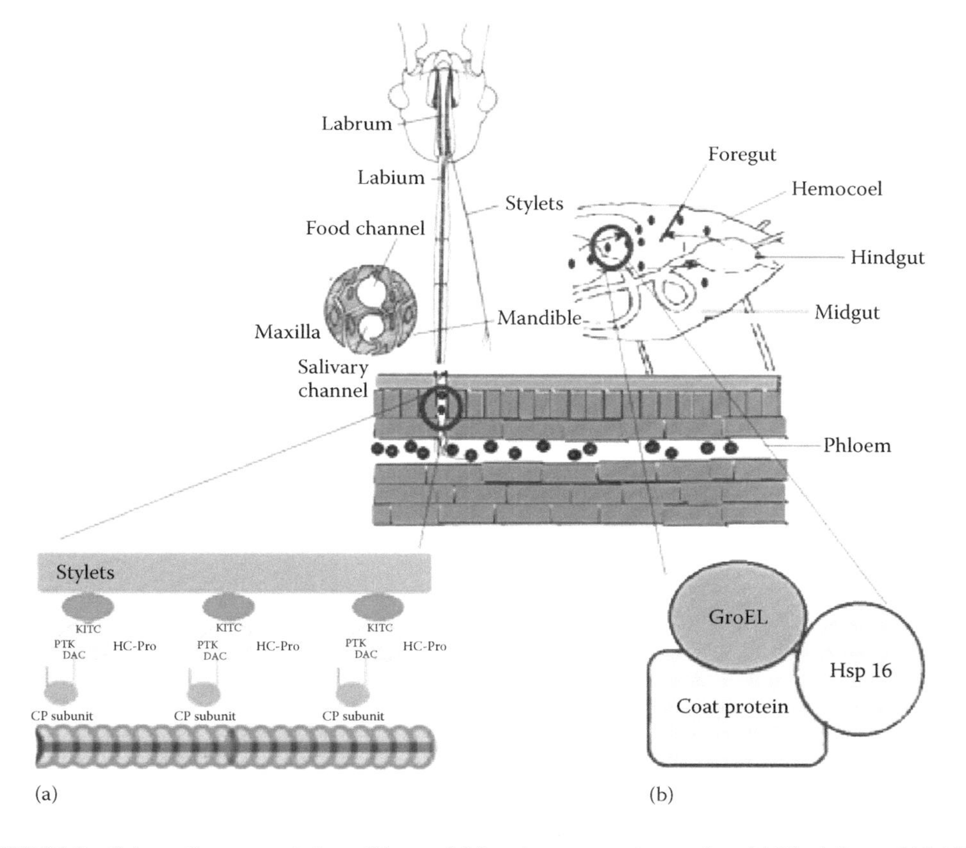

FIGURE 12.2 Schematic representation of the model for virus–vector interaction. (a) The left panel highlights the interaction site of stylets. Noncirculative, nonpersistent viruses are staying in the distal tip of the insect stylet (i.e., aphid). (b) The right panel depicts the interaction in the gutwall. Circulative, propagative viruses replicate and systemically occupy several insect organs and tissues entering the hemolymph to reach the salivary glands for transmission (i.e., whitefly and leafhopper).

12.3 COMPARISON OF WHITEFLIES WITH APHIDS AND LEAFHOPPERS AS VIRUS VECTORS

Aphids and leafhoppers follow non-circulative, non-persistent, and propagative manner transmission of plant viruses. Propagative vectors retained viruses for a longer time in comparison to non-persistent virus carriers (Ng and Falk 2006). Recent research revealed a time period of five minutes for virus acquisition by *Psammotettix alienus* (a leafhopper) for WDV (genus *Mastrevirus*) and movement in the vector (Wang et al. 2014). Further, when *P. alienus* leafhopper was fed on WDV recombinant coat protein, it got contained at the midgut. Even the antibodies raised against wild-type WDV CP resulted in a decline in virus titer in the leafhopper (Wang et al. 2014). On the other hand, 30 different species of aphids are responsible for the transmission of *Cauliflower mosaic virus* (CaMV). CaMV follows a noncirculative and semipersistent strategy of transmission (Uzest et al. 2007). P2 protein encoded by CaMV acts as a helper protein, which mediates the binding of CaMV to the aphid's stylets. The virus is retained in the aphid for several minutes to hours before delivering itself to another healthy host plant (Drucker and Then 2015).

Potyviruses are transmitted through aphids in a non-persistent manner (Pirone and Perry 2002; Bhargava and Khurana 1970; Singh et al. 1988; Verma et al. 2015). In non-persistent type transmission, the vectors can transmit immediately after an acquisition feeding but remain infective for only a short period of time. The stylets of aphids become contaminated with virus during penetration in the infected tissue, and only this fraction of the virus is of significance in transmission (Blanc et al. 2011). Aphid transmission of Potyviruses depends on the presence of specific sequence domains in two virus-encoded proteins, the CP and helper component-proteinase (HC-Pro) (Figure 12.2). The bridge hypothesis adopted by Potyviruses states that the HC-Pro binds one hand to the stylet and the other hand to the virion capsomeres (Kassanis and Govier 1971), providing a bridge to link the virus particles to the aphid (Pirone and Blanc 1996; Mishra et al. 2014). The highly conserved KITC (Lys-Ile-Thr-Cys) motif in the N-terminal domain of HC-Pro is necessary to retain virions in the food canal and foregut of aphids, and PTK (Pro-Thr-Lys) motif located in the HC-Pro C-terminal domain is required for efficient interaction with DAG (Asp-Ala-Gly) motif of CP (Revers and Garcia 2015; Ruiz-Ferrer et al. 2005; Dombrovsky et al. 2005; Verma et al. 2014). All this data support a bridge model in which HC-Pro binds to both viral CP and to a putative receptor within the aphid stylet and facilitates the virus transmission (Brault et al. 2010). However, the receptor of HC-pro in the aphid stylet has not yet been identified. In recent years, studies using *in vitro* binding assay, far-Western blotting strategy, identified several proteins, including cuticle proteins, which were interacting with HC-Pro (Dombrovsky et al. 2007; Fernandez-Calvino et al. 2010).

The transmission of Potyviruses by aphids is also associated with aphid colonizing and probing behavior (Fereres and Moreno 2009; Stafford et al. 2012). Non-colonizer (visiting) aphid species contribute more to Potyvirus spread than colonizing species (Yuan and Ullman 1996; Revers et al. 2015). Virus infections of the host plant also influence vector behavior and performance; for instance, PVY infection on potato plants differentially influenced the aphid feeding behavior of *Myzuspersicae* and *Macrosiphum euphorbiae* (Boquel et al. 2010). In mixed infection of host with a Polerovirus, it increased the fecundity and settling of both aphid species (Srinivasan and Alvarez 2007). Similarly, the aphid *Rhopalo siphummaidis* remains on healthy soybean plants for a longer time period than on the SMV-infected plants (Fereres et al. 1999), and squash plants infected with ZYMV emitted more organic volatile compounds, which were attractive for aphids compared to the healthy plants (Salvaudon et al. 2013). In a recent study, it was also shown that the expression of NIaPro alone in *N. benthamiana* is sufficient to increase *M. persicae* fecundity and settling (Casteel et al. 2014), but the precise role of Nia-Pro in aphid transmission is still unknown.

Leafhoppers and aphids transmit Rhabdoviruses in a persistent, circulative, propagative manner. Rhabdovirus has a spike-like glycoprotein (G) residing on the surface of virions which is predicted to interact with receptors in the midgut, allowing virions to enter epithelial cells by endocytosis (Ammar et al. 2009; Mann and Dietzgen 2014; Jackson et al. 2005). Aphid anchorage with some bacteria has an endosymbiotic relationship belonging to the genus *Buchnera* in specialized cells called mycetocytes located in their abdominal cavity. These endosymbionts generate an abundant quantity of a protein called symbionin that facilitates in protein folding, translocation across membranes, and recovery from stress (Baumann et al. 1995). *In vitro* studies revealed that the symbionin binds to six different purified Luteoviruses (Filichkin et al. 1997). Symbionin may protect the virus from targeting by the aphid immune system, or alternatively, it may function as a chaperonin to preserve or change the structure of the capsid and facilitate virus movement into the accessory salivary gland (van Den Heuvel et al. 1997).

Rice dwarf virus (RDV) is a Phytoreovirus and is transmitted by a leafhopper *Nephotettixc incticeps*. The Phytoreovirus enters the alimentary canal followed by the epithelial cells of the filter chamber region, and further traverses into the anterior midgut. Phytoreovirus pursues the path to the basal lamina to infect the muscle cells, and finally reaches the salivary glands via a hemolymph route, ready to cause infection on a new host. RDV is taken up by clathrin-mediated endocytosis in the midgut and encodes a protein P2 that facilitates virion release from the endocytic vesicle (Wei et al. 2007, 2008). RDV-induced tubule structures are composed primarily of a nonstructural protein Pns10, which is associated with actin-based filopodia, protruding from the surface of cells, and is capable of penetrating neighboring cells. RNAi knockdown of Pns10 in vector feeding experiments inhibited formation of tubules, prevented intercellular spread, and reduced leafhopper transmission efficiency of the virus. These data conclusively show that the Pns10 tubules facilitate the intercellular distribution of RDV in the leafhopper vector (Wei et al. 2009; Chen et al. 2012, 2015).

12.4 DETERMINING VIRUS–VECTOR INTERACTIONS

A polymerase chain reaction (PCR)-based method was established for studying the interactions between the virus and vector. Acquisition access periods (AAPs) were the base for determining the allocation of *Beet mild curly top virus* (BMCTV) in the leafhopper, ranging from 1 to 48 hours, using BMCTV-infected shepherd's purse plants. Upon testing, BMCTV was perceived in the hemolymph (3 hours AAP) and in the salivary glands (3 hours AAP). The quantity of BMCTV found in the hemolymph and salivary glands increased with AAP duration (Soto and Gilbertson 2003). Similarly, the presence of *Maize streak virus* (MSV) in the vector species *Cicadulin ambila* and the non-vector species *C. chinaï* was scrutinized by devising a conventional and real-time quantitative PCR. Both *C. mbila* and *C. chinaï* were subjected to 3-day AAP on MSV-infected plants which had more virus accumulation in *C. chinaï* than *C. mbila*. MSV was detected in the gut, the hemolymph, and also the head of *C. mbila*, but only in the gut of *C. chinaï* (Lett et al. 2002). Virus–vector behavior can also be supervised very intimately using electronic devices such as DC-amplifiers, which differentiate among the intercellular and intracellular surroundings, thus confirming when the plant cell membranes get punctured by the insect stylets (Walker 2000; Backus and Bennett 2009; McLean and Kinsey 1964).

Once the plant cell membrane gets pierced, a different variant of electrical penetration graph (EPG) signal is documented in the form of a potential drop (pd) (Tjallingii and Esch 1993). Geminiviruses are incapable of moving outside phloem tissue and seek the help of phloem-feeding vectors which spread from plant to plant (Johnson and Walker 1999). EPG technique has been employed for studying the transmission of MSV by leafhopper *C. astoreyi*, which showed a continuous and sustained pattern immediately after a sustained potential drop for 60–100 minutes. Numerous EPG wave forms have been depicted for *C. ambila*, MSV vector on maize (Kimmins and Bosque-Perez 1996; Lett et al. 2001).

Microarray, real-time PCR, and Western blot techniques suggested that *B. tabaci* heat shock protein 70 (HSP70) exclusively react in the company of amonopartite TYLCV and the bipartite *Squash leaf curl virus*. Moreover, virus overlay protein-binding assays clearly revealed that protein coimmunoprecipitation and immunocapture PCR an interaction among TYLCV and HSP70 *in vitro*. Immunolocalization and *in situ* fluorescence for the hybridization illustrated colocalization of

TYLCV, and also the bipartite *Watermelon chlorotic stunt* virions with HSP70 in the midgut epithelial cells of *B tabaci* was identified (Götz et al. 2012). Similarly, gas chromatography mass spectrometry (GCMS) was used to assay various plant volatile terpens such thymene, caryophyllene, β-myrcene, α-humulene, (1)-4-carene, and β-phellandrene. Such volatile terpens were released by healthy plants than TYLCV-infected plants. An evidence report suggested preference of *Bemisia tabaci* for infected plants over healthy plants. Infection by TYLCV amends host plant preference of *B. tabaci* B (Middle East-Minor Asia 1) and Q (Mediterranean genetic group). TYLCV-infected tomato plants were preferred as a source of food by TYLCV-free *B. tabaci* Q. This is in contrast to healthy tomato plants that were the choice of TYLCV-free *B. tabaci* B. Finally, TYLCV-infected *B. tabaci*, either B or Q, did not display an inclination of preference between TYLCV-infected and healthy tomato plants. The study showed that TYLCV can amend the host preference of its vector *B. tabaci* B and Q (Fang et al. 2013).

12.5 CONCLUSIONS

To successfully invade new hosts, to break the host resistance, to move the virus particles within and between plants, Geminiviruses have evolved with a coordinated network of viral and cellular protein interactions/relationships/behaviors with the host plants and their insect vectors to complete the viral life cycle. New discoveries on the subject of Geminiviruses and homopterans have increased our understanding of the basic mechanisms of virus transmission, which in turn have enabled the development of numerous management strategies. The literature in the chapter suggests that Geminiviruses are transmitted by a diverse group of arthropods, which employ a similar course for virus movement and circulation in the vector, for which the Geminivirus CP is the chief determinant. Interaction of Geminivirus proteins with GroEL, HSPs, and Cyclophilins is also equally essential for the virions to facilitate their circulation and translocation across membrane barriers, particularly through midgut to hemolymph to salivary glands of the insects. Geminiviruses induce many changes in host plants, especially volatiles (terpenoids), which play an important role in mediating virus-vector interactions/relationships. Protein–protein interactions between Geminiviruses and the insect vectors is an indispensable molecular boundary that aids virus acquisition from infected host plants in transmission to healthy hosts.

ACKNOWLEDGMENTS

The authors are thankful to the Science and Engineering Research Board – Department of Science and Technology, New Delhi, India for financial assistance (File No. YSS/2015/000265) and also to University Grant Commission, New Delhi, for providing financial assistantship under Research Award for Teacher (F.30-1/2014/RA-2014-16-GE-RAJ-4696 (SA-II) and to the Amity University authorities for encouragement and support.

REFERENCES

Adams MJ, King AMQ, Carstens EB (2013). Ratification vote on taxonomic proposals to the International Committee on Taxonomy of Viruses. *Arch Virol* 158(9):2023–2030.

Ammar ED, Tsai CW, Whitfield AE, Redinbaugh MG, Hogenhout SA (2009). Cellular and molecular aspects of Rhabdovirus interactions with insect and plant hosts. *Annu Rev Entomol* 54:447–468.

Andret-Link P, Fuchs M (2005). Transmission of plant viruses by vectors. *J Plant Pathol* 87(3):153–165.

Backus EA, Bennett WH (2009). The AC–DC correlation monitor: New EPG design with flexible input resistors to detect both R and emf components for any piercing sucking hemipteran. *J Insect Physiol* 55:869–884.

Bairoch A (2000). The ENZYME database. *Nucleic Acids Res* 28:304–305.

Baumann P, Baumann L, Lai C, Rouhbakhsh D, Moran NA, Clark MA (1995). Genetics physiology and evolutionary relationships of the genus Buchnera: Intracellular symbionts of aphids. *Annu Rev Microbiol* 49:55–94.

Bhargava KS, Khurana SMP (1970). Insect transmission of papaya viruses with special reference to papaya mosaic. *Zbl Bakt Abt* 124:688–696.

Blanc S, Ammar ED, Garcia-Lampasona S, Dolja VV, Llave C, Baker J, Pirone TP (1998). Mutations in the Potyvirus helper component protein: Effects on interactions with virions and aphid stylets. *J Gen Virol* 79:3119–3122.

Blanc S, Michalakis Y (2016). Manipulation of hosts and vectors by plant viruses and impact of the environment. *Curr Opin Insect Sci.* 16:36–43.

Blanc S, Uzest M, Drucker M (2011). New research horizons in vector-transmission of plant viruses. *Curr Opin Microbiol* 14:483–491.

Boquel S, Giordanengo P, Ameline A (2010). Divergent effects of PVY-infected potato plant on aphids. *Eur J Plant Pathol* 129:507–510.

Bosco D, Mason G, Accotto GP (2004). TYLCSV DNA but not infectivity can be transovarially inherited by the progeny of the whitefly vector Bemisia tabaci (Gennadius). *Virology* 323:276–283.

Brault V, Uzest M, Monsion B, Jacquot E, Blanc S (2010). Aphids as transport devices for plant viruses. *Comptes Rendus Biologies* 333:524–538.

Briddon RW, Pinner MS, Stanley J, Markham PG (1990). Geminivirus coat protein gene replacement alters insect specificity. *Virology* 177:85–94.

Brumfield S, Willits D, Tang L, Johnson JE, Douglas T, Young M (2004). Heterologous expression of the modified coat protein of Cowpea chlorotic mottle bromovirus results in the assembly of protein cages with altered architectures and function. *J Gen Virol* 85(4):1049–1053.

Casteel CL, Yang C, Nanduri AC, De Jong HN, Whitham SA, Jander G (2014). The NIa–Pro protein of Turnip mosaic virus improves growth and reproduction of the aphid vector Myzus persicae (Green Peach Aphid). *Plant J* 77:653–663.

Chen AYS, Walker GP, Carter D, Ng JCK (2011). A virus capsid component mediates virion retention and transmission by its insect vector. *PNAS* 10(40):16777–16782.

Chen Q, Chen H, Mao Q, Liu Q, Shimizu T, Uehara-Ichiki T, Wu Z, Xie L, Omura T, Wei T (2012). Tubular structure induced by a plant virus facilitates viral spread in its vector insect. *PLoS Pathog* 8:e1003032.

Chen Q, Wang H, Ren T, Xie L, Wei T (2015). Interaction between non-structural protein Pns10 of Rice dwarf virus and cytoplasmic actin of leafhoppers is correlated with insect vector specificity. *J Gen Virol* 96: 933–938.

Cicero J, Brown JK (2011). Functional anatomy of whitefly organs associated with Squash leaf curl virus (Geminiviridae:Begomovirus) transmission by the B biotype of Bemisia tabaci (Hemiptera:Aleyrodidae). *Ann Entomol Soc Am* 104:261–279.

De Barro PJ, Liu SS, Boykin LM, Dinsdale AB (2011). Bemisia tabaci: A statement of species status. *Annu Rev Entomol* 56:1–19.

Dietrich CH (2009). Auchenorrhyncha (cicadas spittlebugs leafhoppers treehoppers and plant hoppers) In: *Encyclopedia of Insects*. 2nd edn. Resh V, H Cardé, RT Burlington (Eds). Boston, MA: Elsevier: 56–64.

Dietzgen RG, Mann KS, Johnson KN (2016). Plant virus–insect vector interactions: Current and potential future research directions. *Viruses* 8(303). doi:103390/v8110303.

Dinsdale A, Cook L, Riginos C, Buckley YM, De Barro P (2010). Refined global analysis of Bemisiatabaci (Hemiptera: Sternorrhyncha: Aleyroidae:Aleyrodidae) mitochondrial cytochrome oxidase 1 to identify species level genetic boundaries. *Ann Entomol Soc Am* 103:196–208.

Dombrovsky A, Gollop N, Chen SB, Chejanovsky N, Raccah B (2007). In vitro association between the helper component-proteinase of Zucchini yellow mosaic virus and cuticle proteins of Myzus persicae. *J Gen Virol* 88:1602–1610.

Dombrovsky A, Huet H, Chejanovsky N, Raccah B (2005). Aphid transmission of a Potyvirus depends on suitability of the helper component and the N terminus of the coat protein. *Arch Virol* 150:287–298.

Drucker M, Then C (2015). Transmission activation in non-circulative virus transmission: A general concept? *Curr Opin Virol* 15:63–68.

Fang Y, Jiao X, Xie W et al. (2013). Tomato yellow leaf curl virus alters the host preferences of its vector Bemisiatabaci. *Sci Rep* 3:2876.

Fereres A, Kampmeier GE, Irwin ME (1999). Aphid attraction and preference for soybean and pepper plants infected with Potyviridae. *Ann Entom Soc Am* 92:542–548.

Fereres A, Moreno BA (2009). Behavioural aspects influencing plant virus transmission by homopteran insects. *Virus Res* 141:158–168.

Fernandez-Calvino L, Goytia E, Lopez-Abella D, Giner A, Urizarna M, Vilaplana L et al. (2010). The helper-component protease transmission factor of Tobacco etchpotyvirus binds specifically to an aphid ribosomal protein homologous to the laminin receptor precursor. *J Gen Virol* 91:2862–2873.

Filichkin SA, Brumfield S, Filichkin TP, Young MJ (1997). In vitro interactions of the aphid endosymbioticsymL with Barley yellow dwarf virus. *J Virol* 71:569–577.
Gaur RK, Prajapat R, Marwal A, Sahu A, Rathore MS (2011). First report of a Begomovirus infecting Mimosa pudica in India. *J Plant Pathol* 93(4):480.
Gottlieb Y, Zchori-Fein E, Mozes-Daube N, Kontsedalov S, Skaljac M, Brumin M, Sobol I, Czosnek H, Vavre F, Fleury F, Ghanim M (2010). The transmission efficiency of Tomato yellow leaf curl virus by the whitefly Bemisiatabaci is correlated with the presence of a specific symbiotic bacterium species. *J Virol* 84:9310–9317.
Götz M, Popovski S, Kollenberg M et al. (2012). Implication of Bemisiatabaci heat shock protein 70 in Begomovirus-whitefly interactions. *J Virol* 86(24):13241–13252.
Gray S, Gildow FE (2003). Luteovirus-aphid interactions. *Annu Rev Phytopathol* 41:539–566.
Hanley-Bowdoin L, Bejarano ER, Robertson D, Mansoor S (2013). Geminiviruses: Masters at redirecting and reprogramming plant processes. *Nat Rev Microbiol* 11:777–788.
Hannum G, Srivas R, Guénolé A, van Attikum H, Krogan NJ, Karp RM, Ideker T (2009). Genome-wide association data reveal a global map of genetic interactions among protein complexes. *PLoS Genet* 5: e1000782.
Hogenhout SA, Ammar E-D, Whitfield AE, Redinbaugh MG (2008). Insect vector interactions with persistently transmitted viruses. *Annu Rev Phytopathol* 46:327–359.
Hoh F, Uzest M, Drucker M, Plisson-Chastang C, Bron P, Blanc S, Dumas C (2010). Structural insights into the molecular mechanisms of Cauliflower mosaic virus transmission by its insect vector. *J Virol* 84:4706–4713.
Jackson AO, Dietzgen RG, Goodin MM, Bragg JN, Deng M (2005). Biology of plant Rhabdoviruses. *Annu Rev Phytopathol* 43:623–660.
Johnson DD, Walker GP (1999). Intracellular punctures by the adult whitefly Bemisia argentifolii on DC and AC electronic feeding monitors. *Entomol Exp Appl* 92:257–270.
Kanakala K, Ghanim M (2016). Implication of the Whitefly Bemisiatabaci Cyclophilin B protein in the transmission of Tomato yellow leaf curl virus. *Front Plant Sci* 7:1702.
Kassanis B, Govier DA (1971). The role of the helper virus in aphid transmission of Potato aucuba mosaic virus and potato virus C. *J Gen Virol* 13:221–228.
Kimmins FM, Bosque-Perez NA (1996). Electrical penetration graphs from Cicadulina spp and the inoculation of a persistent virus into maize. *Entomol Exp Appl* 80:46–49.
Kluth S, Kruess A, Tscharntke T (2002). Insects as vectors of plant pathogens: Mutualistic and antagonistic interactions. *Oecologia* 133:193–199.
Leh V, Jacquot E, Geldreich A, Hermann T, Leclerc D, Cerutti M, Yot P, Keller M, Blanc S (1999). Aphid transmission of Cauliflower mosaic virus requires the viral PIII protein. *EMBO J* 18:7077–7085.
Lett JM, Granier M, Grondin M, Turpin P, Molinaro F, Chiroleu F, Peterschmitt M, Reynaud B (2001). Electrical penetration graphs from Cicadulinambila on maize the fine structure of its stylet pathways and consequences for virus transmission efficiency. *Entomol Exp Appl* 101:93–109.
Lett J-M, Granier M, Hippolyte I (2002). Spatial and temporal distribution of Geminiviruses in leafhoppers of the genus Cicadulina monitored by conventional and quantitative polymerase chain reaction. *Phytopathology* 92(1):65–74.
Luan JB, Li JM, Varela N, Wang YL, Li FF, Bao YY, Zhang CX, Liu SS, Wang XW (2011). Global analysis of the transcriptional response of whitefly to Tomato yellow leaf curl China virus reveals the relationship of coevolved adaptations. *J Virol* 85:3330–3340.
Luan JB, Yao DM, Zhang T, Walling LL, Yang M, Wang YJ, Liu SS (2013). Suppression of terpenoid synthesis in plants by a virus promotes its mutualism with vectors. *Ecol Lett* 16:390–398.
Mann KS, Dietzgen RG (2014). Plant Rhabdoviruses: New insights and research needs in the interplay of negative-strand RNA viruses with plant and insect hosts. *Arch Virol* 159:1889–1900.
Martin B, Collar JL, Tjallingii WF, Fereres (1997). A: Intracellular ingestion and salivation by aphids may cause the acquisition and inoculation of non-persistently transmitted plant viruses. *J Gen Virol* 78:2701–2705.
Marwal A, Sahu A, Prajapat R, Choudhary DK, Gaur RK (2012). First report of association of Begomovirus with the leaf curl disease of a common weed Datura inoxia. *Virus Disease* 23(1):83–84.
Marwal A, Sahu A, Sharma P, Gaur RK (2013a). Molecular characterizations of two Begomoviruses infecting Vinca rosea and Raphanus sativus in India. *Virol Sin* 28(1):53–56.
Marwal A, Sahu AK, Choudhary DK, Gaur RK (2013b). Complete nucleotide sequence of a Begomovirus associated with satellites molecules infecting a new host Tagetes patula in India. *Virus Genes* 47(1):194–198.

Marwal A, Sahu AK, Gaur RK (2014). First report of airborne Begomovirus infection in Melia azedarach (Pride of India) an ornamental tree in India. *Aerobiologia* 30(2):211–215.

Mauck KE, De Moraes CM, Mescher MC (2010). Deceptive chemical signals induced by a plant virus attract insect vectors to inferior hosts. *Proc Natl Acad Sci U S A* 107:3600–3605.

McLean DL, Kinsey MG (1964). Technique for electronically recording aphid feeding salivation. *Nature* 202:1358–1359.

Mishra R, Verma RK, Sharma P, Choudhary DK, Gaur RK (2014). Interaction between viral proteins with the transmission of Potyvirus. *Arch Phytopathol Plant Protect* 47(2):240–253.

Moreno A, Tjallingii WF, Fernandez-Mata G, Fereres A (2012). Differences in the mechanism of inoculation between a semi-persistent and a non-persistent aphid-transmitted plant virus. *J Gen Virol* 93:662–667.

Moreno-Delafuente A, Garzo E, Moreno A, Fereres AA (2013). Plant virus manipulates the behaviour of its whitefly vector to enhance its transmission efficiency and spread. *PLoS One* 8:e61543.

Morin S, Ghanim M, Sobol I, Czosnek H (2000). The GroEL protein of the whitefly Bemisiatabaci interacts with the coat protein of transmissible and non-transmissible Begomoviruses in the yeast two hybrid system. *Virology* 276:404–416.

Morin S, Ghanim M, Zeidan M, Czosnek H, Verbeek M, van den Heuvel JFJM (1999). A GroEL homologue from endosymbiotic bacteria of the whitefly Bemisiatabaci is implicated in the circulative transmission of Tomato yellow leaf curl virus. *Virology* 256:75–84.

Nault LR, Ammar ED (1989). Leafhopper and planthopper transmission of plant viruses. *Annu Rev Entomol* 34:503–530.

Ng JC, Falk BW (2006). Virus–vector interactions mediating non-persistent and semi-persistent transmission of plant viruses. *Annu Rev Phytopathol* 44:183–212.

Ng JC, Zhou JS (2015). Insect vector–plant virus interactions associated with non-circulative semi-persistent transmission: Current perspectives and future challenges. *Curr Opin Virol* 15:48–55.

Ohnesorge S, Bejarano ER (2009). Begomovirus coat protein interacts with a small heatshock protein of its transmission vector (Bemisia tabaci) insect. *Mol Biol* 18:693–703.

Omura T, Yan J, Zhong B, Wada M, Zhu Y et al. (1998). The P2 protein of Rice dwarf Phytoreovirus is required for adsorption of the virus to cells of the insect vector. *J Virol* 72:9370–9373.

Osman F, Grantham GL, Rao AL (1997). Molecular studies on Bromovirus capsid protein IV Coat protein exchanges between brome mosaic and Cowpea chlorotic mottle viruses exhibit neutral effects in heterologous hosts. *Virology* 24; 238(2):452–459.

Pan H, Chu D, Ge D, Wang S, Wu Q, Xie W, Jiao X, Liu B, Yang X, Yang N, Su Q, Xu B, Zhang Y (2011). Further spread of and domination by Bemisia tabaci biotype Q on field crops in China. *J Econ Entomol* 104:978–985.

Pieterse CMJ, Dicke M (2007). Plant interactions with microbes and insects: From molecular mechanisms to ecology. *Trends Plant Sci* 12:564–568.

Pirone TP, Blanc S (1996). Helper-dependent vector transmission of plant viruses. *Annu Rev Phytopathol* 34:227–247.

Pirone TP, Perry KL (2002). Aphids: Non-persistent transmission. In: *Advances in Botanical Research 36*. RT Plumb (Ed). San Diego, CA: Academic Press:1–19.

Prajapat R, Marwal A, Gaur RK (2013). Begomovirus associated with alternative host weeds: A critical appraisal. *Arch Phytopathol Plant Protect* 47(2):157–170.

Prajapat R, Marwal A, Sahu A, Gaur RK (2012). Molecular in silico structure and recombination analysis of betasatellite in Calotropis procera associated with Begomovirus. *Arch Phytopathol Plant Protect* 45 (16):1980–1990.

Revers F, Garcia JA (2015). Molecular biology of Potyviruses. *Adv Virus Res* 92:101–199.

Rubinstein G, Czosnek H (1997). Long-term association of Tomato yellow leaf curl virus (TYLCV) with its whitefly vector Bemisia tabaci: Effect on the insect transmission capacity longevity and fecundity. *J Gen Virol* 78:2683–2689.

Ruiz-Ferrer V, Boskovic J, Alfonso C, Rivas G, Llorca O, Lopez-Abella D et al. (2005). Structural analysis of Tobacco etchpotyvirus HC-Pro oligomers involved in aphid transmission. *J Virol* 79:3758–3765.

Salvaudon L, De Moraes CM, Mescher MC (2013). Outcomes of co-infection by two Potyviruses: Implications for the evolution of manipulative strategies. *Proceedings of the Royal Society B: Biological Sciences.* doi: 10.1098/rspb.2012.2959.

Satyanarayana T, Gowda S, Ayllón MA, Dawson WO (2004). Closterovirus bipolar virion: Evidence for initiation of assembly by minor coat protein and its restriction to the genomic RNA 5′ region. *Proc Natl Acad Sci U S A* 101:799–804.

Singh MN, Khurana SMP, Nagaich BB, Agrawal HO (1988). Environmental factors influencing aphid transmission of potato virus Y and leafroll virus. *Potato Res* 31:501–509.

Soto MJ, Gilbertson RL (2003). Distribution and rate of movement of the Curtovirus Beet mild curly top virus (Family Geminiviridae) in the Beet Leafhopper. *Phytopathology* 478–484.

Srinivasan R, Alvarez JM (2007). Effect of mixed viral infections (potato virus Y-Potato leafroll virus) on biology and preference of vectors Myzuspersicae and Macrosiphum euphorbiae (Hemiptera: Aphididae). *J Econ Entomol* 100:646–655.

Stafford CA, Walker GP, Ullman DE (2012). Hitching a ride: Vector feeding and virus transmission. *Commun Int Biol* 5:43–49.

Stout MJ, Thaler JS, Thomma BP (2006). Plant-mediated interactions between pathogenic microorganisms and herbivorous arthropods. *Annu Rev Entomol* 51:663–689.

Tjallingii WF, Esch TH (1993). Fine structure of aphid stylet routes in plant tissues in correlation with EPG signals. *Physiol Entomol* 18:317–328.

Uzest M, Gargani D, Drucker M, Hébrard E, Garzo E, Candresse T, Fereres A, Blanc S (2007). A protein key to plant virus transmission at the tip of the insect vector stylet. *Proc Natl Acad Sci U S A* 104:17959–17964.

van Den Heuvel JFJM, Bruyere A, Hogenhout SA, Ziegler Graff V, Brault V et al. (1997). The N-terminal region of the Luteovirus readthrough domain determines virus binding to Buchnera GroEL and is essential for virus persistence in the aphid. *J Virol* 71:7258–7265.

van Den Heuvel JFJM, Verbeek M, van der Wilk F (1994). Endosymbiotic bacteria associated with circulative transmission of Potato leafroll virus by Myzuspersicae. *J Gen Virol* 75:2559–2565.

Varsani A, Navas-Castillo J, Moriones E et al. (2014). Establishment of three new genera in the family Geminiviridae: Becurtovirus Eragrovirus and Turncurtovirus. *Arch Virol* 159(8):2193–2203.

Varsani A, Roumagnac P, Marc F, Navas-Castillo J, Moriones E, Idris A, Briddon RW, Rivera-Bustamante R, Zerbini FM, Martin DP (2017). Capulavirus and Grablovirus: Two new genera in the family Geminiviridae. *Arch Virol* 162(6):1819–1831.

Verma, RK, Mishra R, Petrov N, Stoyanova M et al. (2015). Molecular characterization and recombination analysis of Indian isolate of Onion yellow dwarf virus. *Eur J Plant Pathol* 143(3):437–445.

Verma RK, Mishra R, Sharma P, Choudhary DK, Gaur RK (2014). Systemic infection of Potyvirus: A compatible interaction between host and viral proteins. In: *Approaches to Plant Stress and Their Management*. RK Gaur, Sharma P (Eds). New Delhi, India: Springer: 353–363.

Walker GP (2000). A beginner's guide to electrical monitoring of homopteran probing behaviour. In: *Principles and Applications of Electronic Monitoring and Other Techniques in the Study of Homopteran Feeding Behavior*. GP Walker, Backus EA (Eds). Annapolis, MD: Thomas Say Publications in Entomology. Entomological Society of America: 14–40.

Wang L, Wei X, Ye X, Xu H, Zhou X, Liu S, Wang X (2014). Expression and functional characterisation of a soluble form of Tomato yellow leaf curl virus coat protein. *Pest Manag Sci* 70:1624–1631.

Wang RY, Ammar ED, Thornbury DW, Lopez-Moya JJ, Pirone TP (1996). Loss of Potyvirus transmissibility and helper component activity correlate with non-retention of virions in aphid stylets. *J Gen Virol* 77:861–867.

Wei T, Chen H, Ichiki-Uehara T, Hibino H, Omura T (2007). Entry of Rice dwarf virus into cultured cells of its insect vector involves clathrin mediated endocytosis. *J Virol* 81:7811–7815.

Wei T, Hibino H, Omura T (2009). Release of Rice dwarf virus from insect vector cells involves secretory exosomes derived from multivesicular bodies. *Commun Integr Biol* 2:324–326.

Wei T, Shimizu T, Omura T (2008). Endo membranes and myosin mediate assembly into tubules of Pns10 of Rice dwarf virus and intercellular spreading of the virus in cultured insect vector cells. *Virology* 372:349–356.

Whitfield AE, Bryce W, Rotenberg FD (2005). Insect vector-mediated transmission of plant viruses. *Virology* 479–480, 278–289.

Yuan C, Ullman DE (1996). Comparison of efficiency and propensity as measures of vector importance in Zucchini yellow mosaic potyvirus transmission by Aphis gossypii and A craccivora. *Phytopathology* 86:698–703.

Zhang T, Luan J-B, Qi J-F, Huang C-J et al. (2012). Begomovirus whitefly mutualism is achieved through repression of plant defences by a virus pathogenicity factor. *Molec Ecol* 21:1294–1304.

Zhou F, Pu Y, Wei T, Liu H, Deng W et al. (2007). The P2 capsid protein of the non-enveloped Rice dwarf Phytoreovirus induces membrane fusion in insect host cells. *Proc Natl Acad Sci U S A* 104:19547–19552.

13 Proteomics in Understanding Host–Virus Interactions

C. Anuradha and R. Selvarajan

CONTENTS

13.1 INTRODUCTION

Plant viruses exploit the host machinery for their replication to establish its infection. Favorable host-virus interactions result in systemic infections that are typically accompanied by the onset of disease symptoms with a yield loss. In contrast, incompatible interactions result in cessation of virus replication and movement at or near the sites of inoculation. In compatible hosts, viral invasion triggers numerous biochemical and physiological changes in cells, tissues, and even whole plants (Maule et al. 2002). Among these are local and systemic changes in host gene expression. Some local changes occur in the cells where viruses are actively replicating and include both induction and shutoff host gene expression (Wang and Maule 1995; Aranda et al. 1996; Escaler et al. 2000; Havelda and Maule 2000). Other local changes in gene expression can occur in advance of or behind the viral replication front. Both localized and systemic infections of plant viruses result in the development of characteristic disease symptoms due to the complex molecular interplay between the host plant and the invading virus.

Dissecting the host gene-expression network and protein expression profile that occurs in response to virus infection should assist in a better understanding of the infection process which would lead to development of sustainable strategies for management of the virus. To develop virus-resistant cultivars, a better understanding of the defense mechanism and essential genes involved in defense responses is the key to circumventing the viral disease. However, little is known about the molecular basis involved in the defense mechanism against many of the viruses. Besides traditional, genetic, molecular, cellular, and biochemical methods for studying plant–virus interactions, both global and specialized proteomic methods are emerging as useful approaches to study the plant's response to different stress factors, including plant–pathogen interactions, and other biotic and abiotic stresses. Global proteome profiling will provide a wealth of information about the impact of virus infection. Proteomics constitute a priority research for all host–pathogen interactions in the post-genomic era. Proteomics studies include measuring differential expression of proteins in virus-infected versus non-infected cells, analysis of viral and host protein components or other virus-induced complexes, as well as proteome-wide screens to identify the interactions of host and viral protein using protein arrays or yeast two-hybrid assays. A related concern is that substantial regulation of cellular events can occur at the protein level with no apparent changes in mRNA abundance. The post-translational modification of proteins can result in a dramatic increase in protein complexity without a concomitant increase in gene expression. Proteomics may be the most promising technique for identification of proteins that are induced, repressed, and post-translationally modified during virus infection in plants. In this chapter, we review different proteomics approaches to study how the plants interact with viruses at the protein level and how this can be correlated with the pathway level to understand the mechanisms of pathogenesis and resistance exerted in the host.

13.2 PROTEOMIC TECHNIQUES USED IN INTERACTION STUDIES

Quantitative proteomic approaches can be classified as either gel-based or gel-free methods as well as label-free or label-based, of which the latter can be further subdivided into the various types of labeling approaches such as chemical and metabolic labeling. There are many gel-based or gel-free protein separation methods followed by mass spectrometry (MS) or MS/MS analysis to qualitatively and quantitatively study the difference in protein abundance during host–virus interactions. Different proteomic approaches (Figure 13.1) are employed to quantitatively determine the proteins during viral infection. The first important step in any proteomic studies is the extraction of proteins from different tissues of the plant, and this is essential for precise results. Different protocols such as trichloroacetic acid (TCA)/acetone precipitation and phenol ammonium acetate precipitation were developed for isolating proteins with less contamination (Carpentier et al. 2005; Jacobs et al. 2001). The isolated proteins need to be separated either by gel-based or gel-free methods and further identified by MS/tandem MS analysis.

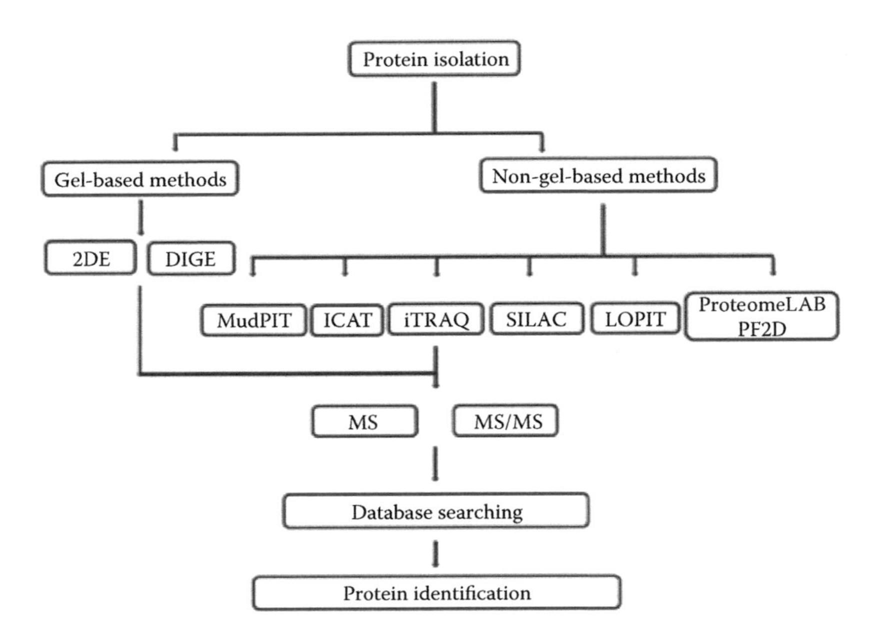

FIGURE 13.1 Proteomic approaches involved in interaction studies.

13.2.1 Gel-Based Methods

Among these preparation methods, two-dimensional gel electrophoresis (2-DE) and two-dimensional differential in gel electrophoresis (2D-DIGE) are well-established gel-based methods with high reproducibility and sensitivity. The 2-DE method is widely used in quantitative proteomics studies and has become the workhorse of protein separation and the method of choice for host-pathogen interaction studies. As a first dimension, the extracted proteins are first separated based on their net charge at different pH values, called isoelectric point focusing (IEF), and in the second dimension further separation is performed based on molecular weight (MW). The gels are then stained by silver, Coomassie, or fluorescent stains such as Sypro Ruby, Lava, and Deep Purple (Nock et al. 2008). For greater reliability of data, biological and technical replicates are compared using various software (e.g., PDQuest, Bionumerics, etc.). Gel spots which are differentially expressed (based on statistical analysis of the gels) are excised, trypsin-digested, and processed for identification by MS analysis. Trypsin cleaves the peptide chain at the carboxyl side of lysine and arginine, except when these are followed by proline. The digested peptides are then introduced into the mass analyzer, with the protein identified either through a technique known as peptide mass fingerprinting (PMF) or via tandem MS analysis by *de novo* sequencing. In PMF, the absolute masses of the peptides originating from the unknown protein are accurately measured with a mass spectrometer (Clauser et al. 1999). The mass of the proteins obtained is then compared by bioinformatics to a database containing known protein sequences or the genome of the organism. This is achieved by using computer programs, such as MASCOT (matrixscience.com/search_form_select.html), Phenyx (phenyx.vital-it.ch/pwi/login/login.jsp), and OMSSA (pubchem.ncbi.nlm.nih.gov/omssa). These programs translate genomes into proteins and then theoretically cleave proteins into peptides and calculate the absolute mass of the peptides for each protein. The peptide masses of the unknown target protein are compared with the theoretical peptide masses of each protein deposited in database. PMF is a high-throughput technique; however, it will only work if the protein sequence is present in the database utilized. Therefore many laboratories prefer to use MS/MS to sequence the peptides.

Differential in gel electrophoresis (DIGE) is an advanced version of 2-DE wherein the proteins are labeled with different fluorescent dyes, and this technology has been exploited successfully for studying the changes that take place during biotic and abiotic stresses. This technique enables protein detection at subpicomolar levels and relies on pre-electrophoretic labeling of samples with one of three spectrally resolvable fluorescent CyDyes (Cy2, Cy3, and Cy5). For quantitative analysis, imaging software is required to align gel spots and measure their intensities once the gel is digitalized by using a scanner recording light either transmitted through or reflected from the stained gel or fluorescent scanner. The images are subsequently analyzed by image analysis software such as DeCyder (GE Healthcare), Proteomweaver (Bio-Rad), PDQuest (Bio-Rad), and Progenesis Same Spots (Nonlinear Dynamics). The software uses multivariate statistical packages such as ANOVA (analysis of variance) based on spot size, and intensity spots are assigned to values and fold changes between groups. Proteins that were highly and significantly changed in expression in both 2-DE and DIGE were selected for identification by MS combined with biometrics. These data will provide a valuable resource for discovering novel proteins involved in host response upon virus invasion.

13.2.2 Non-Gel-Based Methods

Currently, shotgun techniques are slowly taking over, utilizing the rapid expansion and advancement of MS to provide a new toolbox of gel-free quantitative techniques. When coupled to MS, the shotgun proteomic pipeline can fuel new routes in sensitive and high-throughput profiling of proteins, leading to high accuracy in quantification. Some of the non-gel-based proteomic methods used to study the proteomic profiles are multidimensional protein identification technology (MudPIT), isotope-coded affinity tags (ICAT), isotope-coded protein labeling (ICPL), localization of organelle proteins by isotope tagging (LOPIT), isobaric tags for relative and absolute quantificationi (TRAQ), liquid chromatography (LC) of peptides, Proteome LAB PF2D (LC/LC of peptides), and stable isotopic labeling with amino acids in cell culture (SILAC).

In MudPIT, liquid chromatography is directly coupled to tandem mass spectrometry (Washburn et al. 2001). In ICAT, proteins are labeled to the sulfhydryl groups of cysteinyl residues using lighter or heavier isotopes. The labeling takes place at the protein level, allowing samples to be pooled prior to protease treatment, thus eliminating vial-to-vial variations and quantitatively analyzed by tandem MS (Gygi et al. 1999). The ICPL approach is based on isotopic labeling of all free amino groups in proteins. Two protein mixtures are reduced and alkylated to ensure easier access to free amino groups that are subsequently derivatized with the deuterium-free (light) or 4 deuterium-containing (heavy) form, respectively. Light- and heavy-labeled samples are then mixed, fractionated, and digested prior to high throughput MS analysis (Schmidt et al. 2005). In cases of LOPIT, it is used to discriminate organelle proteins in plants. This technique involves partial separation of the organelles by density gradient centrifugation followed by the analysis of protein distributions in the gradient by ICAT and MS (Miles et al. 2009). Unlike ICAT and ICPL, iTRAQ tags are isobarics and are primarily designed for the labeling of peptides rather than proteins (Ross et al. 2004). Another is Proteome Lab PF2D, in which the intact proteins are separated in the first dimension by chromotofocusing followed by reverse phase chromatography in the second dimension, further the proteins are then directly identified by LC-MS/MS (Chang et al. 2016). SILAC is a metabolic labeling technique for MS-based quantitative proteomics (Ong et al. 2002). In SILAC, differentially labeled samples are mixed early in the experimental process and analyzed together by LC-MS/MS. Since the labeling does not affect the chemical properties of the molecules, they co-elute from the LC column and are analyzed together in the mass spectrometer. The peptide peaks of the differentially labeled samples can be very accurately quantified relative to each other to determine the peptide and protein ratios. Although probably the most general and global labeling strategy, SILAC appears to be less suited for quantitative proteomic studies in plants because plants are autotrophic organisms and are capable of synthesizing all the amino acids from inorganic nitrogen, and therefore have lower incorporation efficiency of the exogenously supplied labeled amino acids (Gruhler et al. 2005).

Another commonly used proteomic approach for studying the interaction studies is the yeast two-hybrid system, which is based on the modular structure of transcription factors in yeast (Fields and Song 1989). It can also be used to screen cDNA libraries prepared from plants against different viral proteins. These "omic" approaches, provide a global perspective and useful information to the understanding of the plant host–virus interactome, and may possibly reveal protein targets/markers useful in the design of future diagnosis and/or plant protection strategies.

13.3 PROTEOMIC APPROACHES FOR STUDYING HOST–VIRUS INTERACTION

Comparative proteomic analysis in host-virus interaction has gained momentum due to rapid advancement in bioinformatics and proteomics tools, especially MS analysis. These advancements in proteomic research help in identifying the changes in cellular or tissue plant proteome induced by different viruses, thereby shedding light on the mechanism underlying successful or unsuccessful infection. Some of the proteomics studies conducted using different techniques to find the host virus interactions are furnished in (Table 13.1).

13.3.1 2DE

Xu and his co-workers (2013) reported that *Rice Black-Streaked Dwarf Virus* (RBSDV) produces hydrogen peroxide (H_2O_2) during the plant-virus compatible interaction. The cell responses regulated by the enhanced H_2O_2 in virus-infected plant were investigated through the global proteome changes of rice under long-term RBSDV infection. The increased H_2O_2 in RBSDV-infected plant produced oxidative stress, impaired photosynthesis, disturbed the metabolism and eventually resulted in abnormal growth.

TABLE 13.1
Proteomic Studies on Host–Virus Interactions

Virus	Host	Tissue	Method	Number of Differentially Expressed Proteins	Functional Annotation	Viral Protein Detected	Type of Plant–Virus Interaction	Reference
Pepper mild mottle virus (PMMoV-S)	*Nicotiana benthamiana*	Leaf	Chloroplast proteins prefractionation, 2-DE/ Immunoassay/ N-terminal sequencing	4	PsbP protein of the oxygen evolving complex (OEC)	Not detected	Compatible interaction	Perez-Bueno et al. 2004
Rice yellow mottle virus (RYMV)	*Ir64 Oryza sativa indica*/ Azucena, *Oryza sativa japonica*	Cellular suspensions	2DE/MALDI-MS	IR64 – 19 Azucena - 13	Metabolism, stress-related proteins, and translation	Not detected	Compatible and partially resistant	Ventelon-Debout et al. 2004
Rice yellow mottle virus (RYMV)	*Oryza sativa*	Leaf	Gel exclusion chromatography/ SDS-PAGE/nano LC-MS/MS	171-incompatible cultivar 135-compatible cultivar	Energy, metabolism, translation, protein synthesis, defense, stress transport, and transcription	Not detected	Compatible and partially resistant	Brizard et al. 2006
Tobacco mosaic virus (TMV)	*Solanum lycopersicum*	Fruit	2DE/MALDI-TOF MS/LC-MS/MS	17	Defense related protein, scavenging of reactive oxygen species, pathogenesis-related (PR)	CP	Compatible interaction	Casado-Vela et al. 2006
Plum pox virus (PPV)	*Prunus persica* L.	Leaf	Apoplastic fraction 2-DE/MALDI-TOF	4	PR proteins	Not detected	Compatible interaction	Diaz-Vivancos et al. 2006
Tobacco mosaic virus (TMV)	*Capsicum annum*	Leaf	Nuclei protein prefractionation 2DE/MALDI-TOF	6	Stress-related RNA binding protein proteasome	Not detected	Incompatible interaction	Lee et al. 2006

(*Continued*)

TABLE 13.1 (CONTINUED)
Proteomic Studies on Host–Virus Interactions

Virus	Host	Tissue	Method	Number of Differentially Expressed Proteins	Functional Annotation	Viral Protein Detected	Type of Plant–Virus Interaction	Reference
Cucumber Necrosis Tombusvirus (CNV)	CNV replicase complex in yeast	Enriched membrane fraction of yeast	2DE/MALDI-TOF MS	15	CNV p33 replication protein, the Ssa1/2p molecular chaperones (Hsp70 homologs in yeast), Tdh2/3p (glyceraldehyde-3-phosphate dehydrogenase), and Pdc1p (pyruvate decarboxylase)	p33 and p92 replicase	Yeast interaction study	Serva and Nagy 2006
Citrus tristeza virus (CTV), *Citrus sudden death-associated virus* (CSDaV)	Citrus	Stem bark	2DE/MALDI-TOF/TOF	13	Defense reactions, pr proteins	Not detected	Compatible and incompatible interaction	Cantu et al. 2008
Plum pox virus (PPV)	*Pisumsativum* L.	Leaf	Chloroplast and soluble fraction proteins 2-DE/MALDI TOF and n-ESI–ITMS/MS	12 (chloroplast), 17 (soluble fraction)	Energy metabolism	CP	Compatible interaction	Diaz Vivancos et al. 2008
Pepper mild mottle virus (PMMoV-S PMMoV-I)	*Capsicum chinense*	Leaf	2DE/MALDI-TOF MS/MS-M/nESI-IT MS/MS	17	Defense-related proteins - PR protein	CP	Compatible and incompatible interaction	Elvira et al. 2008

(*Continued*)

TABLE 13.1 (CONTINUED)
Proteomic Studies on Host–Virus Interactions

Virus	Host	Tissue	Method	Number of Differentially Expressed Proteins	Functional Annotation	Viral Protein Detected	Type of Plant–Virus Interaction	Reference
Beet necrotic yellow vein virus (BNYVV)	*Beta vulgaris*	Root	ProteomeLab PF2D/ tandem MALDI-TOF-MS	65	Hormone/stress/ defense-related proteins, oxidative response-related proteins, gene and protein expression, primary/ secondary metabolism, signal transduction, cell development-related proteins, photosynthesis	Not detected	Compatible and incompatible interactions	Larson et al. 2008
Cucumber mosaic virus (CMV)	*Solanum lycopersicum*	Leaf	DIGE/nLC-ESI-IT-MS/MS)	50	Photosynthesis, primary metabolism and defense activity	CP	Compatible interaction	Di Carli et al. 2010
Pepper mild mottle virus (PMMoV-S)	*Nicotiana benthamiana*	Leaf	Chloroplast proteins prefractionation, 2-DE/MS	36	Energy metabolism, nitrogen metabolism, chloroplastidic large ribosomal protein L14	CP	Compatible interaction	Pineda et al. 2010
GLRaV-1, GVA, and RSPaV	*Vitis vinifera*	Berry pulp/skin	2-DE/MALDI-TOF/ TOF	12 (pulp), 7 (skin)	Oxidative stress (skin), cell structure metabolism (pulp) energy metabolism, protein turnover, signal transduction	RSPaV CP	Compatible interaction	Giribaldi et al. 2011

(Continued)

TABLE 13.1 (CONTINUED)
Proteomic Studies on Host–Virus Interactions

Virus	Host	Tissue	Method	Number of Differentially Expressed Proteins	Functional Annotation	Viral Protein Detected	Type of Plant–Virus Interaction	Reference
Rice black-streaked dwarf virus (RBSDV)	*Zea mays* L.	Leaf	PEG prefractionation-2-DE/MS	91	Defense-stress-related, energy metabolism, cell wall formation, carbon metabolism, amino acid metabolism	P9-1	Compatible interaction	Li et al. 2011
Papaya meleira virus (PMeV)	*Carica papaya*	Leaf	2-DE/DIGE/MALDI-TOF-MS/MS/LC-IonTrap-MS/MS	2DE-75 DIGE-79	Metabolism-related proteins, stress-responsive proteins	Not detected	Compatible interaction	Rodrigues et al. 2011
Soybean mosaic virus (SMV)	*Glycine max* L.	Leaf	2-DE/MS	16	Energy metabolism, cell wall formation, defense-stress-related, translation-protein turnover	Not detected	Partially resistant leaf	Yang et al. 2011
Soybean mosaic virus (SMV)	*Glycine max*	Leaf	2DE/; MALDI-TOF/tandem TOF/TOF MS	28	Protein degradation, defense, signal transfer, reactive oxygen, cell wall reinforcement, energy and metabolism regulation	Not detected	Compatible interaction	Yang et al. 2011
Tomato yellow leaf curl virus (TYLCV)	*Solanum lycopersicum*	Leaf	Nano-LC-ESI-MS/MS	68	Stress responsive, pathogenesis-related proteins, wound-induced proteins, chaperones/heat shock proteins, photosynthesis	CP	Compatible and incompatible interaction	Adi et al. 2012

(*Continued*)

TABLE 13.1 (CONTINUED)
Proteomic Studies on Host–Virus Interactions

Virus	Host	Tissue	Method	Number of Differentially Expressed Proteins	Functional Annotation	Viral Protein Detected	Type of Plant–Virus Interaction	Reference
Papaya meleira virus (PMeV)	*Carica papaya*	Fruit latex	1-DE-LC–ESI–MS/MS	10	Plant proteases	Not detected	Compatible interaction	Rodrigues et al. 2012
Mungbean Yellow Mosaic India virus (MYMIV)	*Vigna mungo*	Leaf	2DE/MALDI-TOF/TOF	109	Defense and stress response, energy and metabolism, photosynthesis, protein degradation	Not detected	Compatible and incompatible interactions	Kundu et al. 2013
Papaya ringspot virus (PRSV)	*Carica papaya*	Leaf	2DE/MALDI-TOF MS/LC-MS/MS	43	Photosynthesis, photorespiration, metabolism, gene and biosynthesis, defense-related, stress response, signal transduction, and unknown processes	Not detected	Compatible interaction	Siriwan et al. 2013
Sugarcane mosaic virus (SCMV)	*Zea mays*	Leaf	DIGE/MALDI-TOF/TOF	93	Metabolism, stress and defense responses, photosynthesis, and carbon fixation	Not detected	Compatible and incompatible interactions	Wu et al. 2013a
Sugarcane mosaic virus (SCMV)	*Zea mays*	Leaf	2DE/MALDI-TOF/TOF	96	Energy and metabolism, stress and defense responses, photosynthesis	Not detected	Compatible and incompatible interactions	Wu et al. 2013b

(*Continued*)

TABLE 13.1 (CONTINUED)
Proteomic Studies on Host–Virus Interactions

Virus	Host	Tissue	Method	Number of Differentially Expressed Proteins	Functional Annotation	Viral Protein Detected	Type of Plant–Virus Interaction	Reference
Rice black-streaked dwarf virus (RBSDV)	*Oryza sativa*	Aerial part of rice seedling	2-DE/ MALDI-TOF/TOF-MS	72	Photosynthesis, redox homeostasis, metabolism, energy pathway, and cell wall modification	ND	Compatible interaction	Xu et al. 2013

Yang et al. (2011) studied the molecular basis of defense mechanism against *Soybean Mosaic Virus* (SMV) by employing the 2-DE approach to identify differentially expressed proteins in soybean leaves. Twenty-eight protein spots that showed 2-fold difference in intensity were identified between mock-inoculated and SMV-infected samples and analyzed through MALDI-TOF mass spectrometry and tandem TOF/TOF MS, and were potentially involved in protein degradation, defense signal transfer, reactive oxygen, cell wall reinforcement, and energy and metabolism regulation. Among these, metabolism and photosynthesis genes were downregulated throughout the infection cycle.

Two viral strains of *Pepper Mild Mottle Virus*, PMMoV-S and PMMoV-I, were challenge inoculated onto *Capsicum chinense* PI159236 plants harboring the L3 resistance gene to study the PR proteins induced upon the infection. Various PR protein isoforms belonging to the PR-1, b-1,3-glucanases (PR-2), chitinases (PR-3), osmotin-like protein (PR-5), peroxidases (PR-9), germin-like protein (PR-16), and PRp27 (PR-17) have been identified. Three isoforms of these PR proteins, i.e., an acidic b-1,3-glucanase isoform (PR-2), an osmotin-like protein (PR-5), and a basic PR-1 protein isoform were induced during PMMoV-S activation of *C. chinense* L3 gene-mediated resistance (Elvira et al. 2008).

Significant changes in peptidases, endoglucanase, chitinase, and proteins participating in the ascorbate-glutathione cycle were observed in *Tobacco Mosaic Virus* (TMV)-infected tomato as compared to the control fruits in 2-DE analysis, were compared, which suggests that pathogenesis-related proteins and antioxidant enzymes may play a role in the protection against oxidative damage derived from the virus-induced biotic stress (Casado-Vela et al. 2006).

Ventelon-Debout and co-workers (2004) reported the differential changes in protein expression in *Rice Yellow Mottle Virus* (RYMV)-infected cells of two cultivars, IR64 (*Oryza sativa indica*, susceptible) and Azucena (*O. sativa japonica*, partially resistant). Nineteen and 13 proteins were differentially regulated in IR64 and Azucena, respectively. The study revealed that the virus activated the proteins of abiotic stress response pathway in both cultivars.

Serva and Nagy (2006) used the yeast system to study the interaction of replicase gene of a plant virus, *Cucumber Necrosis Virus* (CNV). 2-DE of CNV replication complex revealed the presence of CNV p33 and p92 replicase proteins along with four major host proteins such as Ssa1/2pmolecular chaperones (yeast homologs of Hsp70 proteins), Tdh2/3p (glyceraldehyde-3-phosphate dehydrogenase, an RNA-binding protein), Pdc1p (pyruvate decarboxylase), and an unknown 35-kDa acidic protein. Further, they have shown that the two identified proteins p33 and Ssa1p found to interact with each other using co-precipitation method.

The *Papaya Ringspot Virus* (PRSV) causes severe economic losses in both papaya and cucurbits throughout the tropical and subtropical regions. The protein profiles of virus-infected and healthy papaya leaves were compared by 2-DE. Many proteins showed differential expression such as those belonging to photosynthesis, photorespiration, metabolism and biosynthesis, defense related, stress response, signal transduction, and unknown processes. Ribulose-1,5-bisphosphate carboxylase, Rieske protein ubiquinol cytochome C, and chlorophyll A/B binding were down-regulated in infected plants. Ubiquitin-like modifiers, vascular processing enzyme, and germin-like protein were up-regulated in infected plants at transcription and translation levels (Siriwan et al. 2013).

Adi et al. (2012) studied differential stress response to *Tomato Yellow Leaf Curl Virus* (TYLCV)-infected resistant (R) and susceptible (S) tomato inbred. Comparison of protein profiles revealed a completely different host stress response. Higher levels of reactive oxygen species (ROS) compounds, anti-oxidative, PR- and wound-induced proteins were predominant in S plants. In TYLCV infected R inbred tomatoes, host defense mechanisms were much less activated, and the protein and chemical chaperones were triggered to maintain tissue homeostasis. Sources of carbon and nitrogen were less affected by TYLCV in R than in S plants, which make R plants more balanced and fit against the virus biotic stress. Moreover, more stable patterns of key cellular regulators, such as SnRK1 and MAPKs, were characteristic in R plants.

The molecular events occurring during compatible and incompatible interactions between *Vigna mungo* and *Mungbean Yellow Mosaic India Virus* (MYMIV) patho system was explored by comparative proteome analyses using 2-DE coupled with mass spectrometry by Kundu et al. (2013). Proteins of several functional categories were differentially changed in abundance during both compatible and incompatible interactions. Among these, photosynthesis-related proteins were mostly affected in the susceptible genotype, resulting in reduced photosynthesis rate under MYMIV-stress. Differential intensities of chlorophyll fluorescence and chlorophyll contents are in congruence with proteomics data, and it was revealed that Photosystem II electron transports are the primary targets of MYMIV during pathogenesis. The network of various cellular pathways that are involved in inducing defense response contains three major proteins, ascorbate peroxidase, rubisco activase, and serine/glycine hydroxymethyl transferase.

Li and colleagues (2011) scrutinized the long-term adjustment of the maize elite but very susceptible line (Ye478), to RBSDV. Fifty days post-infection, accumulation of detoxifying enzymes peroxidase 39, APx2-cytosolicAPX, and catalase isozyme 3 implies an onset of oxidative stress caused by RBSDV, and a subsequent need for regulation of cellular redox state. Higher levels of lipoxygenases and small GTP-binding proteins in virus-infected plants suggest that these proteins modulate the triggering of downstream defense via signaling pathways. Upregulation of UDP-glucosyltransferase BX9 in RBSDV-infected plants indicates that these natural insecticides are very likely a part of maize defense against RBSDV.

Maize rough dwarf disease (MRDD) establishment is also followed by profound adaptation in carbohydrate metabolism, as numerous proteins from glycolysis, TCA cycle, glycogenesis, and pentose phosphate metabolism, were extensively altered in expression. Decrease in transketolase levels in diseased plants, leading to a severely weakened influx of proteins affecting plant growth and development, was evident. Elevated amounts of starch granules that were observed in MRDD-diseased leaves were very likely the result of enhanced expression of ADP-glucose phospohorylase small subunit in virus-infected plants. Photosynthesis, carbon fixation, and assimilation as well as starch synthesis also appeared intensified in virus-infected leaves (Li et al. 2011).

13.3.2 2D-DIGE

Papaya Meleira Virus (PMeV), a laticifer-infecting virus, is the causal agent of papaya sticky disease, and to understand the systemic effects of PMeV in papaya, proteomic analysis was carried out using 2-DE and DIGE followed by MALDI-TOF-MS/MS and LC-ion trap-MS/MS, respectively. PMeV infection of papaya was shown to induce stress-related proteins and repress metabolism-related proteins. This expression pattern was corroborated by the results of the DIGE analysis. Three proteins, calreticulin and the proteasome 20S and RPT5a subunits were confirmed to be induced during this disease and they represent the first disease marker candidates (Rodrigues et al. 2011).

Proteomic analyses of leaf samples from resistant and susceptible ecotypes of maize infected with *Sugarcane Mosaic Virus* (SCMV) was conducted by Wu and co-workers (2013a,b) to explore the molecular mechanisms involved in the plant–SCMV interaction and to identify candidate SCMV resistance proteins. Functional categorization showed that SCMV-responsive proteins were mainly involved in energy and metabolism, stress and defense responses, photosynthesis, and carbon fixation. The study revealed that most of the identified proteins were in chloroplast and cytoplasm. The oxidative burst was more pronounced during incompatible plant–SCMV interactions, as compared to those defined as compatible. There was an increase of enzymes involved in glycolysis and gluconeogenesis pathways in the resistant maize ecotype Siyi, while they decreased in the susceptible maize ecotype Mo17. In addition, there is a marked increase of guanine nucleotide-binding protein beta subunit in the resistant Siyi, which suggests a possible involvement of G-protein associated pathways in the resistant responses of maize to SCMV.

Comparative study on compatible plant-virus interactions between engineered immune protected and susceptible wild type tomato plants to *Cucumber Mosaic Virus* (CMV) helped in understanding the antibody-mediated disease resistance mechanisms in tomato. Most of these proteins identified were related to photosynthesis, primary metabolism, and defense activity, and demonstrated to be actively down-regulated by CMV in infected leaves. The analysis revealed that asymptomatic apical leaves of transgenic inoculated plants had no protein profile alteration as compared to control wild-type uninfected plants, demonstrating that virus infection is confined to the inoculated leaves and systemic spread is hindered by the CMV coat protein (CP)-specific single-chain variable fragment (scFv) G4 molecules (Di Carli et al. 2010).

13.3.3 Multidimensional Liquid Chromatography

Beet Necrotic Yellow Vein Virus (BNYVV) is a devastating sugar beet pathogen. BNYVV-induced differential sugar beet protein expression was evaluated with multidimensional liquid chromatography. Of more than 1000 protein peaks detected in root extracts, 7.4% and 11% were affected by BNYVV in resistant and susceptible genotypes, respectively. Using tandem MALDI-TOF-MS, 65 proteins were identified. Proteomic data suggest involvement of systemic resistance components in Rz1-mediated resistance and phyto-hormones in symptom development (Larson et al. 2008).

From these studies, the common responses observed are the upregulation of defense/stress-related proteins, but whether they are upregulated by disease stress of by virus needs to be confirmed. An overall alteration in protein profile belonging to different functional classes due to virus infection is given in Figure 13.2. Energy, metabolism, protein synthesis, and protein turnover were upregulated due to the redirection of cellular metabolism by the virus for its own replication processes. Another important observation is that photosynthetic activity has been negatively regulated upon virus infection, which may be due to the allocation of energy for general plant defense mechanism. These types of quantification-based analyses may be enriched for proteins far downstream in signaling pathways that are manipulated or perturbed by viruses. Hence to complement these types of experiments, it will be imperative to elucidate the direct interaction of viral proteins with the host.

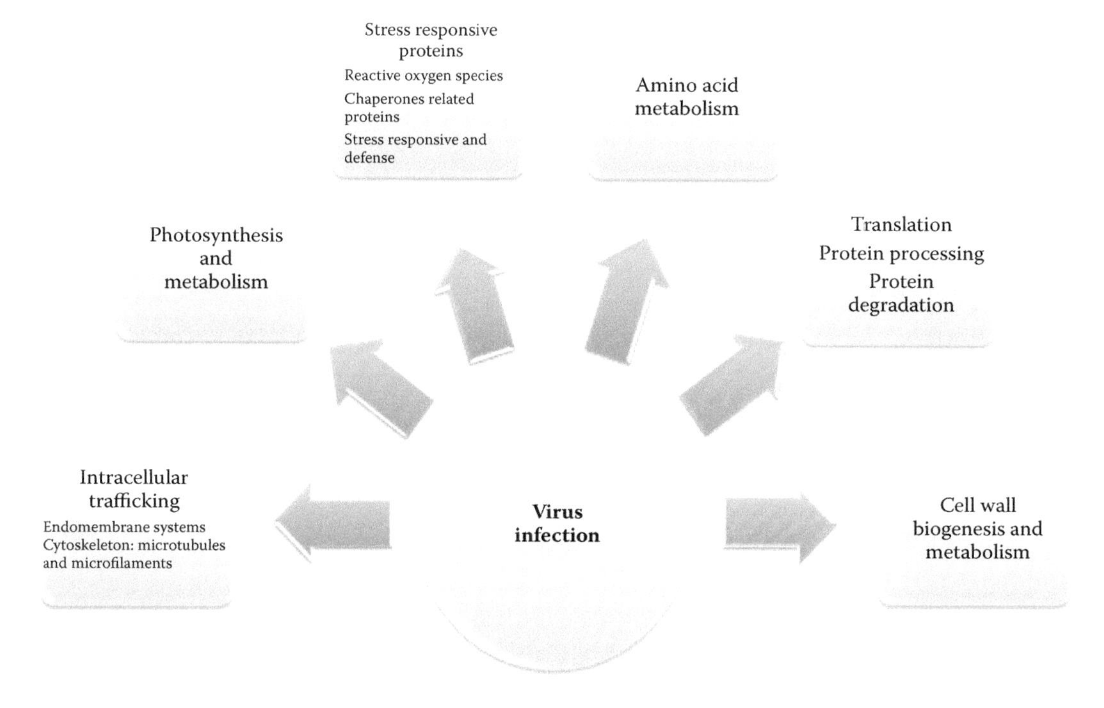

FIGURE 13.2 Protein classes that are manipulated upon viral infection.

13.3.4 Yeast Two-Hybrid System

Interaction between the viral proteins and selected plant protein is critical for establishing the viral infection in plant. Studying the direct interaction between plants and virus is very important for developing effective control measures against any targeted virus. A rapid and economical approach for studying the direct interaction is the yeast two-hybrid (Y2H) system wherein the interaction between two proteins of interest is detected through the reconstitution of a transcription factor and the subsequent activation of reporter genes under the control of this transcription factor (Fields and Song 1989). Using the Y2H system, a number of plant proteins interacting with viral proteins have been identified (Table 13.2), recently, increasing our knowledge of the molecular basis of viral infection and host defense mechanisms.

The Y2H assay has been applied to show interactions between viral replication proteins critical for the assembly of the viral replication complex (RC) for many plant (+)RNA viruses. Examples include interactions between *Brome Mosaic Virus* (BMV) 1a-2apol and 1a-1a, TBSV p33-p92pol and p33-p33 and TMV 126K-183K, and 126K-126K, as well as the P1-P2 proteins of *Alfalfa Mosaic Virus* (Nagy 2008).

The Y2H system has been used to show interactions between the *Plum Pox Virus* CI protein with photosystem I (PSI-K) protein. Downregulation of this gene leads to higher PPV accumulation, suggesting a role for the CI–PSI-K interaction (Jimenez et al. 2006). The CP of *Turnip Crinkle Virus* interacts with TIP, a member of the NAC family of proteins that are involved in plant development and defense (Ren et al. 2000). A single substitution in the interacting domain in CP abolished interaction with TIP, thereby allowing the virus to overcome HR-mediated resistance in *Arabidopsis*. The *Tobacco Etch Virus* HC-Pro interacted in Y2H with rgs-CaM, a calmodulin-related protein (Anandalakshmi et al. 2000) and *Potato Virus A* HC-Pro was found to interact with RING finger protein (HIP1). The P19 of *Tomato Bushy Stunt Virus* interacted with ALY proteins that are likely to be involved in transcriptional activation and export of RNAs from the nucleus in plants (Uhrig et al. 2004). The TMV replicase protein was found to interact with an ATPase protein and a subunit of the oxygen-evolving complex of photosystem II (Abbink et al. 2002). In another study, the TMV replicase protein interacted with PAP1, a regulator of auxin response genes involved in plant development (Padmanabhan et al. 2005). *Tomato Spotted Wilt Virus* NSm protein was found to bind to DnaJ-domain proteins, which are regulators of the Hsp-70 chaperone protein that is also known to be involved in closterovirus movement (Soellick et al. 2000). The movement protein of TMV was found to interact with pectin methylesterase, a cell wall protein (Chen et al. 2000). In many potyviruses, viral genome-linked protein (VPg) was found to interact with host translation initiation factors (eIF4E and eIF(iso)4E) (Wittman et al. 1997; Schaad et al. 2000; Nicaise et al. 2003; Leonard et al. 2004; Michon et al. 2005; Miyoshi et al. 2008; Tavert-Roudet et al. 2012; Moury et al. 2014). The key role of this interaction in the virus life cycle is supported by the finding that a mutation in either VPg or eIF4E abolishes the interaction, thereby preventing virus infection in plants (Ruffel et al. 2002; Gao et al. 2004). Maize Elongin C interacts with the VPg of *Sugarcane Mosaic Virus* and facilitates virus infection (Zhu et al. 2014). In the case of *Tomato Leaf Curl Virus*, a different NAC protein from tomato was found to interact with the replication enhancer protein (Ren) of this virus, and overexpression of the NAC protein increased virus replication (Selth et al. 2005). Capsid protein of *Potato Virus-Y* interacted with tobacco DnaJ-like proteins (Hofius et al. 2007). Protein kinase from tobacco interacts with the *Cucumber Mosaic Virus* 1a methyltransferase domain (Kim et al. 2006). The AL2 protein of *Tomato Golden Mosaic Virus* was found to interact with SNF1-kinase, and silencing of this gene made the plants more susceptible to virus infection (Hao et al. 2003). In another study, AL2 was found to interact with adenosine kinase (ADK) and inactivate this gene in plants (Wang et al. 2003). Geminivirus replication protein was found to interact with many host proteins like retinoblastoma-related protein, proliferative cell nuclear antigen (PCNA), and the host cell sumoylation enzyme, NbSCE1 (homolog of the yeast UBC9) (Ach et al. 1997;

TABLE 13.2
Plant Virus and Host Protein Interactions Confirmed through Yeast Two-Hybrid System

S. No.	Virus	Proteins Interacting in Y2H		Reference
		Virus Protein	**Host Protein**	
	Potyviridae			
1	*Turnip Mosaic Virus*	VPg	eIF(iso)4E	Wittmann et al. 1997
2	*Tobacco Etch Virus*	HC-Pro	rgs-cam, calmodulin related protein	Anandalakshmi et al. 2000
		NIa	eIF4E	Schaad et al. 2000
3	*Potato Virus A*	HC-Pro	RING finger protein (HIP1)	Guo et al. 2003
4	*Turnip Mosaic Virus*	VPg	eIF(iso)4E	Leonard et al. 2004
5	*Plum Pox Virus*	CI (rep, MP)	PSI K protein	Jimenez et al. 2005
6	*Soybean Mosaic Virus*	VPg	eIF4E, eIF(iso)4E Poly A binding protein PABP	Seo et al. 2007
		NIb	PABP, chlorophyll a/b binding preprotein RUBISCO, photosystem I-N subunit STLS1 protein, ubiquitin, translation initiation factor, LRR protein kinase, TIR-NBS type resistant protein, RNA binding protein	
		CP	PABP	
7	*Potato Virus Y*	CP	DnaJ-like (NtCPIPs), NtCPIP2a	Hofius et al. 2007
		VPg	eIF4E	Moury et al. 2014
8	*Lettuce Mosaic Virus*	VPg	eIF(iso)4E	Tavert-Roudet et al. 2012
	Virgaviridae			
1	*Tobacco Mosaic Virus*	Movement protein	Pectin methylesterases	Chen et al. 2000
		RdRp, 3′ UTR	eEF1A	Yamaji et al. 2006
		RNA helicase domain of replicase protein	ATPase protein and a subunit of the oxygen-evolving complex of photosystem II	Abbink et al. 2002
		P50 domain of helicase	N-factor, an R factor	Ueda et al. 2006
2	*Tomato Mosaic Virus*	MP	RIO kinases: Rio1P	Yoshioka et al. 2003
	Secoviridae			
1	*Cow Pea Mosaic Virus*	60K	SNARE-like proteins: VAP27-1 and VAP27-2 (VAP33 family), eEF1-ß	Carette et al. 2002

(Continued)

TABLE 13.2 (CONTINUED)
Plant Virus and Host Protein Interactions Confirmed through Yeast Two-Hybrid System

S. No.	Virus	Proteins Interacting in Y2H		Reference
		Virus Protein	Host Protein	
	Bromoviridae			
1	*Cucumber Mosaic Virus*	CMV1a, methyl transferase domain	Putative protein kinase-Tcoi2 (Tobacco CMV1a interacting protein)	Kim et al. 2006
		2b	2b interacting protein (2bip) Similar to Lut B protein	Ham et al. 1999
2	*Brome Mosaic Virus*	CP	Oxidoreductase enzyme, HCP1	Okinaka et al. 2002
		RdRp	P-body component Lsm 1P	Beckham et al. 2007
3	*Alfalfa Mosaic Virus*	P1, P2	Tonoplast co-localized	Van Der Heijden, 2001
	Bunyaviridae			
1	*Tomato Spotted Wilt Virus*	NSm protein	DnaJ-domain proteins, which are regulators of the Hsp-70 chaperone	Soellick et al. 2000
	Tombusviridae			
1	*Turnip Crinkle Virus*	CP	TIP, a member of the NAC family of proteins	Ren et al. 2000
2	*Tomato Bushy Stunt Virus*	Replicase	Transcription factor Rpb11P, via regulating $P^{33}$and P^{92}pol	Jaag et al. 2007
		P19 silencing suppressor protein	ALY proteins	Uhrig et al. 2004
	Alphaflexiviridae			
1	*Potato Virus X*	TGB (triple gene block) TGB25k, TGB12k, TGB8k	TIP1, TIP2, and TIP3	Fridborg et al. 2002
	Geminiviridae			
1	*Tomato Golden Mosaic Virus*	AL1	RRB1 (retinoblastoma-related protein 1)	Ach et al. 1997
2	*Maize Streak Virus*	RepA and RepB	ZmRb1 (maize retinoblastoma protein)	Horvath et al. 1998
3	*Wheat Dwarf Virus*	Rep-A	GRAB-Geminivirus Rep A-binding protein	Xie et al. 1999
		Replication initiation protein RIP	TmRFC-1	Luque et al. 2002
4	*Tomato Yellow Leaf Curl Sardinia Virus*	REn	PCNA, RBR	Castillo et al. 2003

(*Continued*)

TABLE 13.2 (CONTINUED)
Plant Virus and Host Protein Interactions Confirmed through Yeast Two-Hybrid System

		Proteins Interacting in Y2H		
S. No.	Virus	Virus Protein	Host Protein	Reference
5	*Tomato Golden Mosaic Virus*	Rep-AC1/Rep	PCNA, RFC-1, SCE-SUMO conjugating enzyme	Castillo et al. 2004
6	*Mung Bean Yellow Indian Mosaic Virus*	Rep (AL1/AC1)	PCNA	Bagewadi et al. 2004
7	*Tomato Leaf Curl Virus*	REn (replication enhancer protein)	NAC protein	Selth et al. 2005
8	*Tomato Yellow Leaf Curl Virus*	C3 (replication enhancer protein)	PCNA, RBR	Settlage et al. 2005

Horvath et al. 1998; Xie et al. 1999; Ren et al. 2000; Abbink et al. 2002; Luque et al. 2002; Castillo et al. 2003; Selth et al. 2005; Settlage et al. 2005; Padmanabhan et al. 2005; Lozano-Duran et al. 2011).

The Y2H assay has also been used to screen cDNA libraries prepared from several plants against virus movement, silencing suppressor and CPs to identify possible host factors regulating cell-to-cell movement, RNA encapsidation, and host responses (Carette et al. 2002; Seo et al. 2007; Schaad 2000; Desvoyes et al. 2002; Dunoyer et al. 2004; Fridborg et al. 2003; Guo et al. 2003; Ham et al. 1999; Hofius et al. 2007; Jimenez et al. 2006; Okinaka et al. 2003; Samuels et al. 2007; Schaad et al. 2000; Ueda et al. 2006; Yoshioka et al. 2004). Overall, the results obtained from the yeast model will then be applied to dissect the interactions between plant viruses and their native plant hosts.

13.4 CONCLUSION

The rising number of comparative proteomics and Y2H studies on plant–virus interactions reviewed here proves the validity of this analysis not only in the characterization of resistance mechanisms but also in the identification of possible protein targets to be used in diagnostics of viral pathogens or for the development of novel plant-protection strategies. Proteomics approaches should also accelerate the identification of various post-translational modifications of viral and host proteins that could affect their functions during the replication process. A multidisciplinary approach attempting the integration of data deriving from transcriptomics, proteomics, and metabolomics platforms, each with their own degree of sensitivity, performance, and processivity, could provide further insights into the comprehension of viral pathogenic processes.

REFERENCES

Abbink T E M, Peart J R, Mos T N M, Baulcombe D C, Bol J F and Linthorst H J M (2002). Silencing of a gene encoding a protein component of the oxygen-evolving complex of photosystem II enhances virus replication in plants. *Virology* 295;307–319.

Ach R A, Durfee T, Miller A B et al. (1997). RRB1 and RRB2 encode maize retinoblastoma-related proteins that interact with a plant D-type cyclin and geminivirus replication protein. *Molecular Cell Biology* 17;5077–5086.

Adi M, Jens P, Brotman Y et al. (2012). Stress responses to tomato yellow leaf curl virus (TYLCV) infection of resistant and susceptible tomato plants are different. *Metabolomics* S;1–13.

Anandalakshmi R, Marathe R, Ge X et al. (2000). A calmodulin-related protein that suppresses posttranscriptional gene silencing in plants. *Science* 290;142–144.

Aranda M A, Escaler M, Wang D and Maule A J (1996). Induction of HSP70 and polyubiquitin expression associated with plant virus replication. *Proceedings of the National Academy of Sciences* 93;15289–15293.

Bagewadi B, Chen S, Lal S K, Choudhury N R and Mukherjee S K (2004). PCNA interacts with Indian mung bean yellow mosaic virus rep and downregulates Rep activity. *Journal of Virology* 78;11890–11903.

Beckham C J, Light H R, Nissan T A, Ahlquist P, Parker R and Noueiry A (2007). Interactions between brome mosaic virus RNAs and cytoplasmic processing bodies. *Journal of Virology* 81;9759–9768.

Brizard J P, Carapito C, Delalande F, Van Dorsselaer A and Brugidou C (2006). Proteome analysis of plant–virus interactome: Comprehensive data for virus multiplication inside their hosts. *Molecular Cell Proteomics* 5;2279–2297.

Cantu M D, Mariano A G, Palma M S, Carrilho E and Wulff N A (2008). Proteomic analysis reveals suppression of bark chitinases and proteinase inhibitors in citrus plants affected by the citrus sudden death disease. *Phytopathology* 98(10);1084–1092.

Carette J E, Verver J, Martens J, van Kampen T, Wellink J and van Kammen A (2002). Characterization of plant proteins that interact with cowpea mosaic virus "60K" protein in the yeast two-hybrid system. *Journal of General Virology* 83;885–893.

Carpentier S C, Witters E, Laukens K, Deckers P, Swennen R and Panis B (2005). Preparation of protein extracts from recalcitrant plant tissues: An evaluation of different methods for two-dimensional gel electrophoresis analysis. *Proteomics* 5;2497–2507.

Casado-Vela J, Selles S and Martinez R B (2006). Proteomic analysis of tobacco mosaic virus-infected tomato (*Lycopersicon esculentum* M) fruits and detection of viral coat protein. *Proteomics* 6 (Suppl1);S196–206.

Castillo A, Collinet D, Deret S, Kashoggi A and Bejerano ER (2003). Dual interaction of plant PCNA with geminivirus replication accessory protein (REn) and viral replication protein (Rep). *Virology* 312;381–394.

Castillo A G, Kong L J, Hanley-Bowdoin L and Bejarano E R (2004). Interaction between a geminivirus replication protein and the plant sumoylation system. *Journal of Virology* 78;2758–2769.

Chang Y K, Lai Y H, Chu Y, Lee M C, Huang C Y and Wu S (2016). Haptoglobin is a serological biomarker for adenocarcinoma lung cancer by using the ProteomeLab PF2D combined with mass spectrometry. *American Journal of Cancer Research* 6(8);1828–1836.

Chen M H, Sheng J, Hind G, Handa A K and Citovsky V (2000). Interaction between the tobacco mosaic virus movement protein and host cell pectin methylesterases is required for viral cell-to-cell movement. *EMBO Journal* 19;913–920.

Clauser K R, Baker P and Burlingame A L (1999). Role of accurate mass measurement (+/– 10 ppm) in protein identification strategies employing MS or MS/MS and database searching. *Analytical Chemistry* 71;2871–2882.

Desvoyes B, Faure-Rabasse S, Chen M H, Park J W and Scholthof H B (2002). A novel plant homeodomain protein interacts in a functionally relevant manner with a virus movement protein. *Plant Physiology* 129;1521–1532.

Di Carli M, Villani M E, Bianco L, Lombardi R, Perrotta G, Benvenuto E and Donini M (2010). Proteomic analysis of the plant–virus interaction in cucumber mosaic virus (CMV) resistant transgenic tomato. *Journal of Proteome Research* 9(11);5684–5697.

Diaz-Vivancos P, Clemente-Moreno M J, Rubio M et al. (2008). Alteration in the chloroplastic metabolism leads to ROS accumulation in pea plants in response to plum pox virus. *Journal of Experimental Botany* 59(8);2147–2160.

Dunoyer P, Thomas C, Harrison S, Revers F and Maule A (2004). A cysteine-rich plant protein potentiates Potyvirus movement through an interaction with the virus genome-linked protein VPg. *Journal of Virology* 78;2301–2309.

Elvira M I, Galdeano M M, Gilardi P, García-Luque I and Serra M T (2008). Proteomic analysis of pathogenesis-related proteins (PRs) induced by compatible and incompatible interactions of pepper mild mottle virus (PMMoV) in *Capsicum chinense* L3 plants. *Journal of Experimental Botany* 59(6);1253–1265.

Escaler M, Aranda M A, Thomas C L and Maule A J (2000). Pea embryonic tissues show common responses to the replication of a wide range of viruses. *Virology* 267;318–324.

Fields S and Song O (1989). A novel genetic system to detect protein–protein interactions. *Nature* 340;245–246.

Fridborg I, Grainger J, Page A, Coleman M, Findlay K and Angell S (2003). TIP, a novel host factor linking callose degradation with the cell-to-cell movement of Potato virus X. *Molecular Plant-Microbe Interactions* 16;132–140.

Gao Z, Johansen, E, Eyers S, Thomas C L, Ellis N and Maule A J (2004). The Potyvirus recessive resistance gene, sbm1, identifies a novel role for translation initiation factor eIF4E in cell-to-cell trafficking. *Plant Journal* 40;376–385.

Giribaldi M, Purrotti M, Pacifico D et al. (2011). A multidisciplinary study on the effects of phloem-limited viruses on the agronomical performance and berry quality of *Vitis vinifera* cv Nebbiolo. *Journal of Proteomics* 75;306–315.

Gruhler A, Schulze W X, Matthiesen R, Mann M, and Jensen O N (2005). Stable isotope labeling of Arabidopsis thaliana cells and quantitative proteomics by mass spectrometry. *Molecular and Cellular Proteomics* 4;1697–1709.

Guo D, Spetz C, Saarma M and Valkonen J P (2003). Two potato proteins, including a novel RING finger protein (HIP1), interact with the potyviral multifunctional protein HC-Pro. *Molecular Plant-Microbe Interactions* 16;405–410.

Gygi S P, Rist B, Gerber S A, Turecek F, Gelb M H and Aebersold R (1999). Quantitative analysis of complex protein mixtures using isotope-coded affinity tags. *Nature Biotechnology* 17;994–999.

Ham B K, Lee T H, You J S, Nam Y W, Kim J K and Paek K H (1999). Isolation of a putative tobacco host factor interacting with cucumber mosaic virus-encoded 2b protein by yeast two-hybrid screening. *Molecules and Cells* 9;548–555.

Hao L, Wang H, Sunter G and Bisaro D M (2003). Geminivirus AL2 and L2 proteins interact with and inactivate SNF1 kinase. *Plant Cell* 15;1034–1048.

Havelda Z and Maule A J (2000). Complex spatial responses to cucumber mosaic virus infection in susceptible. *Cucurbita pepo* cotyledons *Plant Cell* 12;1975–1986.

Hofius D, Maier A T, Dietrich C et al. (2007). Capsid protein-mediated recruitment of host DnaJ-like proteins is required for Potato virus Y infection in tobacco plants. *Journal of Virology* 81;11870–11880.

Horvath G V, Pettko-Szandtner A, Nikovics K et al. (1998). Prediction of functional regions of the maize streak virus replication-associated proteins by protein–protein interaction analysis. *Plant Molecular Biology* 38;699–712.

Jaag H M, Stork J and Nagy PD (2007). Host transcription factor Rpb11p affects tombusvirus replication and recombination via regulating the accumulation of viral replication proteins. *Virology* 368;388–404.

Jacobs D I, van Rijssen M S, der Heijden R and Verpoorte R (2001). Sequential solubilization of proteins precipitated with trichloroacetic acid in acetone from cultured *Catharanthu sroseus* cells yields 52% more spots after two-dimensional electrophoresis. *Proteomics* 1;1345–1350.

Jimenez I, Lopez L, Alamillo JM, Valli A and Garcia J A (2006). Identification of a plum pox virus CI-interacting protein from chloroplast that has a negative effect in virus infection. *Molecular Plant-Microbe Interactions* 19;350–358.

Kim M J, Ham B K and Paek K H (2006). Novel protein kinase interacts with the Cucumber mosaic virus 1a methyltransferase domain. *Biochemical and Biophysical Research Communications* 340;228–235.

Kundu S, Chakraborty D and Pal A (2011). Proteomic analysis of salicylic acid induced resistance to Mungbean Yellow Mosaic India Virus in *Vigna mungo*. *Journal of Proteomics* 74(3);337–349.

Larson R L, Wintermantel W M, Hill A, Fortis L and Nunez A (2008). Proteome changes in sugar beet in response to Beet necrotic yellow vein virus. *Physiological and Molecular Plant Pathology* 72(1–3);62–72.

Lee B J, Kwon S J, Kim S K et al. (2006). Functional study of pepper 26S proteosome subunit rpn7 induced by tobacco mosaic virus from nuclear proteome analysis. *Biochemical and Biophysical Research Communications* 351;405–411.

Leonard S, Viel C, Beauchemin C, Daigneault N, Fortin M G and Laliberte J F (2004). Interaction of VPg-Pro of turnip mosaic virus with the translation initiation factor 4E and the poly(A)-binding protein in planta. *Journal of General Virology* 85;1055–1063.

Li K, Xu C and Zhang J (2011). Proteome profile of maize (*Zea Mays* L) leaf tissue at the flowering stage after long-term adjustment to rice black-streaked dwarf virus infection. *Gene* 485(2);106–113.

Lozano-Duran R, Rosas-Diaz T, Gusmaroli G et al. (2011). Geminiviruses subvert ubiquitination by altering csn-mediated derubylation of scf e3 ligase complexes and inhibit jasmonate signaling in Arabidopsis thaliana. *Plant Cell* 23;1014–1032.

Luque A, Sanz-Burgos A P, Ramirez-Parra M, Castellano M M and Gutierrez C (2002). Interaction of geminivirus Rep protein with replication factor C and its potential role during geminivirus DNA replication. *Virology* 302;83–94.

Maule A, Leh V and Lederer C (2002). The dialogue between viruses and hosts in compatible interactions. *Current Opinion Plant Biology* 5;279–284.

Michon T, Estevez Y, Walter J, German-Retana S and LeGall O (2006). The potyviral virus genome-linked protein VPg forms a ternary complex with the eukaryotic initiation factors eIF4E and eIF4G and reduces eIF4E affinity for a mRNA cap analogue. *FEBS Journal* 273;1312–1322.

Miles G P, Samuel M A, Ranish J A, Donohoe S M, Sperrazzo G M, and Ellis B E (2009). Quantitative proteomics identifies oxidant-induced, AtMPK6-dependent changes in Arabidopsis thaliana protein profiles. *Plant Signaling and Behavior* 4(6);497–505.

Miyoshi H, Okade H, Muto S et al. (2008). Turnip mosaic virus VPg interacts with *Arabidopsis thaliana*eIF(iso) 4E and inhibits in vitro translation. *Biochimie* 90(10);1427–1434.

Moury B, Morel C, Johansen E et al. (2004). Mutations in potato virus Y genome-linked protein determine virulence toward recessive resistances in *Capsicum annuum* and *Lycopersicon hirsutum. Molecular Plant-Microbe Interactions* 17;322–329.

Nagy P D (2008). Yeast as a model host to explore plant virus–host interactions. *Annual Review of Phytopathology* 46;217–242.

Nicaise V, German-Retana S, Sanjuan R et al. (2003). The eukaryotic translation initiation factor 4E controls lettuce susceptibility to the potyvirus lettuce mosaic virus. *Plant Physiology* 132;1272–1282.

Nock C M, Ball M S, White I R, Skehel J M, Bill L and Karuso P (2008). Mass spectrometric compatibility of Deep Purple and SYPRO Ruby total protein stains for high-throughput proteomics using large-format two-dimensional gel electrophoresis. *Rapid Communications in Mass Spectrometry* 22(6);881–886.

Okinaka Y, Mise K, Okuno T and Furusawa I (2003). Characterization of a novel barley protein, HCP1, that interacts with the Brome mosaic virus coat protein. *Molecular Plant-Microbe Interactions* 16;352–359.

Ong S, Blagoev B, Kratchmarova I et al. (2002). Stable isotope labeling by amino acids in cell culture, SILAC, as a simple and accurate approach to expression proteomics. *Molecular and Cellular Proteomics* 1;376–386.

Padmanabhan M S, Goregaoker S P, Golem S, Shiferaw H and Culver J N (2005). Interaction of the Tobacco mosaic virus replicase protein with the Aux/IAA protein PAP1/IAA26 is associated with disease development. *Journal of Virology* 79;2549–2558.

Perez-Bueno M L, Rahoutei J, Sajnani C, Garcia-Luque I and Baron M (2004). Proteomic analysis of the oxygen-evolving complex of photosystem II under biotec stress: Studies on *Nicotiana benthamiana* infected with tobamoviruses. *Proteomics* 4;418–425.

Pineda M, Sajnani C and Baron M (2010). Changes induced by the pepper mild mottle tobamovirus on the chloroplast proteome of *Nicotiana benthamiana. Photosynthesis Research* 103;31–45.

Ren T, Qu F and Morris T J (2000). HRT gene function requires interaction between a NAC protein and viral capsid protein to confer resistance to Turnip crinkle virus. *Plant Cell* 12;1917–1925.

Rodrigues S P, Ventura J A, Aguilar C et al. (2011). Proteomic analysis of papaya (Carica papaya L) displaying typical sticky disease symptoms. *Proteomics* 11(13);2592–2602.

Rodrigues S P, Ventura J A, Aguilar C et al. (2012). Label-free quantitative proteomics reveals differentially regulated proteins in the latex of sticky diseased *Carica papaya* L plants *Journal of Proteomics* 75 (11);3191–3198.

Ross P L, Huang Y N, Marchese J N et al. (2004). Multiplexed protein quantitation in *Saccharomyces cerevisiae* using amine-reactive isobaric tagging reagents. *Molecular and Cellular Proteomics* 3;1154–1169.

Ruffel S, Dussault M H, Palloix A et al. (2002). A natural recessive gene against potato virus Y in pepper corresponds to the eukaryotic initiation factor 4E (eIF4E). *Plant Journal* 32;1067–1075.

Samuels T D, Ju H J, Ye C M, Motes C M, Blancaflor E B and Verchot-Lubicz J (2007). Subcellular targeting and interactions among the Potato virus X TGB proteins *Virology* 367;375–389.

Schaad M C, Anderberg R J and Carrington J C (2000). Strain-specific interaction of the tobacco etch virus Nia protein with the translational initiation factor eIF4E in the yeast two hybrid system. *Virology* 273;300–306.

Schmidt A, Kellermann J and Lottspeich F (2005). A novel strategy for quantitative proteomics using isotope-coded protein labels. *Proteomics* 5;4–15.

Selth L A, Dogra S C, Rasheed M S, Healy H, Randles J W and Rezaian M A (2005). A NAC domain protein interacts with tomato leaf curl virus replication accessory protein and enhances viral replication. *Plant Cell* 17;311–325.

Seo J K, Hwang S H, Kang S H et al. (2007). Interaction study of soybean mosaic virus proteins with soybean proteins using the yeast-two hybrid system. *Plant Pathology Journal* 23(4);281–286.

Serva S and Nagy P D (2006). Proteomics analysis of the tombusvirus replicase: Hsp70 molecular chaperone is associated with the replicase and enhances viral RNA replication. *Journal of Virology* 80;2162–2169.

Settlage S B, See R G and Hanley-Bowdoin L (2005). Geminivirus C3 protein: Replication enhancement and protein interactions. *Journal of Virology* 79;9885–9895.

Siriwan W, Roytrakul S, Shimizu M, Takaya N and Chowpongpang S (2013). *Papaya ringspot virus*-infected papaya leaves *Kasetsart. Natural Science* 47;589–602.

Soellick T R, Uhrig J F, Bucher G L, Kellman J W and Schreier PH (2000). The movement protein NSm of Tomato spotted wilt tospovirus (TSWV): RNA binding, interaction with the TSWV N protein, and identification with interacting plant proteins. *Proceedings of the National Academy of Sciences* 97;2373–2378.

Tavert-Roudet G, Abdul-Razzak A, Doublet B et al. (2012). The C terminus of lettuce mosaic potyvirus cylindrical inclusion helicase interacts with the viral VPg and with lettuce translation eukaryotic initiation factor 4E. *Journal of General Virology* 93(1);184–193.

Ueda H, Yamaguchi Y and Sano H (2006). Direct interaction between the tobacco mosaic virus helicase domain and the ATP-bound resistance protein, N factor during the hypersensitive response in tobacco plants. *Plant Molecular Biology* 61;31–45.

Uhrig J, Canto T, Marshall D and MacFarlane S A (2004). Relocalization of nuclear ALY proteins to the cytoplasm by the Tomato bushy stunt virus p19 pathogenicity protein. *Plant Physiology* 135;2411–2423.

Van Der Heijden M W, Carette J E, Reinhoud P J, Haegi A and Bol J F (2001). Alfalfa mosaic virus replicase proteins P1 and P2 interact and colocalize at the vacuolar membrane. *Journal of Virology* 75(4);1879–1887.

Ventelon-Debout M, Tranchant-Dubreuil C, Nguyen T T, Bangratz M, Sire C, Delseny M and Brugidou C (2008). Rice yellow mottle virus stress responsive genes from susceptible and tolerant rice genotypes. *BMC Plant Biology* 8;26.

Wang D and Maule A J (1995). Inhibition of host gene expression associated with plant virus replication. *Science* 267;229–231.

Wang H, Hao L, Shung C Y, Sunter G and Bisaro D M (2003). Adenosine kinase is inactivated by geminivirus AL2 and L2 proteins. *Plant Cell* 15;3020–3032.

Washburn M P, Wolters D and Yates J R (2001). Large-scale analysis of the yeast proteome by multidimensional protein identification technology. *Nature Biotechnology* 19;242–247.

Wittman S, Chatel H, Fortin M G and Laliberte J F (1997). Interaction of the viral protein genome linked of turnip mosaic potyvirus with the translational eukaryotic initiation factor (iso) 4E of Arabidopsis thaliana using the yeast two-hybrid system. *Virology* 234;84–92.

Wu L, Han Z, Wang S, Wang X, Sun A, Zu X and Chen Y (2013a). Comparative proteomic analysis of the plant-virus interaction in resistant and susceptible ecotypes of maize infected with sugarcane mosaic virus. *Journal of Proteomics* 89;124–140.

Wu L, Wang S, Chen X et al. (2013b). Proteomic and phytohormone analysis of the response of maize (*Zea mays* L) seedlings to sugarcane mosaic virus. *PLoS ONE* 8;e70295.

Xie Q, Sanz-Burgos A P, Guo H, Garcia J A and Gutierrez C (1999). GRAB proteins, novel members of the NAC domain family, isolated by their interaction with a geminivirus protein. *Plant Molecular Biology* 39;647–656.

Xu Q, Ni H, Chen Q et al. (2013). Comparative proteomic analysis reveals the cross-talk between the responses induced by H_2O_2 and by long-term rice black-streaked dwarf virus infection in rice *PLoS One* 27;8(11); e81640.

Yang H, Huang Y, Zhi H and Yu D (2011). Proteomics-based analysis of novel genes involved in response toward soybean mosaic virus infection. *Molecular Biology Reports* 38(1);511–521.

Yoshioka K, Matsushita Y, Kasahara M, Konagaya K and Nyunoya H (2004). Interaction of tomato mosaic virus movement protein with tobacco RIO kinase. *Molecules and Cells* 17;223–229.

Zhu M, Chen Y, Ding XS et al. (2014). Maize Elongin C interacts with the viral genome-linked protein, VPg, of *Sugarcane mosaic virus* and facilitates virus infection. *New Phytologist* 203;1291–1304.

14 A Case Study of *Tomato Spotted Wilt Virus*–Pepper Interaction

Asztéria Almási, Katalin Salánki, Katalin Nemes, László Palkovics, and István Tóbiás

CONTENTS

14.1 INTRODUCTION

Tomato spotted wilt virus (TSWV) is the type member of the genus *Tospovirus*, the only plant virus group within the family *Bunyaviridae*. Members of this family are important pathogens of animals and humans. The genus name is derived from its first described member's name ***To**mato **spo**tted wilt **virus***. The first observation of spotted wilt disease of tomato was in Australia in 1915 (Brittlebank 1919); its etiology was later studied and described, and the causal agent was named *Tomato spotted wilt virus* (Samuel et al. 1930). TSWV was found in several countries mainly in protected crops, but the disease did not become significant due to effective control of its vector *Thrips tabaci*. In the 1980s, very severe damage was observed worldwide in horticultural crops caused by TSWV; therefore it was considered one of the ten most economically destructive plant viruses (Tomlinson 1987). In this process, the worldwide distribution of its very efficient vector, western flower thrips (*Frankliniella occidentalis*), and the extremely broad host range of the virus played a crucial role in this process. Tospoviruses are found in every continent and the genus consists of eight accepted and 11 tentative species (Pappu et al. 2009). On the basis of N protein sequences, tospoviruses are divided into two main geographical groups, Asian and American (Dong et al. 2008). Tospoviruses are unique in that they can replicate in both plant and insect vectors (German et al. 1992).

TSWV particles are 80–120 nm in diameter. The viral genome consists of three single-stranded RNA molecules, a large (L RNA), a medium (M RNA), and a small (S RNA) segment, which enables the virus to develop reassortants. The L RNA encodes RNA-dependent RNA polymerase (RdRp) in negative polarity, while the M and S RNAs are ambisense. M RNA encodes the precursor of two glycoproteins (Gn and Gc) that incorporate in the outer envelope membrane of host-origin, and a non-structural protein NSm playing a role in the virus movement. From the S RNA, two proteins are translated: the nucleocapsid (N) and a nonstructural protein (NSs) (Figure 14.1).

The three genomic RNAs are tightly linked with the N protein, forming ribonucleoproteins (RNPs). These RNPs are encased within a lipid envelope consisting of two virus-encoded

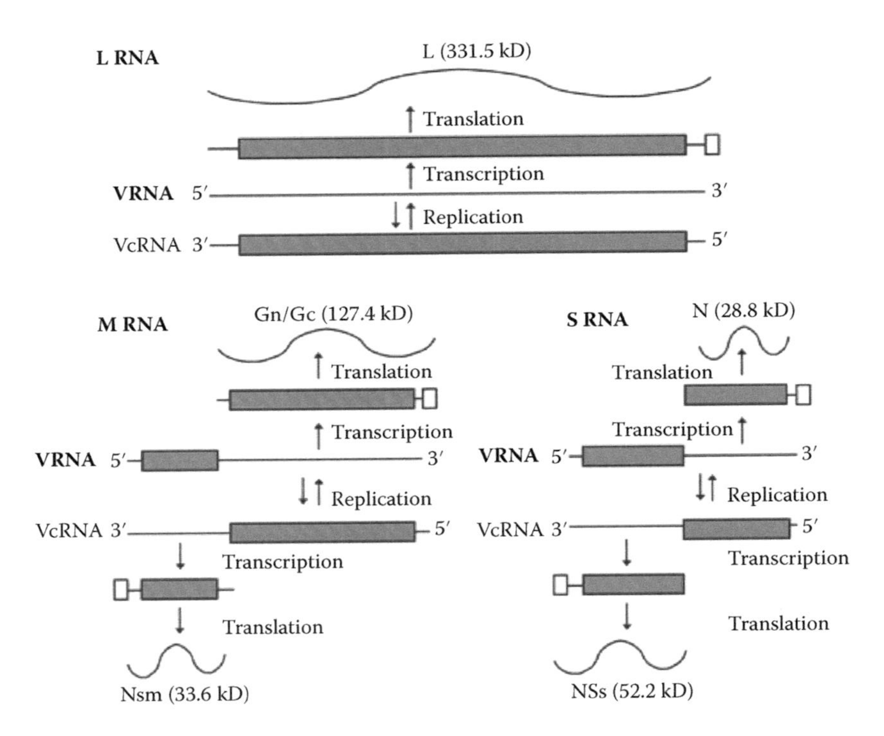

FIGURE 14.1 Genome organization of *Tomato spotted wilt virus*.

glycoproteins and a host-derived membrane. Due to the negative-stranded genome, virions contain several molecules of the RNA-dependent RNA polymerase to initiate rounds of replication of the virion RNAs. Genome expression is facilitated through synthesis of subgenomic RNAs. The 331 kDa RdRp coded by the L RNA serves as a multifunctional, replication-associated protein and is believed to function cooperatively with host-encoded factors. M RNA contains determinants for thrips transmission and host adaptation, as well as overcoming host plant resistance in tomato (Pappu et al. 2009). NSs protein encoded by S RNA functions as a silencing suppressor and an avirulent determinant in TSWV–pepper interaction.

TSWV has an extremely wide host range and infects more than 900 plant species including numerous crops and weeds. The infected crops mainly belong to *Solanaceae* and *Asteraceae* families and show chlorotic or necrotic spots, rings and patterns on leaves on fruits, stunting, and dieback of the plant. Crop losses can even reach 100% in glasshouse production. The weeds in most cases are symptomless hosts and considered as important reservoirs of the TSWV. Virus reservoirs vary between different climatic conditions. Although TSWV has a wide natural host range among weeds, only these species (*Stellaria media*, *Galinsoga parviflora*, *Amarathus sp.*, *Chenopodium sp.*, etc.) play a role in epidemics on which thrips vectors can multiply.

In nature, TSWV is transmitted by nine thrips species (*Frankliniella occidentalis*, *F. fusca*, *F. schultzei*, *F. intonsa*, *F. bispinosa*, *F. cephalica*, *F. gemina*, *Thrips setosus*, and *T. tabaci*) (Rotenberg et al. 2015). The virus is transmitted in a persistent, propagative manner by which the virus replicates in the insect, circulates through the body, and persists through the various developmental stages. Virus acquisition is a developmental stage-dependent phenomenon, as only those thrips are able to transmit the virus that have acquired it at the larvae stage (Rotenberg et al. 2015) (Figure 14.2).

Biologically distinct isolates of TSWV exist in nature, which differ in molecular diversity, thrips transmissibility, pathological properties, symptomatology, and symptom severity, have been described (Jacobson and Kennedy 2013; Qiu et al. 1998; Zhang et al. 2016; Roggero et al. 2002). TSWV

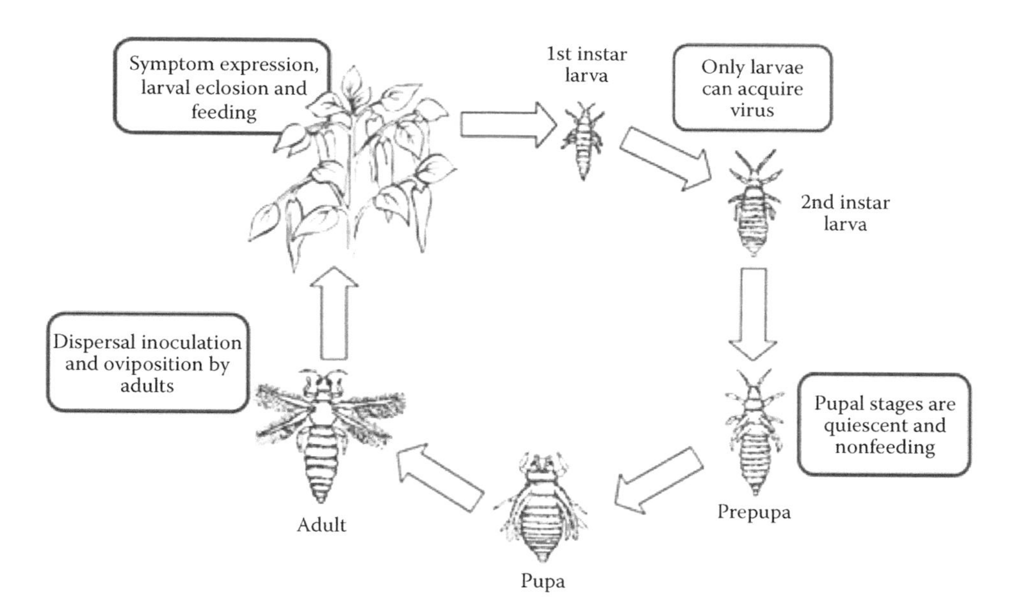

FIGURE 14.2 TSWV transmission cycle.

isolates originating from different hosts cause different types of symptoms on their hosts as well as on indicator plants. The heterogeneity of TSWV was observed and explained by Norris (1946) and it was suggested that disease symptoms on different hosts were caused by the variable ratio of TSWV strains in the viral complex. Biological diversity of TSWV was described in the early 1950s, when Finlay (1952) differentiated ten distinct TSWV strains on five tomato genotypes. The high variability of TSWV was also demonstrated in the case of pepper isolates collected in pepper growing areas, which caused mild, chlorotic, necrotic, and recovery type symptoms on *Nicotiana tabacum* cv. *Xanthi*-nc test plants (Figure 14.3).

Serological properties of TSWV also confirmed the considerable variability of the virus. Polyclonal and monoclonal antibodies were used for serotyping TSWV isolates. According to de Ávila et al. (1990), nucleocapsid proteins are less conserved than the glycoproteins.

In the last 15–20 years in accordance with the development of molecular techniques, molecular diversity of the genome has been demonstrated in cases of different TSWV isolates. The length of TSWV RNA varied to different degrees. The size of L RNA varied between 8897–8917 nt, M RNA 4752–4830 nt, and in the case of S RNA the variation is even larger, expanding to 2916–3364 nt (Zhang et al. 2016). The driving force for the rapid adaptation of TSWV and for its variability is the high frequency of mutation, reassortment, and recombination (Qiu et al. 1998). The requirement for reassortment and/or recombination is that two or more different isolates of the same virus infect the same host cell. First publication of recombination events in the case of TSWV was reported by Best (1961), who was able to isolate three recombinants based on phenotypic markers after co-inoculation with two parental strains. The phylogenetic analysis of the whole-genome sequences showed reassortment of TSWV isolates. According to Tentchev et al. (2011), 17% of the 224 TSWV isolates tested were reassortants. Lian and co-workers (2013) have studied the Korean isolates of TSWV and demonstrated that both the L RNA and M RNA segments originated in Western Europe and North America, but the S RNA segments for all Korean isolates originated in China and Japan.

In laboratory conditions the virus can be transmitted mechanically; transmission efficiency increases by using antioxidants in the extraction buffer. Several repeated passages in tobacco test plants lead to form defective mutants of TSWV. Mutation and degeneration in L RNA leads to attenuation of symptoms on tobacco plant, while in M RNA leads to glycoprotein-deficient mutants

FIGURE 14.3 Different types of symptoms on *Nicotiana tabacum* cv. *Xanthi*-nc plant caused by *Tomato spotted wilt virus* isolated from pepper. (a) Mild strain of TSWV. (b) Chlorotic-necrotic strain of TSWV with recovery type. (c) Chlorotic-necrotic strain of TSWV. (d) Necrotic strain of TSWV.

that lost their vector transmissibility (De Resende et al. 1991). Since symptom phenotype of the glycoprotein-deficient mutants is similar to the wild-type virus, it was suggested that glycoproteins are required for thrips transmission (Wijkamp 1995). Sequence analysis of the GC protein and transmission electron microscopic studies revealed that a specific nonsynonymous mutation (C1375A) in the GN-GC open reading frame (ORF) of the M RNA resulted in the loss of thrips transmissibility without inhibition of virion assembly (Sin et al. 2005).

Effective disease management against TSWV requires multicomponent approaches. The approaches are based mainly on avoidance of sources of infectivity, control of vectors, and use of resistant varieties. Since there are no data for seed transmission of TSWV, the use of virus-free propagation material and effective control of natural weed hosts of TSWV is of primary importance.

Because the different thrips species have different life cycle parameters (Riley et al. 2011), their control is considerably more difficult. Effective protection against thrips is based on biological (predators, parasites, etc.), agrotechnical (reflective mulch, pruning, etc.), and chemical control methods.

Increasing crop resistance, either by using transgenic plants or using resistance gene in a traditional crossing method, can be efficient against viral diseases. *Tsw* resistance was found in *C. chinense* accession PI152225 and PI159236, as described by Black et al. (1996), and it was introduced into several types of pepper varieties. Some years later, rapid adaptation of TSWV to plant resistance was demonstrated, as *Tsw* resistance gene was broken down only a few years after its deployment in pepper crops (Roggero et al. 2002; Thomas-Caroll and Jones 2003; Margaria et al. 2004; Sharman and Persey 2006).

The evidence of the complex crop protection approach is the application of a resistant pepper variety in the TSWV-infected environment. The thrips-transmitted TSWV shows very severe HR symptom not only on the leaves, but also on pepper fruit (Figure 14.4).

FIGURE 14.4 Hypersensitive (HR) symptom on fruits of pepper varieties containing *Tsw* resistance gene after TSWV infestation.

14.2 TSWV–PEPPER INTERACTION

14.2.1 Resistance

Unless gene manipulation techniques in crop cultivation are allowed in the European community and it becomes popular and acceptable for the broad public, the most effective control against TSWV is resistance breeding. Although plant resistance against viruses can be affected by several factors and can be established by distinct mechanisms such as gene silencing, widely studied and applied resistance is based on a general dominant gene. The main efforts are on finding natural resistance genes in different plant species to introduce into plant varieties of horticultural or agricultural significance; still there are only a few of them widespread in production. However, these spp. or cultivars bear one and the same gene against certain virus strains or isolates. In the case of TSWV, there are only two resistance genes—*Sw-5* in *Lycopersicum esculentum* and *Tsw* in *Capsicum* spp. *Tsw* originating from *C. chinense* was incorporated into several pepper cultivars that are broadly used in pepper cultivation.

Soon after extensive propagation of *Tsw*-bearing pepper cultivars, resistance-breaking strains of TSWV have emerged all over the world. First, in the main pepper cultivating countries in Europe and other continents, including Brazil, Italy, Spain, Australia, the United States, China, and Hungary. The rapid appearance of resistance-breaking strains after intensive use of resistant pepper cultivars in homogenous populations could be the consequence of distinct genetic events. Most likely, a population of various virus isolates or mutants is present in a glasshouse or production area. If fitness of one or another mutation alters by selection pressure, that particular isolate is becoming more dominant under those circumstances. Or, as a result of *Tsw* gene-induced selective pressure, a resistance-breaking phenotype has been evolving by mutation. In TSWV–pepper interaction the NSs coding S RNA segment of TSWV genome is responsible for resistance breaking

(Margaria et al. 2007; Takeda et al. 2002; de Ronde et al. 2013, 2014a); therefore mutation(s) in this region is/are important. Phylogenetic analysis based on the NSs protein amino acid sequences of the RB and WT strains showed that there is no common amino acid substitution that could be responsible for the altered phenotype in all cases, while tomato SRB (*Sw-5* resistance breaking) strains comprise amino acid changes at the same two positions, and these strains have evolved several times by convergent processes (López et al. 2011). With pepper, several mutations have been identified, but none of them occurred in all resistance-breaking isolates. Rather, geographical distribution played a role in the evolution of RB strains, which may support the hypothesis that RB strains develop locally from WT strains by point mutations (Almási et al. 2015). The other way to develop RB strain could be by genome reassortment. As TSWV genome consists of three segments (three RNA molecules), interchange is possible between TSWV isolates, which is supported by study (Aramburu et al. 2015) on naturally occurring mixed infections. To this process crop/seedlings in commercial trade may contribute, as distinct virus sources are able to mix with others originating from other countries. According to this conception, if an S RNA of a WT strain is replaced by a foreign S RNA segment of an RB strain, the new reassortant will interact with the *Tsw* gene because the NSs protein is encoded on that region. Recombination could lead to sequence variation at the subgenomic level. Invasion of thrips species as the vector of TSWV has been added to the epidemics and virus evolution as another factor. The bunyavirus ancestor of TSWV could be an insect/arthropod virus that was able to adapt to the plant hosts of the insect and replicate in them too.

TSWV is a unique species among plant viruses not only in respect to its unique genome organization, virion assembly, and life cycle, but in virus evolution and population genetics. However, TSWV has an extremely wide host range; host specificity is not characteristic of the strains. Isolates adapted to given crop plant usually are able to systemically infect other species too. Strains of TSWV isolated from pepper easily infect tomato cultivars, and vice-versa. Resistance in tomato does not give protection against RB pepper isolates; similarly, *Tsw* gene in pepper does not provide resistance against RB tomato strains. Hence the two resistance genes are different, and avirulence factors of TSWV are on different subgenomes—against tomato *Sw-5* gene on NSm (M RNA), while against *Tsw* gene on NSs. In contrast to tomato, where *Sw-5b* (allele of the *Sw-5* gene cluster) provides broad and relatively constant/stable resistance against even distantly related tospoviruses, *Tsw* in pepper only prevents infection of TSWV isolates (Turina et al. 2016). To date, no strains that could break both resistance genes in tomato and pepper have been found.

14.3 POPULATION GENETICS

Tospoviruses among plant viruses are considered emerging viruses, which may explain its very broad molecular diversity and host range. Three strategies that lead to this extreme genetic plasticity are mutation, recombination, and reassortment (Roossinck 1997). Phylogenetic assays were carried out in different countries to study the role of recombination or reassortment in TSWV strain development. In Korea, whole genome of 13 TSWV isolates collected from different regions and various hosts has been analyzed that contained four isolates from pepper, and these were compared to other full-length/complete gene sequences retrieved from the GenBank (Lian et al. 2013). The nucleotide sequences of L RNA were highly variable among the isolates, and the authors highlighted several possible recombination events in this region. It was concluded that geographical origin has a major impact on sequence similarity compared to host origin. Phylogenetic tree composed upon M RNA or S RNA segments showed controversial results, as the former suggested Italian, Spanish, and American ancestor, while the latter indicated a closer relationship with Asian strains, such as from China and Japan. Recombination events could be detected in lower frequency with these segments, only one in the case of S RNA, but the low number of analyzed whole-genome sequences influences the result. All these data suggest that reassortment rather than recombination occurs with higher frequency in case of TSWV and other (mostly negative sense or ambisense RNA) viruses with segmented genome.

A detailed evolution genetic study with partial sequences of the three RNA segments of a large number of southern European TSWV isolates was carried out mostly from pepper and tomato crops (Tentchev et al. 2011). More than 80% of the isolates clustered into three clades, depending on the origin, while the remaining cca. 20% did not cluster according to their geographical origin, indicating their reassortant nature. Phylogenetic study including more American and Asian strains revealed that diversity of European isolates was established locally and independently from American and Asian virus sources. Seedling trade or vector activity could have also contributed to the virus diversification leading to the worldwide resurgence of TSWV. The authors identified and analyzed positive selection in the NSs and N genes, and they found significantly higher selection intensity in NSs gene than in N coding region. Summing up these facts, the possible role of NSs protein in resistance breaking was suggested. Later, *Agrobacterium*-mediated transient expression assays and analysis of silencing suppressor activity using coinfiltration of green fluorescence protein (GFP) encoding construct with *NSs* genes of WT and RB strains (de Ronde et al. 2013) and a great number of NSs mutants generated by direct PCR mutagenesis proved the hypothesis (de Ronde et al. 2014a). At positions 17/18, a GW/WG motif may be essential for RNA binding, since mutation in this motif makes NSs defective in RNA silencing suppression and HR induction. However, in this experiment reverse mutations did not result in recovery of the original phenotype. In the case of Hungarian RB and WT isolates, only two amino acid position differences were recognized in the NSs protein which are unique in all the RB strains from different countries implemented from the GenBank. These differences were sufficient to induce or delete the necrogenic phenotype/characteristic for the WT strain.

14.4 SILENCING SUPPRESSION

Besides functioning as an avirulence factor, NSs protein also acts as a gene-silencing suppressor. Therefore it seemed plausible that resistance breaking could be linked to silencing suppressor activity, and mutation in this region could lead to the loss of suppressor activity. At first, contradictory results were published, as Zhai et al. (2014) demonstrated by site-directed mutations that two conserved motifs of NSs protein (181–183 and 412–413) are required for silencing suppressor activity. Supposedly, contradictory reports on resistance-breaking phenomenon coupled with diminished or complete loss of suppressor function or contrarily retained suppressor activity could be due to the several point mutations establishing RB phenotype; the sequences for avirulence and RNA-silencing suppression (RSS) factor are overlapping and the two functions could alter simultaneously. Not only in local RNA silencing but in systemic silencing suppression strategies, NSs acts as key viral factor as well (Hedil et al. 2015). Virus infection triggers the plant antiviral immune system, and as part of this siRNAs are transported into distant parts of the plant via the phloem source-to-sink transport system. NSs protein counteracts at distinct points with different elements of the plant silencing system, such as RNA-induced silencing complex (RISC) or Dicer-like proteins (DCLs), thus blocking the plant defense machinery and finally preventing the upload of plant silencing signals to the phloem and entering uninfected cells. This counter-defense system enables TSWV to spread systemically in the entire plant. The systemic RNA silencing suppression of TSWV proved to be dose-dependent; i.e., in correlation with the amount of NSs protein, but this feature is unique among closely related tospoviruses. The viral suppressor may bind to different factors like AGO or siRNAs, as it was proved in agroinfiltration tests with presumable AGO-binding sites mutant constructs resulted in abolished or reduced RNA silencing suppressor activity (De Ronde et al. 2013, 2014a). However, when local RSS was demolished, systemic RNA silencing suppression could still be detected. Observations with RB NSs in the same system confirmed the previous results, that RSS and avirulence activities are functionally divided (Hedil et al. 2015).

Molecular mechanisms of dominant *R* gene-mediated resistance in plant defense processes against viruses are the focus of interest, yet not understood in detail. Tospoviruses, as the only plant (and insect)-infecting viruses among arthropod-borne members of the family *Bunyaviridae*, show similarities to animal viruses due to a common insect-infecting ancestor; and co-evolution within their plant and animal host has caused homology in their innate immune sensory system (de Ronde et al. 2014b). Both plants' R proteins and animals' immune receptors contain a nucleotide-binding domain and a leucine-rich repeat (LRR), which may indicate some common feature in foreign pathogen recognition, but downstream response must be different.

14.5 PLANT DEFENSE INTERACTIONS BETWEEN THRIPS AND TSWV

Maris et al. (2004) observed complex relationship among host plant, TSWV, and thrips vector in relation to the effect of virus infection on/and plant preference. Pepper cultivars resistant or sensitive to thrips were investigated to learn if virus infection has an impact on thrips feeding behavior and oviposition. TSWV-infected pepper plants attracted *Frankliniella occidentalis* specimens, as a greater number of them fed on these plants than on non-inoculated ones. Female insects deposited more eggs on TSWV inoculated plants, and larvae developed faster and resulted in larger adults. Abe et al. (2012) further studied the factors and mechanisms playing a role in plant defense regulation against thrips and TSWV. Phytohormone-regulated marker gene expression levels were detected and analyzed. Two well-studied plant hormones that are generally involved in plant defense mechanisms against pathogens were chosen to analyze tripartite association. Salicylic acid (SA) is the signal element of basal resistance responses in plants against microbe pathogens, while jasmonate (JA) acts in plant responses against pathogen attack such as insect feeding, and mechanical wounding manifested this case in thrips resistance. Accumulation of TSWV in plants was measured by NSs expression, SA- and JA-regulated marker gene expression was detected in a 16-day time frame and choice assay, thrips population composition assessed. In both, marker gene expression was up regulated by virus infection. Insect feeding did not alter the expression levels of TSWV infection-induced SA markers. Although JA marker gene expression increased greatly during feeding on non-inoculated plants, feeding on TSWV-infected plants caused only a slight increase. Summing up all these and formerly documented results by other research groups, Abe and co-workers (2012) suggested that TSWV infection boosts thrips feeding on host plant (likely not depending on the plant family of host plant species), because TSWV infection induces antagonistic changes in SA—a regulator of basal immune system—and JA pathways, and this way incidentally amplifies preference of its vector.

14.6 RESISTANCE MANAGEMENT STRATEGIES IN PLANT PROTECTION

Progress to find new resistance genes/sources against TSWV other than already existing dominant *R* genes is more and more crucial.

Hoang et al. (2013) reported a new source of resistance against TSWV in pepper. A total of 487 *Capsicum* spp. and F1 hybrid accessions were tested for resistance against a Korean TSWV strain, and one accession (*Capsicum chinense* "AC09-207") proved to be resistant. Extensive study on morphological, genetic, inheritance, and other characteristics, marker analysis, crossing back, and allelism tests demonstrated that the resistance gene in this accession was located at the same chromosome (10) as the *Tsw* gene, but it may be a distinct allele or a member of tightly linked genes. Since RB strain was not available in Korea, the reaction of "AC09-207" accession in the case of RB strain infection is still in question.

A completely different new scope for crop protection against viruses based on micro RNAs (miRNAs) and the molecular background of this approach was thoroughly discussed by Ramesh et al. (2014). In plants, developmental gene regulation is controlled by miRNAs, but like siRNAs they also play a role in antiviral defense. Consistent with the zigzag model (Jones and Dangl 2006) representing the permanent competition between plant defense system and pathogens, viruses encode miRNAs to bypass host RNA silencing. Analogous to this interaction, viral RSS could be knocked out by expression of transgenic artificial miRNAs. This technique would accelerate reverse genetics, which for TSWV and similarly other viruses that have a genome of negative or ambisense polarity, would be very promising.

REFERENCES

Abe H, Tomitaka H, Shimoda T, Seo S, Sakura T, Kuqimiya S, Tsuda S and Kobayashi M (2012). Antagonistic plant defence system regulated by phytohormones assists interactions among vector insect, thrips and a tospovirus. *Plant Cell Physiology* 53: 204–212.

Almási A, Csilléry G, Csömör Z, Nemes K, Palkovics L, Salánki K and Tóbiás I (2015). Phylogenetic analysis of Tomato spotted wilt virus (TSWV) NSs protein demonstrates the isolated emergence of resistance-breaking strains in pepper. *Virus Genes* 50: 71–78.

Aramburu J, Galipienso L, Soler S, Rubio L and López C (2015). A severe symptom phenotype in pepper cultivars carrying the Tsw resistance gene is caused by a mixed infection between resistance-breaking and non-resistance-breaking isolates of Tomato spotted wilt virus. *Phytoparasitica* 43: 597–605.

Best RJ (1961). Recombination experiments with strains A and E of tomato spotted wilt virus. *Virology* 15: 327–339.

Black LL, Hobbs HA and Kammerlohr DS (1996). Resistance of *Capsicum chinense* lines to tomato spotted wilt virus from Louisiana, USA, and inheritance of resistance. *Acta Horticulturae* 431: 393–401.

Brittlebank CC (1919). Tomato diseases. *Journal of the Department of Agriculture of Victoria* 17: 231–235.

De Ávila AC, Huguenot C, Resende RO, Kitajima EW, Goldbach RW and Peters D (1990). Serological differentiation of 20 isolates of tomato spotted wilt virus. *Journal of General Virology* 71: 2801–2807.

De Resende RO, de Haan P, de Avila AC, Kormelink R, Goldbach R and Peters D (1991). Generation of envelope and defective interfering RNA mutants of tomato spotted wilt virus by mechanical passage. *Journal of General Virology* 72: 2375–2383.

De Ronde D, Butterbach P, Lohuis D, Hedil M, van Lent JWM and Kormelink R (2013). Tsw gene-based resistance is triggered by a functional RNA silencing suppressor protein of the Tomato spotted wilt virus. *Molecular Plant Pathology* 14: 405–415.

De Ronde D, Pasquier A, Ying S, Butterbach P, Lohuis D and Kormelink R (2014a). Analysis of Tomato spotted wilt virus NSs protein indicates the importance of the N-terminal domain for avirulence and RNA silencing suppression. *Molecular Plant Pathology* 15: 185–195.

De Ronde D, Butterbach P and Kormelink R (2014b). Dominant resistance against plant viruses. *Frontiers in Plant Science* 5: 307.

Dong J-H, Cheng X-F, Yin Y-Y, Fang Q, Ding M, Li T-T, Zhang L-Z, Su X-X, McBeath JH and Zhang Z-K (2008). Characterization of tomato zonate spot virus, a new tospovirus in China. *Archives of Virology* 153: 855–864.

Finlay KW (1952). Inheritance of spotted wilt resistance in the tomato I. *Identification of strains of the virus by the resistance or susceptibility of tomato species. Australian Journal of Biological Sciences* 5: 303–315.

German TL, Ullman DE and Moyer JW (1992). Tospoviruses: Diagnosis, molecular biology, phylogeny, and vector relationships. *Annual Review of Phytopathology* 30: 315–348.

Hedil M, Sterken MG, de Ronde D, Lohuis D and Kormelink R (2015). Analysis of Tospovirus NSs proteins in suppression of systemic silencing. *PLoS ONE* 10(8): e0134517.

Hoang NH, Yang H-B and Kang B-C (2013). Identification and inheritance of a new source of resistance against Tomato spotted wilt virus (TSWV) in Capsicum. *Scientia Horticulturae* 161: 8–14.

Jacobson AL and Kennedy GG (2013). Specific insect–virus interactions are responsible for variation in competency of different Thrips tabaci isolines to transmit different Tomato spotted wilt virus. *PLoS ONE* 8(1): e54567.

Jones JDG and Dangl JL (2006). The plant immune system. *Nature* 444: 323–329.

Lian S, Lee J-S, Cho WK, Yu J, Kim M-K, Choi H-S and Kim K-H (2013). Phylogenetic and recombination analysis of Tomato spotted wilt virus. *PLoS ONE* 8(5): e63380.

López C, Aramburu J, Galipienso L, Soler S, Nuez F and Rubio L (2011). Evolutionary analysis of tomato Sw-5 resistance-breaking isolates of Tomato spotted wilt virus. *Journal of General Virology* 92: 210–215.

Margaria P, Ciuffo M and Turina M (2004). Reistance breaking strain of *Tomato spotted wilt virus* (*Tospovirus, Bunyaviridae*) on resistant pepper cultivars in Almería, Spain. *Plant Pathology* 53: 795.

Margaria P, Ciuffo M, Pacifico D and Turina M (2007). Evidence that the nonstructural protein of Tomato spotted wilt virus is the avirulence determinant in the interaction with resistant pepper carrying the TSW gene. *Molecular Plant Microbe Interaction* 20: 547–558.

Maris PC, Joosten NN, Goldbach RW and Peters D (2004). *Tomato spotted wilt virus* infection improves host suitability for its vector *Frankliniella occidentalis*. *Phytopathology* 94: 706–711.

Norris DO (1946). The strain complex and symptom variability of tomato spotted Wilt virus. Australian Council of Scientific and Industrial Research Bulletin no. 202.

Pappu HR, Jones RAC and Jain RK (2009). Global status of tospovirus epidemics in diverse cropping systems: Success achieved and challenges ahead. *Virus Research* 141: 219–236.

Qiu WP, Geske SM, Hickey CM and Moyer JW (1998). Tomato spotted wilt tospovirus genome reassortment and Genome segment-specific adaptation. *Virology* 244: 186–194.

Ramesh SV, Ratnaparkhe MB, Kumawat G, Gupta GK and Husain SM (2014). Plant miRNAome and antiviral resistance: A retrospective view and prospective challenges. *Virus Genes* 48: 1–14.

Riley DG, Joseph SV, Srinivasan R and Diffie S (2011). Thrips vectors of Tospoviruses. *Journal of Integrated Pest Management* 1: 1–10.

Roggero P, Masenga V and Travella L (2002). Field isolates of Tomato spotted wilt virus overcoming resistance in pepper and their spread to other hosts in Italy. *Plant Disease* 86: 950–954.

Roossinck MJ (1997). Mechanisms of plant virus evolution. *Annual Review of Phytopathology* 35: 191–209.

Rotenberg D, Jacobson AL, Schneweis DJ and Whitfield AE (2015). Thrips transmission of tospoviruses. *Current Opinion in Virology* 15: 80–89.

Samuel G, Bald JG and Pittman HA (1930). Investigations on "spotted wilt" of tomatoes. *Council for Scientific and Industrial Research* 44: 1–6.

Sharman M and Persley DM (2006). Field isolates of *Tomato spotted wilt virus* overcoming resistance in Capsicum in Australia. *Australasian Plant Pathology* 35: 123–128.

Sin S-H, McNulty BC, Kennedy GG and Moyer JW (2005). Viral genetic determinants for thrips transmission of Tomato spotted wilt virus. *Proceedings of the National Academy of Sciences* 102: 5168–5173.

Takeda A, Sugiyama K, Nagano H, Mori M, Kaido M, Mise K, Tsuda S and Okuno T (2002). Identification of a novel RNA silencing suppressor, NSs protein of Tomato spotted wilt virus. *FEBS Letters* 532: 75–79.

Tentchev D, Verdin E, Marchal C, Jacquet M, Aguilar JM and Moury B (2011). Evolution and structure of Tomato spotted wilt virus populations: Evidence of extensive reassortment and insights into emergence processes. *Journal of General Virology* 92: 961–973.

Thomas-Caroll ML and Jones RAC (2003). Selection, biological properties and fitness of resistance-breaking strains of *Tomato spotted wilt virus*. *Annals of Applied Biology* 142: 235–243.

Tomlinson JA (1987). Epidemiology and control of virus diseases of vegetables. *Annals of Applied Biology* 110: 661–681.

Turina M, Kormelink R and Resende RO (2016). Resistance to tospoviruses in vegetable crops: Epidemiological and molecular aspects. *Annual Review of Phytopathology* 54: 347–371.

Wijkamp I (1995). *Virus-vector relationships in the transmission of tospoviruses*. PhD Diss. Wageningen, The Netherlands: Agricultural University Wageningen.

Zhai Y, Bag S, Mitter N, Turina M and Pappu HR (2014). Mutational analysis of two highly conserved motifs in the silencing suppressor encoded by tomato spotted wilt virus (genus Tospovirus, family Bunyaviridae). *Archives of Virology* 159: 1499–1504.

Zhang Z, Wang D, Yu C, Dong J, Shi K and Yuan X (2016). Identification of three new isolates of Tomato spotted wilt virus from different hosts in China: Molecular diversity, phylogenetic and recombination analysis. *Virology Journal* 13: 1–18.

15 Virus–Virus Interactions in Plants

Ulrich Melcher and Akhtar Ali

CONTENTS

15.1 INTRODUCTION

No virus "is an island, entire unto itself" (Donne 1624). Other chapters in this section discuss a multitude of interactions of viruses with environmental conditions, insects and other potential vectors, plant tissue components, and other plant-associated organisms. This chapter focuses on interactions between and among viruses. Cultivated plants have often been found to be infected by more than one virus. Beginning with early attempts at discovering the plant virome (Wren et al. 2006; Muthukumar et al. 2008), evidence has been mounting that non-cultivated plants are also frequently infected by multiple viruses (Mascia and Gallitelli 2016). The occupancy of a plant by multiple viruses almost guarantees that there will be interactions among viruses in that plant. Such interactions have been reviewed recently (Syller 2012; Mascia and Gallitelli 2016; Syller and Grupa 2016) and long ago (Marchoux 1988).

Virus–virus interactions can be conceptualized by considering four entities (A, B, C, and D; Figure 15.1). Entities A and B (traditional viruses), when inoculated individually into a naïve susceptible plant, will replicate and make more copies of themselves. Entities C and D are unable to start an infection when introduced into a plant alone. They can, however, replicate when introduced into a plant already infected with an A or B entity. Also, certain C and D entities alone can initiate an infection when both are inoculated together into a plant. The outcomes of these interactions may have positive, negative, or neutral effects. Interactions that do not result in changes in the genomic sequences of the interacting entities will be discussed before those in which changes in genomic nucleotide sequences occur.

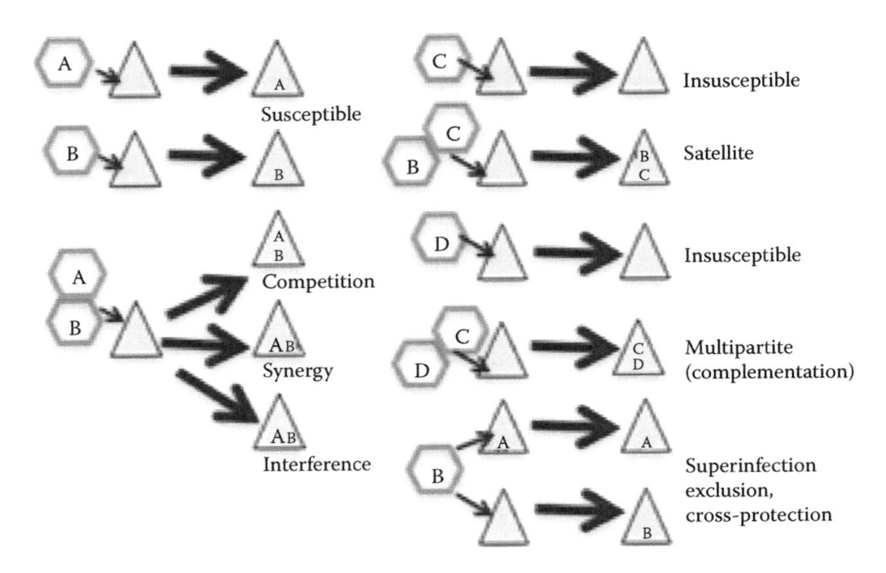

FIGURE 15.1 Modes of virus–virus interactions in plants. Plants are represented by triangles and virus or virus-like entities as hexagons. The comparative magnitude of results of inoculation with a particular entity or entity combination are indicated by the size of the font identifying the entity in the plant.

15.2 CONSERVED SEQUENCE INTERACTIONS

Outcomes of entity A and B interactions depend on the order of inoculation and on the interval of time between inoculation of the first and second entities. Simultaneous inoculation of the plant with a mixture of entities A and B (co-inoculation) sets up a "competition" (Figure 15.1) between them for cellular resources. A reduction or enhancement of virus production or symptom levels relative to levels in single infection by the A or B entities may result. Reductions are termed as "interference," while "synergy" is used for enhancements. A delayed inoculation of an A entity-infected plant with the B entity may result in little or no replication of the B entity, a phenomenon called "superinfection exclusion." Should the reverse also be true (B-infection protecting against replication of the A entity), the phenomenon should be called "cross-protection." However, this latter term is often used also for cases of superinfection exclusion where there is agronomic benefit to the exclusion.

"Multipartite" viruses (Sicard et al. 2016) are examples of co-infection by two or more C and D entities, each incapable of an infection by itself. Usually, each part of the multipartite virus consists of a particle enclosing a specific part of the viral genome, a viral chromosome, or in some cases, dictated by size considerations in packaging, more than one part. The functions of each viral chromosome "complement" the functions of the other chromosomes to create a multipartite virus. The term "satellite virus" describes cases in which a C or D entity replicates in the presence of an A or B entity. The dependent C or D entity is called a satellite virus being complemented by an A or B virus, often thought of as the "helper" virus, itself not benefitting from the presence of the satellite. Satellite RNAs or DNAs are additional RNA or DNA molecules requiring plant viruses for replication and propagation. Some helper viruses also foster the replication of specific RNAs or DNAs incapable of replicating without such fostering ("satellite" nucleic acids).

A study of virus–virus interactions must take account of interactions among their components, including not only the physical parts of the virion and the virion itself, but also a variety of proteins, nucleic acids, and assemblages of the same possibly involving membranes and organelles. When two or more viruses interact, the interaction can be between any of the various manifestations of viral presence in an organism either with each other (direct) or through the medium of host components with which they both interact (Mascia and Gallitelli 2016).

Outcomes of virus–virus interactions depend on temporal and spatial factors. Is the infection with the two interacting viruses simultaneous or does infection with one precede infection with the other? The answer must consider the speeds and pathways of intercellular movement of the infections. Upon rub-inoculation of a plant leaf or through virus delivery by an arthropod or other vector species, single plant cells become infected initially usually with a single virus (Syller 2012). From these initial cells, the infection may travel systemically to neighboring cells via movement proteins (Melcher 2000) to create a "lesion" surrounded by uninfected cells (Melcher et al. 1981). Some virus escapes, traveling longer distances, to infect emerging leaves. Indeed, spreading throughout the plant may also set up patches of infection by single entities.

15.3 COMPETITION

When two related viruses arrive in the same cell, they compete with one another for interaction with host and/or vector (Hall and Little 2013) components. Successive inoculation of zucchini plants with a mild and a severe strain of PRSV (see Table 15.1 footnote for key to virus abbreviations) led to the conclusion that the availability of replication sites is important for the result of the competition (Sansini Freitas and Marques Rezende 2008). That access to host sites required for competition is revealed by the observation that silencing of the *Nicotiana benthamiana* TOM1 gene decreased accumulation of two tobamoviruses (Table 15.1) in a cross-protection situation (Wen et al. 2013). The outcome of competitions provides an estimate of the relative virulence of viruses, a property important for understanding virus evolution. Thus the competition for host or vector components has the effect of a virus–virus interaction even though that interaction is through the intermediary of non-viral components.

Competition may have positive, neutral, or negative effects on virus replication and/or on disease manifestations (Mascia and Gallitelli 2016). The order of inoculation and the length of the interval between inoculations may affect the outcome. Modeling of the infection with classic and emerging strains of WMV suggests that, in most cases (Table 15.1), competition results in a stable equilibrium of the two viruses (Fabre et al. 2010), consistent with the invocation of competition to explain differences in levels of virus strains in natural infections (Tian et al. 2009). Similarly, in BYDV species PAV and PAS infections, PAV accounted for more than half of the viruses present regardless of the order of, or the time interval between, inoculations (Hall and Little 2013). The equilibrium point is related to the efficiency of super-infecting a plant with a strain already infected with another strain (Fabre et al. 2010). It is important to remember that the result of a competition co-inoculation on whole plants need not necessarily reflect two viruses competing at the cellular level. Viruses may enter cells singly, establishing local infections that exclude other infections from that cell and its neighbors (Hall et al. 2001).

15.4 SYNERGY

Synergy is a virus–virus competition, mediated by virus–host interactions, that produces symptoms that are more severe than when occurring with either virus alone. It was first studied in the severe mosaic disease resulting from an interaction between a potyvirus (PVY) and a potexvirus (PVX) (Rochow and Ross 1955). Other examples include Rice Tungro (Hibino et al. 1978), Corn Lethal Necrosis (Niblett and Claflin 1978), Groundnut Rosette (Syller 2012), and Sweet Potato Virus (Untiveros et al. 2007) diseases. Although synergistic effects can be seen after co-inoculation, superinfection of an already infected plant with the second virus is more common in nature (Untiveros et al. 2007).

Synergy is observed more frequently between unrelated viruses (Syller 2012), such as between potexviruses and potyviruses (Table 15.1). In the case of the unrelated PRSV and PapMV, synergy occurs when PRSV is inoculated first, but when the order is reversed, superinfection exclusion is detected (Chávez-Calvillo et al. 2016). CMV and TuMV exhibit a synergistic effect on symptoms when infecting *N. benthamiana* (Takeshita et al. 2012). Tobacco and cucurbits react differently to

TABLE 15.1
Examples of Synergy, Interference, and Cross Protection in Plant Viruses

Hosts	Virus 1	Virus 2	Order[a]	Effect on Virus 2	References
A. thaliana	TMGMV	ORMV	>	No ORMV infection	Aguilar et al. 2000
	ORMV	TMGMV	>	TMGMV infection	Aguilar et al. 2000
A. thaliana	TMV-P	TMV-C	>	Lowered systemic ELISA +ve	Rezende et al. 1992
A. thaliana	TMV-P	TMV-C	<>	No systemic movement	Rezende et al. 1992
A. thaliana	TMGMV	ORMV	>	Systemic spread prevented	Aguilar et al. 2000
A. thaliana	ORMV	TMGMV	>	No effect	Aguilar et al. 2000
A. thaliana	TCV	CMV	>, =	Protection	Aguilar et al. 2000
Black soybean	SMV attenuated	SMV	>	Seed transmitted cross-protection	Kosaka and Fukunishi 1994
Brach podium	PMV	Sat	=	vRNA and symptoms increased	Mandadi and Scholthof 2012
Cantaloupe, zucchini	ZYMV-Wk	ZYMV-Ca	>	Protection of yield	Perring et al. 1995; Lecoq et al. 1991
Cassava	EACMV-U			Cross-protection	Owor et al. 2004
Chili	ChiLCDV	β-sat	=	Virus and symptoms increased	Kumar et al. 2015
Chinese yam	CYNMV				Kondo et al. 2015
	PMMV				Leclerc and Abouhaidar 1995
Chinese yam	JYMV-mild	JYMV-severe	>	Protection	Kajihara et al. 2008
Citrus	CTV				Cook et al. 2016
Cocoa	EACMV-U	Severe strain			
Cucumber	WMV-CL	WMV-EM	>	WMV-EM less likely than <	Fabre et al. 2010
Datura metel	PVM I-38	PVM-Uran	>	Superinfection exclusion	Grupa and Syller 2016
Grapefruit	CTV 3 mild	CTV (wild)	>	New aphid vector decreases effectiveness	Powell et al. 2003
Green pepper	PMMoV L3	PMMoV	>	Protection	Ogai et al. 2013
Groundnut	GRV + sat	GRAV	=	Groundnut Rosette Disease	Murant 1990
Maize	MCMV	MDMV	>	Corn Lethal Necrosis Disease	Niblett and Claflin 1978
Maize	MDMV	MCDV-S	>	Synergistic disease	Morales et al. 2014
Maize	MDMV	MCDV-M1	>	No symptom enhancement	Morales et al. 2014
N. benthamiana	CMV	TuMV	=	Synergistic symptoms	Takeshita et al. 2012
N. benthamiana	CMV	TNV	=	CMV RNA, capsid, and symptom increase	Xi et al. 2007
N. benthamiana TOM1	HLSV	TMV-U1	>	TOM1 needed for max protection	Wen et al. 2013
N. benthamiana	TCV	TNV	=	Reduction in levels of both	Xi et al. 2010

(Continued)

TABLE 15.1 (CONTINUED)
Examples of Synergy, Interference, and Cross Protection in Plant Viruses

Hosts	Virus 1	Virus 2	Order[a]	Effect on Virus 2	References
N. benthamiana	TCV	CMV	>, =	No protection	Yang et al. 2010
N. clevelandii	recCyRSV	CyRSV	>	Interference	Burgyan et al. 1991
Papaya	PRSV-mild	PRSV-severe	>	1–5 month delay in symptoms	Sheen et al. 1998
Papaya	PRSV	PapMV	>	Synergy	Chávez-Calvillo et al. 2016
Papaya	PRSV	PapMV	<	Superinfection exclusion	Chávez-Calvillo et al. 2016
	TNV	TCV			Yang et al. 2010
Pepper, melon	CMV-S & sat	CMVD		3-week interval, near complete protection	Montasser et al. 1998
Potato	PAMV	Various	=	Increased aphid transmission	Syller 2012
Protoplast	PVX CP	TMV	>	Trans rotection	Bazzini et al. 2006
	ORMV	TMGMV	>	no TMGMV infection	Bazzini et al. 2006
Rice	RSV	RTBV	=	Rice Tungro Disease	Hibino et al. 1978
Sugar beet	BSBMV	BNYVV	>, <	Cross-protection	Mahmood and Rush 1999
Sweet potato	SPCSV	SPFMV	>	Sweet potato Virus Disease	Untiveros et al. 2007
Tobacco	PVM	PAMV	>	Mutated CP protects	Lecoq et al. 1991
Tobacco	CMV	ZYMV, WMV	=	Strong synergy	Wang et al. 2002
Tobacco	TMV-P	TMV-C	<>	Superinfection exclusion	Rezende et al. 1992
Tobacco	ORMV	TMGMV	<, >	Superinfection exclusion	Rezende et al. 1992
Tobacco	PVX	PVY	=	Synergy; PVX –ve strand increased	Rochow and Ross 1955; Vance 1991
Tobacco	satTMV5	satTMV6	> only	Protection against satTMV6	Kurath and Dodds 1994
Tobacco	TMV-C	TMV-P			Rezende et al. 1992
Tobacco	TMGMV	ORMV	>	No ORMV infection	Rezende et al. 1992
Tobacco	TMGMV-S	TMGMV-L	>	> 3-day gap needed for protection	Bodaghi et al. 2004
Tobacco	TMGMV-S	TMGMV-L	>	No delay needed for protection	Bodaghi et al. 2004
Tomato	CMV	Sat	=	Synergistic symptoms;	Taliansky et al. 1998
Turnip	CaMV	TVCV	=	Synergistic symptoms	Hii et al. 2002
Turnip	CaMV-UM130	CaMV – CabbS	=	CabbS	Zhang and Melcher 1989
Turnip	CaMV-UM130	CaMV – CabbS	>	8-day delay no protection	Zhang and Melcher 1989

(*Continued*)

TABLE 15.1 (CONTINUED)
Examples of Synergy, Interference, and Cross Protection in Plant Viruses

Hosts	Virus 1	Virus 2	Order[a]	Effect on Virus 2	References
Wheat	BYDV-PAV	BYDV-PAS	><	Set point [PAV}/[PAS] > 1.0	Hall and Little 2013
Zucchini	PRSV-W-1	PRSV-W-C	>	6–9-day interval needed to avoid symptoms	Sansini Freitas and Marques Rezende 2008

Note: Virus abbreviations: ACMV, African cassava mosaic virus; AMV, Alfalfa mosaic virus; ArMV, Arabis mosaic virus; ASG, Apple Stem Grooving Virus; BCM, Bean common mosaic virus; BNYVV, beet necrotic yellow vein virus; BSBMV, Beet soilborne mosaic virus; BYDV, Barley yellow dwarf virus; CaMV, Cauliflower mosaic virus; ChiLCDV, Chili leaf curl disease virus; CMV, Cucumber mosaic virus; CPMVV, Cowpea mosaic virus; CPSMV, Cowpea severe mosaic virus; CTV, Citrus tristeza virus; CYNMV, Chinese yam necrotic mosaic virus; CyRSV, Cymbidium ringspot virus; EACMV-U, East African cassava mosaic virus-Uganda; GFLV, Grapevine fanleaf virus; JYMV, Japanese yam mosaic virus; MCDV, Maize chlorotic dwarf virus; MDMV, Maize dwarf mosaic virus; ORMV, Oilseed rape mosaic virus; PAMV, Potato aucuba mosaic virus; PanMV, Panicum mosaic virus; PapMV, Papaya mosaic virus; PepMV, Pepino mosaic virus; PMMo, Pepper mild mottle virus; PRSV, Papaya ringspot virus; PVMV, Potato virus M; PVX, Potato virus X; PVY, Potato virus Y; satTMV, Satellite tobacco mosaic virus; SMV, Soybean mosaic virus; SPCSV, Sweet potato chlorotic stunt virus; SPFMV, Sweetpotato Feathery Mottle Virus; TCV, Turnip crinkle virus; TMGMV, Tobacco mild green mosaic virus; TMV, Tobacco mosaic virus; TNV, Tumor necrosis virus; ToMV, Tomato mosaic virus; TRV, Tobacco rattle virus; TrYMV, Triumfetta yellow mosaic virus; TuMV, Turnip mosaic virus; TVCV, Turnip vein clearing virus; WMV, Watermelon mosaic virus; ZYMV, Zucchini yellow mosaic virus

[a] > Virus 1 inoculation precedes virus 2 inoculation
< Virus 1 inoculation follows virus 2 inoculation
= coinfection with viruses 1 and 2.

mixed infections by CMV and potyviruses (Wang et al. 2002), indicating a complex series of interactions with hosts during dual infection. In a tomato infected with CMV and a satellite, the satellite RNA acts synergistically on symptoms (Taliansky et al. 1998).

Commonly, not all strains of a synergizing virus are equally effective. In sweet potato virus disease (Table 15.1), SPCSV is regarded as the active agent since it can synergize with a variety of viruses, while the original partner, SPFMV, does not result in synergy with the other viruses (Untiveros et al. 2007). Considerable variation in effectiveness of synergy among strains of the active agent exists. In another example, MDMV and MCD-S (severe strain) exhibit symptom synergy, but a mild strain (MCDV-M1) does not enhance symptoms (Morales et al. 2014).

Many cases of symptom synergy occur in interactions in which some virus components are elevated in level. However, the degree of symptom synergy is not necessarily correlated with levels of virus components (Tsuda et al. 2007). Synergistic interactions include increases in virus levels, increased symptom severity, and increased movement throughout the host plant, among other phenomena. For example (Table 15.1), CMV and TNV coinfecting *N. benthamiana* result in a synergistic interaction in symptom enhancement, and increased levels of CMV(+) RNA and capsid protein (Xi et al. 2007). Further, in the PVX-PVY synergy, most viral components are not differentially increased except for the negative strand PVX template, which is increased substantially (Vance 1991), suggesting that PVX is the active partner in this synergy. However, more complex interactions are observed when simultaneous infection with more than two virus species occurs. Upon mixed infections by viruses PVX, PVY and PVS, PVX concentrations were reduced in potato plants by the presence of PVS (Singh and Khurana 1995). The nature of the shifts in virus levels is complex, being dependent on host species and variety (Singh and Singh 1995).

Determinants of synergy are identifiable by mapping, using strains that differ in synergistic potential (Table 15.1). A synergistic severe necrotic response was encountered in turnip plants being used for experimentation with CaMV (Hii et al. 2002). The extra severity was traced to the presence of TVCV (Lartey et al. 1994). Synergy was dependent on the isolate of CaMV used. Using strain chimeric CaMV genomes, the synergy-inducing sequence was mapped to the ORF IV-V region of CaMV DNA (Hii et al. 2002). In PAMV infection, its HC-Pro region helps aphid transmission of a broad range of potyvirus particles (Syller 2012). Mutation of a region of the HC-Pro gene of the potyvirus destroys synergism inducing ability (Shi et al. 1997). One mechanism for this loss is suppression of gene silencing, discussed below.

Some synergistic viral diseases involve not only two viruses or viral entities, but also require a satellite virus or satellite RNA. Satellites may alter the result of synergistic coinfection of CMV with a potyvirus (Cillo et al. 2007). Association of the *Begomovirus* (ChiLCD) with beta-satellites increased ChiLCD accumulation and severity of symptoms in the host (Kumar et al. 2015). The satellite of PanMV exacerbates symptoms and increases viral RNA levels when present with its helper virus (Mandadi and Scholthof 2012). Similar disastrous effects of satellite presence are seen in cassava infections by ACMVor EACMV-U (Legg et al. 2006).

15.5 INTERFERENCE AND CROSS-PROTECTION

Complementary to synergy, *interference* describes negative interactions between viruses in the same plant. Experimentally, interference occurs when a plant harboring one virus is challenged with a second virus, and there is no, or much reduced, infection by the second virus, relative to inoculation of uninfected plants. In some situations (Table 15.1), the challenge virus accumulates but fails to induce normal symptoms (Cassells and Herrick 1977). In other cases, such as that of PRSV and PapMV, an interaction, synergistic with one inoculation order, becomes an interference when the order is reversed (Chávez-Calvillo et al. 2016). Some examples of interference occur with an inoculum that contains both viruses. “Concurrent protection” describes this mixed inoculum approach (Bruening et al. 2000). Interference is the basis of cross-protection, used in agriculture and horticulture to protect plants against infection by viruses that produce significant yield losses. A popular practical application of interference occurs when the first virus produces a mild infection, agronomically tolerable, while infection by the second, highly pathogenic on naïve plants, does not produce major disease.

Cross-protection was identified nearly a century ago (McKinney 1929; Fulton 1986; Pennazio et al. 2001). Reciprocity (the second virus interfering with the first virus) is only seen occasionally (Table 15.1), such as in consideration of systemic, but not local, symptom spread of two strains of TMV in *Arabidopsis thaliana* (Rezende et al. 1992). However, in *Nicotiana* species, the initially inoculated strain prevented establishment of the second strain. Similarly, in *A. thaliana*, TMGMV prevented systemic infection by ORMV, but ORMV was inactive in preventing superinfection by TMGMV, whereas in *N. tabaccum*, superinfection was prevented in both directions (Aguilar et al. 2000). Differences in plant host reaction exist. TCV protects *A. thaliana* against CMV even when co-inoculated (Yang et al. 2010). However, the protection is not seen when *N. benthamiana* is used as a host (Chen et al. 2014). Yet the same host, co-inoculated with TNV and TCV, produces levels of both RNAs that are less than in singly-inoculated plants (Xi et al. 2010). TMGMV-S, an isolate of TMGMV with a shorter coat protein chain, excluded TMGMV-L if at least three days were allowed between inoculations of *N. glauca* (Bodaghi et al. 2004). No such delays were needed in the reciprocal case (TMGMV-L protecting against TMGMV-S). Cross-protection was also observed for DNA viruses. Infection with the Cabbage S isolate of CaMV-protected turnip plants against subsequent inoculation with a derivative of Cabbage S (Zhang and Melcher 1989). Cross-protection was also documented for a satellite virus (*Satellite tobacco mosaic virus*) (Kurath and Dodds 1994; Montasser et al. 1998).

Results are often interpreted as due to differential effects on plant defense systems. However, interference with wild-type CyRSV infection resulted from defective RNAs or virions generated by RNA recombination (Burgyan et al. 1991). Virus strains belonging to different genera also sometimes generate cross-protection. For example, BSBMV and BNYVV are serologically distinct but related cross-protecting viruses (Rush 1999).

Interference and crop protection have significantly impacted world agriculture (Table 15.1). Cross-protection is a major part of strategies to fight cassava mosaic diseases in Africa (Legg et al. 2006). Natural infection of cassava in Uganda by a mild strain of *EACMV-U* has favored the natural spread of this protective strain that prevents later infection by a severe strain (Owor et al. 2004). Although use of mild strains of a virus often provide a needed level of protection, the protective ability in long-lived plants such as cocoa trees decreases over time (Ameyaw et al. 2016). Protection of grapefruit trees against *Citrus tristeza virus* (CTV) by mild isolates has been highly productive but complex due to the mixed nature of natural isolates (Cook et al. 2016). Japanese yams are protected against infection from JYMV by the presence in the yams of a mild strain (Kajihara et al. 2008). Similarly, in Chinese yams, an attenuated strain of CYNMV protects against infection with virulent strains (Kondo et al. 2015). Papaya in Taiwan have been protected against PRSV using the cross-protection strategy (Sheen et al. 1998). In Japan, the use of soil fumigation with methyl bromide was targeted to be replaced by pre-inoculation with an attenuated strain of PMMoV (Ogai et al. 2013). Cross-protection has been used to produce crops of cantaloupes despite the presence of a severe strain of ZYMV (Perring et al. 1995). Cross-protection may break down after introduction of new vectors, as observed with CTV-protected grapefruit (Powell et al. 2003). PRSV HA5-1, highly effective in control of PRSV disease in papaya, does not provide protection against PRSV strain W in cucurbits (You et al. 2005). Incorporation of certain sequence stretches of the W strain into HA5-1 results in viruses that do provide protection. Importantly, cross-protection of black soybean against SMV leads to the production of cross-protected seeds (Kosaka and Fukunishi 1994).

Instances of trans-protection, such as of PVX coat protein against TMV infection of protoplasts (Bazzini et al. 2006), are known. New examples of cross-protection continue to be discovered, for example between carlavirus isolates in *Datura metel* (Grupa and Syller 2016). Coat protein genes with deletions in the core region for the protein were able, nevertheless, to cross-protect tobacco against PAMV (Leclerc and Abouhaidar 1995). Prior infection of zucchini squash plants with a weak strain of ZYMV (ZYMV-WK) infection made ZYMV disease in this squash much less severe than without treatment (Lecoq et al. 1991). Mild disease was also the result of testing protection against a variety of ZYMV challenges (Wang et al. 1991). CMV coat protein expression in *N. tabaccum* produced a significant degree of protection against a variety of CMV isolates (Quemada et al. 1991). In sweet potato, a mild SPFMV infection is able to control russet crack disease of sweet potato (Yamasaki et al. 2009). Modeling aids planning the utilization of cross-protection in agricultural practice (Zhang et al. 2000; Neofytou et al. 2016). It revealed that in a field situation, maintaining a certain level of infection by the protecting virus was important to maintain protection in the whole field.

15.6 COAT PROTEIN-MEDIATED PROTECTION

Since the initial description of cross-protection (McKinney 1929), a multitude of mechanisms underlying protection have been proposed (Urban et al. 1990). *Coat protein-mediated protection* (Hackland et al. 1994) arose from exploration of cross-protection. The latter was rediscovered among specific tobamoviruses about the same time as the process of uncoating of the viral RNA prior to RNA replication was being studied. It was argued that the extra concentration of capsid subunits that would be present in a transgenic plant, transgenic for the synthesis of the TMV capsid protein, would shift the uncoating equilibrium toward coating, thus preventing late viral RNAs from being uncoated. This was consistent with the observation that small host-specific differences in coat protein sequence influence the ability to provide cross-protection (Nakazono-Nagaoka et al. 2009).

Cross-protection motivated creation of transgenic plants that could synthesize viral proteins (Ziebell and Carr 2010). It must be remembered that natural isolates likely are mixtures of multiple related strains that can be distinguished by sequencing or by SSCP (Sambade et al. 2002). Preference for a genetically uniform protectant motivated using mutagenesis of the coat protein gene to create a mild PepMV to protect tomato (Chewachong et al. 2015). The resulting virus protected as expected. However, also as expected, resistance to ToMV in tomatoes is more effective with a ToMV coat protein transgene than with a transgene from the close relative, TMV (Sanders et al. 1992). A PVX vector construct that expressed the TMV coat protein gene was able to protect *N. benthamiana* from subsequent infection by TMV (Culver 1996), lending support to the concept that packaging and unpackaging may be involved in cross-protection. Highly ordered states of TMV coat protein subunits, generated by mutation, were implicated as the strongest inducers of interference in a protoplast study (Bazzini et al. 2006).

15.7 RNA SILENCING

The coat protein gene is the major viral determinant of cross-protection (Valkonen et al. 2002). However, coat proteins were gradually realized not to play as large a role in these phenomena as originally thought (Morange 2012). Today, we believe that much of this resistance is due to interaction through *RNA silencing* and its suppression (Palukaitis 2011). This shift was engendered in part by the observation that non-coat protein viral genes may also produce cross-protection. In some cases, movement proteins were implicated. The RNA polymerase domain of TMV replicase cross-protects against TMV in *N. benthamiana* (Goregaoker et al. 2000). Mutations in the ZYMV HC-Pro generate cross-protecting mild strains (Lin et al. 2007). Presumably, this region interacts with the RNA silencing system in the tested plants. In other cases, in addition to the viral genes, satellite viruses or nucleic acids presented as transgenes lead to protection. Plants transgenic for production of wild-type satellite CMV have good superinfection-exclusion activities against CMV and TAV (McGarvey et al. 1994).

The protection offered by the *N. tabaccum* M(IC)1,3 variant of TMV against infection by wild type TMV correlated with the appearance of TMV specific siRNAs, suggesting that RNA silencing plays a major role in this cross-protection (Ding et al. 2004). Similar results were obtained with *N. tabaccum* plants made transgenic for production of TMV RNA12 with an extra insertion from a mobile element (Donson et al. 1993). Further, in a TMV system lacking a coat protein gene, superinfection-exclusion could nevertheless be seen (Manuel Julve et al. 2013). In vitro exchange of selected regions of viral genomes between superinfection exclusion and inactive strains of a virus have been used to map genetic determinants needed for this exclusion in CTV (Atallah et al. 2016). For this virus, reviewed by Folimonova (2013), different isolates of a strain of the virus provide protection against infection by other isolates, but not by isolates of other strains (Folimonova et al. 2010). The ability of CTV to make a 33kD protein is critical for the protective effect (Folimonova 2012). The likely role of RNA silencing in cross-protection was strengthened by the observation that TMV double-stranded RNA from the 126K gene induced protection when used as leaf inoculum (Konakalla et al. 2016). However, studies with turnip mosaic virus (TuMV) on several model host plants suggest that while RNA silencing must, in part, be responsible for cross-protection, a variety of other induced immunity responses are likely also involved (Kung et al. 2014).

15.8 SILENCING SUPPRESSION

Another mutation that seems effective in producing a protecting strain is the deletion of the 2b gene of CMV (Ziebell et al. 2007). This gene encodes a silencing suppressor. Supposedly, allowing RNA silencing to take place will allow protection over a range of related sequences.

Silencing suppression has been implicated as a factor responsible for synergy (MacDiarmid 2005; Untiveros et al. 2007; Nishiguchi and Kobayashi 2011), and provides an intermediary for interaction between viruses. Silencing suppression (Qu and Morris 2005) may lead to the exacerbated expression of symptoms of infection. Protection against infection by virulent isolates of three genotypes of PepMV depends on the mild isolate being of the same genotype as the challenge virus (Hanssen et al. 2010). This sequence dependence suggests an RNA silencing related mechanism. The roles of the 2b protein of CMV and TuMV in synergy between the two viruses has been explored (Takeshita et al. 2012). However, silencing has been ruled out as a factor in the case of cross-protection afforded by isolates of CTV strains (Folimonova et al. 2014). In *A. thaliana*, cross-protection does not require activity of RNA-dependent RNA polymerase SDE1/SGS2, enzymes critical for RNA silencing. That examples of cross-protection include only limited similarity in sequence between protecting and challenge virus suggests that some cross-protection involves mechanisms other than sequence-specific RNA silencing (Li et al. 2016).

15.9 COMPLEMENTATION

A defective virus or viral genome may nonetheless produce a viral infection when in the presence of another virus (or other entity) that provides the function missing from the virus of interest. Such rescue of a mutant virus is termed *complementation*. Complementation can be reciprocal such that two viruses, each defective in a different gene, can complement each other, producing an infection, while inoculation with either viral genome alone does not lead to infection (Choe et al. 1985; Melcher et al. 1986a). However, successful infection can also be due to a concomitant recombination between the two mutant genomes, or in the case of multipartite viruses, due to reassortment of segments. Satellite viruses require complementation by the helper virus. Multipartite viruses are examples of complementation since individual entities cannot propagate in the absence of other parties. Complementation may even be responsible for the increased transmission by aphids of PVY from plants infected with three strains relative to those infected with fewer strains (Mondal et al. 2017).

15.10 INTER- AND INTRA-VIRAL COMPLEMENTATION

Transmission of viruses is complicated by packaging constraints. Genetic transmission of genomes in most forms of life proceeds by packaging at least one complete set of chromosomes in a unit for transmission. Many viruses resemble the first-described virus, TMV, in that all genetic information is stored in one continuous nucleic acid, the genome. However, viruses whose genomes are distributed between or among multiple segments exist. The interactions between and among segments and particles, respectively, of multisegmented and multipartite viruses are analogous to complementation between defective viruses, discussed above, and are therefore considered in this chapter.

Some viruses, such as the influenza viruses, are multisegmented, yet manage to package all segments into a single particle and are thus designated monopartite viruses. Other multisegmented viruses are multipartite, meaning that multiple particles are required for an infection.

For filamentous or rod-shaped viruses, such as TMV, the RNA genome is packaged in a coat of identical protein subunits. The number of subunits and the lengths of the particles are determined by the length of the RNA. This flexibility is absent in icosahedral particles and constrains the genome size of nonsegmented viral genomes. CaMV DNA with extra DNA inserted loses the extra DNA rapidly during infections, presumably due to packaging constraints (Pennington and Melcher 1993).

In some multipartite viruses, the ability to interact synergistically with infection by another virus resides on a single component of the multipartite virus, as in the interaction between CPMV and CPSMV where RNA-1, but not RNA-2 containing, CPMV particles reduced infection by CPSMV (Eastwell and Kalmar 1997; Bruening et al. 2000).

If two viruses invade a stand of host plants by co-inoculation, evolution is accelerated. Viral fitness has been predicted to increase more quickly than after sequential inoculation (Miralles et al. 2001). Despite the demonstration that inter-strain recombination occurred between GFLV and ArMV, leading to emergence of a new strain, such recombination is not a frequent result (Vigne et al. 2009).

15.11 SEQUENCE CHANGE THROUGH INTERACTION

The above-described interactions between viruses do not result directly in any changes in the genomic sequence of nucleotides of the participating viral entities. There are, however, three ways in which genomic sequences have changed as a result of interaction. These are reassortment, allele loss, and recombination.

15.11.1 Reassortment

Reassortment is a documented outcome of coinfection by related multipartite viruses. It can be thought of as a type of genomic sex and is only possible for multisegmented viruses regardless of whether they are also multipartite. These interactions take place when two or more different viruses or strains of the same virus are present in the same tissues in the same organism. Phylogenetic analyses of the genomes of segmented plant viruses often show different evolutionary patterns for different segments. This suggests that reassortment is a common phenomenon in plant viruses. Reassortment has been reported to occur in RNA viruses (+ssRNA, -ssRNA, and dsRNA genomes) and also DNA viruses, both naturally as well as experimentally.

Plant viruses in the family *Geminiviridae* and *Nanoviridae* are multipartite single-stranded DNA viruses. DNA viruses in *Geminiviridae* are bipartite particularly in the genus *Begomovirus* while in *Nanoviridae*, the genus is subdivided into two genera: *Babuvirus* and *Nanovirus*. DNA viruses in *Babuvirus* and *Nanovirus* genera have six to twelve ssDNA molecules encapsidated into individual virions (Vetten et al. 2012). In addition, satDNA molecules (beta- and delta-satellites) are frequently associated with DNA virus infection. Reassortment events have been detected in ssDNA viruses belonging to genera *Babuvirus* and *Nanovirus* (Hu et al. 2007; Fu 2009; Hyder 2011; Stainton et al. 2012; Grigoras et al. 2014; Kumar et al. 2015) as well as begomoviruses (Kumar et al. 2015). Interestingly, with pairs of the bipartite geminiviruses, it has been shown that when both components of one virus are present, then the DNA-A component of another virus can be replicated (without its B-component) (Reddy et al. 2012). The relative frequencies of the multiple segments of a reassorting virus (gene copy number; Sicard et al. 2016) appear to be highly regulated. The mechanism(s) governing this gene copy number set point remain to be elucidated.

15.11.2 Recombination

Complementation, as mentioned above, can occur when a defective virus is co-inoculated with another virus defective in a different function than in which the first virus is defective. Originally, it was hoped that such complementation would allow easy introduction of foreign genetic material into plants by using one DNA as carrier of one necessary factor (for selection of transformed recipient plant cells) and a second bearing a second accessory factor and a payload. However, when distinct pairs of mutant CaMV DNAs (Melcher et al. 1986a) were used and the resulting virus particles characterized, it was determined that neither of the two original mutations was present in the progeny (Choe et al. 1985). Examination of nucleotide sequences of a variety of CaMV isolates suggested that recombination was not just a laboratory curiosity, but also had occurred during the divergence of strains of CaMV (Chenault and Melcher 1994). Mechanistic studies produced evidence that the predominant path for recombination was the reverse transcription of RNA into DNA (Grimsley et al. 1986; Vaden and Melcher 1990). This led to the concept that plant virus genomes can undergo recombination, a concept soon verified also for RNA viruses (Bujarski and Kaesberg 1986). Recombination observations complicated plans for plant genetic engineering using viruses.

Recombination is the most important possible consequence of coinfection by two or more viruses. It is possible between related genomes present in the same tissue. Homology is required. Recombination of double-stranded DNA in plant protoplasts requires a minimum of 0.5 kbp of similar sequence, and its frequency increases with increasing length of overlap (Puchta and Hohn 1991).

Perhaps the simplest example of recombination at work is the use of CaMV DNA constructs that should have, when introduced in pairs, the opportunity to generate circular viral DNA by recombination between duplicated sequences at the arms (Walden and Howell 1983). Central to many of these interactions are plant defense mechanisms that condition host plant resistance and susceptibility: initiation factors in translation, RNA silencing, molecular pathogen recognition patterns, various signaling pathways, and apoptosis.

In recent decades with the advent of increased rates of nucleotide sequencing and software analysis tools, new strains of viruses derived from recombination of pairs of old strains are increasingly being described. For example, Tomato yellow leaf curl virus (Anfoka et al. 2016), Plum pox virus (James et al. 2016), and other begomoviruses (Khan and Samad 2016). Of newly analyzed isolates of ASGV, 29% were identified as recombinant (Jo et al. 2016). A virus of azuki beans, BCMV-Az, has been shown to be a recombinant derived from a pair of BCMV strains (Li et al. 2016).

Recombination was implicated in the restoration of infectivity of an RNA3 deletion mutant of AMV. The restored RNA3 was suggested to have arisen by recombination events between RNAs 1 and 3 (Bujarski and Dzianott 1991). Further studies characterized multiple AMV genomes that resulted from recombination between P3 and CP deletion mutants (Vanderkuyl et al. 1991). RNA recombination was also implicated in other cases, for example in the generation of the TCM isolate of *Tobacco rattle virus* in which the RNA-2 apparently had acquired 3′ and 5′ ends from the RNA-2 of *Pea early browning virus* (Goulden et al. 1991). Begomoviruses from a pair of distinct host species are reported to parent the formation of TrYMV (Nascimento et al. 2016). Intra- and interspecies recombination events have been detected in ssDNA viruses belonging to genera *Babuvirus* and *Nanovirus* (Hu et al. 2007; Fu 2009; Hyder 2011; Stainton et al. 2012; Grigoras et al. 2014) as well as begomoviruses (Kumar et al. 2015). In general, for begomoviruses, the frequency of recombination events decreases as the parental sequence diversity increases (Padidam et al. 1999). Recombination and reassortment have also been reported to occur in satellites, such as the satellite RNA of GFLV (Cepin et al. 2016).

15.11.3 Allele Loss

In schemes to describe recombination in nuclear DNA, situations are created in which strand invasion around a sequence difference occurs. The mismatch is corrected by DNA repair systems. Such gene conversion was invoked to have occurred during the generation of infectious CaMV DNA from inocula containing pairs of mutant DNAs (Melcher et al. 1986a). This view was supported by placing pairs of CaMV DNAs tandemly in the plant chromosome (Gal et al. 1991). Infectious DNA and virions result from recombination between homologous sites in the viral genomes. These can also result from recombination between a viral gene introduced into the chromosome by agrobacterial transformation and viral DNA lacking the transgene. Both RNA and DNA recombination events were implicated (Gal et al. 1992). Changes suggestive of mismatch repair in the recombination region were detected.

15.12 CONCLUSION

The most important outcomes of research on interactions between viruses have been superinfection-exclusion, coat protein-mediated protection, and an understanding of virus evolution resulting from reassortment and recombination. Knowledge of the co-occurrence of multiple viruses in a variety of plants led to the investigation of the interactions and the identification of cross-protection (superinfection-exclusion) and the revelation that previously undescribed viruses interact synergistically with known viruses to create devastating diseases. Not only have studies of virus interactions in synergy and interference led to advances in the protection of crops against virus infection, but they have also provided tools for the molecular exploration of plant molecular biology. Both recombination and reassortment are driving forces in the evolution of the multipartite plant viruses which can create new strains, expand host ranges, and increase the virulence of these viruses.

ACKNOWLEDGMENTS

Ulrich Melcher acknowledges the support of the USDA National Institute of Food and Agriculture, Hatch OKL01789, and the Division of Agricultural Sciences and Natural Resources at Oklahoma State University. Akhtar Ali thanks the support of the Office of Research and Sponsored Programs, The University of Tulsa, Oklahoma.

REFERENCES

Aguilar I, Sanchez F, Ponz F (2000). Different Forms of Interference between Two Tobamoviruses in Two Different Hosts. *Plant Pathology* 49: 659–665.

Ameyaw GA, Domfeh O, Dzahini-Obiatey H, Ollennu LAA et al. (2016). Appraisal of Cocoa Swollen Shoot Virus (CSSV) Mild Isolates for Cross Protection of Cocoa against Severe Strains in Ghana. *Plant Disease* 100: 810–815.

Anfoka G, Al-Talb M, Haj-Ahmad F (2016). A New Isolate of Tomato Yellow Leaf Curl Axarquia Virus Associated with Tomato Yellow Leaf Curl Disease in Jordan. *Journal of Plant Pathology* 98: 145–149.

Atallah OO, Kang SH, El-Mohtar CA, Shilts T et al. (2016). A 5′-Proximal Region of the Citrus Tristeza Virus Genome Encoding Two Leader Proteases Is Involved in Virus Superinfection Exclusion. *Virology* 489: 108–115.

Bazzini AA, Asurmendi S, Hopp HE, Beachy RN (2006). Tobacco Mosaic Virus (TMV) and Potato Virus X (PVX) Coat Proteins Confer Heterologous Interference to PVX and TMV Infection, Respectively. *Journal of General Virology* 87: 1005–1012.

Bodaghi S, Mathews DM, Dodds JA (2004). Natural Incidence of Mixed Infections and Experimental Cross Protection between Two Genotypes of Tobacco Mild Green Mosaic Virus. *Phytopathology* 94: 1337–1341.

Bruening G, Buzayan JM, Ferreiro C, Lim W (2000). Evidence for Participation of RNA 1-Encoded Elicitor in Cowpea Mosaic Virus-Mediated Concurrent Protection. *Virology* 266: 299–309.

Bujarski JJ, Dzianott AM (1991). Generation and Analysis of Nonhomologous RNA-RNA Recombinants in Brome Mosaic-Virus: Sequence Complementarities at Crossover Sites. *Journal of Virology* 65: 4153–4159.

Bujarski JJ, Kaesberg P (1986). Genetic-Recombination between RNA Components of a Multipartite Plant-Virus. *Nature* 321: 528–531.

Burgyan J, Rubino L, Russo M (1991). Denovo Generation of Cymbidium Ringspot Virus Defective Interfering RNA. *Journal of General Virology* 72: 505–509.

Cassells AC, Herrick CC (1977). Cross Protection between Mild and Severe Strains of Tobacco Mosaic-Virus in Doubly Inoculated Tomato Plants. *Virology* 78: 253–260.

Cepin U, Gutierrez-Aguirre I, Ravnikar M, Pompe-Novak M (2016). Frequency of Occurrence and Genetic Variability of Grapevine Fanleaf Virus Satellite RNA. *Plant Pathology* 65: 510–520.

Chávez-Calvillo G, Contreras-Paredes CA, Mora-Macias J, Noa-Carrazana JC et al. (2016). Antagonism or Synergism between Papaya Ringspot Virus and Papaya Mosaic Virus in Carica Papaya Is Determined by Their Order of Infection. *Virology* 489: 179–191.

Chen YJ, Zhang J, Liu J, Deng XG et al. (2014). The Capsid Protein P38 of Turnip Crinkle Virus Is Associated with the Suppression of Cucumber Mosaic Virus in Arabidopsis Thaliana Co-Infected with Cucumber Mosaic Virus and Turnip Crinkle Virus. *Virology* 462: 71–80.

Chenault KD, Melcher U (1994). Phylogenetic Relationships Reveal Recombination among Isolates of Cauliflower Mosaic Virus. *Journal of Molecular Evolution* 39: 496–505.

Chewachong GM, Miller SA, Blakeslee JJ, Francis DM et al. (2015). Generation of an Attenuated, Cross-Protective Pepino Mosaic Virus Variant through Alignment-Guided Mutagenesis of the Viral Capsid Protein. *Phytopathology* 105: 126–134.

Choe IS, Melcher U, Richards K, Lebeurier G et al. (1985). Recombination between Mutant Cauliflower Mosaic-Virus DNAs. *Plant Molecular Biology* 5: 281–289.

Cillo F, Pasciuto MM, De Giovanni C, Finetti-Sialer MM et al. (2007). Response of Tomato and Its Wild Relatives in the Genus Solanum to Cucumber Mosaic Virus and Satellite RNA Combinations. *Journal of General Virology* 88: 3166–3176.

Cook G, van Vuuren SP, Breytenbach JHJ, Steyn C et al. (2016). Characterization of Citrus Tristeza Virus Single-Variant Sources in Grapefruit in Greenhouse and Field Trials. *Plant Disease* 100: 2251–2256.

Culver JN (1996). Tobamovirus Cross Protection Using a Potexvirus Vector. *Virology* 226: 228–235.

Ding XS, Liu JZ, Cheng NH, Folimonov A et al. (2004). The Tobacco Mosaic Virus 126-kDa Protein Associated with Virus Replication and Movement Suppresses RNA Silencing. *Molecular Plant-Microbe Interactions* 17: 583–592.

Donne J (1624). Meditation 17 from *Devotions upon Emergent Occasioms.*

Donson J, Kearney CM, Turpen TH, Khan IA et al. (1993). Broad Resistance to Tobamoviruses Is Mediated by a Modified Tobacco Mosaic-Virus Replicase Transgene. *Molecular Plant-Microbe Interactions* 6: 635–642.

Eastwell KC, Kalmar GB (1997). Characterizing the Interference between Two Comoviruses in Cowpea. *Journal of the American Society for Horticultural Science* 122: 163–168.

Fabre F, Chadoeuf J, Costa C, Lecoq H et al. (2010). Asymmetrical Over-Infection as a Process of Plant Virus Emergence. *Journal of Theoretical Biology* 265: 377–388.

Folimonova SY (2012). Superinfection Exclusion Is an Active Virus-Controlled Function That Requires a Specific Viral Protein. *Journal of Virology* 86: 5554–5561.

Folimonova SY (2013). Developing an Understanding of Cross-Protection by Citrus Tristeza Virus. *Frontiers in Microbiology* 4.

Folimonova SY, Harper SJ, Leonard MT, Triplett EW et al. (2014). Superinfection Exclusion by Citrus Tristeza Virus Does Not Correlate with the Production of Viral Small RNAs. *Virology* 468: 462–471.

Folimonova SY, Robertson CJ, Shilts T, Folimonov AS et al. (2010). Infection with Strains of Citrus Tristeza Virus Does Not Exclude Superinfection by Other Strains of the Virus. *Journal of Virology* 84: 1314–1325.

Fu HC, Hu JM, Hung TH, Su HJ, Yeh HH (2009). Unusual Events Involved in Banana Bunchy Top Virus Strain Evolution. *Phytopathology* 99: 812–822.

Fulton RW (1986). Practices and Precautions in the Use of Cross Protection for Plant-Virus Disease-Control. *Annual Review of Phytopathology* 24: 67–81.

Gal S, Pisan B, Hohn T, Grimsley N et al. (1991). Genomic Homologous Recombination in Planta. *Embo Journal* 10: 1571–1578.

Gal S, Pisan B, Hohn T, Grimsley N et al. (1992). Agroinfection of Transgenic Plants Leads to Viable Cauliflower Mosaic-Virus by Intermolecular Recombination. *Virology* 187: 525–533.

Goregaoker SP, Eckhardt LG, Culver JN (2000). Tobacco Mosaic Virus Replicase-Mediated Cross-Protection: Contributions of RNA and Protein-Derived Mechanisms. *Virology* 273: 267–275.

Goulden MG, Lomonossoff GP, Wood KR, Davies JW (1991). A Model for the Generation of Tobacco Rattle Virus (TRV) Anomalous Isolates: Pea Early Browning Virus RNA-2 Acquires TRV Sequences from Both RNA-1 and RNA-2. *Journal of General Virology* 72: 1751–1754.

Grigoras I, Ginzo AID, Martin DP, Varsani A et al. (2014). Genome Diversity and Evidence of Recombination and Reassortment in Nanoviruses from Europe. *Journal of General Virology* 95: 1178–1191.

Grimsley N, Hohn T, Hohn B (1986). Recombination in a Plant-Virus: Template Switching in Cauliflower Mosaic-Virus. *Embo Journal* 5: 641–646.

Grupa A, Syller J (2016). Cross-Protection between a Naturally Occurring Mild Isolate of Potato Virus M (PVM) and a More Virulent Isolate in Datura Metel Plants. *Journal of Phytopathology* 164: 69–73.

Hackland AF, Rybicki EP, Thomson JA (1994). Coat Protein-Mediated Resistance in Transgenic Plants. *Archives of Virology* 139: 1–22.

Hall GS, Little DP (2013). Within-Host Competition between Barley Yellow Dwarf-PAV and -PAS. *Virus Research* 174: 148–151.

Hall JS, French R, Hein GL, Morris TJ et al. (2001). Three Distinct Mechanisms Facilitate Genetic Isolation of Sympatric Wheat Streak Mosaic Virus Lineages. *Virology* 282: 230–236.

Hanssen IM, Gutierrez-Aguirre I, Paeleman A, Goen K et al. (2010). Cross-Protection or Enhanced Symptom Display in Greenhouse Tomato Co-Infected with Different Pepino Mosaic Virus Isolates. *Plant Pathology* 59: 13–21.

Hibino H, Roechan M, Sudarisman S (1978). Association of 2 Types of Virus-Particles with Penyakit Habang (Tungro Disease) of Rice in Indonesia. *Phytopathology* 68: 1412–1416.

Hii G, Pennington R, Hartson S, Taylor CD et al. (2002). Isolate-Specific Synergy in Disease Symptoms between Cauliflower Mosaic and Turnip Vein-Clearing Viruses. *Archives of Virology* 147: 1371–1384.

Hu JM, Lin CH, Su HJ, Yeh HH (2007). Reassortment and Concerted Evolution in Banana Bunchy Top Virus Genomes. *Journal of Virology* 81: 1746–1761.

Hyder MZ, Shah SH, Hameed S, Naqvi SM (2011). Evidence of Recombination in the Banana Bunchy Top Virus Genome. *Infection, Genetics and Evolution* 11: 1293–1300.

James D, Sanderson D, Varga A, Sheveleva A et al. (2016). Genome Sequence Analysis of New Isolates of the Winona Strain of Plum Pox Virus and the First Definitive Evidence of Intrastrain Recombination Events. *Phytopathology* 106: 407–416.

Jo Y, Choi H, Kim SM, Kim SL et al. (2016). Integrated Analyses Using RNA-Seq Data Reveal Viral Genomes, Single Nucleotide Variations, the Phylogenetic Relationship, and Recombination for Apple Stem Grooving Virus. *BMC Genomics* 17.

Julve JM, Gandia A, Fernandez-del-Carmen A, Sarrion-Perdigones A et al. (2013). A Coat-Independent Superinfection Exclusion Rapidly Imposed in Nicotiana Benthamiana Cells by Tobacco Mosaic Virus Is Not Prevented by Depletion of the Movement Protein. *Plant Molecular Biology* 81: 553–564.

Kajihara H, Kameya-Iwaki M, Oonaga M, Kimura I et al. (2008). Field Studies on Cross-Protection against Japanese Yam Mosaic Virus in Chinese Yam (Dioscorea Opposita) with an Attenuated Strain of the Virus. *Journal of Phytopathology* 156: 75–78.

Khan A, Samad A (2016). Molecular and Biological Characterization of Begomoviruses Infecting Andrographis Paniculata and Their Genetic Recombination Lineage. *European Journal of Plant Pathology* 146: 177–189.

Khurana SMP, Singh MN (1988). Yield Loss Potential of Potato Viruses X and Y in Indian Potatoes. *Journal of the Indian Potato Association* 15: 27–29.

Konakalla NC, Kaldis A, Berbati M, Masarapu H et al. (2016). Exogenous Application of Double-Stranded RNA Molecules from TMV P126 and CP Genes Confers Resistance against TMV in Tobacco. *Planta* 244: 961–969.

Kondo T, Kogawa K, Ito K (2015). Evaluation of Cross Protection by an Attenuated Strain of Chinese Yam Necrotic Mosaic Virus in Chinese Yam. *Journal of General Plant Pathology* 81: 42–48.

Kosaka Y, Fukunishi T (1994). Application of Cross-Protection to the Control of Black Soybean Mosaic Disease. *Plant Disease* 78: 339–341.

Kumar RV, Singh AK, Singh AK, Yadav T et al. (2015). Complexity of Begomovirus and Betasatellite Populations Associated with Chilli Leaf Curl Disease in India. *Journal of General Virology* 96: 3157–3172.

Kung YJ, Lin PC, Yeh SD, Hong SF et al. (2014). Genetic Analyses of the FRNK Motif Function of Turnip Mosaic Virus Uncover Multiple and Potentially Interactive Pathways of Cross-Protection. *Molecular Plant-Microbe Interactions* 27: 944–955.

Kurath G, Dodds JA (1994). Satellite Tobacco Mosaic-Virus Sequence Variants with Only 5 Nucleotide Differences Can Interfere with Each Other in a Cross Protection-Like Phenomenon in Plants. *Virology* 202: 1065–1069.

Lartey RT, Lane LC, Melcher U (1994). Electron Microscopic and Molecular Characterization of Turnip Vein-Clearing Virus. *Archives of Virology* 138: 287–298.

Leclerc D, Abouhaidar MG (1995). Transgenic Tobacco Plants Expressing a Truncated Form of the PAMV Capsid Protein (CP) Gene Show CP-Mediated Resistance to Potato Aucuba Mosaic-Virus. *Molecular Plant-Microbe Interactions* 8: 58–65.

Lecoq H, Lemaire JM, Wipfscheibel C (1991). Control of Zucchini Yellow Mosaic-Virus in Squash by Cross Protection. *Plant Disease* 75: 208–211.

Legg JP, Owor B, Sseruwagi P, Ndunguru J (2006). Cassava Mosaic Virus Disease in East and Central Africa: Epidemiology and Management of a Regional Pandemic. *Advances in Virus Research* 67: 355–418.

Li Y, Cao Y, Fan Z, Wan P (2016). Identification of a Naturally Occurring Bean Common Mosaic Virus Recombinant Isolate Infecting Azuki Bean. *Journal of Plant Pathology* 98: 129–133.

Li Y, Zhang J, Zhao F, Ren H et al. (2016). The Interaction between Turnip Crinkle Virus p38 and Cucumber Mosaic Virus 2b and Its Critical Domains. *Virus Research* 222: 94–105.

Lin SS, Wu HW, Jan FJ, Hou RF et al. (2007). Modifications of the Helper Component-Protease of Zucchini Yellow Mosaic Virus for Generation of Attenuated Mutants for Cross Protection against Severe Infection. *Phytopathology* 97: 287–296.

MacDiarmid R (2005). RNA Silencing in Productive Virus Infections. *Annual Review of Phytopathology* 43: 523–544.

Mahmood T, Rush CM (1999). Evidence of Cross-Protection between Beet Soilborne Mosaic Virus and Beet Necrotic Yellow Vein Virus in Sugar Beet. *Plant Disease* 83: 521–526.

Mandadi KK, Scholthof KBG (2012). Characterization of a Viral Synergism in the Monocot *Brachypodium Distachyon* Reveals Distinctly Altered Host Molecular Processes Associated with Disease. *Plant Physiology* 160: 1432–1452.

Marchoux G (1988). Interactions between Viruses in Plants. 2. Synergy, Complementation, Assistance. *Agronomie* 8: 471–490.
Mascia T, Gallitelli D (2016). Synergies and Antagonisms in Virus Interactions. *Plant Science* 252: 176–192.
McGarvey PB, Montasser MS, Kaper JM (1994). Transgenic Tomato Plants Expressing Satellite RNA Are Tolerant to Some Strains of Cucumber Mosaic-Virus. *Journal of the American Society for Horticultural Science* 119: 642–647.
McKinney HH (1929). Mosaic Diseases in the Canary Islands, West Africa, and Gibraltar. *Journal of Agricultural Research* 39: 556–578.
Melcher U (2000). The 30k Superfamily of Viral Movement Proteins. *Journal of General Virology* 81: 257–266.
Melcher U, Choe IS, Lebeurier G, Richards K et al. (1986a). Selective Allele Loss and Interference between Cauliflower Mosaic Virus DNAs. *Molecular and General Genetics* 203: 230–236.
Melcher U, Gardner CO Jr, Essenberg RC (1981). Clones of Cauliflower Mosaic Virus Identified by Molecular Hybrdization in Turnip Leaves. *Plant Molecular Biology* 1: 63–73.
Melcher U, Steffens DL, Lyttle DJ, Lebeurier G et al. (1986b). Infectious and Non-Infectious Mutants of Cauliflower Mosaic Virus DNA. *Journal of General Virology* 67: 1491–1498.
Miralles R, Ferrer R, Solé RV, Moya A et al. (2001). Multiple Infection Dynamics Has Pronounced Effects on the Fitness of RNA Viruses. *Journal of Evolutionary Biology* 14: 654–662.
Mondal S, Lin YH, Carroll J, Wenninger EJ et al. (2017). Potato Virus Y Transmission Efficiency from Potato Infected with Single or Multiple Virus Strains. *Phytopathology* 107: 491–498.
Montasser MS, Tousignant ME, Kaper JM (1998). Viral Satellite RNAs for the Prevention of Cucumber Mosaic Virus (CMV) Disease in Field-Grown Pepper and Melon Plants. *Plant Disease* 82: 1298–1303.
Morales K, Luis Zambrano J, Stewart LR (2014). Co-Infection and Disease Severity of Ohio Maize Dwarf Mosaic Virus and Maize Chlorotic Dwarf Virus Strains. *Plant Disease* 98: 1661–1665.
Morange M (2012). What History Tells Us XXIX. Transfers from Plant Biology: From Cross Protection to RNA Interference and DNA Vaccination. *Journal of Biosciences* 37: 949–952.
Murant AF (1990). Dependence of Groundnut Rosette Virus on Its Satellite RNA as Well as on Groundnut Rosette Assistor Luteovirus for Transmission by Aphis Craccivora. *Journal of General Virology* 71: 2163–2166.
Muthukumar V, Melcher U, Pierce ML, Wiley GB et al. (2008). Non-Cultivated Plants of the Tallgrass Prairie Preserve of Northeastern Oklahoma Frequently Contain Virus-Like Sequences in Particulate Fractions. *Virus Research* 141: 169–173.
Nakazono-Nagaoka E, Takahashi T, Shimizu T, Kosaka Y et al. (2009). Cross-Protection against Bean Yellow Mosaic Virus (BYMV) and Clover Yellow Vein Virus by Attenuated BYMV Isolate M11. *Phytopathology* 99: 251–257.
Nascimento LD, Silva SJC, Sobrinho RR, Ferro MMM et al. (2016). Complete Nucleotide Sequence of a New Begomovirus Infecting a Malvaceous Weed in Brazil. *Archives of Virology* 161: 1735–1738.
Neofytou G, Kyrychko YN, Blyuss KB (2016). Mathematical Model of Plant-Virus Interactions Mediated by RNA Interference. *Journal of Theoretical Biology* 403: 129–142.
Niblett CL, Claflin LE (1978). Corn Lethal Necrosis: New Virus-Disease of Corn in Kansas. *Plant Disease Reporter* 62: 15–19.
Nishiguchi M, Kobayashi K (2011). Attenuated Plant Viruses: Preventing Virus Diseases and Understanding the Molecular Mechanism. *Journal of General Plant Pathology* 77: 221–229.
Ogai R, Kanda-Hojo A, Tsuda S (2013). An Attenuated Isolate of Pepper Mild Mottle Virus for Cross Protection of Cultivated Green Pepper (Capsicum Annuum L.) Carrying the L-3 Resistance Gene. *Crop Protection* 54: 29–34.
Owor B, Legg JP, Okao-Okuja G, Obonyo R et al. (2004). Field Studies of Cross Protection with Cassava Mosaic Geminiviruses in Uganda. *Journal of Phytopathology* 152: 243–249.
Padidam M, Sawyer S, Fauquet CM (1999). Possible Emergence of New Geminiviruses by Frequent Recombination. *Virology* 265: 218–225.
Palukaitis P (2011). The Road to RNA Silencing is Paved with Plant-Virus Interactions. *Plant Pathology Journal* 27: 197–206.
Pennazio S, Roggero P, Conti M (2001). A History of Plant Virology. Cross Protection. *Microbiologica* 24: 99–114.
Pennington R, Melcher U (1993). In Planta Deletion of DNA Inserts from the Large Intergenic Region of Cauliflower Mosaic Virus DNA. *Virology* 192: 188–196.
Perring TM, Farrar CA, Blua MJ, Wang HL et al. (1995). Cross Protection of Cantaloupe with a Mild Strain of Zucchini Yellow Mosaic Virus: Effectiveness and Application. *Crop Protection* 14: 601–606.

Powell CA, Pelosi RR, Rundell PA, Cohen M (2003). Breakdown of Cross-Protection of Grapefruit from Decline-Inducing Isolates of Citrus Tristeza Virus Following Introduction of the Brown Citrus Aphid. *Plant Disease* 87: 1116–1118.

Puchta H, Hohn B (1991). A Transient Assay in Plant-Cells Reveals a Positive Correlation between Extra-chromosomal Recombination Rates and Length of Homologous Overlap. *Nucleic Acids Research* 19: 2693–2700.

Qu F, Morris TJ (2005). Suppressors of RNA Silencing Encoded by Plant Viruses and Their Role in Viral Infections. *FEBS Letters* 579: 5958–5964.

Quemada HD, Gonsalves D, Slightom JL (1991). Expression of Coat Protein Gene from Cucumber Mosaic-Virus Strain-C in Tobacco: Protection against Infections by CMV Strains Transmitted Mechanically or by Aphids. *Phytopathology* 81: 794–802.

Reddy RVC, Dong W, Njock T, Rey MEC et al. (2012). Molecular Interaction between Two Cassava Geminiviruses Exhibiting Cross-Protection. *Virus Research* 163: 169–177.

Rezende JAM, Urban L, Sherwood JL, Melcher U (1992). Host Effect on Cross Protection between Two Strains of Tobacco Mosaic Virus. *Journal of Phytopathology* 136: 147–153.

Rochow WF, Ross AF (1955). Virus Multiplication in Plants Doubly Infected by Potato Viruses X and Y. *Virology* 1: 10–27.

Sambade A, Rubio L, Garnsey SM, Costa N et al. (2002). Comparison of Viral RNA Populations of Pathogenically Distinct Isolates of Citrus Tristeza Virus: Application to Monitoring Cross-Protection. *Plant Pathology* 51: 257–265.

Sanders PR, Sammons B, Kaniewski W, Haley L et al. (1992). Field-Resistance of Transgenic Tomatoes Expressing the Tobacco Mosaic-Virus or Tomato Mosaic-Virus Coat Protein Genes. *Phytopathology* 82: 683–690.

Sansini Freitas DM, Marques Rezende JA (2008). Protection between Strains of Papaya Ringspot Virus - Type W in Zucchini Squash Involves Competition for Viral Replication Sites. *Scientia Agricola* 65: 183–189.

Sheen TF, Wang HL, Wang DN (1998). Control of Papaya Ringspot Virus by Cross Protection and Cultivation Techniques. *Journal of the Japanese Society for Horticultural Science* 67: 1232–1235.

Shi XM, Miller H, Verchot J, Carrington JC et al. (1997). Mutations in the Region Encoding the Central Domain of Helper Component-Proteinase (HC-Pro) Eliminate Potato Virus X/Potyviral Synergism. *Virology* 231: 35–42.

Sicard A, Michalakis Y, Gutiérrez S, Blanc S (2016). The Strange Lifestyle of Multipartite Viruses. *PLoS Pathogens* 12: e1005819.

Singh M, Singh RP (1995). Host-Dependent Cross-Protection between PVY^N, PVY^O, and PVA in Potato Cultivars and Solanum-Brachycarpum. *Canadian Journal of Plant Pathology-Revue Canadienne De Phytopathologie* 17: 82–86.

Stainton D, Kraberger S, Walters M, Wiltshire EJ et al. (2012). Evidence of Inter-Component Recombination, Intra-Component Recombination and Reassortment in Banana Bunchy Top Virus. *Journal of General Virology* 93: 1103–1119.

Syller J (2012). Facilitative and Antagonistic Interactions between Plant Viruses in Mixed Infections. *Molecular Plant Pathology* 13: 204–216.

Syller J, Grupa A (2016). Antagonistic Within-Host Interactions between Plant Viruses: Molecular Basis and Impact on Viral and Host Fitness. *Molecular Plant Pathology* 17: 769–782.

Takeshita M, Koizumi E, Noguchi M, Sueda K et al. (2012). Infection Dynamics in Viral Spread and Interference under the Synergism between Cucumber Mosaic Virus and Turnip Mosaic Virus. *Molecular Plant-Microbe Interactions* 25: 18–27.

Taliansky ME, Ryabov EV, Robinson DJ, Palukaitis P (1998). Tomato Cell Death Mediated by Complementary Plant Viral Satellite RNA Sequences. *Molecular Plant-Microbe Interactions* 11: 1214–1222.

Tian Z, Qiu J, Yu J, Han C et al. (2009). Competition between Cucumber Mosaic Virus Subgroup I and II Isolates in Tobacco. *Journal of Phytopathology* 157: 457–464.

Tsuda S, Kubota K, Kanda A, Ohki T et al. (2007). Pathogenicity of Pepper Mild Mottle Virus is Controlled by the RNA Silencing Suppression Activity of Its Replication Protein but Not the Viral Accumulation. *Phytopathology* 97: 412–420.

Untiveros M, Fuentes S, Salazar LF (2007). Synergistic Interaction of Sweet Potato Chlorotic Stunt Virus (Crinivirus) with Carla-, Cucumo-, Ipomo-, and Potyviruses Infecting Sweet Potato. *Plant Disease* 91: 669–676.

Urban LA, Sherwood JL, Rezende JAM, Melcher U (1990). Examination of Mechanisms of Cross Protection with Non-Transgenic Plants. In *Recognition and Response in Plant-Virus Interactions*. RSS Fraser (ed.). Berlin, Springer Verlag: 415–426.

Vaden VR, Melcher U (1990). Recombination Sites in Cauliflower Mosaic-Virus DNAs: Implications for Mechanisms of Recombination. *Virology* 177: 717–726.
Valkonen JPT, Rajamaki ML, Kekarainen T (2002). Mapping of Viral Genomic Regions Important in Cross-Protection between Strains of a Potyvirus. *Molecular Plant-Microbe Interactions* 15: 683–692.
Vance VB (1991). Replication of Potato Virus X RNA Is Altered in Coinfections with Potato Virus Y. *Virology* 182: 486–494.
Vanderkuyl AC, Neeleman L, Bol JF (1991). Complementation and Recombination between Alfalfa Mosaic-Virus RNA3 Mutants in Tobacco Plants. *Virology* 183: 731–738.
Vetten HJ, Dale JL, Grigoras I, Gronenborn B et al. (2012). *Family Nanoviridae. Virus Taxonomy. Ninth Report of the International Committee on Taxonomy of Viruses*. MJ Adams, AMQ King, EC Carstens and EJ Lefkowitz. London, Elsevier/Academic Press: 395–404.
Vigne E, Marmonier A, Komar V, Lemaire O et al. (2009). Genetic Structure and Variability of Virus Populations in Cross-Protected Grapevines Superinfected by Grapevine Fanleaf Virus. *Virus Research* 144: 154–162.
Walden RM, Howell SH (1983). Uncut Recombinant Plasmids Bearing Nested Cauliflower Mosaic-Virus Genomes Infect Plants by Intragenomic Recombination. *Plant Molecular Biology* 2: 27–31.
Wang HL, Gonsalves D, Provvidenti R, Lecoq HL (1991). Effectiveness of Cross Protection by a Mild Strain of Zucchini Yellow Mosaic-Virus in Cucumber, Melon, and Squash. *Plant Disease* 75: 203–207.
Wang YZ, Gaba V, Yang J, Palukaitis P et al. (2002). Characterization of Synergy between Cucumber Mosaic Virus and Potyviruses in Cucurbit Hosts. *Phytopathology* 92: 51–58.
Wen Y, Lim GXY, Wong SM (2013). Profiling of Genes Related to Cross Protection and Competition for *NbTOM1* by HLSV and TMV. *Plos One* 8: 1–12.
Wren JD, Roossinck MJ, Nelson RS, Scheets K et al. (2006). Plant Virus Biodiversity and Ecology. *PLoS Biology* 4: e80.
Xi D, Feng H, Lan L, Du J et al. (2007). Characterization of Synergy between Cucumber Mosaic Virus and Tobacco Necrosis Virus in Nicotiana Benthamiana. *Journal of Phytopathology* 155: 570–573.
Xi D, Yang H, Jiang Y, Xu M et al. (2010). Interference between Tobacco Necrosis Virus and Turnip Crinkle Virus in Nicotiana Benthamiana. *Journal of Phytopathology* 158: 263–269.
Yamasaki S, Sakai J, Kamisoyama S, Goto H et al. (2009). Control of Russet Crack Disease in Sweetpotato Plants Using a Protective Mild Strain of Sweet Potato Feathery Mottle Virus. *Plant Disease* 93: 190–194.
Yang H, Wang S, Xi D, Yuan S et al. (2010). Interaction between Cucumber Mosaic Virus and Turnip Crinkle Virus in Arabidopsis Thaliana. *Journal of Phytopathology* 158: 833–836.
You BJ, Chiang CH, Chen LF, Su WC et al. (2005). Engineered Mild Strains of Papaya Ringspot Virus for Broader Cross Protection in Cucurbits. *Phytopathology* 95: 533–540.
Zhang XS, Holt J, Colvin J (2000). Mathematical Models of Host Plant Infection by Helper-Dependent Virus Complexes: Why Are Helper Viruses Always Avirulent? *Phytopathology* 90: 85–93.
Zhang XS, Melcher U (1989). Competition between Isolates and Variants of Cauliflower Mosaic Virus in Infected Turnip Plants. *Journal of General Virology* 70: 3427–3437.
Ziebell H, Carr JP (2010). Cross-Protection: A Century of Mystery. *Natural and Engineered Resistance to Plant Viruses, Volume 76* Part II. JP Carr and G Loebenstein. London, Elsevier/Academic Press: 211–264.
Ziebell H, Payne T, Berry JO, Walsh JA et al. (2007). A Cucumber Mosaic Virus Mutant Lacking the 2b Counter-Defence Protein Gene Provides Protection against Wild-Type Strains. *Journal of General Virology* 88: 2862–2871.

16 Viruses and Plant Development

Flora Sánchez and Fernando Ponz

CONTENTS

16.1 INTRODUCTION

As obligate intracellular parasites, viruses rely on their host cellular components to complete their full cycle of own gene expression, genome replication, accumulation, encapsidation, movement, and transport to another cell in the same or in another host. The depletion of cellular factors exploited for viral purposes, the production of viral proteins in the cell, and the combined effects of these two processes on the host metabolism frequently have the consequence of host disease, which can take many forms and be expressed through different symptoms induced by the viral infection. In the case of plants, viral disease symptoms usually affect global plant size (dwarfism), photosynthetic ability (mottles and mosaics, yellows), and photosynthate delivery (vein yellows). In comparison with our knowledge about virus replication cycles and viral molecular biology in general, the mechanisms underlying plant viral diseases are much less well known (Pallás and García, 2016). In addition, for many years, the main focus of studies on the host side of plant/virus interactions has been that of resistance to the pathogen, in close connection with breeding programs for cultivated plant species (Palloix and Ordon, 2011). This has led to a present situation in which many studies concerning the plant response to the infection are biased toward the defense response, a response that normally takes place even in susceptible hosts. However, some of the effects of viral infections on the biology of the infected plant are not obviously connected to a defense reaction; rather, they are just the consequence of the infection by a viral pathogen.

Although not abundant, good studies can be found in the literature about impacts of viruses on plant physiology and plant metabolism. These issues have been recently reviewed by several authors (Pallás and García, 2016; Culver and Padmanabhan, 2007). In general, there is increasingly precise information about the viral determinants of the different disease symptoms affecting physiology and metabolism. The plant side has received less attention comparatively, possibly reflecting the fact that the majority of scientists studying these aspects are virologists. As to the viral determinants, it is not really possible to depict generalities about the types of viral proteins more directly involved, since many different ones have been pointed out as mediators of certain specific effects (García and Pallás, 2015).

Even taking into account the limitations of this area of research, it can be said that it has generously shown its potential as a source of extremely useful tools for plant biology studies. Two major examples can be sufficient to illustrate the benefits to the field of plant physiology and plant molecular biology. One is the macromolecular transport within and between plant cells. Viruses must move their genomes from one infected cell to the next to propagate the infection, and for this purpose they have developed proteins with this specific role. The so-called viral movement proteins (MPs) mediating short-distance cell-to-cell movement of the viral infections have been extensively used in studies of macromolecular transport between cells. Many MPs have the ability to increase the size exclusion limit of plasmodesmata, so they have been used to study the macromolecular transport process (Heinlein, 2015). The other plant research area in which viral proteins have proven their value is gene silencing. The phenomenon was already approached in connection with viral studies from the very beginning (Ratcliff et al., 1997), and the subsequent discovery that viruses avoid or fight the silencing defensive mechanism through specific silencing suppressors encoded in their genomes allowed the exploitation of these proteins in the characterization of the process, thus facilitating enormously the progress in its understanding (Csorba et al., 2015). These two instances exemplify the notion that studying plant/virus interactions in their intimacy not only pays off in viral studies, but also in plant biology ones.

In this context, an aspect of plant/virus interactions which has not received a lot of attention so far is the effect of viral infections on specific plant developmental traits. Globally considered, most typical symptoms of viral infections can in fact be seen as impacts on development, since aspects like dwarfisms or diminished flower and seed production are nothing but developmental defects. However, if one pays attention to particular symptoms involving specific alterations of the plant developmental plan, studies are scarce. There is little doubt that focusing research from this perspective will be as beneficial to plant development studies as it has been to other areas of plant biology, as shown above. In fact, virus-induced alterations of specific developmental traits should be complementary to the classical approach of plant genetic mutants. Thus, opposed to the more permanent and stable nature of mutations, the alterations induced by viruses are dynamic and can evolve over time as the infection progresses, potentially providing a new view to the alteration and its control. Moreover, development is a sensitive matter for the plant life cycle and some of the genes involved in its control may not be so amenable to mutation. Virus infections affecting development can be started at will in the development program, potentially providing the opportunity to study processes not so easily addressable with mutants.

This approach should not only benefit plant studies, but also virus-related ones. As virus infections can affect plant development, the stage of plant development affects the outcome of the infection. Examples that infections are different depending on the time when they are started can be found in old literature, showing that the effects can be quite dramatic, even raising the notion of resistance to infection. How the developmental stage of the plant can be so influential to the virus life cycle is an issue far from being well understood, but some clues are available. The characterization of these mutual effects may well have the indirect effect of providing new, still unexplored, biological venues for resistance to virus infections.

This chapter provides a brief panorama of the studies linking virus infections with development-related plant traits. It is necessarily of a fragmented nature, since no systematic studies on the issue have been really approached. One of its aims is to promote them.

16.2 VIRUSES AND VIRAL ELEMENTS AFFECTING PLANT DEVELOPMENT

Not only virus infections have an effect on plant development, but also some of their individual elements. A consequence of the extensive studies on viral genomes to study functions of viral genes was the identification of many of these functions. A typical approach in this type of studies is the subsequent production of the specific viral products in transgenic plants. The pioneer work of Roger Beachy expressing the coat protein of *Tobacco mosaic virus* (TMV) in transgenic tobacco led to the discovery of the virus-derived strong transgenic resistance to the homologous virus and closely related ones (Power-Abel et al., 1986). Similar approaches with other viral-encoded proteins uncovered their involvement in the alteration of plant developmental traits. Among these, two main types of viral proteins have been studied in this context: the viral suppresors of RNA silencing (VSRs) and the previously alluded to MPs.

16.2.1 VSRs and Plant Development

The so-called small RNAs (smRNAs) are major regulators of plant development (Borges and Martienssen, 2015; Chen, 2009). There are several classes of these smRNAs described in plants, such as microRNAs (miRNAs), trans-acting small interfering RNAs (ta-siRNAs), natural-antisense RNAs (nat-siRNAs), repeat-associated siRNAs (ra-siRNAs), viral siRNAs (vsiRNAs), and virus-activated siRNAs (vasiRNAs). Each of these has its own specifity in terms of origin, biosynthesis, or mode of action. However, they all share some common features. Either encoded by the plant genome or originated from a viral pathogen, the generation of these RNAs involves certain dedicated enzymatic activities. A larger precursor RNA which must have at least certain regions in a double-stranded form (dsRNA) is cleaved by an RNase III-like activity called Dicer-like (DCL) in plants. Products of the cleavage are incorporated into Argonaute (Ago) proteins, which act as effectors of an RNA-induced silencing complex (RISC) to produce post-transcriptional silencing (PTGS), or an RNA-induced transcriptional silencing complex (RITS), which mediates transcriptional silencing (TGS). The whole process of RNA-induced silencing in plants has been reviewed in recent excellent reviews (Borges and Martienssen, 2015; Martínez de Alba et al., 2013; Weiberg et al., 2015).

Among the many biological roles of RNA-induced silencing in plants is antiviral defense, mostly in the form of PTGS. The replication process of RNA viruses (the vast majority of plant viruses) implies a step involving the generation of a dsRNA, which can trigger the process against the viral genome. In the case of DNA viruses, overlapping bidirectional read-through transcripts or highly structured viral transcripts may play the triggering role. As expected, within the never-ending battle between hosts and parasites viruses have developed mechanisms against this potent plant defense route. The most relevant of them is the incorporation of genes encoding VSRs into their genomes. Their roles, mechanistic details, and biotechnological applications have been thoroughly reviewed recently (Csorba et al., 2015). These proteins form a family, but only from the standpoint of its function, i.e., interfering with the silencing process, since they do not display any obvious sequence homology or a common way of action. Conversely, different viruses have very different VSRs, altogether acting at almost any step in the complex silencing process.

Soon after the discovery of the VSRs, they were stably incorporated through transgenesis into the genomes of several plant species, to facilitate their study (Chapman et al., 2004; Chellappan et al., 2005; Chen et al., 2004; Dunoyer et al., 2004; Kasschau et al., 2003). This approach uncovered their implication in the symptoms of virus-induced diseases affecting developmental traits. Aspects such as flower defects, large degree of or complete sterility, leaf shape defects, cell size, etc. were reported in association with transgenic VSR expression. Considering the extensive involvement of the different smRNAs in the control and regulation of the plant developmental plan, it is not surprising that these defects appeared in the different transgenic lines. A first notion put forward after this finding was that symptoms of virus infections affecting development were largely or exclusively due to the fact that viruses encode VSRs, which upon expression from the viral genome would alter the developmental program through their effect on smRNAs. However, this straightforward view was soon questioned as the only reason for the alterations. On one hand, some transgenically expressed VSRs do not induce developmental symptoms (Dunoyer et al., 2004; Chapman et al., 2004), and on the other hand some complementary results directly question this view. For example, transgenic expression of DCL1 in Arabidopsis could alleviate the developmental defects found in transgenic plants expressing *Turnip mosaic virus* (TuMV) VSR, but it did not have an effect on the defects induced in smRNA pathways, suggesting that not only these pathways are responsible for the induced developmental defects (Mlotshwa et al., 2005). Furthermore, when different TuMV strains were used to analyze the developmental defects induced, it was found that another viral protein (P3) is a more determinant factor than the VSR (Sánchez et al., 2015). These results are discussed further in the next section.

Another important consideration in the use of transgenic lines is that transgenic gene expression is not always as precisely regulated, and it is often the subject of silencing itself. This type of situation has been found in the incorrect assignment to auxin response factor 8 of a role in the mediation of VSR-induced developmental defect (Mlotshwa et al., 2016). Thus, since transgenic VSR expression is not an exact reproduction of viral infections, extrapolations to the effects of true infections must be very carefully considered. It seems clear that VSRs have a strong potential to play a significant role in the induction of plant development alterations. Whether or not a particular VSR does so in a specific infection, will require the precise characterization of the particular instance, since other viral-encoded factors can modulate this potential significantly.

16.2.2 MPs and Plant Development

All viral genomes sequenced so far encode one or several MPs. These viral proteins are critical to allow the infection to propagate from one infected cell to the next. Like the VSRs, they do not form a compact group of phylogenetically related proteins sharing large stretches of homology sequences. Rather, it is possible to find a good number of different biological solutions to the problem of overcoming the formidable barrier to intercellular viral movement which the cell wall means. Viruses have developed many forms to get around it in order to facilitate viral transport within ground tissue cells (Heinlein, 2015). Viruses move cell-to-cell through plasmodesmata, which under normal conditions present a size exclusion limit (SEL) incompatible with the passage of the infectious material. This traveling material is not the same in all plant virus infections. Some viruses move their genome in a truly encapsidated form (moving virions), but others do it through not fully characterized complexes involving viral nucleic acid and viral and cellular proteins. Most likely, this solution is not unique, meaning that the moving complexes may take different forms for different viruses.

In the few cases in which the process has been studied in detail, the plasmodesma SEL is only temporarily altered, for the time required for the infection to propagate intercellularly. After the new cell is infected, the SEL usually returns to normal. Thus there is not really much opportunity for this aspect of the infection process to play a significant role in the alteration of plant developmental traits. When MPs are transgenically expressed constitutively, things can be different. The most extensively studied cases deal with MPs of tobamoviruses, mostly TMV (Lucas et al., 1993; Heinlein et al., 1998; Olesinski et al., 1996). One major effect reported for solanaceous plants transgenically expressing TMV-MP is the alteration of sugar storage and biomass partitioning. This is an important developmental trait affecting the final architecture of the plant from the point of view of some of its components. For example, root mass can be half of a normal plant (Balachandran et al., 1997), and the availability of sugars can also be different. An immediate theoretical explanation for these altered traits comes from the fact that constitutively expressing MP-transgenic plants should have many plasmodesmata with a permanently higher SEL. In fact, this has been shown in several cases (Wolf et al., 1989; Lucas, 2006). Sugar loading from source tissues and unloading in sink tissues should be significantly affected in ground and vascular tissues with altered plasmodesma SEL. However, the phenomenon is not a universal one for transgenic tobamoviral MPs. For instance, upon expression in Arabidopsis of the MP of *Oilseed rape mosaic virus* (ORMV), a tobamovirus, no significant alteration in biomass partitioning was found (Mansilla et al., 2006). In this case, an environmentally dependent flowering time alteration was noticed (our unpublished results), leading to early flowering plants, also an important developmental trait. In fact, given the mobile nature of many smRNAs and transcription factors which are important development regulators, moving through plasmodesmata, it is somewhat surprising that not so many other developmental trait alterations have been reported in association with transgenic tobamoviral MP-expression.

Possibly more dramatic effects can be found in cases in which the MPs are derived from other types of MPs. Transgenic expression of polerovirus MP *Potato leaf roll virus* (PLRV) in tobacco led in some instances to more obvious alterations of sugar accumulation in source and sink leaves, and especially of biomass partitioning, than those reported for tobamoviral transgenic MPs (Herbers et al., 1997). Maybe the fact that PLRV is a phloem-limited virus has some relationship to this higher symptom intensity. In the case of the DNA-encapsidating begomovirus *Bean dwarf mosaic virus* (BDMV), also a phloem-limited virus, the expression of one of its two MPs led to strong developmental alterations in transgenic tomatoes (Hou et al., 2000).

So, as discussed for the transgenic VSRs, transgenesis does not fully reflect the effects of viral infections. However, viral MP transgenesis is a useful system to study links between development and macromolecular transport, an issue of growing importance in the study of plant development regulation.

16.2.3 Plant Developmental Traits Affected by Virus Infections

The effects of true viral infections on plant developmental traits have not reached a degree of characterization comparable to the previously discussed individual viral elements. That virus infections often affect the normal development of infected plants was noticed long ago, along with the discovery of viruses as plant pathogens (Hull, 2002), but insights into processes and mechanisms governing the effects have proven more difficult to study, especially if focused on plant traits more directly connected to the sequential steps of the plant developmental plan. An interesting recent example is the alteration in stomatal development induced by infections of two different susceptible hosts by two different tobamoviruses (Murray et al., 2016). The reduction in the number of stomata was significantly different compared to non-infected or resistant hosts. No clues as to the underlying processes for this alteration are yet available. Other examples affecting different developmental traits can be found in the literature (e.g., Culver and Padmanabhan, 2007, and other specific examples mentioned in this chapter).

We have concentrated our characterization efforts on the pathosystem *Arabidopsis thaliana*/TuMV. This potyvirus displays a wide host range compared to the majority of other viruses in the family. The variability and evolution of this virus has received attention over the last decade or so (Gibbs et al., 2015; Sánchez et al., 2003; Tomimura et al., 2004). The studies have uncovered a high degree of genetic variability leading to the separation of the different sequenced virus isolates into true genetic strains. Representatives of two of the major strains induce differential symptoms when infecting Arabidopsis, several of which are true specific developmental traits (Figure 16.1) (Sánchez et al., 2015) such as flower stalk elongation, flower whorl morphology and flower fertility, or root growth. This differential behavior provides a very convenient ground for detailed studies of viral and plant factors involved. The lack of flower stalk elongation in plants infected by one of isolates (TuMV-UK 1) is major developmental trait affected, since a non-elongated flower stalk leads to plants not developing an inflorescence and all the consequences derived (almost total absence of flowers, infertility, lack of siliques and seeds, etc.). A more detailed characterization of the viral determinant(s) involved was subsequently carried out. As discussed above, transgenically expressed VSRs can mimic the developmental defects induced by a certain infection. In the case of TuMV, the VSRs of the two isolates, i.e. UK 1, the stalk arresting, and JPN 1, a non-arresting isolate, induced the same type of symptoms in the transgenic plants: a phenocopy of the UK 1-infected plants. This result indicated that, even though the VSR (called HC-Pro in the potyviruses) is likely playing a role in mediating the non-elongated stalk trait, other viral factors should be the main responsible ones, since JPN 1-infected plants do not show stalk arrest. A detailed study involving the generation of chimeras between the two isolates allowed mapping the responsible elongation determinant to the C-terminal region of the viral protein P3, a membrane protein, not a VSR. Transgenic plants expressing the P3 proteins of either isolate rendered asymptomatic plants. Possibly, when similar studies are carried out in other viral pathosystems, non-VSR viral proteins will be uncovered as major determinants of

FIGURE 16.1 Differential effects on Arabidopsis (ecotype Col-0) flower stalk development induced by different *Turnip mosaic virus* (TuMV) strains. (a) Buffer-inoculated plant. (b) Plant inoculated with TuMV isolate JPN 1. (c) Plant inoculated with TuMV isolate UK 1. (Courtesy of Silvia López-González, CBGP.)

developmental defects. We propose that VSRs are potent effectors of developmental alterations, but that often other viral-encoded proteins expressed during the infection process can, and will, regulate these alterations.

A final comment in this section concerns the relationship of viruses, developmental traits, and plant hormones. Hormones are major players in controlling development, and the likely role of altered hormones acting as mediators of virus-induced developmental alterations is an emerging area of research in the study of plant/virus interactions. So far, a global view relating the three factors is still lacking. There are several reports about alterations in plant hormones after virus infections, which have been recently reviewed (Alazem and Lin, 2015; Collum and Culver, 2016). Auxins, brassinosteroids, cytokinins, abscisic acid, salicylic acid, and gibberellins have all been found altered in specific virus infections in several hosts. The role they may play in affecting specific developmental traits clearly needs further characterization. In addition, a good number of the reported studies focusing on plant hormones and viruses use transgenic plants expressing viral genomic elements. As discussed above, results from this approach must be taken with care when applied to true virus infections.

16.3 PLANT DEVELOPMENT AFFECTING VIRUS INFECTIONS

Not only virus infections can alter plant development significantly. The stage of development of a plant can influence the outcome of the infection. Two major developmental stages have been identified as strong influencers. They are briefly reviewed and discuss in this section.

Over 25 years ago the concept of "developmental resistance" was put forward during studies of infections by *Cauliflower mosaic virus* (CaMV) (Leisner et al., 1993), although the association of certain stages of development and resistance to infection is also active in other types of phytopathogens (Panter and Jones, 2002). In the case of viruses, studies with CaMV infecting Arabidopsis and turnip showed that the well-known relationship between virus long distance movement through the phloem and photosynthate distribution throughout the plant was also at the basis of developmental resistance. Since most plant parts undergo a transition from photosynthate sinks to phosynthate sources, a parallel process takes place in the vascular-mediated distribution of viruses from infected tissues to still uninfected ones. This form of resistance is not a genetic, one in the sense that the tissue is unable to support virus replication. If the uninvaded tissue is directly inoculated with the virus, it will support the infection. It is development-associated, because the virus in the vasculature will mostly escape tissues that have become photosynthate sources, and will be unloaded only in sink tissues. As a consequence, older tissues that have already undergone the transition will be uninfected or apparently resistant. This is probably the main reason for the widely observed phenomenon in the field that virus infections are much more aggressive in plants infected young than in older plants. The same applies to young tissues of fully grown plants, in comparison with older plant parts. Interestingly, plasmodesma architecture and frequency also change dramatically during the sink-to-source transition (Fitzgibbon et al., 2013; Oparka et al., 1999), possibly also influencing the capability of viruses to invade older tissues.

An unexpected link was uncovered between virus infections and the transition period from vegetative to reproductive growth in Arabidopsis infected by either of two quite different viruses (Lunello et al., 2007). Both TuMV and ORMV had their infection pattern severely altered within this period. The transition period, a major developmental trait, was identified as the time for inflorescence bud formation and appearance. Quantitative measurements of virus titers revealed the alterations in the patterns of virus accumulation. Before the transition period, virus titers grew at good rate in systemically infected tissues, both roots and leaves. During transition, no increase of virus accumulation was detected in leaves, which restored accumulation afterward. It is possible that phloem unloading is greatly diminished in leaves at this time, at which a high demand of energy (sugars) is required for the formation of the new reproductive organs, and that the decreased photosynthate unloading affects virus unloading too. Whether this effect is simply derived from a straight competition in the strength of photosynthate sinks or some other anatomical change precluding temporary virus unloading is also taking place remains to be investigated. The effect was more dramatic in roots. Here the combined results of quantitative measurements of coat protein and viral RNA accumulation revealed that not only viral unloading was halted during the transition period, but also a degradative environment affected the levels of virus titer previously achieved, a phenomenon more drastic for TuMV than for ORMV. There are still no mechanistic clues of the process relating transition to reproductive stage and viral replication, accumulation, movement, and/or stability. However, the phenomenon is a major example of the intimate relationship between plant development and virus infections.

16.4 CONCLUSION

Plant virus infections and specific plant developmental traits are processes intimately interconnected and mutually influenced. The overall effects of virus infections on the global developmental plan of plants ranges from minor to dramatic depending on the specific virus/host combination. For the study of the mutual influence it is convenient to focus on specific developmental traits, rather than global assessments, since not all traits involved in the developmental plan can be similarly affected by a certain infection, and different virus infections may affect different developmental traits. Studies on the effects of viruses on development can be approached by the dissection of the virus into components tested individually. When approached this way, care must be taken to take a reductionistic view of the influence, since even though the overall effect on development may seem similar to the effect by a virus infection; several studies have shown that the actual effect of single viral components may be modulated by other factors. When the effect of the plant developmental stage on the virus infection is studied, no such reductionistic approach can be taken, because a conclusion would be mostly meaningless, virus-wise.

REFERENCES

Alazem M, Lin NS (2015). Roles of plant hormones in the regulation of host–virus interactions. *Molecular Plant Pathology* 16:529–540.

Balachandran S, Hull RJ, Rosemary AM, Yoash V, Lucas WJ (1997). Influence of environmental stress on biomass partitioning in transgenic tobacco plants expressing the movement protein of Tobacco mosaic virus. *Plant Physiology* 114:475–481.

Borges F, Martienssen RA (2015). The expanding world of small RNAs in plants. *Nature Reviews Molecular Cell Biology* 16:727–741.

Chapman EJ, Prokhnevsky AI, Gopinath K, Dolja VV, Carrington JC (2004). Viral RNA silencing suppressors inhibit the microRNA pathway at an intermediate step. *Genes and Development* 18:1179–1186.

Chellappan P, Vanitharani R, Fauquet CM (2005). MicroRNA-binding viral protein interferes with Arabidopsis development. *Proceedings of the National Academy of Sciences of the United States of America* 102:10381–10386.

Chen J, Li WX, Xie DX, Peng JR, Ding SW (2004). Viral virulence protein suppresses RNA silencing-mediated defense but upregulates the role of MicroRNA in host gene expression. *Plant Cell* 16:1302–1313.

Chen XM (2009). Small RNAs and their roles in plant development. *Annual Review of Cell and Developmental Biology* 25:21–44.

Collum TD, Culver JN (2016). The impact of phytohormones on virus infection and disease. *Current Opinion in Virology* 17:25–31.

Csorba T, Kontra L, Burgyan J (2015). Viral silencing suppressors: Tools forged to fine-tune host-pathogen coexistence. *Virology* 479-480:85–103.

Culver JN, Padmanabhan MS (2007). Virus-induced disease: Altering host physiology one interaction at a time. *Annual Review of Phytopathology* 45:221–243.

Dunoyer P, Lecellier CH, Parizotto EA, Himber C, Voinnet O (2004). Probing the microRNA and small interfering RNA pathways with virus-encoded suppressors of RNA silencing. *Plant Cell* 16:1235–1250.

Fitzgibbon J, Beck M, Zhou J, Faulkner C, Robatzek S, Oparka K (2013). A developmental framework for complex plasmodesmata formation revealed by large-scale imaging of the Arabidopsis leaf epidermis. *Plant Cell* 25:57–70.

García JA, Pallás V (2015). Viral factors involved in plant pathogenesis. *Current Opinion in Virology* 11:21–30.

Gibbs AJ, Nguyen HD, Ohshima K (2015). The "emergence" of turnip mosaic virus was probably a "gene-for-quasi-gene" event. *Current Opinion in Virology* 10:20–26.

Heinlein M (2015). Plant virus replication and movement. *Virology* 479-480:657–671.

Heinlein M, Padgett HS, Gens JS, Pickard BG, Casper SJ, Epel BL, Beachy RN (1998). Changing patterns of localization of the tobacco mosaic virus movement protein and replicase to the endoplasmic reticulum and microtubules during infection. *Plant Cell* 10:1107–1120.

Herbers K, Tacke E, Hazirezaei M, Krause KP, Melzer M, Rohde W, Sonnewald U (1997). Expression of a luteoviral movement protein in transgenic plants leads to carbohydrate accumulation and reduced photosynthetic capacity in source leaves. *Plant Journal* 12:1045–1056.

Hou YM, Sanders R, Ursin VM, Gilberston RL (2000). Transgenic plants expressing geminivirus movement proteins: Abnormal phenotypes and delayed infection by Tomato mottle virus in transgenic tomatoes expressing the Bean dwarf mosaic virus BV1 or BC1 proteins. *Molecular Plant-Microbe Interactions* 13:297–308.

Hull R (2002). *Matthew's Plant Virology*. San Diego, CA: Academic Press.

Kasschau KD, Xie Z, Allen E, Llave C, Chapman EJ, Krizan KA, Carrington JC (2003). P1/HC-Pro, a viral suppressor of RNA silencing, interferes with Arabidopsis development and miRNA function. *Developmental Cell* 4:205–217.

Leisner SM, Turgeon R, Howell SH (1993). Effects of host plant development and genetic determinants on the long distance movement of Cauliflower mosaic virus in Arabidopsis. *Plant Cell* 5:191–202.

Lucas WJ (2006). Plant viral movement proteins: Agents for cell-to-cell trafficking of viral genomes. *Virology* 344:169–184.

Lucas WJ, Olesinski A, Hull RJ, Haudenshicld JS, Deom CM, Beachy RN, Wolf S (1993). Influence of the tobacco mosaic virus 30-kDa movement protein on carbon metabolism and photosynthate partitioning in transgenic tobacco plants. *Planta* 190:88–96.

Lunello P, Mansilla C, Sanchez F, Ponz F (2007). A developmentally linked, dramatic, and transient loss of virus from roots of Arabidopsis thaliana plants infected by either of two RNA viruses. *Molecular Plant-Microbe Interactions* 20:1589–1595.

Mansilla C, Aguilar I, Martínez-Herrera D, Sánchez F, Ponz F (2006). Physiological effects of constitutive expression of Oilseed Rape Mosaic Tobamovirus (ORMV) movement protein in Arabidopsis thaliana. *Transgenic Research* 15:761–770.

Martínez De Alba, AE, Elvira-Matelot E, Vaucheret H (2013). Gene silencing in plants: A diversity of pathways. *Biochimica et Biophysica Acta* 1829:1300–1308.

Mlotshwa S, Pruss GJ, Macarthur JL, Reed JW, Vance V (2016). Developmental defects mediated by the P1/HC-Pro Potyviral silencing suppressor are not due to misregulation of Auxin Response Factor 8. *Plant Physiology* 172:1853–1861.

Mlotshwa S, Schauer SE, Smith TH, Mallory AC, Herr JM, Roth B, Merchant DS, Ray A, Bowman LH, Vance VB (2005). Ectopic DICER-LIKE1 expression in P1/HC-Pro Arabidopsis rescues phenotypic anomalies but not defects in microRNA and silencing pathways. *Plant Cell* 17:2873–2885.

Murray RR, Emblow MS, Hetherington AM, Foster GD (2016). Plant virus infections control stomatal development. *Scientific Reports* 6:34507.

Olesinski AA, Almon E, Navot N, Perl A, Galun E, Lucas WJ, Wolf S (1996). Tissue-specific expression of the tobacco mosaic virus movement protein in transgenic potato plants alters plasmodesmal function and carbohydrate partitioning. *Plant Physiology* 111:541–550.

Oparka KJ, Roberts AG, Boevink P, Santa Cruz S, Roberts L, Pradel KS, Imlau A, Kotlizky G, Sauer N, Epel B (1999). Simple, but not branched, plasmodesmata allow the nonspecific trafficking of proteins in developing tobacco leaves. *Cell* 97:743–754.

Pallás V, García JA (2016). Viral pathogenesis: Still many missing pieces in the puzzle. *Current Opinion in Virology* 17:v–vii.

Palloix A, Ordon F (2011). Advanced Breeding for Virus Resistance in Plants. In: *Recent Advances in Plant Virology*. Caranta C, Aranda MA, Tepfer M, López-Moya JJ (eds). 195–218. Norwich: Caister Academic Press.

Panter SN, Jones DA (2002). Age-related resistance to plant pathogens. *Advances in Botanical Research* 38:251–280.

Power-Abel P, Nelson RS De B, Hoffmann N, Rogers SG, Fraley RT, Beachy RN (1986). Delay of disease development in transgenic plants that express the tobacco mosaic virus coat protein gene. *Science* 232:738–743.

Ratcliff F, Harrison BD, Baulcombe DC (1997). A similarity between viral defense and gene silencing in plants. *Science* 276:1558–1560.

Sánchez F, Manrique P, Mansilla C, Lunello P, Wang X, Rodrigo G, López-González S, Jenner C, González-Melendi P, Elena SF, Walsh J, Ponz F (2015). Viral strain-specific differential alterations in Arabidopsis developmental patterns. *Molecular Plant-Microbe Interactions* 28:1304–1315.

Sánchez F, Wang X, Jenner CE, Walsh JA, Ponz F (2003). Strains of Turnip mosaic potyvirus as defined by the molecular analysis of the coat protein gene of the virus. *Virus Research* 94:33–43.

Tomimura K, Spak J, Katis N, Jenner CE, Walsh JA, Gibbs AJ, Ohshima K (2004). Comparisons of the genetic structure of populations of Turnip mosaic virus in West and East Eurasia. *Virology* 330:408–423.

Weiberg A, Bellinger M, Jin H (2015). Conversations between kingdoms: Small RNAs. *Current Opinion in Biotechnology* 32:207–215.

Wolf S, Deom CM, Beachy RN, Lucas WJ (1989). Movement protein of tobacco mosaic virus modifies plasmodesmatal size exclusion limit. *Science* 246:377–379.

Section III

Management

17 Host miRNAs and Virus-Derived Small RNAs in Plants Infected with Certain Potyviruses

Zhimin Yin

CONTENTS

17.1 INTRODUCTION

In plants, two main categories of small regulatory RNAs, i.e., microRNAs (miRNAs) and small interfering RNAs (siRNAs), are distinguished based on their biogenesis and function (Chapman and Carrington 2007; Chen 2009; Vazquez et al. 2010). Although miRNAs are encoded by endogenous *MIR* genes, siRNAs can arise from either exogenous nucleic acids such as viruses or endogenous transcripts (Bartel 2004; Meins et al. 2005; Vaucheret 2006). The accumulation of exogenous virus-derived siRNAs (vsiRNAs) is a hallmark in plant antiviral defense (Hamilton and Baulcombe 1999; Waterhouse and Fusaro 2006; Ding and Voinnet 2007; Pantaleo et al. 2007). On the other hand, recent studies demonstrated that plant endogenous small RNAs (sRNAs), including miRNAs and siRNAs, play an important role in plant immunity (Jin 2008; Voinnet 2008; Padmanabhan et al. 2009).

The first zigzag model for illustrating plant immunity was outlined based on antibacterial/antifungal immune system (Jones and Dangl 2006). In this model, plants detect microbial/pathogen-associated molecular patterns (MAMPs/PAMPs) via pattern recognition receptors (PRRs) to trigger PAMP-triggered immunity (PTI), which is a basal and broad-spectrum defense. As a counterdefense, pathogens have evolved molecules known as effectors to suppress PTI, resulting in effector-triggered susceptibility (ETS). To counter counterdefense, plants have coevolved resistance (*R*) genes to detect the pathogen effectors that overcame PTI. If one effector is recognized by a corresponding R protein, the effector-triggered immunity (ETI) is activated. The recognized effector is termed an avirulence (Avr) protein. Nature selection drives pathogens to avoid or to suppress ETI by modifying Avr proteins or acquiring new Avr proteins, which results in new *R* specificities and triggers ETI again.

Since RNA silencing against viruses is reminiscent of basal resistance against fungi and bacteria, by regarding RNA silencing as a type of PTI, viruses are readily superimposed on the classic zigzag model, in which the viral double-stranded RNA (dsRNA) act as a PAMP (Moffett 2009; Zvereva and Pooggin 2012; Nakahara and Masuta 2014). In the modified zigzag model of virus–host interactions (Nakahara and Masuta 2014), viruses produce dsRNAs in the infected plants, which activates RNA silencing as a PTI-like phase to target the viral RNAs. Then, the viruses produce RNA silencing suppressors (RSSs) as viral effectors to suppress RNA silencing. In turn, plants activate an *R* gene-encoded protein that specifically recognizes a viral RSS as the Avr protein, leading to the ETI-like phase. Unlike the non-viral pathogens, when one of the viral proteins is recognized by a host R protein, viruses usually modify it, instead of replacing it, by changing the amino acid sequence while retaining the protein structure necessary for the function (Moffett 2009; Donze et al. 2014). The viral Avr proteins usually are also virulence factors, which might suppress innate immune responses in the susceptible hosts, and are recognized by R protein in the resistant hosts (Soosaar et al. 2005; Moffett 2009; Zvereva and Pooggin 2012).

In this chapter, host miRNAs and vsiRNAs in plant-*Potyvirus* interaction are summarized, with highlights on plant-PVY phytosystems.

17.2 HOST miRNAs AND VIRAL vsiRNAs IN PLANT–VIRUS INTERACTION

Plant miRNAs are endogenous non-coding RNAs of 20–24 nucleotides (nt) that regulate eukaryotic gene expression post-transcriptionally by targeting specific messenger RNAs (mRNAs) for cleavage or translational inhibition (Bartel 2004; Voinnet 2009). Some species of miRNA may regulate gene expression transcriptionally by directing DNA methylation at their own loci in *cis* or at their target genes in *trans* (Wu et al. 2010; Hu et al. 2014). Plant miRNAs play essential roles in plant growth and development as well as in the regulation of the miRNA pathway itself (Jones-Rhoades et al. 2006; Chen 2010; Jin et al. 2013). They also regulate plant responses to biotic and abiotic stresses (Shukla et al. 2008; Ruiz-Ferrer and Voinnet 2009; Sunkar 2010; Khraiwesh et al. 2012; Sunkar et al. 2012; Ramesh et al. 2014). In virus–host interaction, recent studies provide evidence that several miRNAs, i.e., miR482, miR6019, and miR6020, are involved in antiviral immunity by regulation of *R* genes that encode proteins with nucleotide binding (NB) and leucine-rich repeat (LRR) domains (Li et al. 2012; Shivaprasad et al. 2012; Permar et al. 2014). In the local inoculated leaves of tomato plants infected with *Turnip crinkle virus* (TCV), *Cucumber mosaic virus* (CMV), and *Tobacco rattle virus* (TRV), the levels of miR482 was decreased compared to that of the healthy plants, which resulted in enhanced expression of the *NB-LRR* transcripts (Shivaprasad et al. 2012). The authors proposed that the RNA silencing system is a first layer of defense, the viral RSSs are a counterdefense system, and miR482 is then a counter-counterdefense system that is dependent on the counterdefense system (Shivaprasad et al. 2012). In susceptible plants, the experimentally evidenced changes in the levels of host miRNAs following viral infection were reviewed (Yin et al. 2014). A common set of miRNAs, mainly those related to plant development or described as responsive to biotic and abiotic stresses, was summarized. Although the miR168 was up-regulated by virus infection in a plant- and virus-independent manner, many miRNAs and their host mRNA targets respond to viral infection in a virus-, strain-, plant-, and tissue-specific manner (Yin et al. 2014). The symptoms caused by viral infection might be in part the consequence of misregulation of host miRNAs (Kasschau et al. 2003; Silhavy and Burgyán 2004; Xu et al. 2014).

RNA silencing is a eukaryotic surveillance mechanism against invasive nucleic acids (Waterhouse et al. 2001). The antiviral RNA silencing is induced by viral dsRNA molecules (Ding and Voinnet 2007). These viral dsRNAs are recognized by the plant and are processed into the primary vsiRNAs of 21-nt by DCL4 and of 22-nt by DCL2 (Deleris et al. 2006; Waterhouse and Fusaro 2006; Qu et al. 2008; Llave 2010). The secondary vsiRNAs are produced from the aberrant viral transcription byproducts, which are generated by the primary vsiRNAs (Garcia-Ruiz et al. 2010; Wang et al. 2010). The vsiRNA is incorporated into Argonaute 1 (AGO1)-, AGO2-, or AGO7-containing RNA-induced silencing complex (RISC) and degrades the viral RNA with sequence complementary to it (Morel et al. 2002; Bortolamiol et al. 2007; Takeda et al. 2008; Harvey et al. 2011; Alvarado and Scholthof 2012; Wang et al. 2012). The vsiRNAs are the main components of RNA silencing based antiviral response, and they are involved in the regulation of host gene expression and mediate disease symptoms in plants as well (Llave 2010; Pantaleo 2011; Shimura et al. 2011; Smith et al. 2011). The molecular and cellular bases for antiviral silencing have been proposed (Pumplin and Voinnet 2013).

17.3 HOST miRNAs IN PLANT–POTYVIRUS INTERACTION

The potyvirus-responsive host miRNAs and their mRNA targets showed common and strain-specific regulation. Some PVY-responsive host miRNAs were differentially regulated depending on plant species, symptom severity, or showing strain-specificity; their host mRNA targets showed antagonistic expression, up-regulation, or strain-specific alteration. Potato miRNAs with the potential to target the PVY viral genome have been predicted.

17.4 HOST miRNAs AND mRNA TARGETS RESPONDING TO *POTYVIRUS*

The genus *Potyvirus* belong to the family *Potyviridae*. The potyviruses are widespread in cultivated plants worldwide (Ivanov et al. 2014). *Potato virus Y* (PVY), the type member of the genus *Potyvirus*, is among the top 10 plant viruses with significant economic and scientific importance (Scholthof et al. 2011). The plant miRNAs responsive to potyviruses have been studied in different plant species. It included PVY-infected *Nicotiana tabacum*, *N. benthamiana* and *Solanum tuberosum* (Bazzini et al. 2007; Pacheco et al. 2012; Yin et al. 2017), *Turnip mosaic virus* (TuMV)-infected *Brassica rapa* and Chinese cabbage (He et al. 2008; Wang et al. 2015), and *Soybean mosaic virus* (SMV)-infected soybean (Yin et al. 2013; Chen et al. 2016, 2017) (Table 17.1). The potyvirus-responsive host miRNAs and their mRNA targets showed common and strain-specific regulation. In soybean following infections by three different SMV isolates, i.e., isolates L and LRB (G2 strain) and isolate G7 (G7 strain), the miRNAs and mRNA targets responsive to all isolates were found to be involved in protein synthesis and modification (Chen et al. 2016). The strain/isolate-specific responsive ones may associate with the distinct pathogenesis of SMV strains/isolates (Chen et al. 2016).

17.5 HOST miRNAs RESPONDING TO PVY

Alteration of host miRNAs upon PVY infection was studied in *N. tabacum* and *N. benthamiana* (Bazzini et al. 2007; Pacheco et al. 2012). Although expression of potato (*S. tuberosum*) miRNAs in healthy plants has been confirmed (Xie et al. 2011; Zhang et al. 2013; Lakhotia et al. 2014), there is no published data on miRNA expression in the PVY-infected potato except several selected miRNAs (Yin et al. 2017). In PVY-infected plants, some conserved miRNAs exhibit either a decrease or an increase in different species, while certain miRNAs are differentially regulated depending on the plant species, virus strain, and symptom severity (Table 17.2).

In PVY-infected *N. tabacum*, *N. benthamiana*, and/or *S. tuberosum*, the expression levels of miR168, miR162, miR166, miR171, miR172, miR398, miR159, miR167, and miR482 were up-regulated, while that of miR164 and miR169 was down-regulated (Bazzini et al. 2007; Pacheco et al.

TABLE 17.1
Host miRNAs and Their mRNA Target Expression in Different Plant Species in Response to *Potyvirus*

Plant Species	*Potyvirus* (Strain)	Methodology	Host miRNA/ Host mRNA Target	References
Nicotiana tabacum	PVY	Northern blot	10 miRNAs	Bazzini et al. 2007
Brassica rapa	TuMV	Sequencing Northern blot	9 conserved miRNA 3 novel miRNAs *TIR-NBS-LRR* targets	He et al. 2008
N. benthamiana	PVY	Northern blot RT-PCR	4 miRNAs 3 mRNA targets	Pacheco et al. 2012
Solanum tuberosum	PVY (PVY^{NTN}, PVY^{N-Wi}, PVY^{Z}-NTN)	Stem-loop RT-qPCR RT-qPCR	10 miRNAs 14 mRNA targets Strain-specific regulation	Yin et al. 2017
Soybean	SMV	Deep sequencing	179 miRNAs 346 predicted targets	Yin et al. 2013
Soybean	SMV (G2 and G7)	sRNA-seq Degradome-Seq Transcriptome-Seq	164 up-regulated DEMs/115 down-regulated DEMs 1432 up-regulated DEGs/1264 down-regulated DEGs Common or strain-specific regulation	Chen et al. 2016
Soybean	SMV (G7)	sRNA-seq Degradome-Seq Transcriptome-Seq	37 up-regulated DEMs/47 down-regulated DEMs 318 up-regulated DEGs/588 down-regulated DEGs	Chen et al. 2017
Chinese cabbage	TuMV	High-throughput sequencing	69 TuMV responsive Bra-miRNAs 271 predicted targets involving in growth and stress resistance	Wang et al. 2015

Note: DEGs, differentially expressed genes; DEMs, differentially expressed miRNAs; PVY, *Potato virus Y*; SMV, *Soybean mosaic virus*; TuMV, *Turnip mosaic virus*.

2012; Yin et al. 2017; Table 17.2). These PVY-altered miRNAs are among those previously described stress-responsive miRNAs (Liu et al. 2008; Trindade et al. 2010; Bazzini et al. 2011; Khraiwesh et al. 2012). The levels of miR156 and miR160 decreased in PVY-infected *N. tabacum*, but increased in PVY-infected *N. benthamiana* or *S. tuberosum*. Previous studies suggested that the regulation of miRNA expression appears to vary between plant species (He et al. 2008; Sunkar et al. 2012; Khraiwesh et al. 2012). Although some miRNA-target modules are conserved within or beyond angiosperms, the biologic functions of the regulatory modules may vary in different species (Chen 2010).

Strain-specific alteration of miRNAs was observed in potato-PVY interaction (Yin et al. 2017). Potato cv. Etola exhibits hypersensitive response (HR) resistance to PVY^{NTN} strain, partial HR resistance to PVY^{N-Wi} with severe leaf symptoms, and partial HR resistance to PVY^{Z}-NTN with mild leaf symptoms. Increase of miR162, miR168a, miR172e, and miR482 are detected only in the upper leaves of Etola plants infected with the severe PVY^{N-Wi} strain but not in Etola plants infected with the mild PVY^{Z}-NTN isolate, nor in the PVY^{NTN}-inoculated Etola plants that are resistant to the PVY^{NTN} (Yin et al. 2017). The FRNK motif, which was suggested to bind the siRNA and miRNA duplex (Shiboleth et al. 2007), is present in the helper component-proteinase (HC-Pro) protein of the PVY^{N-Wi}, PVY^{Z}-NTN, and PVY^{NTN} isolates (Yin et al. 2017). However, the levels of PVY *HC-Pro RNA* in

TABLE 17.2
Experimentally Validated Host miRNAs and Their mRNA Target Expression in Different Plant Species in Response to PVY

miRNAs and/or *mRNA Targets*[a]	*Nicotiana tabacum* (Bazzini et al. 2007)	*N. benthamiana* (Pacheco et al. 2012)	*Solanum tuberosum* (Yin et al. 2017)		
	PVY	PVY	$PVY^{N\text{-}Wi}$ Strain	PVY^{Z}-NTN Strain	PVY^{NTN} Strain
miR162	nd	nd	↑	–	–
DCL1	nd	nd	↑	–	–
miR168	nd	↑	↑	–	–
AGO1	nd	nd	↑	–	–
miR156	↓	↑	nd	nd	nd
SPL9	nd	↑	nd	nd	nd
miR159	↑	nd	–	–	–
14-3-3	nd	nd	↑	–	–
miR160	↓	nd	↑	–	–
ARF16	nd	nd	↓/–	–	–
miR166	↑	nd	–	–	–
ATHB-9	nd	nd	–	–	–
ATHB-14	nd	nd	–	–	–
miR171	↑	↑	nd	nd	nd
miR171*	↑	nd	nd	nd	nd
SCL6	nd	↑	nd	nd	nd
miR172	nd	nd	↑	–	–
TOE3	nd	nd	↑	–	–
RAP2-7	nd	nd	–	–	–
miR398	nd	↑	nd	nd	nd
CSD	nd	↑	nd	nd	nd
miR482	nd	nd	↑	–	–
Gpa2	nd	nd	↑	↑	–
R	nd	nd	–	–	–
miR482	nd	nd	↑	–	–
CC-NBS-LRR	nd	nd	↑	–	–
NBS-LRR	nd	nd	–	–	–
NBS	nd	nd	–	–	–
miR164	↓	nd	nd	nd	nd
miR165	–	nd	nd	nd	nd
miR167	↑	nd	nd	nd	nd
miR169	↓	nd	nd	nd	nd

Note: ↑, up-regulated; ↓, down-regulated; –, no change; nd, not determined; PVY, *Potato virus Y*; AGO1, Argonaute 1; ARF16, auxin response factor 16; CSD, copper superoxide dismutase; DCL1, RNAse III-like DICER-like I endonuclease; ATHB-9, homeobox-leucine zipper protein ATHB-9 like; ATHB-14, hHomeobox-leucine zipper protein ATHB-14 like; 14-3-, 14-3-3 protein 7; SCL6, scarecrow-like transcription factor 6; SPL9, squamosa promoter binding protein-like transcription factor 9; RAP2-7, ethylene-responsive transcription factor RAP2-7 like; TOE3, apetala2-like ethylene-responsive transcription factor TOE3 like; Gpa2, disease resistance protein Gpa2; R, resistance protein N; CC-NBS-LRR, CC-NBS-LRR resistance protein, JHL06P13.14 protein; NBS-LRR, NBS-LRR protein, Rx protein; NBS, NBS-coding resistance gene analog.

[a] mRNA targets are in *italic*.

PVY$^{N\text{-}Wi}$-infected Etola plants are 40-fold of that of PVY^{Z}-NTN-infected ones (Yin et al. 2017). The different levels of PVY *HC-Pro* RNA, which encodes an RSS, expressed in different strains might be related to the strain-specific alteration of miRNAs in Etola-PVY interaction.

The induction of miR168 is ubiquitous in plant-virus interactions, and the increased miR168 accumulation is accompanied by *AGO1* mRNA induction (Várallyay et al. 2010). Induction of *AGO1* mRNA during viral infections is a plant defense response, because AGO1 protein, which is translated from its mRNA, guides vsiRNAs against viral RNAs. However, virus-induced accumulation of miR168, a counterdefense action of the invading virus, is involved in repression of AGO1 protein accumulation (Várallyay et al. 2010, 2014). Failure of viruses to cause miR168 accumulation promotes accumulation of AGO1 protein and results in a stronger antiviral response (Várallyay et al. 2010).

Besides, in tobacco, the potyviruses PVY and *Tobacco etch virus* (TEV) inducing mild symptoms altered miRNA accumulation to a less extent than *Tobacco mosaic virus* (TMV) and *Tomato mosaic virus* (ToMV) inducing severe symptoms (Bazzini et al. 2007). In potato cv. Etola, the strain-specific alteration of certain host miRNAs might be correlated with the severe symptoms induced by PVY$^{N\text{-}Wi}$ strain (Yin et al. 2017). The severity of symptoms induced by viruses seems to be correlated with miRNA accumulation (Bazzini et al. 2007; Naqvi et al. 2010; Amin et al. 2011; Lang et al. 2011; Yin et al. 2017).

17.6 HOST mRNA TARGETS RESPONDING TO PVY

Plant miRNAs perfectly or near perfectly complement their mRNA targets, leading to target cleavage and degradation (Bartel 2004). The miRNAs and their mRNA targets essentially show mutually antagonistic expression levels in a virus-infected plant (Naqvi et al. 2010; Bazzini et al. 2011; Du et al. 2011), whereas a parallel increase in expression of target mRNAs and the corresponding miRNA species was also frequently observed in virus-infected plants (Cillo et al. 2009; Naqvi et al. 2010; Várallyay et al. 2010; Feng et al. 2012; Hu et al. 2011; Pacheco et al. 2012). In PVY-infected plants, the mRNA targets and their corresponding miRNAs show antagonistic expression, up-regulation, or strain-specific alteration (Table 17.2). In some cases, the altered miRNAs did not cause any changes in levels of the corresponding mRNA targets, which contains the predicted miRNA binding site. The functions of the PVY-responsive mRNA targets are summarized in Table 17.3.

The antagonistic expression of miR160 and its target were detected in PVY-infected potato. The enhanced expression of miR160 resulted in reduced expression of its targeting transcripts' auxin response factor 16 (*ARF16*) (Yin et al. 2017; Table 17.2). Li et al. (2010) demonstrated that miR160 plays an important role in PTI against bacteria. Treatment of *Arabidopsis* plants with Flg22, a bacterial flagellin-derived peptide, induced miR160 accumulation and repressed the expression of the miR160 targets *ARF10*, *ARF16*, and *ARF17*, which led to greater callose deposition, a PTI response induced by PAMP (Li et al. 2010). Likewise, in SMV-infected soybean, the up-regulation of miR160 suggests that the defense responses triggered miRNA-mediated suppression of auxin-signaling pathways (Yin et al. 2013). Infection of *S. lycopersicum* with *Tomato leaf curl New Delhi virus* (ToLCNDV) resulted in up-regulation of miR160 and down-regulation of its target mRNA *ARF17* (Pradhan et al. 2015). The up-regulation of miR160 has also been observed in CMV-infected tomato, *Southern rice black-streaked dwarf virus* (SRBSDV)-infected rice and in *N. benthamiana* co-infected with *Tomato yellow leaf curl China virus* (TYLCCNV) and its betasatellite (Feng et al. 2014; Xiao et al. 2014; Xu et al. 2014).

Conversely, a parallel increase in mRNA levels of *AGO1* and *DCL1*, along with the corresponding miR168 and miR162, were observed in PVY-infected *S. tuberosum* and/or *N. benthamiana* (Pacheco et al. 2012; Yin et al. 2017; Table 17.2). Plant DCL1 and AGO1 are the two key enzymes in the miRNA biogenesis pathway, and undergo sophisticated homeostatic regulations through the action of miR162 and miR168, respectively (Xie et al. 2003; Vaucheret et al. 2004; Voinnet 2009). Changes in expression levels of miR162 and miR168 may influence the global levels of miRNA production (Voinnet 2009).

TABLE 17.3
Function of Host miRNAs and Their mRNA Targets Responsive to PVY[a]

miRNA	Target Gene	Target Protein Class	Function	References
miR162	*DCL1*	Endoribonuclease Dicer homolog 1	Regulation of miRNA	Zhang et al. 2013; Khraiwesh et al. 2012; Yin et al. 2014; Xie et al. 2003
miR168	*AGO1*	Argonaute 1	Regulation of miRNA Leaf polarity Adaptive response to stress	Xie et al. 2011; Khraiwesh et al. 2012; Yin et al. 2014; Liu et al. 2008; Vaucheret et al. 2004
miR156	*SPL9*	Squamosa promoter binding protein-like transcription factor	Shoot maturation	Pacheco et al. 2012
miR159	*14-3-3*	14-3-3 protein7, phosphoserine/ threonine binding modules	Primary metabolism Signaling Stress response Plant-pathogen interaction Immunity-associated programmed cell death	Xie et al. 2011; TAIR
miR160	*ARF16*	Auxin response factor Transcription factor	Auxin signaling Floral organ identity Adaptive response to stress	Xie et al. 2011; Khraiwesh et al. 2012; Yin et al. 2014; Liu et al. 2008; Navarro et al. 2006
miR166	*ATHB-14*	Homeobox-leucine zipper protein ATHB-14 like transcription factor	Leaf polarity Leaf morphogenesis Meristem formation Vascular development Phase change and flowering Response to freezing	TAIR; Yin et al. 2017
miR166	*ATHB-9*	Homeobox-leucine zipper protein ATHB-9 like transcription factor	Leaf polarity Meristem formation Vascular development Root development Response to water deprivation Response to salt stress	Xie et al. 2011; TAIR
miR171	*SCL6*	Scarecrow-like transcription factor	Gibberellin signaling	Pacheco et al. 2012
miR172	*TOE3*	Apetala2-like ethylene-responsive transcription factor TOE3 like	Leaf development Floral organ identity Flowering time	Xie et al. 2011; Khraiwesh et al. 2012; Yin et al. 2014
miR172	*RAP2-7*	Ethylene-responsive transcription factor RAP2-7-like	Organ morphogenesis Phase change Response to abscisic acid stimulus Response to salt stress	TAIR; Yin et al. 2017
miR398	*CSD*	Copper superoxide dismutases	Oxidative stress tolerance	Pacheco et al. 2012

(Continued)

TABLE 17.3 (CONTINUED)
Function of Host miRNAs and Their mRNA Targets Responsive to PVY[a]

miRNA	Target Gene	Target Protein Class	Function	References
miR482	*Gpa2*	Disease resistance protein Gpa2	Defense response Incompatible interaction Hypersensitive response	Zhang et al. 2013; TAIR
miR482	*R*	Resistance protein N	Defense response Signal transduction	Zhang et al. 2013; TAIR; PGR
miR482	*CC-NBS-LRR*	CC-NBS-LRR resistance protein, HL06P13.14 protein	Defense response	Zhang et al. 2013; TAIR; PGR
miR482	*NBS-LRR*	NBS-LRR protein, Rx protein	Defense response Incompatible interaction Hypersensitive response	Zhang et al. 2013; PGR
miR482	*NBS*	NBS-coding resistance gene analog	Defense response	Zhang et al. 2013

Note: TAIR, The Arabidopsis Information Resource (http://www.arabidopsis.org) PGR, Potato Genomic Resource (potato.plantbiology.msu.edu/cgi-bin/annotation_report.cgi).

[a] See Table 17.2.

In addition, elevated accumulation of some transcription factor (TF) and non-TF mRNA targets together with the corresponding miRNAs was observed in PVY-infected potato and/or *N. benthamiana* (Pacheco et al. 2012; Yin et al. 2017; Tables 17.2 and 17.3). The TF targets included squamosa promoter binding protein-like TF (*SPL9*), scarecrow-like TF (*SCL6*), and apetala 2-like ethylene-responsive TF TOE3 like (*TOE3*), and the non-TF target was copper superoxide dismutase (*CSD*). The parallel increase in expression of certain miRNAs and their host mRNA targets might be the effect of the viral RSS on miRNA-guided cleavage of endogenous transcripts. Previous studies demonstrated that the potyvirus multifunctional HC-Pro protein is an RSS (Anandalakshmi et al. 1998; Brigneti et al. 1998; Kasschau and Carrington 1998; Urcuqui-Inchima et al. 2001). The siRNA binding feature of the PVY HC-Pro was confirmed by the surface plasmon resonance (SPR) assay (Shimura et al. 2008). The FRNK motif, a probable point of contact with siRNA and miRNA duplexes, which is highly conserved in the potyvirus HC-Pro proteins, is also presented in the PVY HC-Pro (Shiboleth et al. 2007). According to the model suggested by Shiboleth et al. (2007), in uninfected plants, miRNA is incorporated into the RISC, and miRNA* is degraded. RISC then down-regulates the expression of a target mRNA, complementary to the miRNA, through cleavage, whereas in the potyvirus-infected plants, HC-Pro can bind the miRNA/miRNA* duplex, effectively sequestering miRNA and causing ectopic expression of target mRNA (Shiboleth et al. 2007).

Strain-specific alteration of plant mRNA targets of miRNAs was observed in potato-PVY HR type interactions. The parallel increase of the target transcripts *AGO1*, *DCL1*, *TOE3*, and *Gpa2/CC-NBS-LRR*, together with their corresponding miRNAs miR168, miR162, miR172, and miR482, respectively, were observed only in potato cv. Etola infected with strain $PVY^{N\text{-}Wi}$ showing partial HR resistance and severe symptoms (Yin et al. 2017). The same miRNA/target sets were not altered in Etola plants infected with strain PVY^{Z}-NTN showing partial HR resistance and mild symptoms, nor in Etola inoculated with strain PVY^{NTN} showing HR resistance to PVY^{NTN}.

By contrast, in PVY-infected potato, increased expression of miR172 did not induce any changes in the levels of one of its mRNA targets, ethylene-responsive transcription factor *RAP2-7* (Yin et al. 2017; Table 17.2). *RAP2-7* is an abiotic-responsive TF, which may not responsive to biotic stress such as PVY infection. Alternatively, the miRNA enrichment may occur in specific tissues in which the majority of the corresponding targets are not expressed (Hu et al. 2011). The enriched miRNAs may also function in miRNA-guided translational repression (Brodersen et al. 2008; Brodersen and Voinnet 2009; Várallyay et al. 2010) or mediate DNA methylation (Wu et al. 2010; Hu et al. 2014).

17.7 HOST miRNAs PREDICTED TO TARGET PVY GENOME

Potato miRNAs with the potential to target PVY genome have been predicted (Iqbal et al. 2016). In total, 86 potato miRNAs are predicted to target the PVY genome at 151 different sites. Among them, the miRNA families of *S. tuberosum*, such as miR166c-3p, miR482e-5p, miR5303a, miR5303d, miR8004, miR8032b-5p, miR8032c, miR8032e-5p, miR162b-3p, miR164-3p, miR160a-5p, miR8011a-5p, and miR8018, were found to target PVY at multiple loci. The *CI* gene was targeted by 32 different miRNAs, followed by *NIb*, *HC-Pro*, *NIa-Pro*, *VPg*, and *CP*, which were targeted by 26, 19, 18, 16, and 13 miRNAs, respectively. The predicted miRNAs have the potential to be used for the development of PVY-resistant potato.

17.8 VIRAL vsiRNAs IN PLANT–POTYVIRUS INTERACTION

The potyvirus-derived vsiRNAs have shown GC preference. They may target different regions of the viral genomes and the host mRNAs as well. The PVY-derived vsiRNAs exhibit strain-specificity. PVY^{NTN}-infected potato plants produced the highest population of vsiRNA accumulation in comparison to that of PVY^{N} and PVY^{O} infected ones. The PVY vsiRNAs are biased to the positive strand of the genome, with the 21-nt class vsiRNAs being the most prevalent. *In silico* prediction revealed the putative host mRNAs targeted by PVY vsiRNAs, which include TFs having a known role in leaf development and symmetry. The vsiRNA-based deep sequencing has been used for the study of within-host PVY population structure.

17.9 *POTYVIRUS*-DERIVED SMALL INTERFERING RNAs

Among many characterized RNA viruses, the vsiRNAs derived from potyviruses were studied in plants infected with TuMV, *Watermelon mosaic virus* (WMV), *Sugarcane mosaic virus* (SCMV), and PVY (Ho et al. 2007, 2008; Donaire et al. 2009; Xia et al. 2014, 2016; Catalano et al. 2012; Nie and Molen 2015; Naveed et al. 2014) (Table 17.4). The potyvirus-derived vsiRNAs have shown GC preference. They are predicted to target different regions of the viral genomes and the host mRNAs as well. The host mRNAs targeted by potyvirus vsiRNAs are involved in metabolic process, regulation of transcription, protein phosphorylation, and oxidation-reduction process, among others (Xia et al. 2014, 2016).

17.10 PVY-DERIVED SMALL INTERFERING RNAs

The profiles of PVY-derived vsiRNAs were analyzed in the infected potato (*S. tuberosum*) by deep sequencing (Naveed et al. 2014), in the infected tomato (*S. lycopersicum*) by *in silico* prediction (Catalano et al. 2012), and in the infected tobacco and potato by Northern blot (Nie and Molen 2015). The characteristics of the PVY-derived vsiRNAs are summarized in Table 17.5. Naveed et al. (2014) investigated the levels of vsiRNAs in potato cv. Russet Burbank infected with three strains of PVY, i.e., the ordinary strain PVY^{O}, the tobacco veinal-necrotic strain PVY^{N}, and the tuber necrotic strain PVY^{NTN}. The three PVY strains interact differently in the same host genetic background. PVY^{NTN}-infected plants produced the highest population of vsiRNA accumulation in comparison to that of PVY^{N} and PVY^{O} infected plants. The plants developed stronger RNA interference mechanism towards necrotic strains PVY^{N} and PVY^{NTN}, which resulted in an increased vsiRNA accumulation (Naveed et al. 2014). However, plants exhibiting a weak gene-silencing response upon PVY^{O} infection resulted in less vsiRNA accumulation. The 21- and 22-nt vsiRNAs were relatively abundant, indicating the importance of DCL4 and DCL2 in antiviral RNA silencing (Ding 2010; Llave 2010; Naveed et al. 2014). The PVY-derived vsiRNAs were biased to the positive strand of the genome

TABLE 17.4
Virus-Derived Small Interfering RNAs (vsiRNA) in Potyvirus-Infected Plants

Virus	Host	Methodology	vsiRNA/Host mRNA Target	References
TuMV	*Brassica juncea*	Sequencing	842 TuMV-siRNA with GC bias	Ho et al. 2007
CSV	*Dactylis glomerata*	Sequencing	1057 CSV-siRNA with GC bias	Ho et al. 2008
WMV	*Cucumis melon*	Deep sequencing	1473 vsRNA	Donaire et al. 2009
TuMV	*Arabidopsis thaliana*		497 vsRNA	
SCMV	Maize	Deep sequencing	6220433 vsiRNA reads Predicted maize mRNA targets involving in gene expression, energy metabolism, signal transduction, transcriptional regulation, cell defense	Xia et al. 2014
SCMV	Maize	Deep sequencing	6740592 vsiRNA reads 1560 predicted maize mRNA targets	Xia et al. 2016
PVY	*Solanum lycopersicum*	*In silico* prediction	11430 vsiRNA reads 381 tomato mRNA targets	Catalano et al. 2012
PVY	*S. tuberosum*	Deep sequencing	359884 vsiRNA reads	Naveed et al. 2014
PVY	*S. tuberosum* *Nicotiana tabacum*	Northern blot	PVY-derived small RNAs were detected in both tobacco and potato plants, correlating with PVY RNA levels	Nie and Molen 2015

Note: CSV, *Cocksfoot streak virus*; PVY, *Potato virus Y*; SMV, *Soybean mosaic virus*; SCMV, *Sugarcane mosaic virus*; TuMV, *Turnip mosaic virus*; vsRNA, viral small RNA; vsiRNA, virus-derived small interfering RNA; WMV, *Watermelon mosaic virus*.

(Naveed et al. 2014). PVY is a single-stranded positive sense RNA virus. Bias toward the positive sense strand of the viral genome might give the host an advantage to suppress the virus at an early stage of its life cycle and stop transcription of the viral genome (Naveed et al. 2014). Qi et al. (2009) suggested that the nascent viral RNA strands generated by RNA-dependent RNA polymerases (RDRs) are chemically modified to prevent the negative strand of vsiRNA duplex from entering into an AGO complex. Nie and Molen (2015) demonstrated that PVY-derived vsiRNAs were a population of vsiRNAs that might have been produced from all three sections (i.e., 5′ end to nt 3300, nt 3200–6500, and nt 6400-3′ end) of PVY genome in $PVY^{N:O}$-Mb58-infected tobacco and potato cvs Ranger Russet and Cal White. The estimated length of PVY-derived vsiRNAs was 21–24 nt (Nie and Molen 2015).

17.10.1 vsiRNAs Targeting Host Transcripts

Several studies provided experimental evidence that some of the vsiRNAs target host transcripts for post-transcriptional regulation (Smith et al. 2011; Miozzi et al. 2013; Xia et al. 2014). In potyvirus SCMV-infected maize, the vsiRNAs target dozens of host transcripts, some of which encode proteins involved in ribosome biogenesis and in biotic and abiotic stresses (Xia et al. 2014). In PVY-infected tomato, *in silico* prediction revealed that the putative host mRNAs targeted by vsiRNAs include transcription factors having a known role in leaf development and symmetry (Catalano et al. 2012). Modulation of these vsiRNA-targeted host genes would mediate disease symptoms in PVY-infected tomato (Catalano et al. 2012). Recently, two studies reported simultaneously that the yellowing symptoms induced in *N. tabacum* by CMV Y-satellite RNA (Y-Sat) is the consequence of the

TABLE 17.5
Characteristics of Virus-Derived Small Interfering RNAs (vsiRNA) in PVY-Infected Plants

Host	***Solanum lycopersicum* (Catalano et al. 2012)**		***S. tuberosum* (Naveed et al. 2014)**			***S. tuberosum Nicotiana tabacum* (Nie and Molen 2015)**
Virus isolate/ strain	PVYc-to	PVY-SON41	PVY^{O}	PVY^{N}	PVY^{NTN}	$PVY^{N:O}$-Mb58
Symptoms in host	Severe leaf distortion	Mild mosaic in leaves	Mosaic and mottling in leaves	No discernible symptoms in leaves	Mosaic and mottling in leaves	Severe vein clearing and malformation, chlorosis in tobacco; mosaic or no symptoms in potato
Methods for vsiRNAs analysis	*In silico* prediction		Deep sequencing			Northern blot
Size of vsiRNAs	21 nt	21 nt	18 to 26 nt, with the 21 nt predominant			21 nt
Total vsiRNAs (% of total small RNAs)	nd	nd	0.335%	0.406%	0.493%	nd
vsiRNAs of 21 nt (reads)	1750 identical 9671 dissimilar	1750 identical 9680 dissimilar	236545	204841	359884	nd
vsiRNAs of 22 nt (reads)	nd	nd	36087	35102	71741	nd
5' nucleotide on vsiRNAs of 21 nt	nd	nd	5′-U 5′-A	5′-U 5′-A	5′-U 5′-A	nd
Hotspots for vsiRNAs of 21 nt	nd	nd	*CI* *NIb*	*NIb* *CI* *CP*	*P1* *NIb*	3′ end–nt 3300 nt 3200–6500 nt 6400–3′ end
PVY genome regions produced higher number of vsiRNAs	nd	nd	***CI***[a] *NIb* *HC-Pro*	***NIb***[a] *CI* *HC-Pro*	***CI***[a] *NIb* *HC-Pro*	nd
PVY genome regions produced the least number of vsiRNAs	nd	nd	*6K1* *6K2*	*6K1* *6K2*	*6K1* *6K2*	nd
The ratio of sense to antisense strand vsiRNAs	nd	nd	1.2:1	1.25:1	1.1:1	nd

(*Continued*)

TABLE 17.5 (CONTINUED)
Characteristics of Virus-Derived Small Interfering RNAs (vsiRNA) in PVY-Infected Plants

Host	*Solanum lycopersicum* (Catalano et al. 2012)		*S. tuberosum* (Naveed et al. 2014)			*S. tuberosum Nicotiana tabacum* (Nie and Molen 2015)
Putative host-targets of vsiRNAs	315 tomato mRNAs	381 tomato mRNAs	nd	nd	nd	nd
	Mainly transcription factors related to leaf development and symmetry					

Note: 6K1, 6K1 peptide; 6K2, 6K2 peptide; CI, cytoplasmic inclusion protein; CP, coat protein; HC-Pro, helper component-proteinase; nd, not determined; NIb, nuclear inclusion b protein; nt, nucleotides; P1, P1 serine proteinase; PVY, Potato virus Y; vsiRNA, virus-derived small interfering RNAs.
[a] The region produced the highest number of vsiRNAs.

tobacco chlorophyll biosynthetic gene (*CHL1* mRNA) down-regulation mediated by CMV Y-Sat-derived siRNAs (Shimura et al. 2011; Smith et al. 2011).

17.10.2 vsiRNA-Based Deep Sequencing Used for Viral Genomic Study

vsiRNA-based deep sequencing was used for the construction of virus population and discovery of novel viruses or stains. Kutnjak et al. (2015) provided the first study to compare the within-host PVY population structure by deep sequencing obtained from two coexisting PVY sequence pools, i.e., RNA isolated from viral particles and vsiRNA. Both pools give a similar mutational landscape. More unique single-nucleotide polymorphisms (SNPs) occurred in the vsiRNA pool, while nonhomologous recombinations were commonly detected in the viral particle pool. A novel potyvirus tentatively named *Pecan mosaic-associated virus* (PMaV) was discovered in pecan (*Carya illinoensis*) by deep sequencing of vsiRNA (Su et al. 2016).

17.11 PLANT-INFECTING VIRUS-ENCODED miRNAs

The virus-encoded miRNAs were first identified in the *Epstein-Barr virus* (EBV), a causative agent of infectious mononucleosis (Pfeffer et al. 2004). Many animal-infecting virus-encoded miRNAs have evolved to regulate viral transcripts or networks of host genes that are unique to viral miRNAs, and some may mimic host miRNAs and regulate numerous host transcripts that comprise the network of genes regulated by a particular host miRNA (Grundhoff and Sullivan 2011). However, there is little evidence of the existence of plant-infecting virus-derived miRNAs. The possible reason may be that many plant viruses are RNA viruses, and RNA viruses are not known to enter the plant cell nucleus, where the synthesis of miRNAs starts (Ramesh et al. 2014).

According to the available data published in the peer-reviewed journals, the plant virus-encoded miRNAs were first identified in a single-stranded positive-sense RNA virus named *Hibiscus chlorotic ringspot virus* (HCRSV), which belongs to the genus *Carmovirus* (Gao et al. 2012). In HCRSV-infected kenaf (*Hibiscus cannabilis* L.) leaves, the HCRSV RNA was detected in the nucleus of the infected cells, where the viral microRNAs (vir-miRNAs) are generated (Gao et al. 2012). One of the five predicted vir-miRNAs from HCRSV, i.e., hcrsv-miR-H1-5p, was experimentally demonstrated. Since the total RNA used for detection was extracted from highly purified nuclei, the involvement of siRNAs was excluded. In addition, co-infiltration assay showed that hcrsv-miR-H1-5p could target and down-regulate p23 gene of HCRSV, which is essential for HCRSV replication. Hence the presence of hcrsvmiR-H1-5p may inhibit viral replication. A mechanism has been suggested that a single RNA transcript containing both the protein-coding region and the miRNA coding sequence can be potentially expressed coordinately to produce a protein and to generate miRNAs (Allen et al. 2004; Cai et al. 2004). Similarly, in HCRSV-infected kenaf plants, the p23 coding region might be able to generate both the p23 protein and vir-miRNAs (Gao et al. 2012).

The plant virus-encoded miRNAs were also identified in potyvirus TuMV, poacevirus *Sugarcane streak mosaic virus* (SCSMV), and the begomoviruses *African cassava mosaic virus* (ACMV) and *East African cassava mosaic virus-Uganda* (EACMV-UG) (Shazia 2010; Maghuly et al. 2014; Viswanathan et al. 2014). In addition, the miRNA length variants, called isomiRs, which exhibit variations in a few nucleotides at the 3′ or 5′ end leading to the production of multiple mature variants, were identified in ACMV and EACMV-UG by *in silico* prediction and stem-loop reverse transcription-polymerase chain reaction (RT-PCR) (Maghuly et al. 2014). Some predicted viral isomiRs even share a common region (nucleotides 2–8 at the 5′ end), and these isomiRs also share the same targets (Maghuly et al. 2014), consistent with those reported by Cloonan et al. (2011) that isomiRs are biologically relevant and target pathways of functionally related genes. However, no published data were available on miRNAs encoded by PVY.

The newly identified plant viral miRNAs are involved in regulation of viral replication or predict to target host genes that involve in biotic response, metabolic pathways, and transcription factors (Shazia 2010; Gao et al. 2012; Maghuly et al. 2014; Viswanathan et al. 2014). Although the animal-infecting virus-encoded miRNAs mainly utilize the same processing and effector machinery as host miRNAs for their own biogenesis (Grundhoff and Sullivan 2011), the pathway for plant viral miRNA biogenesis varies at certain steps from the miRNA biogenesis pathway known to exist in plants, as analyzed by using *Arabidopsis* RNAi pathway mutants (Shazia 2010).

The experimentally verified plant virus-encoded miRNAs are summarized in Table 17.6.

17.12 CONCLUSION

This chapter summarizes the host miRNAs, the viral vsiRNAs, and vir-miRNAs in plants infected with certain potyviruses, with highlights on the plant-PVY phytosystems. The potyvirus-responding host miRNAs and their host mRNA targets showed common and strain-specific regulation, e.g., in TuMV-infected *Brassica rapa* and Chinese cabbage and in SMV-infected soybean. PVY infection caused alteration of a set of conserved and stress-related host miRNAs in *N. tabacum*, *N. benthamiana*, and/or *Solanum tuberosum*. Some PVY-responding host miRNA/target sets are differentially regulated depending on plant species, PVY strain, and symptom severity. Alteration of miRNAs, i.e., miR162, miR168, miR172, and miR482, and their targets showed strain-specificity in potato-PVY HR reactions, and might be related to symptom severity. In PVY-infected plants, the mRNA targets and their corresponding miRNAs show an antagonistic expression or up-regulation. Bioinformatics study predicted that 86 potato miRNAs have the potential to target the PVY genome at 151 different sites, and these potato miRNAs can be used for development of PVY-resistant potato. The potyvirus-derived vsiRNAs have shown GC preference and may target different regions of viral genomes. The PVY-derived vsiRNAs exhibit strain-specificity. PVY^{NTN}-infected plants produced the highest population of vsiRNA accumulation in comparison to that of PVY^{N} and PVY^{O}-infected *S. tuberosum* plants. The PVY vsiRNAs are biased to the positive strand of the genome, with the 21-nt class vsiRNA being the most prevalent. *In silico* prediction revealed the putative host mRNAs targeted by PVY vsiRNAs, which include transcription factors having a known role in leaf development. The vsiRNA-based deep sequencing has been used for the study of within-host PVY population structure. The HC-Pro encoded by PVY, an RSS, may influence the miRNA pathway or the antiviral RNA silencing pathway by small RNA binding. Although not found in PVY, the plant-infecting virus-encoded miRNAs have been identified in several DNA and RNA viruses, including the potyvirus TuMV. The vir-miRNAs were predicted to regulate the expression of viral- and/or host-encoded genes.

TABLE 17.6
Experimentally Verified Plant Virus-Encoded miRNAs (vir-miRNA) in Virus-Infected Plants

Virus	Family/Genus	Host	Methodology	vir-miRNA/Target	References
TuMV	*Potyviridae/ Potyvirus*	*Arabidopsis thaliana*	*In silico* prediction Northern blot	TuMV-mir-S1 TuMV-mir-S2 Predicted to target *Arabidopsis* stress responsive gene *HVA22D*	Shazia 2010[a] PhD thesis
HCRSV	*Tombusviridae/ Carmovirus*	*Hibiscus cannabilis*	*In silico* prediction Northern blot Stem-loop real-time PCR Co-infiltration assay	5 vir-miRNAs predicted One vir-miRNA named hcrsv-miR-H1-5p was detected and it can downregulate p23 gene HSRSV	Gao et al. 2012
SCSMV	*Potyviridae/ Poacevirus*	Sugarcane	Computationally prediction Stem-loop RT-qPCR	SCSMV miR16 19 target genes were predicted in sugarcane and other plant species such as *Arabidopsis*, rice, sorghum involving cellular component, molecular function, and biological process	Viswanathan et al. 2014
ACMV EACMV-UG	*Geminiviridae/ Begomovirus*	*Jatropha* Cassava	Computationally prediction Stem-loop RT-qPCR	84 out of 111 miRNAs/ miRNAs* corresponded to ACMV and 27 to EACMV-UG isomiRs were predicted in both viruses In *Jatropha*: 234 targets were predicted for 78 ACMV- and 27 EACMV-UG miRNAs/miRNAs* by RNAhybrid analyses 621 targets were predicted for 79 ACMV and 26 EACMV-UG-miRNAs/miRNAs by psRNATarget	Maghuly et al. 2014

(Continued)

TABLE 17.6 (CONTINUED)
Experimentally Verified Plant Virus-Encoded miRNAs (vir-miRNA) in Virus-Infected Plants

Virus	Family/Genus	Host	Methodology	vir-miRNA/Target	References
				In cassava: 370 targets were predicted for 84 ACMV- and 27 EACMV-UG-miRNAs/miRNAs* by RNAhybrid analyses 688 targets were predicted for 81 ACMV and 26 EACMV-UG-miRNAs/miRNAs by psRNATarget The predicted target genes in *Jatropha* and cassava may involve in biotic response, metabolic pathways and transcription factors	

Note: ACMV, *African cassava mosaic virus*; EACMV-UG, *East African cassava mosaic virus-Uganda*; HCRSV, *Hibiscus chlorotic ringspot virus*; isomiR, microRNA length variant; SCSMV, *Sugarcane streak mosaic virus*; TuMV, *Turnip mosaic virus*; vir-miRNA, viral microRNA.

[a] espace.library.uq.edu.au/view/UQ:242468.

17.13 ABBREVIATIONS

6K1	6K1 peptide
6K2	6K2 peptide
14-3-3	14-3-3 protein 7
ACMV	*African cassava mosaic virus*
AGO1	Argonaute 1
ARF	auxin response factor
ATHB-9	homeobox-leucine zipper protein ATHB-9 like
ATHB-14	homeobox-leucine zipper protein ATHB-14 like
Avr	avirulence protein
CC-NBS-LRR	proteins with coiled coil (CC) - nucleotide binding site (NBS) - leucine rich repeat (LRR) domains
CHL1	tobacco chlorophyll biosynthetic gene
CI	cytoplasmic inclusion protein
CMV	*Cucumber mosaic virus*
CP	coat protein
CSD	copper superoxide dismutase
CSV	*Cocksfoot streak virus*
DCL1	RNAse III-like Dicer-like I endonuclease

DEM	differentially expressed miRNA
DEG	differentially expressed gene
dsRNA	double-stranded RNA
EACMV-UG	*East African cassava mosaic virus-Uganda*
EBV	*Epstein-Barr virus*
ETI	effector-triggered immunity
ETS	effector-triggered susceptibility
Gpa2	disease resistance protein Gpa2
HC-Pro	helper-component proteinase
HCRSV	*Hibiscus chlorotic ringspot virus*
HR	hypersensitive response
isomiR	microRNA length variant
MAMPs/PAMP	microbial/pathogen-associated molecular pattern
miRNA	microRNA
mRNA	messenger RNA
NB-LRR	proteins with nucleotide binding (NB) and leucine-rich repeats (LRR) domains
NBS	NBS-coding resistance gene analog
nd	not determined
NIa	nuclear inclusion a protein
NIb	nuclear inclusion b protein
nt	nucleotides
P1	P1 serine proteinase
p23	p23 protein of *Hibiscus chlorotic ringspot virus*
PGR	potato genomic resource
PMaV	*Pecan mosaic-associated virus*
PRR	pattern recognition receptor
PTI	PAMP-triggered immunity
PVY	*Potato virus Y*
RAP2-7	ethylene-responsive transcription factor RAP2-7 like
RDR	RNA-dependent RNA polymerase
RISC	RNA-induced silencing complex
RSS	RNA silencing suppressor
RT-PCR	reverse transcription - polymerase chain reaction
SCL	scarecrow-like transcription factor
SCMV	*Sugarcane mosaic virus*
SCSMV	*Sugarcane streak mosaic virus*
siRNA	small interfering RNA
SMV	*Soybean mosaic virus*
SNP	single nucleotide polymorphism
SPL	squamosa promoter binding protein-like transcription factor
SPR	surface plasmon resonance
SRBSDV	*Southern rice black-streaked dwarf virus*
sRNA	small RNA
TAIR	arabidopsis information resource
TCV	*Turnip crinkle virus*
TEV	*Tobacco etch virus*
TF	transcription factor
TMV	*Tobacco mosaic virus*
TOE3	apetala2-like ethylene-responsive transcription factor TOE3 like
ToLCNDV	*Tomato leaf curl New Delhi virus*

ToMV	*Tomato mosaic virus*
TRV	*Tobacco rattle virus*
TuMV	*Turnip mosaic virus*
TYLCCNV	*Tomato yellow leaf curl China virus*
vir-miRNA	viral microRNA
VPg	viral genome-linked protein
vsiRNA	virus-derived small interfering RNA
vsRNA	viral small RNA
WMV	*Watermelon mosaic virus*
Y-Sat	*Cucumber mosaic virus* Y-satellite RNA

ACKNOWLEDGMENTS

The author would like to thank Professor Ewa Zimnoch-Guzowska (Młochów Research Center, Plant Breeding and Acclimatization Institute—National Research Institute, Poland) for her helpful suggestions on the manuscript. The author would like to thank the editor Professor Khurana SMP's helpful comments on improving the manuscript. The research was supported in part by a Statutory Project 1-3-00-3-03 from the Polish Ministry of Agriculture and Rural Development.

REFERENCES

Allen E, Xie Z, Gustafson AM, Sung GH, Spatafora JW, Carrington JC (2004). Evolution of microRNA genes by inverted duplication of target gene sequences in *Arabidopsis thaliana. Nature Genetics* 36: 1282–1290.

Alvarado VY, Scholthof HB (2012). AGO2: A new agonaute compromising plant virus accumulation. *Frontiers in Plant Science* 2: 112.

Amin I, Patil BL, Briddon RW, Mansoor S, Fauquet CM (2011). A common set of developmental miRNAs are upregulated in *Nicotiana benthamiana* by diverse begomoviruses. *Virology Journal* 8: 143.

Anandalakshmi R, Pruss GJ, Ge X, Marathe R, Mallory AC, Smith TH, Vance VB (1998). A viral suppressor of gene silencing in plants. *Proceedings of the National Academy of Sciences of the United States of America* 95: 13079–13084.

Bartel DP (2004). MicroRNAs: Genomics biogenesis mechanism and function. *Cell* 116: 281–297.

Bazzini AA, Hopp HE, Beachy RN, Asurmendi S (2007). Infection and co-accumulation of tobacco mosaic virus proteins alter microRNA levels correlating with symptom and plant development. *Proceedings of the National Academy of Sciences of the United States of America* 104: 12157–12162.

Bazzini AA, Manacorda CA, Tohge T, Conti G, Rodriguez MC, Nunes-Nesi A, Villanueva S, Fernie AR, Carrari F, Asurmendi S (2011). Metabolic and miRNA profiling of TMV infected plants reveals biphasic temporal changes. *PLoS One* 6: e28466.

Bortolamiol D, Pazhouhandeh M, Marrocco K, Genschik P, Ziegler-Graff V (2007). The *Polerovirus* F box protein P0 targets ARGONAUTE1 to suppress RNA silencing. *Current Biology* 17: 1615–1621.

Brigneti G, Voinnet O, Li WX, Ji LH, Ding SW, Baulcombe DC (1998). Viral pathogenicity determinants are suppressors of transgene silencing in *Nicotiana benthamiana. The EMBO Journal* 17: 6739–6746.

Brodersen P, Sakvarelidze-Achard L, Bruun-Rasmussen M, Dunoyer P, Yamamoto YY, Sieburth L, Voinnet O (2008). Widespread translational inhibition by plant miRNAs and siRNAs. *Science* 320: 1185–1190.

Brodersen P, Voinnet O (2009). Revisiting the principles of microRNA target recognition and mode of action. *Nature Reviews Molecular Cell Biology* 10: 141–148.

Cai X, Hagedorn CH, Cullen BR (2004). Human microRNAs are processed from capped polyadenylated transcripts that can also function as mRNAs. *RNA* 10: 1957–1966.

Catalano D, Cillo F, Finetti-Sialer M (2012). *In silico* prediction of virus-derived small interfering RNAs and their putative host messenger targets in *Solanum lycopersicum* infected by different potato virus Y isolates. *EMBnet Journal* 18A: 83–84.

Chapman EJ, Carrington JC (2007). Specialization and evolution of endogenous small RNA pathways. *Nature Reviews Genetics* 8: 884–896.

Chen X (2009). Small RNAs and their roles in plant development. *Annual Review of Cell and Developmental Biology* 25: 21–44.
Chen X (2010). Small RNAs-secrets and surprise of the genome. *The Plant Journal* 61: 941–958.
Chen H, Arsovski AA, Yu K, Wang A (2016). Genome-wide investigation using sRNA-seq degradome-seq and transcriptome-seq reveals regulatory networks of microRNAs and their target genes in soybean during *Soybean mosaic virus* infection. *PLoS One* 11: e0150582.
Chen H, Arsovski AA, Yu K, Wang A (2017). Deep sequencing leads to the identification of eukaryotic translation initiation factor 5a as a key element in *Rsv1*-mediated lethal systemic hypersensitive response to *Soybean mosaic virus* infection in soybean. *Molecular Plant Pathology* 18: 391–404.
Cillo F, Mascia T, Pasciuto MM, Gallitelli D (2009). Differential effects of mild and severe *Cucumber mosaic virus* strains in the perturbation of microRNA-regulated gene expression in tomato map to the 3′ sequence of RNA 2. *Molecular Plant-Microbe Interactions* 22: 1239–1249.
Cloonan N, Wani S, Xu Q, Gu J, Lea K et al. (2011). MicroRNAs and their isomiRs function cooperatively to target common biological pathways. *Genome Biology* 12: R126.
Deleris A, Gallego-Bartolome J, Bao J, Kasschau KD, Carrington JC, Voinnet O (2006). Hierarchical action and inhibition of plant Dicer-like proteins in antiviral defense. *Science* 313: 68–71.
Ding SW (2010). RNA-based antiviral immunity. *Nature Reviews Immunology* 10: 632–644.
Ding SW, Voinnet O (2007). Antiviral immunity directed by small RNAs. *Cell* 130: 413–426.
Donaire L, Wang Y, Gonzalez-Ibeas D, Mayer KF, Aranda MA, Llave C (2009). Deep-sequencing of plant viral small RNAs reveals effective and widespread targeting of viral genomes. *Virology* 392: 203–214.
Donze T, Qu F, Twigg P, Morris TJ (2014). *Turnip crinkle virus* coat protein inhibits the basal immune response to virus invasion in *Arabidopsis* by binding to the NAC transcription factor TIP. *Virology* 449: 207–214.
Du P, Wu J, Zhang J, Zhao S, Zheng H, Gao G, Wei L, Li Y (2011). Viral infection induces expression of novel phased microRNAs from conserved cellular microRNA precursors. *PLoS Pathogens* 7: e1002176.
Feng J, Lai L, Lin R, Jin C, Chen J (2012). Differential effects of *Cucumber mosaic virus* satellite RNAs in the perturbation of microRNA-regulated gene expression in tomato. *Molecular Biology Reports* 39: 775–784.
Feng J, Liu S, Wang M, Lang Q, Jin C (2014). Identification of microRNAs and their targets in tomato infected with *Cucumber mosaic virus* based on deep sequencing. *Planta* 240: 1335–1352.
Gao R, Liu P, Wong SM (2012). Identification of a plant viral RNA genome in the nucleus. *PLoS One* 7: e48736.
Garcia-Ruiz H, Takeda A, Chapman EJ, Sullivan CM, Fahlgren N, Brempelis KJ, Carrington JC (2010). *Arabidopsis* RNA-dependent RNA polymerases and dicer-like proteins in antiviral defense and small interfering RNA biogenesis during *Turnip Mosaic Virus* infection. *The Plant Cell* 22: 481–496.
Grundhoff A, Sullivan CS (2011). Virus-encoded microRNAs. *Virology* 411: 325–343.
Hamilton AJ, Baulcombe DC (1999). A species of small antisense RNA in posttranscriptional gene silencing in plants. *Science* 286: 950–952.
Harvey JJ, Lewsey MG, Patel K, Westwood J, Heimstädt S, Carr JP, Baulcombe DC (2011). An antiviral defense role of AGO2 in plants. *PLoS One* 6: e14639.
He XF, Fang YY, Feng L, Guo HS (2008). Characterization of conserved and novel microRNAs and their targets including a TuMV-induced TIR-NBS-LRR class *R* gene-derived novel miRNA in Brassica. *FEBS Letters* 582: 2445–2452.
Ho T, Wang H, Pallett D, Dalmay T (2007). Evidence for targeting common siRNA hotspots and GC preference by plant Dicer-like proteins. *FEBS Letters* 581: 3267–3272.
Ho T, Rusholme Pilcher RL, Edwards ML, Cooper I, Dalmay T, Wang H (2008). Evidence for GC preference by monocot Dicer-like proteins. *Biochemical and Biophysical Research Communications* 368: 433–437.
Hu Q, Hollunder J, Niehl A, Kørner CJ, Gereige D, Windels D, Arnold A, Kuiper M, Vazquez F, Pooggin M, Heinlein M (2011). Specific impact of tobamovirus infection on the *Arabidopsis* small RNA profile. *PLoS One* 6: e19549
Hu W, Wang T, Xu J, Li H (2014). MicroRNA mediates DNA methylation of target genes. *Biochemical and Biophysical Research Communications* 444: 676–681.
Iqbal MS, Hafeez MN, Wattoo JI, Ali A, Sharif MN, Rashid B, Tabassum B, Nasir IA (2016). Prediction of host-derived miRNAs with the potential to target PVY in potato plants. *Frontiers in Genetics* 7: 159.
Ivanov KI, Eskelin K, Lõhmus A, Mäkinen K (2014). Molecular and cellular mechanisms underlying potyvirus infection. *Journal of General Virology* 95: 1415–1429.
Jin D, Wang Y, Zhao Y, Chen M (2013). MicroRNAs and their cross-talks in plant development. *The Journal of Genetics and Genomics* 40: 161–170.
Jin H (2008). Endogenous small RNAs and antibacterial immunity in plants. *FEBS Letters* 582: 2679–2684.
Jones JD, Dangl JL (2006). The plant immune system. *Nature* 444: 323–329.

Jones-Rhoades MW, Bartel DP, Bartel B (2006). MicroRNAs and their regulatory roles in plants. *Annual Review of Plant Biology* 57: 19–53.

Kasschau KD, Carrington JC (1998). A counterdefensive strategy of plant viruses: Suppression of posttranscriptional gene silencing. *Cell* 95: 461–470.

Kasschau KD, Xie Z, Allen E, Llave C, Chapman EJ, Krizan KA, Carrington JC (2003). P1/HC-Pro a viral suppressor of RNA silencing interferes with *Arabidopsis* development and miRNA function. *Developmental Cell* 4: 205–217.

Khraiwesh B, Zhu JK, Zhu J (2012). Role of miRNAs and siRNAs in biotic and abiotic stress responses of plants. *Biochimica et Biophysica Acta* 1819: 137–148.

Kutnjak D, Rupar M, Gutierrez-Aguirre I, Curk T, Kreuze JF, Ravnikar M (2015). Deep sequencing of virus-derived small interfering RNAs and RNA from viral particles shows highly similar mutational landscapes of a plant virus population. *Journal of Virology* 89: 4760–4769.

Lakhotia N, Joshi G, Bhardwaj AR, Katiyar-Agarwal S, Agarwal M, Jagannath A, Goel S, Kumar A (2014). Identification and characterization of miRNAome in root stem leaf and tuber developmental stages of potato (*Solanum tuberosum* L) by high-throughput sequencing. *BMC Plant Biology* 14: 6.

Lang Q, Jin C, Lai L, Feng J, Chen S, Chen J (2011). Tobacco microRNAs prediction and their expression infected with *Cucumber mosaic virus* and *Potato virus X*. *Molecular Biology Reports* 38: 1523–1531.

Li F, Pignatta D, Bendix C, Brunkard JO, Cohn MM, Tung J, Sun H, Kumar P, Baker B (2012). MicroRNA regulation of plant innate immune receptors. *Proceedings of the National Academy of Sciences of the United States of America* 109: 1790–1795.

Li Y, Zhang Q, Zhang J, Wu L, Qi Y, Zhou JM (2010). Identification of microRNAs involved in pathogen-associated molecular pattern-triggered plant innate immunity. *Plant Physiology* 152: 2222–2231.

Liu HH, Tian X, Li YJ, Wu CA, Zheng CC (2008). Microarray-based analysis of stress-regulated microRNAs in *Arabidopsis thaliana*. *RNA* 14: 836–843.

Llave C (2010). Virus-derived small interfering RNAs at the core of plant-virus interactions. *Trends in Plant Science* 15: 701–707.

Maghuly F, Ramkat RC, Laimer M (2014). Virus versus host plant microRNAs: Who determines the outcome of the interaction? *PLoS One* 9: e98263.

Meins F Jr, Si-Ammour A, Blevins T (2005). RNA silencing systems and their relevance to plant development. *Annual Review of Cell and Developmental Biology* 21: 297–318.

Miozzi L, Gambino G, Burgyan J, Pantaleo V (2013). Genome-wide identification of viral and host transcripts targeted by viral siRNAs in *Vitis vinifera*. *Molecular Plant Pathology* 14: 30–43.

Moffett P (2009). Mechanisms of recognition in dominant *R* gene mediated resistance. *Advances in Virus Research* 75: 1–33.

Morel JB, Godon C, Mourrain P, Béclin C, Boutet S, Feuerbach F, Proux F, Vaucheret H (2002). Fertile hypomorphic ARGONAUTE (ago1) mutants impaired in post-transcriptional gene silencing and virus resistance. *The Plant Cell* 14: 629–639.

Nakahara KS, Masuta C (2014). Interaction between viral RNA silencing suppressors and host factors in plant immunity. *Current Opinion in Plant Biology* 20: 88–95.

Naqvi AR, Haq QM, Mukherjee SK (2010). MicroRNA profiling of tomato leaf curl New Delhi virus (tolcndv) infected tomato leaves indicates that deregulation of mir159/319 and mir172 might be linked with leaf curl disease. *Virology Journal* 7: 281.

Navarro L, Dunoyer P, Jay F, Arnold B, Dharmasiri N, Estelle M, Voinnet O, Jones JD (2006). A plant miRNA contributes to antibacterial resistance by repressing auxin signaling. *Science* 312: 436–439.

Naveed K, Mitter N, Harper A, Dhingra A, Pappu HR (2014). Comparative analysis of virus-specific small RNA profiles of three biologically distinct strains of *Potato virus Y* in infected potato (*Solanum tuberosum*) cv Russet Burbank. *Virus Research* 191: 153–160.

Nie X, Molen TA (2015). Host recovery and reduced virus level in the upper leaves after *Potato virus Y* infection occur in tobacco and tomato but not in potato plants. *Viruses* 7: 680–698.

Pacheco R, García-Marcos A, Barajas D, Martiáñez J, Tenllado F (2012). PVX-potyvirus synergistic infections differentially alter microRNA accumulation in *Nicotiana benthamiana*. *Virus Research* 165: 231–235.

Padmanabhan C, Zhang X, Jin H (2009). Host small RNAs are big contributors to plant innate immunity. *Current Opinion in Plant Biology* 12: 465–472.

Pantaleo V (2011). Plant RNA silencing in viral defence. *Advances in Experimental Medicine and Biology* 722: 39–58.

Pantaleo V, Szittya G, Burgyán J (2007). Molecular bases of viral RNA targeting by viral small interfering RNA-programmed RISC. *Journal of Virology* 81: 3797–3806.

Permar V, Singh A, Pandey V, Alatar AA, Faisal M, Jain RK, Praveen S (2014). Tospoviral infection instigates necrosis and premature senescence by micro RNA controlled programmed cell death in Vigna unguiculata. *Physiological and Molecular Plant Pathology* 88: 77–84.

Pfeffer S, Zavolan M, Grässer FA, Chien M, Russo JJ, Ju J, John B, Enright AJ, Marks D, Sander C, Tuschl T (2004). Identification of virus-encoded microRNAs. *Science* 304: 734–736.

Pradhan B, Naqvi AR, Saraf S, Mukherjee SK, Dey N (2015). Prediction and characterization of *Tomato leaf curl New Delhi virus* (ToLCNDV) responsive novel microRNAs in *Solanum lycopersicum. Virus Research* 195: 183–195.

Pumplin N, Voinnet O (2013). RNA silencing suppression by plant pathogens: Defence counter-defence and counter-counter-defence. *Nature Reviews Microbiology* 11: 745–760.

Qi X, Bao FS, Xie Z (2009). Small RNA deep sequencing reveals role for *Arabidopsis thaliana* RNA-dependent RNA polymerases in viral siRNA biogenesis. *PLoS One* 4: e4971.

Qu F, Ye X, Morris TJ (2008). Arabidopsis DRB4 AGO1 AGO7 and RDR6 participate in a DCL4-initiated antiviral RNA silencing pathway negatively regulated by DCL1. *Proceedings of the National Academy of Sciences of the United States of America* 105: 14732–14737.

Ramesh SV, Ratnaparkhe MB, Kumawat G, Gupta GK, Husain SM (2014). Plant miRNAome and antiviral resistance: A retrospective view and prospective challenges. *Virus Genes* 48: 1–14.

Ruiz-Ferrer V, Voinnet O (2009). Roles of plant small RNAs in biotic stress responses. *Annual Review of Plant Biology* 60: 485–510.

Scholthof KB, Adkins S, Czosnek H, Palukaitis P, Jacquot E, Hohn T, Hohn B, Saunders K, Candresse T, Ahlquist P, Hemenway C, Foster GD (2011). Top 10 plant viruses in molecular plant pathology. *Molecular Plant Pathology* 12: 938–954.

Shazia I (2010). *Role of viral and host MicroRNAs in plant virus interaction. PhD diss.* School of Biological Sciences, The University of Queensland (http://espacelibraryuqeduau/view/UQ:242468).

Shiboleth YM, Haronsky E, Leibman D, Arazi T, Wassenegger M, Whitham SA, Gaba V, Gal-On A (2007). The conserved FRNK box in HC-Pro a plant viral suppressor of gene silencing is required for small RNA binding and mediates symptom development. *Journal of Virology* 81: 13135–13148.

Shimura H, Fukagawa T, Meguro A, Yamada H, Oh-Hira M, Sano S, Masuta C (2008). A strategy for screening an inhibitor of viral silencing suppressors which attenuates symptom development of plant viruses. *FEBS Letters* 582: 4047–4052.

Shimura H, Pantaleo V, Ishihara T, Myojo N, Inaba J, Sueda K, Burgyán J, Masuta C (2011). A viral satellite RNA induces yellow symptoms on tobacco by targeting a gene involved in chlorophyll biosynthesis using the RNA silencing machinery. *PLoS Pathogens* 7: e1002021.

Shivaprasad PV, Chen HM, Patel K, Bond DM, Santos BA, Baulcombe DC (2012). A microRNA superfamily regulates nucleotide binding site-leucine-rich repeats and other mRNAs. *The Plant Cell* 24: 859–874.

Shukla LI, Chinnusamy V, Sunkar R (2008). The role of microRNAs and other endogenous small RNAs in plant stress responses. *Biochimica et Biophysica Acta* 1779: 743–748.

Silhavy D, Burgyán J (2004). Effects and side-effects of viral RNA silencing suppressors on short RNAs. *Trends in Plant Science* 9: 76–83.

Smith NA, Eamens AL, Wang MB (2011). Viral small interfering RNAs target host genes to mediate disease symptoms in plants. *PLoS Pathogen* 7: e1002022.

Soosaar JL, Burch-Smith TM, Dinesh-Kumar SP (2005). Mechanisms of plant resistance to viruses. *Nature Reviews Microbiology* 3: 789–798.

Su X, Fu S, Qian Y, Zhang L, Xu Y, Zhou X (2016). Discovery and small RNA profile of *Pecan mosaic-associated virus* a novel potyvirus of pecan trees. *Scientific Reports* 6: 26741.

Sunkar R (2010). MicroRNAs with macro-effects on plant stress responses. *Seminars in Cell and Developmental Biology* 21: 805–811.

Sunkar R, Li YF, Jagadeeswaran G (2012). Functions of microRNAs in plant stress responses. *Trends in Plant Science* 17: 196–203.

Takeda A, Iwasaki S, Watanabe T, Utsumi M, Watanabe Y (2008). The mechanism selecting the guide strand from small RNA duplexes is different among argonaute proteins. *Plant & Cell Physiology* 49: 493–500.

Trindade I, Capitão C, Dalmay T, Fevereiro MP, Santos DM (2010). miR398 and miR408 are up-regulated in response to water deficit in *Medicago truncatula. Planta* 231: 705–716.

Urcuqui-Inchima S, Haenni AL, Bernardi F (2001). Potyvirus proteins: A wealth of functions. *Virus Research* 74: 157–175.

Várallyay É, Válóczi A, Agyi A, Burgyán J, Havelda Z (2010). Plant virus-mediated induction of miR168 is associated with repression of ARGONAUTE1 accumulation. *The EMBO Journal* 29: 3507–3519.

Várallyay É, Oláh E, Havelda Z (2014). Independent parallel functions of p19 plant viral suppressor of RNA silencing required for effective suppressor activity. *Nucleic Acids Research* 42: 599–608.

Vaucheret H (2006). Post-transcriptional small RNA pathways in plants: Mechanisms and regulations. *Genes & Development* 20: 759–771.

Vaucheret H, Vazquez F, Crété P, Bartel DP (2004). The action of ARGONAUTE1 in the miRNA pathway and its regulation by the miRNA pathway are crucial for plant development. *Genes & Development* 18: 1187–1197.

Vazquez F, Legrand S, Windels D (2010). The biosynthetic pathways and biological scopes of plant small RNAs. *Trends in Plant Science* 15: 337–345.

Viswanathan C, Anburaj J, Prabu G (2014). Identification and validation of sugarcane streak mosaic virus-encoded microRNAs and their targets in sugarcane. *Plant Cell Reports* 33: 265–276.

Voinnet O (2008). Post-transcriptional RNA silencing in plant-microbe interactions: A touch of robustness and versatility. *Current Opinion in Plant Biology* 11: 464–470.

Voinnet O (2009). Origin biogenesis and activity of plant microRNAs. *Cell* 136: 669–687.

Wang XB, Wu Q, Ito T, Cillo F, Li WX, Chen X, Yu JL, Ding SW (2010). RNAi-mediated viral immunity requires amplification of virus-derived siRNAs in *Arabidopsis thaliana. Proceedings of the National Academy of Sciences of the United States of America* 107: 484–489.

Wang MB, Masuta C, Smith NA, Shimura H (2012). RNA silencing and plant viral diseases. *Molecular Plant-Microbe Interactions* 25: 1275–1285.

Wang Z, Jiang D, Zhang C, Tan H, Li Y, Lv S, Hou X, Cui X (2015). Genome-wide identification of turnip mosaic virus-responsive microRNAs in non-heading Chinese cabbage by high-throughput sequencing. *Gene* 571: 178–187.

Waterhouse PM, Wang MB, Lough T (2001). Gene silencing as an adaptive defence against viruses. *Nature* 411: 834–842.

Waterhouse PM, Fusaro AF (2006). Viruses face a double defense by plant small RNAs. *Science* 313: 54–55.

Wu L, Zhou H, Zhang Q, Zhang J, Ni F, Liu C, Qi Y (2010). DNA methylation mediated by a microRNA pathway. *Molecular Cell* 38: 465–475.

Xia Z, Peng J, Li Y, Chen L, Li S, Zhou T, Fan Z (2014). Characterization of small interfering RNAs derived from *Sugarcane mosaic virus* in infected maize plants by deep sequencing. *PLoS One* 9: e97013

Xia Z, Zhao Z, Chen L, Li M, Zhou T, Deng C, Zhou Q, Fan Z (2016). Synergistic infection of two viruses MCMV and SCMV increases the accumulations of both MCMV and MCMV-derived siRNAs in maize. *Scientific Reports* 6: 20520.

Xiao B, Yang X, Ye CY, Liu Y, Yan C, Wang Y, Lu X, Li Y, Fan L (2014). A diverse set of miRNAs responsive to begomovirus-associated betasatellite in *Nicotiana benthamiana. BMC Plant Biology* 14: 60.

Xie F, Frazier TP, Zhang B (2011). Identification characterization and expression analysis of microRNAs and their targets in the potato (*Solanum tuberosum*). *Gene* 473: 8–22.

Xie Z, Kasschau KD, Carrington JC (2003). Negative feedback regulation of Dicer-Like1 in *Arabidopsis* by microRNA-guided mRNA degradation. *Current Biology* 13: 784–789.

Xu D, Mou G, Wang K, Zhou G (2014). MicroRNAs responding to southern rice black-streaked dwarf virus infection and their target genes associated with symptom development in rice. *Virus Research* 190: 60–68.

Yin X, Wang J, Cheng H, Wang X, Yu D (2013). Detection and evolutionary analysis of soybean miRNAs responsive to soybean mosaic virus. *Planta* 237: 1213–1225.

Yin Z, Chrzanowska M, Michalak K, Zimnoch-Guzowska E (2014). Alteration of host-encoded miRNAs in virus infected plants—Experimentally verified. In: *Plant Virus-Host Interaction*. eds RK Guar, T Hohn and P Sharma, 17–55. San Diego, CA: Elsevier.

Yin Z, Xie F, Michalak K, Pawełkowicz M, Zhang B, Murawska Z, Lebecka L, Zimnoch-Guzowska E (2017). Potato cultivar Etola exhibits hypersensitive resistance to PVY^{NTN} and partial resistance to PVY^{Z}-NTN and $PVY^{N\text{-}Wi}$ strains and strain-specific alterations of certain host miRNAs might correlate with symptom severity. *Plant Pathology* 66: 539–550. doi:101111/ppa12599.

Zhang R, Marshall D, Bryan GJ, Hornyik C (2013). Identification and characterization of miRNA transcriptome in potato by high-throughput sequencing. *PLoS One* 8: e57233.

Zvereva AS, Pooggin MM (2012). Silencing and innate immunity in plant defense against viral and non-viral pathogens. *Viruses* 4: 2578–2597.

18 Possible Approaches for Developing Different Strategies to Prevent Transmission of Geminiviruses to Important Crops

Avinash Marwal, R.K. Gaur, and Khurana SMP

CONTENTS

18.1 INTRODUCTION

There are a large number of threats to agricultural crops across the world. Many insect pests harm the crops and reduce the yield, but apart from these there are other microscopic pathogens that cause a great annual loss to the agriculture sector and the livelihood of Indian farmers and world agricultural economy. Worldwide expansion of agriculture has also resulted in the emergence and spread of numerous diseases and insect pests. Of particular importance are insect-transmitted viruses, especially in tropical and subtropical regions. Geminiviruses are insect-transmitted viruses that have emerged over the past 20 years as the largest group of plant viruses (322 species) and one of the most economically important (CRSP 2012). Several viruses affect crop plants, but the major contribution is from the viruses belonging to the family *Geminiviridae*. Geminiviruses are characterized by their twin icosahedral capsids and small, single-stranded DNA (ssDNA) genome (~2.7 kb) (Brown et al. 2012; Marwal et al. 2014). A study examining genome-wide pairwise sequence identity, genome organization, host range, and insect transmission vector recently classified the family *Geminiviridae* into seven genera: *Mastrevirus*, *Begomovirus*, *Curtovirus*, *Topocuvirus*, *Becurtovirus*, *Turncurtovirus*, and *Eragrovirus*. In nature, geminiviruses are transmitted by phloem-feeding insects, including various species of leafhoppers, a treehopper, and whiteflies of the species *Bemisia tabaci*. Geminiviruses are not transmitted through seeds, whereas many are graft-transmissible and some are mechanically (sap) transmissible. Plants infected with geminiviruses show a wide range of symptoms including stunting, distorted growth, and leaf streaking and striations in monocotyledonous plants and leaf crumpling, curling, distortion, golden-light green-yellow mosaic/mottle, interveinal yellowing, yellow spots, and vein swelling, purpling, and yellowing in dicotyledoneous plants. White flies (*Bemisia tabaci*) are responsible for transmission of geminiviruses from one plant to another (Razavinejad et al. 2013; Halley-Stott et al. 2007; Fauquet et al. 2008; Wang et al. 2004); other vectors such as leafhopper and treehopper also contribute to infection (Gaur et al. 2011).

Bacterial and fungal diseases can be managed by the use of chemicals, while sacrificing or eradication of plants has been the only way to control virus diseases. Viral diseases cause lower yields and reduced quality of plant products, leading to economic losses. In perennial crops, damage is profound in comparison to annuals. Tropical and subtropical regions are the most favorable regions for the emergence of new Geminiviruses which cause severe disease epidemics in cash crops or staple foods such as legumes, oil seeds, cotton, grains, tomato, and others (Khan 2000). Pests and diseases are major threats to the crop yield. In a broad variety of plant species, Geminiviruses are the causal agents of most important crop diseases (Amudha et al. 2011) which are responsible for tremendous economic (yield) losses in crops that provide nutrition to millions of people in the developing regions such as sub-Saharan Africa (Marwal et al. 2013a). Some of these Geminiviruses are *Maize streak mastrevirus* (MSV) and *African cassava mosaic begomovirus* (ACMV) (Palmer and Rybicki 1998; Harrison and Robinson 1999). Other economically important crops are also affected by Geminiviruses (*Lycopersicon esculentum*, *Capsicum annuum*, *Carica papaya*, *Manihot esculenta*, etc.), with variable levels of loss. In a *Science* interview, Dr. Claude Fauquet declared "it is not unusual to see destruction of whole crops, such as tomato, cotton, and cassava, with the viruses causing serious plant disease in at least 39 nations" (Moffat 1999, p. 1835). Climate change can impact the relationship between pests and crop plants by two distinct mechanisms: firstly, changes in climate that have a direct impact on the biology of insects, including vectors, leading to differences in their survival, reproduction, and spread (Kumar and Khurana 2016); and secondly, changes in agricultural practices that will take place as a result of climate change, and the influence of these changes on the availability of host plants for the pest species; e.g. the introduction of new crop species and plant genotypes, and changes in husbandry practice (Harrington 1994; Harrington et al. 2007). Plant viruses face special problems in initiating an infection. The outer surfaces of the plants are composed of protective layers of waxes and pectin, but more significantly, each cell is surrounded by a thick wall of cellulose overlying the cytoplasmic membrane. To date, no plant virus is known to use a specific cellular receptor of the type which animal and bacterial viruses use to attach to host cells. After virus replication in an initial cell, the lack of receptor poses special problems for plant viruses in recruiting new cells for the infection.

There are a number of routes by which plant viruses get transmitted—infected seeds, vegetative propagation/grafting, and vector-mediated transmissions. Mechanical transmission of viruses is also an important natural method of transmission. Virus particles may contaminate soil for long periods and may be transmitted to the leaves of new host plants through wind-blown dust or rain-splashed mud (Marwal et al. 2013b). Extensive areas of monoculture and injudicious use of pesticides that kill natural predators can result in massive populations of insects such as aphids. Insects cause damage and crop loss in many ways, mostly associated with the impact of their feeding in the form of yield loss and fall in harvest quality due to cosmetic damage (Yamamura and Kiritani 1998). However, sucking insects, such as white flies and aphids, are also associated with the transmission of viruses, causing major economic crop losses. The insect transmission of plant viruses can be classified as persistent, semi-persistent, or non-persistent. Persistent transmission requires sustained feeding by the insect, while non-persistent transmission is due to test feed or a more superficial relationship between the insect and the plant (Newman et al. 2003). Breeders working toward the incorporation of Geminivirus resistance genes must take into account the high degree of genetic diversity present within geminiviruses. A given resistance gene might be extremely effective against a particular Geminiviruses species and totally ineffective against distinct, unrelated species (Marwal et al. 2013c). Also, the rate of evolution of Geminiviruses seems to be inordinately fast (Padidam et al. 1999), leading to the quick emergence of new strains or species that might overcome resistance genes. An important question in terms of developing management strategy of Geminivirus disease is to determine the distribution of the virus in terms of geographic location and host plants (host range). Many strategies have been extensively used worldwide for control of Geminiviruses with great success. This chapter focuses on the various strategies used for management of Geminiviruses in agriculturally important crops.

18.2 TRADITIONAL/CONVENTIONAL/CULTURAL APPROACH

Geminiviruses threaten food security and agriculture, infecting key crop species, especially in tropical and subtropical regions. A number of cultural practices have been adopted, although these have either only limited effect or have been difficult for farmers to utilize. If possible, new plantings should be established during the periods when whitefly populations are low. This varies from region to region and needs to be established experimentally. It is possible that suppression of whitefly populations, either via biological control or with other natural or traditional methods, may help reduce the spread of Geminiviruses in certain situations. Many predators like *Chrysopa pallens*, *Typhlodromips swirskii*, *Orius sauteri*, *Chrysopa formosa*, *Harmonia axyridis*, *Erigonnidium graminicolum*, *Scymnus hoffmanni*, *Coccinellid septempunctata*, *Propylaea japonica*, *Euseius scutalis*, and *Neoscona doenitzi* feed on *B. tabaci* (Zhang et al. 2007; Nomikou et al. 2005). These native predators can be efficiently used in sustainable management of *B. tabaci* (Meng et al. 2006).

Chemiecological mechanisms involve honeydews excreted by whitefly adults and nymphs, which provide important chemical cues (kairomone) for host searching by parasitoids such as *E. formosa* adults (Liu et al. 2007). Most predators of *B. tabaci* are polyphagous and can be determined using TaqMan PCR assay. These predators can be used in devising sustainable management measures against *B. tabaci* (Nomikou et al. 2005). This strategy has been successfully used in the Mediterranean regions with *Encarsa formosa*, *E. lutea*, and *Eretmocerus mundus* to control the vector population and virus spread (Gerling 1986; Naratajan 1990; Rapisarda 1990; Henneberry and Bellows 1995).

Plant age is an important factor, especially for tomato plants, as their first five weeks is very critical for establishment since this is the period when whitefly transmits Geminiviruses in healthy tomatoes. Keeping this in mind, an anticipatory approach was taken to diminish contact between the vector and the host tomato plant by masking the crop with living ground covers. This was carried out at the critical period and was performed on a small scale by Costa Rican farmers. Living covers consisting of perennial peanuts, cinquillo, and coriander reduced the number of incoming whitefly adults, therefore delaying the onset of tomato yellow mottle virus (ToYMoV) and decreasing disease severity, resulting in higher yields and profits compared to bare soil control. Coriander provided additional economic returns when sold, and was easier to establish and remove than the other living covers. Therefore living ground covers appeared to offer a viable and economical management alternative for resource-poor farmers (Hilje and Stansly 2008).

Virus reservoirs can be eliminated for virus management in tomato plots. Weeds and other solanaceous crops that are hosts of the virus should be eliminated. However, knowledge of both the wild and domesticated hosts of the virus is essential for a purposeful and result-oriented implementation of this strategy. In Tanzania, the age-old practice of farm sanitation, which involves the clearing of weeds, is a common strategy currently adopted by tomato farmers (Ioannou 1987; Cohen et al. 1988). Another effective method is the manipulation of sowing dates adopted by the growers in Tanzania by delaying tomato seed sowing to avoid periods of peak vector populations. This has been adopted with recorded successes in greenhouse and low-tunnel tomato cultures (Makkouk 1978).

The practice of mulching for transplanted tomato seedlings benefits the late spread of tomato yellow leaf curl geminivirus (TYLCV) disease by two weeks, thus preventing the landing of vector on the crops. Other mulching materials that have been used are yellow polyethylene materials and ultraviolet-absorbing films (Suwwan et al. 1988; Antignus et al. 1995). Intercropping of tomato with other crops such as cucumbers, peppers, and eggplants, is useful in diverting the whiteflies, resulting in delay of tomato infection by TYLCV (El-Serwiy et al. 1987).

Tomatoes are grown across the world and are prone to Geminiviruses (Czosnek et al. 1988). Efforts have been made to find tomato varieties with resistance to TYLCV. Around 1200 breeding lines, cultivars, and wild species have been screened for resistance to TYLCV in India, but only five lines of *L. hirsutum* and two of *L. peruvianum* were found resistant to Geminiviruses (Saikia and Muniyappa 1989). Some accessions of *L. chilense*, a green-fruited wild tomato species, have been reported to be immune to the virus (AVRDC 1994; Kashina et al. 2003; Lapidot and Friedmann 2002; Briddon 2003; Tarr 1951).

It was found that leaf curl disease has higher severity rate in winter-grown crops (83%) than in the summer season (14%). If information is available on the seasonal patterns of whitefly populations and virus pressure, planting times can be modified to avoid periods of high pressure. The disease incidence may be reduced further when polythene mulch (PM) and perforated polythene cover (PC) are used during planting. It was observed that at the same stages of crops, the winter-grown crops had a larger population of whiteflies than the summer crops, suggesting the favorable conditions for whiteflies. The use of polythene bags to cover tomato plants (PC) and polythene sheets as mulch (PM) were found most effective in reducing the disease incidence and promoting the growth and yield of tomato plants as compared to other treatments (Tripathi and Varma 2002).

When aluminum painted mulch was used, the Geminiviruses infections on pepper plant were fewer than on tomato plants, possibly due to the shorter height of pepper plants, which cover less mulch than the tomato plants. Though aluminum painted mulch had no influence on the numbers of leaf mines and mite colonies on tomato foliage, fewer tomato plants got infected with aphid-transmitted virus as compared to non-painted mulch (Kring and Schuster 1992; Smith et al. 2000).

A number of initial techniques of phytosanitation applied to fields result in lower incidence of disease in a crop. For Cauliflower mosaic disease management in Africa, three phytosanitation techniques have been used: tissue culture, selection of disease-free stems, and removing diseased plants from the fields (known as roguing). Roguing is most helpful if the incidence of the virus is low. After roguing, it is also important that there is a minimal level of virus spread in the field. If nymphs are present, rogued plants should be removed in plastic bags. The tissue culture technique engages meristem tip excision, thermotherapy, and virus indexing for producing virus-free plants (Legg et al. 2006). Row covers in Israel and Guatemala have been used to protect plants from whiteflies in crops such as cucurbits, pepper, and tomato (Natwick and Durazo 1985; Gilbertson et al. 2011). Another interesting approach that has been applied in Israel is to grow vegetables (like tomatoes) in greenhouses (glass, plastic, and double doors with positive pressure), protecting from whitefly damage and virus infection by physical means. Excellent management of whiteflies can be achieved with the use of these greenhouses (Ausher 1997; Berlinger et al. 2002).

18.3 INSECTICIDAL MANAGEMENT OF GEMINIVIRUSES

Geminivirus vectors are quite difficult to control in agronomic and horticultural production systems. In the past 10 years, insecticides have been introduced that provide a diversity of novel modes of action and routes of activity to effectively control various virus vectors. This includes the nicotinoids and insect growth regulators (IGRs). The nicotinoids are systemic neurotoxins that target acetylcholine receptors in the insect nervous system. The non-neurotoxic IGR pyriproxyfen, a juvenile hormone analog, has also played an important role in controlling *B. tabaci*, particularly on cotton in North America and Israel.

The most effective group of insecticides is the neonicotinoids, having systemic and translaminar properties with high residual activity (Palumboa et al. 2011; Elbert et al. 1990). Their mode of action is against sucking insects such as whiteflies, aphids, and leafhoppers, and various coleopteran pests such as the Colorado potato beetle, *Leptinotarsa decemlineata*. A thiadizine-like compound, i.e., buprofezin, has a long systemic persistent activity which functions as a chitin synthesis inhibitor. It affects the nymphal stages of sucking insects, particularly whiteflies, through its contact and vapor activity (De Cock and Degheele 1998). Due to the lack of chitin deficiency, the whitefly nymph loses its elasticity and the insect is unable to molt (Horowitz et al. 2011).

For disease management, control of vectors in initial stages is done by seed treatment with systemic insecticides. A number of compounds have been verified, such as Thiomethoxam 70 WS, Imidacloprid 600 FS, Imidacloprid 70 WS, and Carbosulfan 25 DS against cotton leaf curl virus (CLCuV). This results in the reduction of disease severity caused by whitefly population and increase in yield at the end of season. Even the vector population in the later stages can be minimized by careful application of triazophos and Ethion. Ali et al. (2005) showed that buprofezin is effective against nymphs. Similarly, imidacloprid and Acephate 95 SG were effective against the whitefly adults. For cotton pest management insecticide spirotetramat, applied in mixture with imidacloprid, associated or not with methylated soy oil, was efficient on the control of the whitefly (Papa et al. 2008).

For whitefly management in cotton crops an effective pesticide, Diafenthiuron (a thiourea derivative) has been used, particularly in Israel and Europe, since 1998 (Horowitz et al. 1998). It has both insecticidal and acaricidal properties against some species of hemipterans and phytophagous mites. Its remarkable activity is its conversion into carbodiimide in the presence of sunlight (phytochemically), thus enhancing its insecticidal activity (Steinemann et al. 1990; Horowitz et al. 2011). Imidacloprid belongs to a new group of insecticides called nitroguanidines. Imidacloprid, the first nicotinoid registered, has been largely responsible for the sustained management of *B. tabaci* in horticultural production systems worldwide. The insecticide is systemic when applied to soil and is therefore particularly efficient against sucking insects such as leafhoppers, whiteflies, aphids, and thrips (Rubinstein et al. 1999). Thiamethoxam is a neonicotinoid insecticide, and is used in preventing transmission of TYLCV by the whitefly *B. tabaci*. Its effect has been studied through experimental transmissions to new tomato seedlings. At the six-leaf stage, the source plants were foliar treated, whereas the test plants were both foliar and drench treated. Foliar treatment of source plants made infected tomatoes totally ineffective as a virus source for at least eight days. Viruliferous whiteflies exposed to thiamethoxam-treated plants stopped feeding before acquiring enough viruses to subsequently inoculate plants (Mason et al. 2000).

Ryanodine is a plant alkaloid used as a natural botanical insecticide that targets insect ryanodine receptors and is more potent or better than other chemical insecticides (Sattelle et al. 2008). An insecticide called Spirotetramat is systemic with phloem and xylem mobility, and is useful against sucking insects, including aphids, whiteflies, psyllids, and scales. It is particularly effective against juvenile stages of sucking pests and it significantly reduces fecundity and fertility of *B. tabaci* females (Brück et al. 2009). Fuog et al. (1998) suggested that Pymetrozine, which is an azomethine pyridine insecticide, is extremely precise against sucking insect pests. Pymetrozine causes insect death due to starvation by affecting the nerves which control the salivary pump of insects, causing immediate and irreversible cessation of feeding due to an obstruction of stylet penetration (Kayser et al. 1994). The compound is a powerful toxicant against aphids, whiteflies (including *B. tabaci* and *Trialeurodes vaporariorum*), and plant hoppers (Horowitz et al. 2011).

18.4 UNDERSTANDING MANAGEMENT THROUGH RECOMBINATION STUDIES

Recombination is an important source of genetic variability and is one of the important factors in evolutionary processes implicated in determining the structural design of genomes and resulting phenotypes. Understanding the occurrence of recombinants among viruses can help in their management

(Prajapat et al. 2012). A study for analysis of recombination phenomenon among Geminiviruses was carried out, highlighting virus strains reported from India and neighboring countries. Around 1000 sequences such as *Ageratum enation virus*, *Bell pepper leaf curl virus*, *Pumpkin yellow vein mosaic virus*, *Tobacco leaf curl virus*, etc. from India, Bangladesh, China, Nepal, Pakistan, and Sri Lanka were analyzed, suggesting *Tomato yellow leaf curl China viruses* and *Mung bean yellow mosaic India virus* had the highest number of recombination events. This *in silico* analysis has helped in greater understanding of the recombination phenomenon in Geminiviruses, which can help in devising strategic planning for management of plant viral diseases (Morya et al. 2014).

The whitefly-transmitted begomovirus comprises the largest number of species (288 of 325 total species) in the family *Geminiviridae* (Brown et al. 2012). Remarkable emergence of a high degree of recombinations in these viruses has been driven by its vector, *B. tabaci*, especially the polyphagous biotype B that has spread throughout the world (Gilbertson et al. 2015). High intra- and interspecific diversity of begomoviruses facilitates their adaptation to new climates and novel hosts (Monci et al. 2002). Recombination is the most intensively studied population genetic process in begomoviruses, and has been considered more significant than mutation by many researchers (Lefeuvre et al. 2007; Martin et al. 2005; Monci et al. 2002; Padidam et al. 1999; Pita et al. 2001). It appears to contribute heavily to genetic diversity of begomovirus, increasing their evolutionary potential and local adaptation of strains (Graham et al. 2010). Besides the apparent importance of recombination in begomovirus evolution, the marks that it has left on currently sampled Begomovirus genome sequences also have major implications when attempting to use these sequences to infer the evolutionary histories of begomoviruses (Marwal et al. 2012; Prajapat et al. 2011). While analyzing the betasatellite molecule associated with leaf curl disease of ornamental Marigold plants (KC589700), a total of four recombination events were detected in the genome. The evidence is given in Figure 18.1, where breakpoint begins from 717th (position 842 in alignments) position and ending at 783th (position 910 in alignments) position. Approximate *p*-value for this region was 2.947×10^{-4}. The beginning breakpoint probability was found to be 0.048.

The region probability (multiple comparison [MC] uncorrected) was 2.607 E-08 and region probability (MC corrected) was 2.972 E-05. The open reading frame (ORF) beta C1 is devoid of any recombination, but recombination was observed in A rich region of the betasatellite sequence. The major parent was identified as Okra leaf curl virus satellite DNA beta (FN432358) identified in Pakistan and found infecting *Sonchus arvensis*, whereas the minor parent was Pepper leaf curl virus satellite beta DNA (GQ330541), found infecting pumpkin in India (Marwal et al. 2016a).

As per the result of sequence display, only one recombination event was detected in *Marigold leaf curl alphasatellite* (MLCuA: KC206078). Sequences used for the comparison were obtained from the GenBank database. The detection of potential sequences, identification of likely parent sequences, and localization of possible recombination breakpoints were carried out using the recombination detection program (RDP) method (Martin et al. 2015). Remarkably, the recombination was detected in a small portion of the replicase (Rep) ORF of the alphasatellite. The beginning breakpoint position was 936th [position 1101 in the alignment] and ending breakpoint position was 974th [position 1147 in the alignment]. The region probability (MC uncorrected) was 9.393 E-09 and region probability (MC corrected) was 2.686 E-06. Approximate *p*-value for this region was 2.686×10^{-6}. In this case the major parent was identified as nanovirus-like particle Rep gene for truncated replication-associated protein, clone UK8 (AM930248) identified in Pakistan and found infecting *S. arvensis* (Figure 18.2). The minor parent in the RDP plot was found to be Tobacco curly shoot virus-associated DNA 1 Rep gene for replication-associated protein (AJ579349) found in association with tobacco plants in China (Marwal et al. 2016a).

In a similar study by Godara et al (2016), RDP4 was used for detection of recombination events in CLCuD-begomovirus isolates at New Delhi, IARI-34 (CLCuMuV-Rajasthan strain, KJ959629) and IARI-45 (CLCuKoV-Burewala strain, KM070821). Recombination events were detected in IR, C1, and C4 regions with high probability in isolate IARI-34. A recombinant event involving a large sequence at nt 2746-874 (containing IR, V1, V2, C5 genes) was observed but the probability was low.

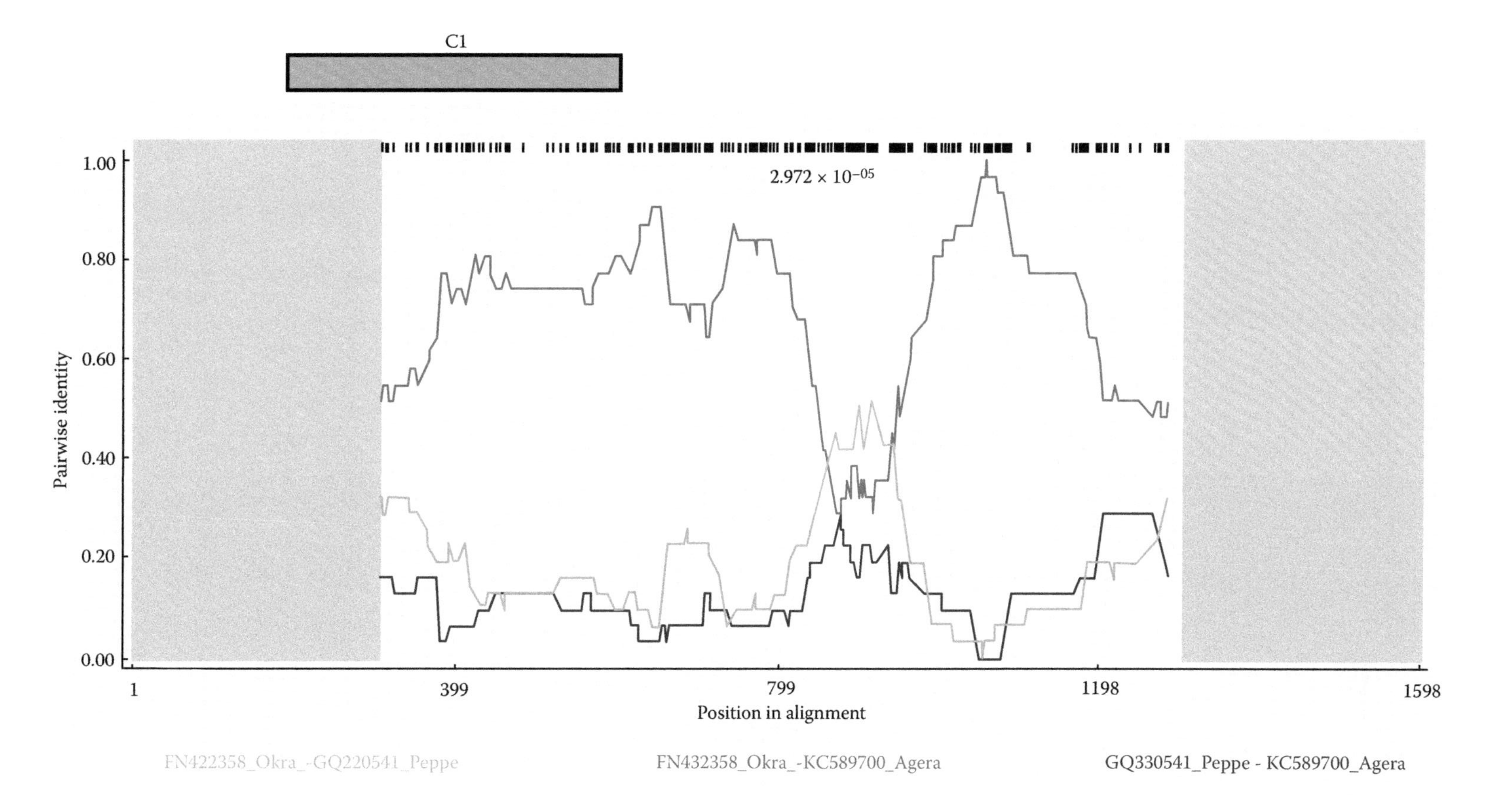

FIGURE 18.1 An RDP pairwise identity plot for the piece of sequence from the major parent (FN432358_Okra). Uppermost bars indicating positions of informative sites; white indicates breakpoint positions suggested by the RDP software method. The pairwise identity plot has major parent: minor parent plot (FN432358_ Okra: GQ330541_Peppe; light gray), major parent: recombinant plot (FN432358_ Okra: KC589700_ Agera; medium gray), and minor parent: recombinant plot (GQ330541_ Peppe: KC589700_ Agera; dark gray)

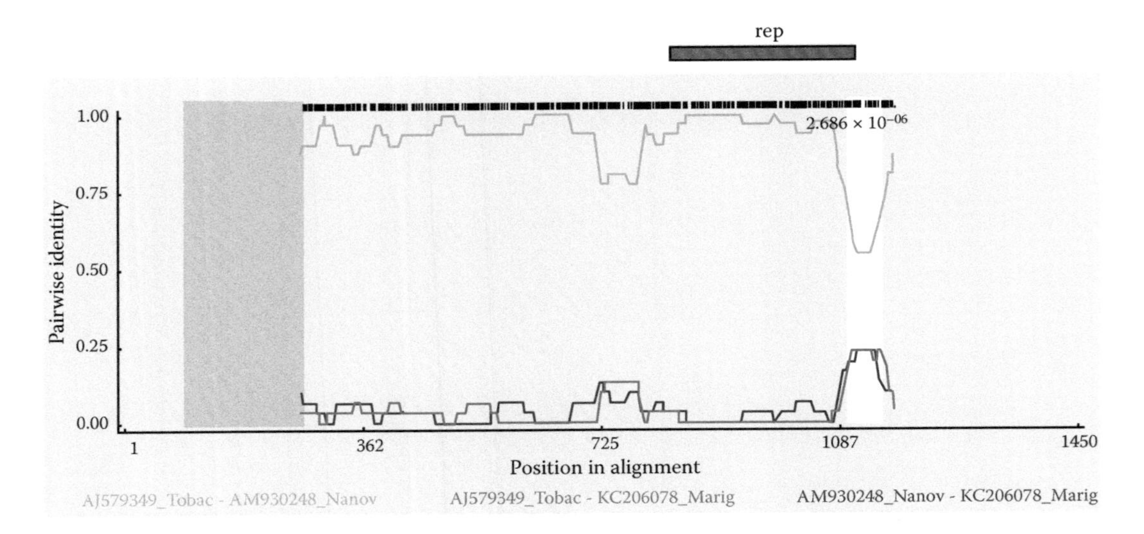

FIGURE 18.2 An RDP pairwise identity plot for the piece of sequence from the major parent (AM930248_Nanov). Uppermost bars indicating positions of informative sites; white indicates breakpoint positions suggested by the RDP software method. The pairwise identity plot has major parent: minor parent plot (AM930248_Nanov: AJ579349_T.obac; light gray), major parent: recombinant plot (AM930248_Nanov: KC206078_ Marig; medium gray), and minor parent: recombinant plot (AJ579349_Tobac: KC206078_ Marig; dark gray)

CLCuMuV-Rajasthan strain (AY795605) and CLCuMuV-Hisar strain (AY765256) were reported from IARI experimental farm during the years 2003 and 2004, respectively. It is hypothesized that due to natural selection only the CLCuD-begomoviruses that evolved by recombination between CLCuMuV and CLCuKoV could survive (Saleem et al. 2016). The previously reported CLCuMuV-Hisar strain (AY765256) was detected as non-recombinant (Godara et al. 2016). As the non-recombinant CLCuMuV-Hisar strain (AY765256) could not survive at IARI, no isolate of CLCuMuV-Hisar was found at IARI, cotton field during the years 2012 and 2013 by Godara et al. (2016).

For many viruses, the generation of recombinant variants has been associated with important moments in the processes of adaptation, gain of pathogenic potential, or increased spreading. For instance, the appearance of CLCuKoV-Burewala strain in Pakistan is associated with the CLCuD epidemic in Pakistan in 2001–2002 leading to 100% crop losses (Amrao et al. 2010; Rajagopalan et al. 2012). This CLCuKoV-Burewala strain is a resistance-breaking recombinant strain that has been shown to infect all of the previously known resistant and tolerant varieties of the cultivated *Gossypium hirsutum* in Pakistan (Amrao et al. 2010; Rajagopalan et al. 2012).

The evolution of viral genome is believed to be a result of natural selection, virus–vector–host interaction and factors affecting recombination and nucleotide substitution (Saleem et al. 2016). In this regard, recombinations among different CLCuD begomoviruses is an important process that involves the combination of divergent genetic backgrounds with an ability to break down the resistance and also adapt to the insect vector.

18.5 *IN SILICO* APPROACHES FOR MANAGEMENT OF GEMINIVIRUSES

Software like RaptorX and 3DLigandSite tactic are beneficial for developing inhibitor compounds for providing viral resistance to plants. Such resistant plants will have imminent advantage to augment production of standard, nongenetically modified varieties while avoiding concerns of transgene flow. This presents confirmation that ligand site prediction is able to bestow resistance to multiple Geminiviruses. Results of these techniques can be effectively applied for virus disease management, crop protection, and development of quarantine strategies at both state and national levels.

Online resources RaptorX and 3DLigandSite are websites for the prediction of ligand-binding sites. These are based upon successful manual methods used in the eighth round of the critical assessment of techniques for protein structure prediction (CASP8). 3DLigandSite and RaptorX utilize protein-structure prediction to provide structural models for proteins that have not been solved (Wass et al. 2010). Ligands bound to structures similar to the query are superimposed onto the model and used to predict the binding site. The website enables users to submit either a query sequence or structure. Predictions are displayed via an interactive Jmol applet. 3DLigandSite is available at sbg.bio.ic.ac.uk/3dligandsite (Wass et al. 2010).

RaptorX distinguishes itself from other websites by the quality of the alignment between a target sequence and one or multiple distantly related template proteins (especially those with sparse sequence profiles) and by a novel nonlinear scoring function and a probabilistic-consistency algorithm. Consequently, RaptorX delivers high-quality structural models for many targets with only remote templates (raptorx.uchicago.edu) (Källberg et al. 2012).

18.5.1 Binding Site Prediction of Begomovirus and Its Satellite Proteins Infecting Marigold

In order to find the binding site of the proteins of *Ageratum enation virus* (KC589699), *Ageratum enation betasatellite* (KC589700), and MLCuA (KC206078), predicted structure models of all the proteins were uploaded on the websites and the results were obtained. The website provides the detail of the amino acids responsible for binding as well as the list of ligand molecules as heterogens provided by Uniprot that are unlikely to be present in protein structures as solvent. The protein and predicted binding site can be shown in cartoon, spacefill, or wireframe formats. The structure of the protein can be differently colored to show the predicted binding site in blue or residue conservation (Marwal et al. 2016b).

During the analysis of AC1 protein it was found that asparagine, glutamine, and glycine residues at 90, 92, and 93 positions are responsible for binding site of AC1 protein. The heterogen/ligand present in binding site were NAG (N-acetyl-D-glucosamine). Glycine at position 37 and phenylalanine at position 41 were found responsible for binding sites in AC2 protein molecule. FAD (flavin-adenine dinucleotide) acts as a ligand molecule/heterogen, very much present in the vicinity of the protein (Figure 18.3). Investigation of AC3 protein revealed that residues responsible for binding and their position were found to be isoleucine at 62, cysteine at 65, isoleucine again at position 71, tryptophan at 72, methionine at 73, threonine at 74, threonine again at 79, leucine at 83, valine at 112, and again at position 116 valine was accountable. The heterogen/ligands present in binding site were ADP (adenosine-5′-diphosphate).

The binding sites exhibit chemical specificity, a measure of the types of ligand that bond, and the affinity that measures strength of the chemical bond (Balakrishnan et al. 2010). When the model of the AC4 protein PDB file was uploaded on the website, it was established that the residues glutamic acid and asparagine at position 72 and 75, respectively, were in charge for binding site, whereas the ligand was found to be MDO (4-methylidene-5-one peptide derived chromophore). While studying the coat protein AV1, it was instituted that heterogen CA (calcium ion) was responsible for binding at glycine residue in the 125th position of AV1 protein model. AV2 protein is the most active protein in begomovirus, hence remarkable findings were observed while analyzing the AV2 protein molecule. Surprisingly, 16 amino acid residues are present at the binding site of the AV2 protein molecule, including glycine, cysteine, methionine, alanine, valine, lysine, leucine, arginine at the 15th, 18th, 19th, 21st, 22nd, 23rd, 25th, 50th positon respectively, and asparagine, tyrosine, valine, glutamine, alanine, serine at the 53rd, 54th, 55th, 56th, 57th, 59th position, further including arginine and tyrosine at the 60th and 61st position, respectively. The ligand responsible at the binding site was HEA (HEME-A). It was found that only one binding site was predicted in *Ageratum leaf curl betasatellite* C1 protein at position 69, conquered by methionine, having MG (magnesium ion) as the ligand molecule. Whereas in the case of Rep protein of MLCuA, five residues are functional at the binding

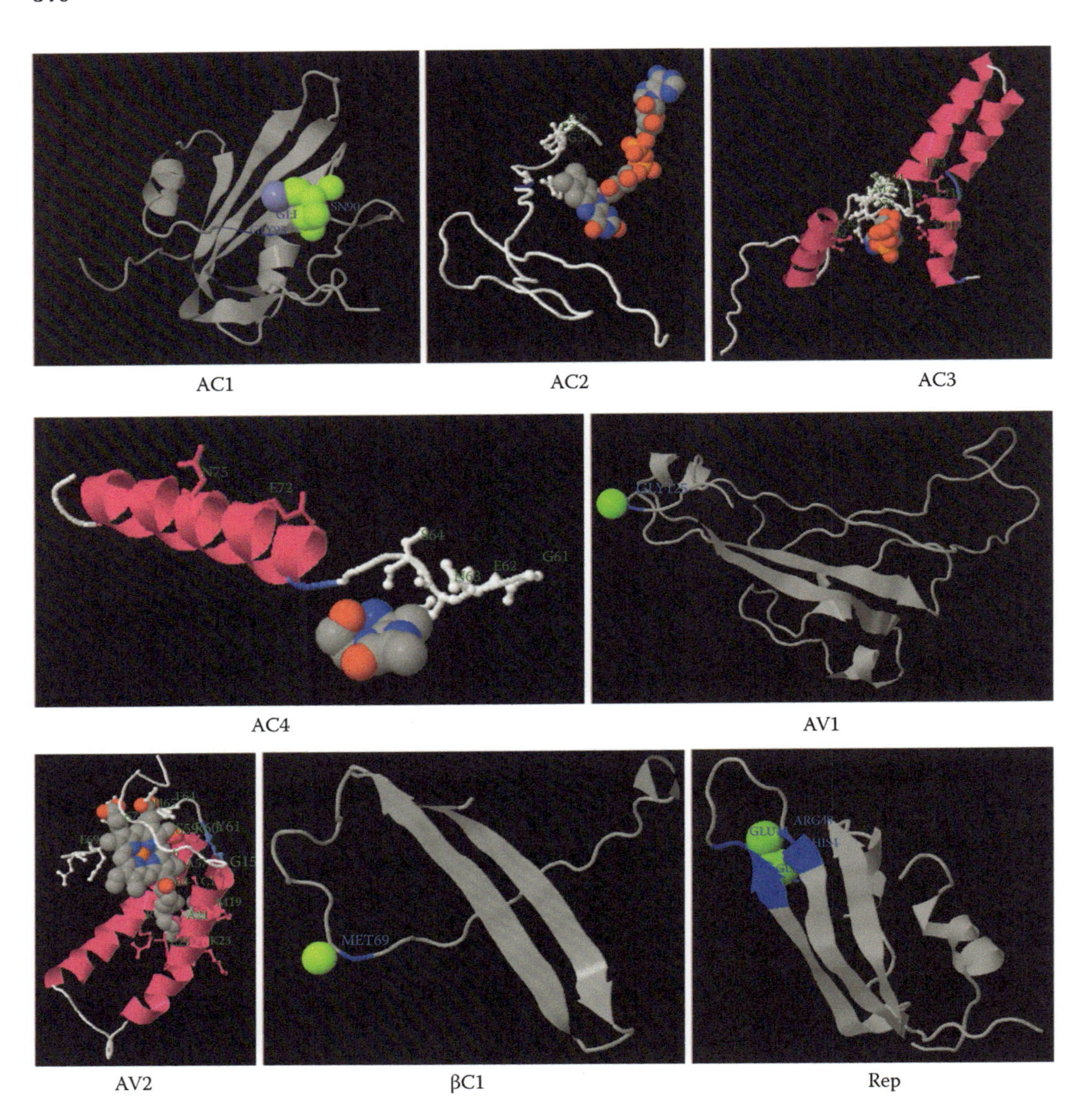

FIGURE 18.3 3DLigandSite and RaptorX website-generated visualization of prediction for target proteins of *Ageratum enation virus* (AC1, AC2, AC3, AC4, AV1, and AV2), *Ageratum enation betasatellite* (βC1) and MLCuA (Rep). The Jmol applet displays the protein structure with predicted binding site colored blue. The ligands in the cluster used to make the prediction are displayed with ions in spacefill and organic molecules in wireframe formats.

site, which includes glutamine at the 39th, 40th, and 41st position, arginine and histidine at the 48th and 49th position, respectively. The heterogen molecule was revealed to be NI (nickel (II) ion).

18.5.2 Binding Site Prediction of Begomovirus and Its Betasatellite Proteins Infecting Rose

Protein–protein interactions are a pivotal component of many biological processes and mediate a diverse variety of functions that include signal transduction, transport, and gene expression regulation, to name a few (Russell et al. 2004; Aloy and Russell 2004). Although detailed structural information of a protein complex is necessary for the understanding of the underlying mechanisms, such details are often difficult to obtain through traditional experimental means (i.e., x-ray crystallography, NMR). Knowing the tertiary structure of a protein complex is therefore essential for understanding the interaction mechanism. However, experimental techniques to solve the structure of the complex are often difficult (Salwinski and Eisenberg 2003; Szilagyi et al. 2005). A similar type of approach has been applied for *Rose leaf curl virus* (KF584008) and its associated *Rose leaf curl betasatellite* (KF584009) molecule infecting rose plant. Investigation of AC1 protein revealed that residues responsible for binding and their positions were found to be asparagine at 88th position, histidine at 90th, and glycine at 91st position was accountable. The heterogen/ligand present in binding site were NAG. When the model of AC2 protein was uploaded on the website, it was established that the residues cysteine, histidine, asparagine, histidine, and glycine were present at the 39th, 44th, 46th, 59th, and 67th position, respectively, for binding site activity, whereas the ligand was found to be MG.

During the analysis of AC3 protein it was clearly found that 12 amino acid residues are responsible for the binding activity of AC3, which includes isoleucine at 50th, glutamine at 51st, phenylalanine at 52nd, histidine at 54th, alanine at 68th, phenylalanine again at 69th, arginine at 70th, isoleucine at 71st, tryptophan at 72nd, mehionine at 73rd, threonine at 74th, and phenylalanine once again at 82nd position are responsible for binding site of AC3 protein. The heterogen/ligand present in binding site was FRU (fructose). Leucine at position 42 and Arginine at position 45 were found responsible for binding site in AC4 protein molecule. PO4 (phosphate ion) acts as a ligand molecule/heterogen, very much present in the vicinity of the protein (Figure 18.4).

When the model of AV1 protein was uploaded on the website, it was established that only one residue glycine at 189th position was in charge for binding site, whereas the heterogen/ligand molecule was found to be CA. It was found that two binding sites were predicted in AV2 protein at positions 59 and 63, conquered by arginine and histidine, respectively, having GOL (glycerol) as the ligand molecule, whereas in the case of beta C1 protein of *Rose leaf curl betsatellite*, two residues are functional at the binding site: glycine at the 70th and asparagine at 71st position, respectively. The heterogen molecule was revealed to be ZN (zinc ion).

Proteomes constitute the backbone of cellular function by carrying out the tasks encoded in the genes expressed by a given cell type. It does, however, remain challenging to efficiently classify the operational role of the individual protein entities identified in such procedures (Bairoch 2000). Functional properties of a protein domain, such as enzymatic activity or the ability to interact with other proteins, can often be derived from the approximate spatial arrangement of its amino acid chain in the folded state (Hannum et al. 2009). Knowledge of the structure of a newly discovered protein is thus highly valuable in determining the role it plays in biological processes, and it can serve as an important stepping stone in generating hypotheses or suggesting experiments to further explore the protein's nature.

18.6 RNAI TECHNOLOGY

Plant viruses are one of the major yield-reducing factors for agricultural and horticultural crops. In India, many destructive diseases are caused by the geminiviruses. Virus-resistant transgenic plants (VRTPs), developed by the transfer of transgenes from virus, plant, or other origins, have been found resistant to a wide range of viruses. The most successful approach is the viral coat protein-mediated resistance (CPMR). Other transgenes of viral origin having promise are Rep protein, movement protein, proteases, and antisense sequences. 'R'-genes from plants, plantibodies, and yeast RNase genes are also useful for developing VRTPs (Huttner et al. 2001; Vazquez-Rovere et al. 2001; Maki-Valkama et al. 2000). In a large number of VRTPs developed using transgenes of viral origin, resistance is conferred by post-transcriptional gene silencing (PTGS); in some cases, PTGS has been overcome when plants are infected by a heterologous virus, indicating the need for a cautious approach. Overall, the bio-safety concerns in the use of VRTPs are insignificant, but they must be

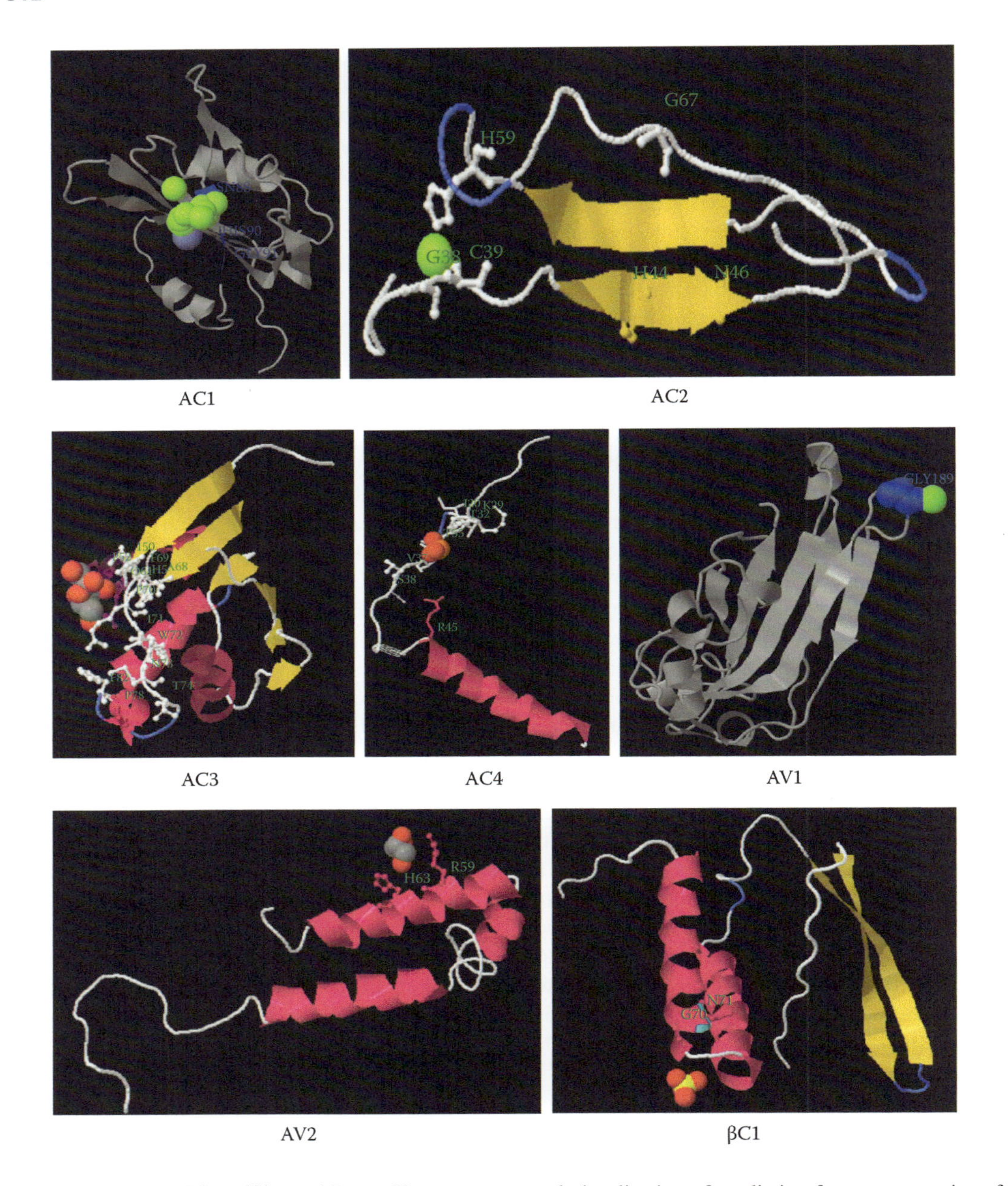

FIGURE 18.4 3DLigandSite and RaptorX server-generated visualization of prediction for target proteins of *Rose leaf curl virus* (AC1, AC2, AC3, AC4, AV1, and AV2) and *Rose leaf curl betasatellite* (βC1). The Jmol applet displays the protein structure with predicted binding site colored blue. The ligands in the cluster used to make the prediction are displayed with ions in spacefill and organic molecules in wireframe formats.

addressed scientifically. In India, initiatives have been taken for developing VRTPs to manage important plant viral diseases (Varma et al. 2002).

RNA silencing is a sequence-specific RNA degradation process that is triggered either by the formation of dsRNA or alternatively by aberrant RNAs associated with transgene viruses and transposons (Vaucheret 2006). RNAs with hairpin-like structures are the actual inducers of PTGS in plants (Ikegami et al. 2011). Transgene-induced silencing in plants is usually associated with methylation of nuclear DNA corresponding to the transcribed region of the target RNA despite transcription levels of the transgene remaining unaffected (Saunders et al. 2004). RNAi technology when used against a *Geminivirus* (ACMV) showed 99% decrease of Rep transcripts and 66% reduction in viral DNA. Herein the siRNA was transiently transferred into the protoplast, making it effective against the Rep-coding sequence of the ACMV (Sanjaya et al. 2005). Targeting Rep and AV_2 gene by antisense technology is found to be quite successful (Dasgupta et al. 2003). To make PTGS an effective method, both sense and antisense RNAs are a prerequisite, and a transgenic tobacco (*Nicotiana benthamiana*) using RNAi was developed (Singh et al. 2007).

Using RNA silencing approach, AC2 protein of *Mungbean yellow mosaic India virus* (MYMIV) was effectively suppressed (Chilakamarthi et al. 2007). Reduction in Rep mRNA of MYMIV was achieved by the catalytic activity of both active and inactive ribozyme, which is primarily due to the of Rep-siRNA formation in presence of the ribozymes (Kumari and Malathi 2012). RNAi-mediated silencing of geminiviruses using transient protoplast assay where protoplast was co-transferred with an siRNA designed to Rep coding sequence of ACMV and the genomic DNA of ACMV resulted in 99% reduction in Rep transcripts and 66% reduction in viral DNA (Ruiz and Voinnet 2007). It was observed that siRNA was able to silence a closely related strain of ACMV but not a more distantly related virus. In plants, PTGS has been widely studied using virus-encoded proteins as transgenes, and viruses are both initiators and targets of suppression of gene silencing. Viruses induced an RNA-mediated defense in plants similar to PTGS that is distinguished by sequence-specific resistance against virus infection. siRNA is an important method for evaluating gene functionality and is being exploited for the development of new approaches to control plant viruses (Mourrain et al. 2000; Covey et al. 1997; Ratcliff et al. 1997).

To develop papaya and tomato crop resistance to geminiviruses, putative siRNA sequences were designed to target essential genes of geminiviruses. This was done by using a software siRNA finder (Ambion), which selects conserved sequences to fulfill all the basic criteria required as per the algorithm. Finally, a cross-search using basic local alignment search tool (BLAST) was performed to confirm that the designed siRNAs do not have any homology to plant genome sequences (Saxena et al. 2011). Sense and antisense constructs of the movement protein genes (BC1 and BV1) of *Tomato mottle geminivirus* (TMoV) were used through Agrobacterium for transformation of tobacco. Such transgenic plants were then tested for virus resistance by exposing viruliferous whiteflies carrying TMoV or *Cabbage leaf curl geminivirus* (CabLCV) for a 72-hour inoculation period. Geminivirus was detected only in inoculated leaves but was not beyond the inoculation sites in the highly resistant plants (Duan et al. 1997).

Replication-associated protein gene, intergenic region, and part of the movement protein gene of CpCDPKV were used to construct a hairpin (hp)RNAi. This construct was placed under the control of the *Cauliflower mosaic virus* 35S promoter. It was triged against a legume infecting mastrevirus, i.e., *Chickpea chlorotic dwarf Pakistan virus* (CpCDPKV) that harms chickpea and other leguminous crops in Pakistan. This construct was later transformed into *N. benthamiana* plants. In transgenic *N. benthamiana* plants, a very low level of viruses was detected, suggesting RNAi technology to be useful in protecting plants from mastrevirus (Nahid et al. 2011). *Maize streak reunion virus* (MSRV) is a member of the *Mastrevirus* genus in the family *Geminiviridae*. A mastrevirus MSRV-YN collected in Yunnan Province, China, was identified by small RNA deep sequencing. This vsiRNA profile derived from MSRV-YN was characterized, which might contribute to insight into the host RNA silencing defense induced by MSRV-YN and provide guidelines on designing antiviral strategies using RNAi against MSRV-YN (Chen et al. 2015).

18.7 CRISPR: THE LATEST APPROACH FOR DISEASE MANAGEMENT

Geminivirus causes infections in numerous economically important crops like cassava, cotton, wheat, maize, and many horticultural plants, resulting in a serious threat to agriculture worldwide by decreasing crop area per capita and production to feed the rapidly increasing world population. Genome engineering strategies have recently emerged as promising tools to introduce desirable traits in many eukaryotic species, including plants (Khatodia et al. 2016; Zaidi et al. 2016). Virus-based guide RNA (gRNA) delivery system for clustered regularly interspaced short palindromic repeats (CRISPR)/CRISPR associated systems (Cas)9 mediated plant genome editing (virus-induced genome editing [VIGE]) cause mutations in target genome locations. VIGE was executed using a modified *Cabbage leaf curl virus* (CaLCuV) vector to express gRNAs in stable transgenic plants expressing Cas9. These results demonstrated that Geminivirus-based VIGE is a powerful tool in plant genome editing (Yin et al. 2015).

With respect to agriculture, genetic engineering has tremendous application by offering an alternative approach through genome editing using sequence-specific nucleases (SSNs). The available SSN techniques are transcription activator-like effector nucleases (TALENs), zinc finger nucleases (ZFNs) and CRISPR/Cas, which have been used in a number of plant species for targeted mutagenesis through non-homologous end joining (NHEJ). In a study, CRISPR/Cas reagents expressing two single-guide RNA (sgRNA) targeted toward potato acetolactate synthase1 (StALS1) gene. This was done by inducing targeted mutations in diploid and tetraploid potato using Agrobacterium-mediated transformation having a conventional T-DNA or a modified Geminivirus T-DNA (Ali et al. 2015). Results showed that single targeted mutations were inherited through the germline of both diploid and tetraploid. This manifestation of CRISPR/Cas in potato enlarges the range of plant species modified by means of CRISPR/Cas and provides a structure for prospective research (Butler et al. 2015). In another approach sgRNAs specific for coding and non-coding sequences of TYLCV into *N. benthamiana* plants stably overexpressed the Cas9 endonuclease, thus targeting TYLCV genome for degradation and introducing mutations at the target sequences. All tested sgRNAs exhibit interference activity, but those targeting the stem-loop sequence within the TYLCV origin of replication in the intergenic region (IR) are the most effective. *N. benthamiana* plants expressing CRISPR/Cas9 exhibit delayed or reduced accumulation of viral DNA, abolishing or significantly attenuating infection symptoms. Moreover, this system could simultaneously target multiple DNA viruses (Ali et al. 2015).

CRISPR/Cas is an adaptive immune system in many archaea and bacteria that cleaves foreign DNA on the basis of sequence complementarity. In order to tackle Geminiviruses, a study was carried out by Baltes et al. (2015) in *Bean yellow dwarf virus* (BeYDV) genome which was targeted for change with the CRISPR/Cas system. By using BeYDV-based replicons, transient assays were performed, disclosing that CRISPR/Cas reagents introduced mutations within the viral genome and thus reduced virus copy number. Transgenic plants infected with BeYDV showed less virus load and symptoms by expressing CRISPR/Cas reagents. This novel strategy demonstrates a promising engineering for resistance to Geminiviruses (Baltes et al. 2015). Similarly, using the Geminivirus, *Beet severe curly top virus* (BSCTV), transient assays performed in *N. benthamiana* demonstrated that the sgRNA–Cas9 constructs inhibited virus accumulation and introduced mutations at the target sequences. In addition, transgenic Arabidopsis and *N. benthamiana* plants overexpressing sgRNA–Cas9 are highly resistant to virus infection (Ji et al. 2015; Chaparro-Garcia et al. 2015).

18.8 CONCLUSION

For better crop production and yield, a combination of various management strategies should be used to control the Geminiviruses in addition to other cultural practices that have been effective so far. Thus collectively, the use of these new technologies coupled with cultural and biological pest management practices presently provides the best model for combating Geminiviruses. Area-wide management programs and other approaches are being used to reduce the risk of exporting pests, helpful in achieving quarantine security. Whatever strategies have been applied, it is necessary for the government and other agencies to introduce greatly improved methods whereever applicable. There is also a need for further information on the epidemiology and control of Geminiviruses and on the prevalence of different strains, especially in parts of world where there have been no previous studies. This chapter highlights important aspects to consider when developing management strategies for Geminiviruses and improving agricultural productivity.

ACKNOWLEDGMENTS

The authors are thankful to the Science and Engineering Research Board, Department of Science and Technology, New Delhi, India, for financial assistance (File No. YSS/2015/000265) and also to University Grant Commission, New Delhi, for providing financial assistantship under the Research Award for Teacher [F.30-1/2014/RA-2014-16-GE-RAJ-4696 (SA-II)].

REFERENCES

Ali MA, Rafiq-ur-Rehman R, Tatla YH, Ali Z (2005). Evaluation of different insecticides for the control of whitefly on cotton crop in Karor district Layyah. *Pak Entomol.* 27:5–8.

Ali Z, Abulfaraj A, Idris A, Ali S, Tashkandi M, Mahfouz MM (2015). CRISPR/Cas9-mediated viral interference in plants. *Genome Biol.* 16:238.

Aloy P, Russell RB (2004). Ten thousand interactions for the molecular biologist. *Nat Biotechnol.* 22:1317–1321.

Amrao L, Amin I, Shahid S, Briddon RW, Mansoor S (2010). Cotton leaf curl disease in resistant cotton is associated with a single begomovirus that lacks an intact transcriptional activator protein. *Virus Res.* 152:153–163.

Amudha J, Balasubramani G, Malathi VG, Monga D, Kranthi KR (2011). *Cotton leaf curl virus* resistance transgenics with antisense coat protein gene (AV1). *Curr Sci.* 101:10.

Antignus Y, Ben-Joseph R, Mor N, Cohen S (1995). The use of UV absorbing films for the protection of different crops against virus diseases vectored by *Bemisia tabaci*. *Phytoparasitica.* 23:3.

Ausher R (1997). Implementation of integrated pest management in Israel. *Phytoparasitica.* 25:119–141.

AVRDC (1994). Tomato diseases in Tanzania: Identification, disease incidence and distribution. *Annual Progress Report.* 478–482.

Bairoch A (2000). The ENZYME database in 2000. *Nucleic Acids Res.* 28:304–305.

Balakrishnan M, Srivastava RC, Pokhriyal M (2010). Homology modeling and docking studies between HIV-1 protease and carbamic acid. *IJBT.* 9:96–100.

Baltes NJ, Hummel AW, Konecna E, Cegan R, Bruns AN, Bisaro DM, Voytas DF (2015). Conferring resistance to geminiviruses with the CRISPR–Cas prokaryotic immune system. *Nature Plants.* 1:15145.

Berlinger MJ, Taylor RAJ, Lebiush-Mordechi J, Shalhevet S, Spharim I (2002). Efficiency of insect exclusion screens for preventing whitefly transmission of *Tomato yellow leaf curl virus* of tomatoes in Israel. *Bull Entomol Res.* 92:367–373.

Briddon RW (2003). Cotton leaf curl disease, a multicomponent begomovirus complex. *Mol Plant Pathol.* 4:427–434.

Brown JK, Fauquet CM, Briddon RW, Zerbini M, Moriones E, Navas-Castillo J (2012). Geminiviridae. In: King AMQ, Adams MJ, Carstens EB, Lefkowitz EJ (eds). *Virus Taxonomy. Ninth Report of the International Committee on Taxonomy of Viruses.* London: Elsevier, pp. 351–373.

Brück E, Elbert A, Fischer R et al. (2009). Movento, an innovative ambimobile insecticide for sucking insect pest control in agriculture: Biological profile and field performance. *Crop Prot.* 28:838–844.

Butler NM, Atkins PA, Voytas DF, Douches DS (2015). Generation and inheritance of targeted mutations in potato (*Solanum tuberosum* L.) using the CRISPR/Cas system. *PLoS ONE.* 10(12):e0144591.

Chaparro-Garcia A, Kamoun S, Nekrasov V (2015). Boosting plant immunity with CRISPR/Cas. *Genome Biol.* 16:254.

Chen S, Huang Q, Wu L, Qian Y (2015). Identification and characterization of a maize-associated mastrevirus in China by deep sequencing small RNA populations. *Virol J.* 12:156.

Chilakamarthi UK, Mukherjee SJ, Deb K (2007). Intervention of geminiviral replication in yeast by ribozyme mediated downregulation of its Rep protein. *FEBS Lett.* 581:2675–2683.

Cohen S, Kern J, Harpaz I, Ben-Joseph R (1988). Epidemiological studies of the tomato yellow leaf curl in the Jordan valley, Israel. *Phytoparasitica.* 16:259–270.

Covey SN, Al-Kaff NS, Lángara A, Turner DS (1997). Plants combat infection by gene silencing. *Nat (Lond).* 385:781–782.

CRSP Digest (2012). Feeding more people with healthier crops: Virology symposium hosted by IPM CRSP in India. http://crsps.net/2012/08/feeding-more-people-with-healthier-crops-virology-symposium-hosted-by-ipm-crsp-in-india (accessed June 8, 2012).

Czosnek H, Ber R, Navot N, Zamir D (1988). Detection of *Tomato yellow leaf curl virus* in lysates of plants and insects by hybridization with a viral DNA probe. *Plant Dis.* 72:949–951.

Dasgupta I, Malathi VG, Mukherjee SK (2003). Genetic engineering for virus resistance. *Curr Sci.* 84:341–354.

De Cock A, Degheele D (1998). Buprofezin: A novel chitin synthesis inhibitor affecting specifically planthoppers, whiteflies and scale insects. In: Ishaaya I, Degheele D (eds.) *Insecticides with Novel Modes of Action: Mechanism and Application.* New York: Springer, pp. 74–91.

Duan YP, Powel CA, Webb SE, Purcifull DE, Hiebert E (1997). Geminivirus resistance in transgenic tobacco expressing mutated BC1 protein. *MPMI.* 10:617–623.

Elbert A, Overbeck H, Iwaya K, Tsuboi S (1990). Imidacloprid, a novel systemic nitromethylene analogue insecticide for crop protection. In: *Proceedings of 1990 Brighton Crop Protection Conference – Pests and Diseases*, Brighton, pp. 21–28.

El-Serwiy SA, Ali AA, Razoki IA (1987). Effect of intercropping of some host plants with tomato on population density of tobacco whitefly, *Bemisia tabaci* (Genn.), and the incidence of *Tomato yellow leaf curl virus* (TYLCV) in plastic houses. *J Agric Water Res Plant Prod.* 6:71–79.

Fauquet CM, Briddon RW, Brown JK, Moriones E, Stanley J, Zerbini M, Zhou X (2008). Geminivirus strain demarcation and nomenclature. *Arch Virol.* 153:783–821.

Fuog D, Fergusson SJ, Flückiger C (1998). Pymetrozine: A novel insecticide affecting aphids and whiteflies. In: Ishaaya I, Degheele D (eds.) *Insecticides with Novel Modes of Action: Mechanism and Application.* New York: Springer, pp. 40–49.

Gaur RK, Prajapat R, Marwal A, Sahu A, Rathore MS (2011). First Report of a *Begomovirus* Infecting *Mimosa Pudica* in India. *J Plant Pathol.* 93(S4):80.

Gerling D (1986). Natural enemies of *Bemisia tabaci*, biological characteristics and potential as biological control agents: A review. *Agric Ecosys Environ.* 17:99–110.

Gilbertson RL, Rojas M, Natwick E (2011). Development of integrated pest management (IPM) strategies for whitefly (*Bemisia tabaci*)-transmissible Geminiviruses. In: WMO Thompson (ed.). *The Whitefly, Bemisia tabaci (Homoptera: Aleyrodidae). Interaction with Geminivirus-Infected Host Plants.* New York: Springer, pp. 323–356.

Godara S, Paul Khurana SM, Biswas KK (2016). Three variants of cotton leaf curl begomoviruses with their satellite molecules are associated with cotton leaf curl disease aggravation in New Delhi. *J Plant Biochem Biotechnol.* 26:97–105.

Graham AP, Martin DP, Roye ME (2010). Molecular characterization and phylogeny of two begomoviruses infecting Malvastrum americanum in Jamaica: Evidence of the contribution of inter-species recombination to the evolution of malvaceous weed associated begomoviruses from the Northern Caribbean. *Virus Res.* 40:256–266.

Halley-Stott RP, Tanzer F, Martin DP, Rybicki EP (2007). The complete nucleotide sequence of a mild strain of *Bean yellow dwarf virus*. *Arch Virol.* 152:1237–1240.

Hannum G et al. (2009). Genome-wide association data reveal a global map of genetic interactions among protein complexes. *PLoS Genet.* 5:e1000782.

Harrington R (1994). Aphid layer. *Antenna.* 18:50–51.

Harrington R, Clark SJ, Welham SJ, Verrier PJ, Denholm CH, Hullé M, Maurice D, Rounsevell MD, Cocu N (2007). Environmental change and the phenology of European aphids. *Global Change Biol.* 13:1550–1564.

Harrison BD, Robinson DJ (1999). Natural genomic and antigenic variation in whitefly-transmitted geminiviruses (begomoviruses). *Annu Rev Phytopathol.* 37:369–398.

Henneberry TJ, Bellows TS (1995). Sweetpotato whitefly. In: JR Nichols, LR Andres, JW Beradsley, RD Goeden, CG Jackson (eds.). *Biological Control in the Western United States.* University of California, Division of Agriculture and Natural Resources, pp. 115–117.

Hilje L, Stansly PA (2008). Living ground covers for management of *Bemisia tabaci* (Gennadius) (Homoptera: *Aleyrodidae*) and Tomato yellow mottle virus (ToYMoV) in Costa Rica. *Crop Prot.* 27:10–16.

Horowitz AR, Antignus Y, Gerling D (2011). Management of *Bemisia tabaci* whiteflies. In: WMO Thompson (ed.). *The Whitefly, Bemisia tabaci (Homoptera: Aleyrodidae). Interaction with Geminivirus-Infected Host Plants.* New York: Springer, pp. 293–322.

Horowitz AR, Mendelson Z, Weintraub PG, Ishaaya I (1998). Comparative toxicity of foliar and systemic applications of two chloronicotinyl insecticides, acetamiprid and imidacloprid, against the cotton whitefly, *Bemisia tabaci*. *Bull Entomol Res.* 88:437–442.

Huttner E, Tucker W, Vermeulen A, Jgnart F, Sawayer B, Birch R (2001). Ribozyme genes protecting transgenic melon plants against potyvirueses. *Curr Issues Mol Biol.* 3:27–34.

Ikegami M, Kon T, Sharma P (2011). RNA silencing and viral encoded silencing suppressors. In: Gaur RK, Gafni Y, Gupta VK, Sharma P (eds). *RNAi Technology*. Boca Raton, FL: CRC Press, pp. 209–240.

Ioannou M (1987). Cultural management of tomato yellow leaf curl disease in Cyprus. *Plant Pathol.* 36:367–373.

Ji X, Zhang H, Zhang Y, Wang Y, Gao C (2015). Establishing a CRISPR–Cas-like immune system conferring DNA virus resistance in plants. *Nature Plants.* 1:15144.

Källberg M, Wang H, Wang S, Peng J, Wang Z, Lu H, Xu J (2012). Template-based protein structure modeling using the RaptorX web server. *Nature Protocols.* 7:1511–1522.

Kashina BD, Mabagala RB, Mpunami A (2003). *Tomato Yellow Leaf Curl Begomovirus* disease in Tanzania: Status and strategies for sustainable management. *J Sustain Agric.* 22:23–41.

Kayser H, Kaufmann L, Schürmann F (1994). Pymetrozine (CGA 215′944): A novel compound for aphid and whitefly control. In: *An Overview of its Mode of Action. Proceedings of 1994 Brighton Crop Protection Conference – Pests and Diseases*. Vol. 2, Brighton, pp. 737–742.

Khan JA (2000). Detection of tomato leaf curl geminivirus in its vector *Bemisia tabaci. Ind J Exper Biol.* 38:512–515.

Khatodia S, Bhatotia K, Passricha N, Khurana SMP, Tuteja N (2016). The CRISPR/Cas genome-editing tool: Application in improvement of crops. *Front Plant Sci.* 7:506.

Kring JB, Schuster DJ (1992). Management of insects on pepper and tomato with uv-reflective mulches. *Fla Entomol.* 75:119–129.

Kumar N, Khurana SMP (2016). Phytochemistry and medicinal values of Glue berry (*Cordia dichotoma* G. Forst). medicinal plants. *Int J Phytomed Rel Ind.* 8:199–206.

Kumari A, Malathi VG (2012). RNAi-mediated strategy to develop transgenic resistance in grain legumes targeting the *Mungbean Yellow Mosaic India Virus* coat protein gene. *International Conference on Plant Biotechnology for Food Security: New Frontiers*. February 21–24, 2012, New Delhi, India.

Lapidot M, Friedmann M (2002). Breeding for resistance to whitefly-transmittted geminiviruses. *Ann Appl Biol.* 140:109–127.

Lefeuvre P, Martin DP, Hoareau M, Naze F, Delatte H, Thierry M, Varsani A, Becker N, Reynaud B, Lett JM (2007). Begomovirus "melting pot" in the south-west Indian Ocean islands: Molecular diversity and evolution through recombination. *J Gen Virol.* 88:3458–3468.

Legg JP, Owor B, Sseruwagi P, Ndunguru J (2006). *Cassava mosaic virus* disease in East and Central Africa: Epidemiology and management of a regional pandemic. *Adv Virus Res.* 67:355–418.

Liu WX, Yang Y, Wan FH et al. (2007). Comparative analysis of carbohydrates, amino acids and volatile components of honeydew produced by two whiteflies *Bemisia tabaci* B-biotype and *Trialeurodes vaporariorum* (Homoptera: *Aleyrodidae*) feeding cabbage and cucumber. *Acta Entomol Sin.* 50:850–857.

Maki-Valkama T, Valkonen JP, Kreuze JF, Pehu E (2000). Transgenic reistance to PVY (O) associated with post transcriptional silencing of PI transgene is overcome by PVY (N) strains that carry highly homologous PI sequences and recover transgene expression at infection. *Mol Plant Microbe Interact.* 13:366–373.

Makkouk KM (1978). A study on tomato viruses in the Jordan valley with special emphasis on tomato yellow leaf curl. *Plant Dis Report.* 64:259–262.

Martin DP, Murrell B, Golden M, Khoosal A, Muhire B (2015). RDP4: Detection and analysis of recombination patterns in virus genomes. *Virus Evol.* 1:vev003. doi: 10.1093/ve/vev003.

Martin DP, Posada D, Crandall KA, Williamson C (2005). A modified Bootscan algorithm for automated identification of recombinant sequences and recombination breakpoints. *AIDS Res Hum Retroviruses.* 21:98–102.

Marwal A, Prajapat R, Gaur RK (2016a). First report of recombination analysis of betasatellite and aplhasatellite sequence isolated from an ornamental plant Marigold in India: An *in silico* approach. *Int J Virol.* 12:10–17.

Marwal A, Prajapat R, Gaur RK (2016b). Prediction of binding site in eight protein molecules of Begomovirus and its satellite components i.e. betasatellite and alphasatellite isolated from infected ornamental plant. *Plant Pathol J.* 15:1–4.

Marwal A, Sahu A, Choudhary DK, Gaur RK (2013a). Complete nucleotide sequence of a begomovirus associated with satellites molecules infecting a new host *Tagetes patula* in India. *Virus Genes.* 47:194–198.

Marwal A, Sahu A, Gaur RK (2014). First report of airborne begomovirus infection in *Melia azedarach* (Pride of India), an ornamental tree in India. *Aerobiologia.* 30:211–215.

Marwal A, Sahu A, Prajapat R, Choudhary DK, Gaur RK (2012). First report of association of *Begomovirus* with the leaf curl disease of a common weed, *Datura inoxia*. *Virus Dis.* 23:83–84.

Marwal A, Sahu A, Sharma P, Gaur RK (2013b). Molecular characterizations of two begomoviruses infecting *Vinca rosea* and *Raphanus sativus* in India. *Virolog Sin.* 28:53–56.

Marwal A, Sahu A, Sharma P, Gaur RK (2013c). Transmission and host interaction of geminivirus in weeds. In: RK Gaur, T Hohn, P Sharma (eds.). *Plant Virus-Host Interaction: Molecular Approaches and Viral Evolution.* Boston, MA: Elsevier, pp. 143–161.

Mason G, Rancati M, Bosco D (2000). The effect of thiamethoxam, a second-generation neonicotinoid insecticide, in preventing transmission of *Tomato yellow leaf curl geminivirus* (TYLCV) by the whitefly *Bemisia tabaci* (Gennadius). *Crop Prot.* 19:473–479.

Meng RX, Janssen A, Nomikou M et al. (2006). Previous and present diets of mite predators affect antipredator behaviour of whitefly prey. *Exp Appl Acarol.* 38:113–124.

Moffat AS (1999). Geminiviruses emerge as serious crop threat. *Science.* 286:1835.

Monci F, Sanchez-Campos S, Navas-Castillo J, Moriones E (2002). A natural recombinant between the geminiviruses Tomato yellow leaf curl Sardinia virus and Tomato yellow leaf curl virus exhibits a novel pathogenic phenotype and is becoming prevalent in Spanish populations. *Virology.* 303:317–326.

Morya VK, Singh Y, Singh BK, Thomas G (2014). Ecogenomics of geminivirus from India and neighbor countries: An in silico analysis of recombination phenomenon. *Interdiscip Sci Comput Life Sci.* 6:1–9.

Mourrain P, Beclin C, Elmayan T, Feuerbach F, Godon C, Morel JB, Jouette D, Lacombe AM, Nikic S, Picault N, Remoue K, Sanial M, Vo TA, Vaucheret H (2000). Arabidopsis SGS2 and SGS3 genes are required for posttranscriptional gene silencing and natural virus resistance. *Cell.* 101:533–542.

Nahid N, Amin I, Briddon RW, Mansoor S (2011). RNA interference-based resistance against a legume mastrevirus. *Virol J.* 8:499.

Naratajan K (1990). Natural enemies of *Bemisia tabaci* (Gennadius) and effect of insecticides on their activity. *J Biol Control.* 4:86–88.

Natwick ET, Durazo A (1985). Polyester covers protect vegetables from whiteflies and virus disease. *Calif Agric.* 39:21–22.

Newman JA, Gibson DJ, Parsons AJ, Thornley JHM (2003). How predictable are aphid population responses to elevated CO_2? *J Anim Ecol.* 72:556–566.

Nomikou M, Meng RX, Schraag R et al. (2005). How predatory mites find plants with whitefly prey. *Exp Appl Acarol.* 36:263–275.

Padidam M, Sawyer S, Fauquet CM (1999). Possible emergence of new geminiviruses by frequent recombination. *Virology.* 265:218–225.

Palmer KE, Rybicki EP (1998). The molecular biology of mastreviruses. *Adv Virus Res.* 50:183–234.

Palumboa JC, Horowitz AR, Prabhaker N (2001). Insecticidal control and resistance management for *Bemisia tabaci*. *Crop Prot.* 20:739–765.

Papa G, Furlan R, Takao W, Fernando J Celoto FJ, Gerlack G (2008). Effect of new insecticide (spirotetramat) in mixture with neonicotinoid on the control of whitefly, *Bemisia tabaci* b-biotype (hemiptera: *Aleyrodidae*), in cotton. *Beltwide Cotton Conferences*, Nashville, TN, January 8–11, 1398–1402.

Pita JS, Fondong VN, Sangare A, Otim-Nape GW, Ogwal S, Fauquet CM (2001). Recombination, pseudorecombination and synergism of geminiviruses are determinant keys to the epidemic of severe cassava mosaic disease in Uganda. *J Gen Virol.* 82:655–665.

Prajapat R, Marwal A, Bajpai V, Gaur RK (2011). Genomics and proteomics characterization of alphasatellite in weed associated with *Begomovirus*. *Int J Plant Pathol.* 2:1–14.

Prajapat R, Marwal A, Sahu A, Gaur RK (2012). Molecular *in silico* structure and recombination analysis of betasatellite in *Calotropis procera* associated with *Begomovirus*. *Arch Phytopathol Plant Prot.* 45:1980–1990.

Rajagopalan PA, Naik A, Katturi P, Kurulekar M, Kankanallu RS, Anandalakshmi R (2012). Dominance of resistance-breaking cotton leaf curl Burewala virus (CLCuBuV) in northwestern *India Arch Virol.* 157:855–868.

Rapisarda C (1990). La *Bemisia tabaci* vetorre del TYLCV in Sicilia. *Inform Fitopatol.* 6:27–31.

Ratcliff FG, Harrison BD, Baulcombe DC (1997). A similarity between viral defense and gene silencing in plants. *Science.* 276:1558–1560.

Razavinejad S, Heydarnejad J, Kamali M, Massumi H, Kraberger S, Varsani A (2013). Genetic diversity and host range studies of *Turnip curly top virus*. *Virus Genes.* 46:345–353.

Rubinstein G, Morin S, Czosnek H (1999). Transmission of *Tomato Yellow Leaf Curl Geminivirus* to imidacloprid treated tomato plants by the whitefly *Bemisia tabaci* (Homoptera: *Aleyrodidae*). *Horticult Entomol.* 92:658–662.

Ruiz FV, Voinnet O (2007). Roles of plant small RNAs in biotic stress responses. *Annu Rev Plant Biol.* 60:485–510.

Russell RB, Alber F, Aloy P, Davis FP, Korkin D, Pichaud M, Topf M, Sali A (2004). A structural perspective on protein–protein interactions. *Curr Opin Struct Biol.* 14:313–324.

Saikia AK, Muniyappa V (1989). Epidemiology and control of *Tomato leaf curl virus* in southern India. *Trop Agric (Trinidad).* 66:350–354.

Saleem H, Nahid N, Shakir S, Ijaz S, Murtaza G, Khan AA et al. (2016). Diversity, mutation and recombination analysis of cotton leaf curl geminiviruses. *PLoS ONE.* 11:e0151161.

Salwinski L, Eisenberg D (2003). Computational methods of analysis of protein-protein interactions. *Curr Opin Struct Biol.* 13:377–382.

Sanjaya VVS, Prasad V, Kirthi N, Maiya SP, Savithri HS, Sita GL (2005). Development of cotton transgenics with antisense AV2 gene for resistance against *Cotton leaf curl virus* (CLCuD) via *Agrobacterium tumefaciens*. *Plant Cell Tiss Organ Cult.* 81:55–63.

Sattelle DB, Cordova D, Cheek TR (2008). Insect ryanodine receptors: Molecular targets for novel pest control chemicals. *Invertebr Neurosci.* 8:107–119.

Saunders K, Norman A, Gucciardo S, Stanley J (2004). The DNA-b satellite component associated with Ageratum yellow vein disease encodes an essential pathogenicity protein (bC1). *Virology.* 324:37–47.

Saxena S, Singh N, Ranade SA, Babu SG (2011). Strategy for a generic resistance to geminiviruses infecting tomato and papaya through in silico siRNA search. *Virus Genes.* 43:409–434.

Singh DK, Karjee S, Malik P, Islam N, Mukherjee SK (2007). DNA replication and pathogenecity of MYMIV. In: *Communicating Current Research and Educational Topics and Trends in Applied Microbiology.* Mendez-Vilas A. (ed.). Formatex, USA, pp. 155–162.

Smith HA, Koenig RL, McAuslane HJ, McSorley R (2000). Effect of silver reflective mulch and a summer squash trap crop on densities of immature *Bemisia argentifolii* (Homoptera: *Aleyrodidae*) on organic bean. *J Econ Entomol.* 93:726–731.

Steinemann A, Stamm E, Frei B (1990). Chemodynamics in research and development of new plant protection agents. *Pestic Outlook.* 1:3–7.

Suwwan MA, Akkawi M, Al-Musa AM, Mansour A (1988). Tomato performance and incidence of Tomato yellow leaf curl (TYLC) virus as affected by type of mulch. *Sci Hortic.* 37:39–45.

Szilagyi A, Grimm V, Arakaki AK, Skolnick J (2005). Prediction of physical protein-protein interactions. *Phys Biol.* 2:S1–16.

Tarr SAJ (1951). *Leaf Curl Disease of Cotton.* The Commonwealth Mycological Institute, Kew, Surrey, England.

Tripathi S, Varma A (2002). Eco-friendly management of leaf curl disease of tomato. *Indian Phytopath.* 55:473–478.

Varma A, Jain RK, Bhat AI (2002). Virus resistant transgenic plants for environmentally safe management of viral diseases. *Indian J Biotechnol.* 1:73–86.

Vaucheret H (2006). Post-transcriptional small RNA pathways in plants: Mechanisms and regulations. *Genes Dev.* 20:759–771.

Vazquez-Rovere C, Asumendi S, Hopp HE (2001). Transgenic resistance in potato plants expressing potato leaf roll virus (PLRV) replicase gene sequences is RNA-mediated and suggests the involvement of post-transcriptional gene silencing. *Arch Virol.* 146:1337–1353.

Wang X, Xie Y, Zhou X (2004). Molecular characterization of two distinct begomovirus from papaya in China. *Virus Genes.* 29:303–309.

Wass MN, Kelley LA, Sternberg MJE (2010). 3DLigandSite: Predicting ligand-binding sites using similar structures *Nucleic Acids Res.* 38:W469–W473.

Yamamura K, Kiritani K (1998). A simple method to estimate the potential increase in the number of generations under global warming in temperate zones. *Appl Entomol Zool.* 33:289–298.

Yin K, Han T, Liu G, Chen T, Wang Y, Yu ALY, Liu Y (2015). A geminivirus-based guide RNA delivery system for CRISPR/Cas9. *Scientific Reports.* 5:14926.

Zaidi SS, Tashkandi M, Mansoor S, Mahfouz MM (2016). Engineering plant immunity: Using CRISPR/Cas9 to generate virus resistance. *Front Plant Sci.* 7:1673.

Zhang GF, Lü ZC, Wan FH (2007). Detection of *Bemisia tabaci* remains in predator guts using a sequence-characterized amplified region marker. *Entomol Exp Appl.* 123:81–90.

19 On-Farm Management of *Papaya Ringspot Virus* in Papaya by Modifying Cultural Practices

Sunil Kumar Sharma and Savarni Tripathi

CONTENTS

19.1 INTRODUCTION OF HOST AND PATHOGEN

Papaya (*Carica papaya* L.) is one of the major tropical fruit crops. It is cultivated in all continents, but the major share of its total production of 11.57 million tonnes from 433,057 hectares in 2014 came from Asia, Central America, and Africa. India, Brazil, Nigeria, Indonesia, and Mexico are among the major papaya producing countries (FAO, 2014). Commercial cultivation of papaya is unable to achieve its full potential due to widespread incidence of viral diseases. Among various viral diseases affecting papaya cultivation, *Papaya ringspot virus* strain papaya (PRSV-P) is the most devastating in all major papaya-growing areas of the world (Gonsalves et al., 2010; Tripathi et al., 2008). The natural spread of PRSV-P is rapid; therefore the virus may infect up to 100% of plants. The disease is so devastating that farmers have stopped growing papaya in severely affected areas. The use of transgenic cultivars, a successful strategy to manage the virus in Hawaii, has so far not been scaled up in other papaya cultivating areas for some technical reasons (virus strain homology specificity of the PRSV-P resistance; Gonsalves et al., 2006) and environmental activism (Herring, 2008). Other approaches to managing PRSV-P have had only limited success. Therefore the approach of management of cultivation practices to minimize yield losses remains the only viable option in the present scenario.

PRSV was first reported from Hawaii and was shown to be viral in nature by Jensen (1949). The name of the disease, ringspot, is taken from the occurrence of ring spots on the fruit of infected plants. Other symptoms produced by the infections are mosaic and chlorosis of the leaf lamina, water-soaked oily streaks on the petiole and upper part of the trunk, and distortion of young leaves that sometimes results in shoestring-like symptoms. Infected plants lose vigor and become stunted. Fruits from infected trees are of poor quality and generally have lower sugar concentrations. Plants subjected to the early infection (before flowering) with the severe strain of the virus usually do not produce marketable fruits. The virus spreads mainly in the field by aphid vectors in a non-persistent manner. It is not transmitted by seeds. Although papaya is the most important primary and secondary source for spread of the virus, it may also be spread by host plants of Cucurbitaceae. PRSV, a member of genus *Potyvirus*, is further classified into two types: Type P (PRSV-P), which infects cucurbits and papaya, and type W (PRSV-W) which infects cucurbits but not papaya (Purcifull et al., 1984; Tripathi et al., 2008; Gonsalves et al., 2010). Both biotype P and W are serologically indistinguishable. The virions are non-enveloped, flexious filamentous in shape and measure 760-800 × 12 nm (Gonsalves and Ishii, 1980). Virus particles contain 94.5% protein and 5.5% nucleic acid. The genome of PRSV-P consists of ssRNA with positive polarity and has the typical array of genes as present in potyviruses (Yeh and Gonsalves, 1985; Yeh et al., 1992; Shukla et al., 1994). The genome is monocistronic and is expressed via a large polypeptide of 381 kDa that is subsequently cleaved by the viral encoded proteinases to yield functional proteins. Like other potyviruses, the functional proteins are produced for PRSV-P and occur by a combination of cotranslational, post-translational, autoproteolytic, and transproteolytic processing by the three virus-encoded endoproteases P1, HC-Pro, and NIa (Yeh and Gonsalves, 1985; Yeh et al., 1992). Phylogenetic studies showed that within the PRSV coat protein gene, sequences can diverge up to 14% at nucleotide level and 10% at amino acid level (Jain et al., 2004).

19.2 INTEGRATED MANAGEMENT OF *PAPAYA RINGSPOT VIRUS*

As there is no prophylactic or therapeutic control measure for PRSV-P infection, the major emphasis is on optimizing yield by management of the disease. The current approach of disease management is to avoid infection at an early stage of plant growth, for the yield losses are inversely proportional to the age of the plant at the time of infection, and proportional to the severity of the infection. It is mainly transmitted through aphid vectors. Therefore avoiding and reducing the aphid population is the main strategy for PRSV management. It has been established that simply adjusting one factor of cultivation or another cannot prevent PRSV-P infection or reduce its further spread effectively. However, adopting a strategic integrated management of cultural practices can help; e.g., using healthy (virus-free) seedlings of a tolerant cultivar, selecting season of transplanting when the vector population is low, planting border crops, selective roguing of infected plants, and controlling the population of aphid vectors. These practices can be integrated in such a way that the infection is avoided in young plants (Sharma et al., 2010). The following approaches can be applied for disease management.

19.2.1 Site Selection

Climatic and edaphic conditions of the site should be suitable for papaya cultivation. The plantation site should not be prone to water stagnation. Sites with conditions favorable for initial PRSV infection should be avoided, such as high population of aphid vectors and proximity to infected plants (papaya and cucurbits) (Sharma and Tripathi, 2014).

19.2.2 Use of Tolerant Cultivars

Although no commercial papaya cultivar is reported to be resistant, a few cultivars provide commercial fruit production despite PRSV infection. A cultivar that provides a better yield under PRSV infection should be selected. Kudada and Prasad (2000) screened 26 commercial varieties of papaya for resistance against PRSV under natural conditions. None was found to be resistant; however, CO-2 showed high yield potential under disease conditions. IARI, Regional station, Pune has successfully developed two high-yielding PRSV-tolerant dioecious lines with pink (Pune Selection-3) and yellow (Pune Selection-1) pulp. These lines performed better than other commercial cultivars under high disease pressure (Datar et al., 2014; Sharma et al., 2016).

19.2.3 Use of Healthy Seedlings

Papaya production is affected drastically if papaya seedlings are infected by PRSV at nursery stage. Often, infected seedlings do not produce any fruit. Therefore, use of virus-free and otherwise healthy seedlings is an integral part of PRSV management. Virus-free papaya seedlings can be raised only in insect-proof polyhouses (Figure 19.1). Care should be taken to avoid entry of aphid vectors, and there should not be host plants inside the nursery that may act as a source of inoculum in case of accidental entry of aphid vectors.

19.2.4 Inoculum Avoidance

19.2.4.1 Shifting Cultivation

Cultivating papaya in an area for a short period and shifting to other areas is an effective strategy by commercial farmers in those areas where agricultural lands are available for short term. In some pockets of western and southern India, climatic conditions are suitable for papaya cultivation and unfavorable for aphid population. Agents of commercial farmers keep track of available land and hire land for short leases of two to three years, cultivate papaya until PRSV infection is mild, and move to a new similar location for further cultivation. A break from the papaya cultivation in the previous location destroys PRSV inoculum. Eventually papaya can be cultivated in that location again (personal discussion with farmers). A similar strategy was applied when PRSV was detected in Brazil and Hawaii, where the papaya cultivation was shifted to PRSV-free areas (Gonsalves et al., 2006). This approach can be applied while taking care of the following considerations:

- PRSV infection usually occurs in the new areas eventually, therefore a second option should be kept open.
- The selected areas should not be far from the main markets so as to avoid increased shipping cost and deterioration in the quality of the fruits.
- The shifting cultivation strategy cannot be applied on small islands.

19.2.4.2 Isolation

An adequate isolation distance of new papaya plantation from the existing PRSV-infected crop can delay the infection. Wolfenbarger (1996) suggested that the new plantation should be at least 100 meters away from the infected ones. Maintaining a safe isolation distance in commercial cultivation is not possible where papaya growers are small landholders. In such cases, all diseased plants should be removed from the vicinity, and a gap of a few months should be maintained before new plantation. Border crops around papaya plantations, use of intercrops, and positioning crops upwind from severe inoculum sources may reduce transmission by aphid vectors even at a closer isolation.

FIGURE 19.1 Insect-proof net house (top) and healthy papaya nursery (bottom).

19.2.4.3 Roguing of Infected Papaya Plants and Other Host Plants

PRSV-infected papaya plants work as potent sources for further spread of infection in the neighboring plants. It is generally recommended to rogue (uproot and destroy) infected plants at an early stage to check further spread of infection. It is presumed that roguing of infected plants will delay PRSV infection (and reduce yield losses) in neighboring plants. However, it is observed that infected plants, too, give some yield. Roguing them will result in loss of their potential yield. Therefore farmers are reluctant to uproot infected plants. Roguing will be economically viable only when the presumed increase in the yield in the neighboring plants would be more than the loss of potential yield of uprooted infected plants. Roguing of plants showing mild PRSV symptoms with potential for fruit production may not be economically viable. It is therefore recommended that only severely infected plants with little or no yield potential should be rogued only until the fruit sets. In conformity

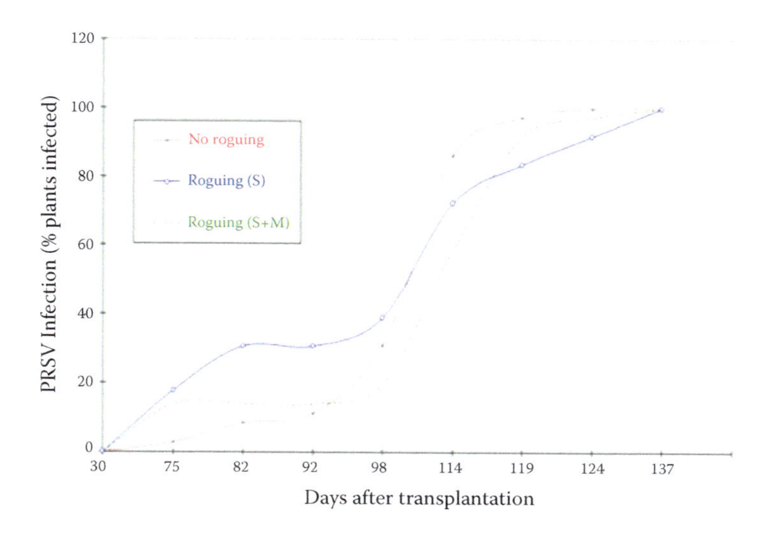

FIGURE 19.2 Effect of roguing on PRSV infection. M = mildly infected plants, S = severely.

with that assumption, roguing of plants with initial symptoms lowered and delayed damage caused by PRSV infection in cv. Maradol Roja in Mexico (Hernandez-Castro et al., 2003). At Pune, spread of PRSV infection showed a pattern according to the different treatments of roguing. The number of infected plants increased in the treatment without roguing (control) 100 days after transplanting (Figure 19.2). Rate of PRSV spread was minimum in plants under roguing of both mild and severely infected plants. Time required for infection in adjoining plants was the shortest in control (32 days), and the longest in roguing of severely infected plants (41 days) (Sharma et al., 2010).

19.2.5 Vector Control

19.2.5.1 Vector Avoidance

19.2.5.1.1 Annual Cultivation

Owing to heavy infestation in the first year of cultivation, the second or third years' crops are no longer economically viable. Farmers cultivate papaya as an annual crop in most parts of India. Under Pune climate, papaya is recommended for transplantation in February. The harvesting begins in October and completes by January. Plants are uprooted, and new crop can be transplanted in February again (Sharma et al., 2010). Farmers in Taiwan are growing papaya as an annual crop. This strategy ensures that damage caused by infection are not carried forward to the next season.

19.2.5.1.2 Season for Transplanting

The aphid vector population follows a seasonal trend; therefore papaya should be planted in the season when the aphid population is naturally low so that it gets a longer PRSV-free period. Under climatic conditions of Pune, spring season (February) transplantation showed a longer PRSV-free period than summer (June) or autumn (October) seasons (Ram and Singh, 1999). However, in central and eastern parts of India, October transplanting gave better yield than other seasons of transplanting (Kudada and Prasad, 1999; Ray et al., 1999).

19.2.5.1.3 Net House Cultivation

Aphid vector population can be reduced by cultivating papaya trees under protective netting. However, under net cultivation fruits do not develop well and have lower sugar content due to limited sunlight. When nets are removed to allow better growth of fruits, trees become infected. Moreover,

FIGURE 19.3 Papaya cultivation in a net house.

net cultivation adds additional cost to papaya cultivation (Gonsalves, 1998). Therefore net cultivation has not become popular among farmers (Figure 19.3).

19.2.5.1.4 Use of Barrier Crops

Raising a border crop around a papaya plantation can reduce entry of aphid vectors inside the main plantation, thereby minimizing contact of papaya plants with virus-carrying aphid vectors, which is likely to delay PRSV infection (Sharma et al., 2010). In addition, border crops can ameliorate the microclimate of the papaya plantation. Their effect is more pronounced during peak summer and winter seasons. Their third role is to work as windbreak under windy conditions. Though border crops may vary from location to location, there are certain parameters that will help in the selection of border crops (Table 19.1). Five rows of banana, planted four months before papaya transplantation at a spacing of 1×1 meter acts as an excellent border crop (Figures 19.4 and 19.5) (Sharma et al., 2005).

TABLE 19.1
Ideal Characteristics of Border Crops

S. No.	Characteristics	Significance
1	It should not be a primary or secondary host to PRSV	No source of PRSV inoculum in the vicinity
2	It should not be a primary or secondary host to aphid vectors transmitting PRSV	Reduced aphid population
3	It should grow at least 2 meters tall	Effective barrier to aphid-vectors entry
4	It should be perennial	Protection throughout the cropping cycle
5	It should be deep rooted	Able to withstand high velocity winds
6	It should be acceptable to the local farming system, and should not compete with papaya crop for general cultivation	High synergy
7	It should supply a commercial product	Loss of area in border crop and expenditure in raising them is justified

FIGURE 19.4 Successful papaya cultivation under border crop of banana.

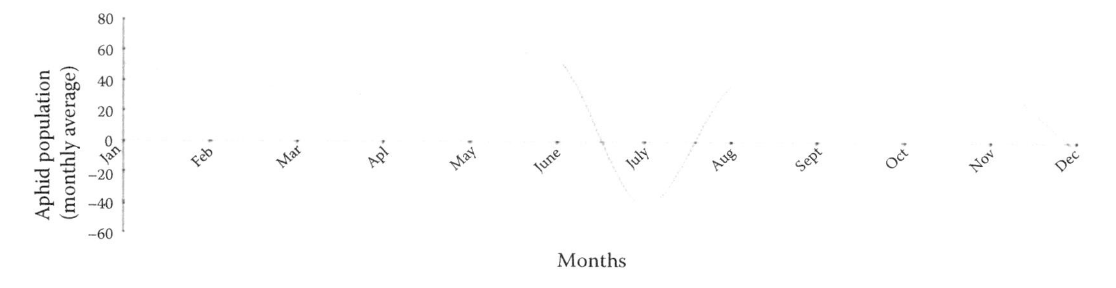

FIGURE 19.5 Effect of banana border crop on aphid population (percent reduction inside the border crop).

Perennial pigeon pea, sorghum, and castor were tried as border crops around papaya by Prasad and Kudada (2005). Perennial pigeon pea proved to be most effective in reducing the incidence of PRSV and providing a higher fruit yield.

19.2.5.2 Vector Eradication

19.2.5.2.1 Chemical and Biological Control

Aphid vectors do not breed or colonize on papaya. Though the disease is transmitted through aphid vectors, the application of insecticides has no conclusive evidence of reduction of PRSV infection. Some reduction in infection is reported by spraying 1% white oil emulsion. Seven percent of plants became infected by PRSV when treated with white oil as compared to 44% among untreated plants (Verghese et al., 2001). Reductions of 15% and 12% were reported in PRSV incidence in integrated pest management (IPM) and IPM without citroline, respectively, as compared to 100% incidence with conventional management (Hernandez Castro et al., 2000). No correlation could be established between foliar application of various doses of dimethoate (alternated with neem oil) and the incidence of PRSV (Sharma, 2005). However, judicious use of botanical and chemical insecticides is recommended up to fruit set only. It will also keep papaya plants free from other harmful insect pests. Papaya mosaic disease had very high incidence at the nursery but no control was known. Bhargava and Khurana (1969) found that weekly sprays of 1% water emulsion of groundnut oil on papaya plants in the nursery completely prevented the aphid transmission of the mosaic virus (PMV=PRSV) for one week.

19.2.6 Good Agricultural Practices (Nutrition and Plant Protection)

Many commercial cultivated papaya cultivars in India are F_1 hybrids. Such plants have higher yield potential. Their yield potential can be exploited only when adequate nutrients are available in the soil and physicochemical properties support the uptake of these nutrients. Soil nutrient status and physicochemical properties of the soil can be enhanced by applying manure, raising "green manuring" crops, and by sheep folding. Subsequent application of manure, fertilizers, and micronutrients at the critical stage of growth will ensure good growth and quality fruit production. Papaya plants require drenching of the rooting zone with copper-based fungicide to avoid mortality of transplanted seedlings due to soil-borne fungal infection. Some fungal diseases are reported on papaya plants during the fruit development and harvesting stages. They should be treated therapeutically. Foliar spray of broad-spectrum fungicides can reduce loss in the fruit quality during transport and storage. Khurana (1968) tested 27 Indian and exotic papaya varieties by mechanical sap inoculation with Papaya mosaic/Ringspot/Distortion potyvirus, but all were susceptible or highly susceptible to the mosaic virus in nursery. Prasad and Kudada (2000) found the same results on field screening of papaya commercial varieties. They then recommended cultivating papaya under nethouses.

19.2.7 Development of Resistance

19.2.7.1 Cross-Protection

When sap from the leaves of the plants infected with a mild strain of PRSV is applied on the leaves of healthy papaya seedlings at the nursery stage, they are presumed to develop immunity against a severe strain of PRSV. When these seedlings are transplanted in the field they show temporary resistance to the disease. This technique is called "cross-protection." Two mild isolates of PRSV (PRSV-HA 5-1 and PRSV-HA 6-1) were developed by mutation of an isolate of severe PRSV from Hawaii. When tried on Hawaiian solo cultivars, cross-protection yielded good results in line 8 and Kamiya, but it showed adverse effects on Sunrise (Gonsalves and Garsney, 1989; Yeh and Gonsalves, 1984). When these mild strains were tried in large-scale field trials in Taiwan, treated plants showed only a delay in the appearance of severe symptoms of PRSV (Yeh et al., 1998). In India, Ram et al. (2006) applied cross-protection using a naturally occurring mild strain. The pre-immunized seedlings exhibited up to 66% protection against severe isolates of the PRSV in the field. The cross-protection technique is not popular because of the following limitations:

- The technique is dependent on the availability of a mild strain homologous to the severe strain of PRSV for each geographical area.
- There is a concern that the mild virus may regain potency and become severe.
- Cross-protection with an effective mild strain does not show consistent results with all cultivars over a period of time.

19.2.7.2 Breeding for Resistance

There is no known source of genes for resistance against PRSV in the genus *Carica* that can be used for resistance breeding. There are claims and counterclaims about resistance/tolerance against PRSV in intergeneric hybrids between *C. papaya* and *Vasconcellea* species resistant to PRSV. *V. cundinamarcensis* has been identified as immune to PRSV, but due to incompatibility with *C. papaya*, development of PRSV-resistant intergeneric is yet to be achieved (Sharma and Tripathi, 2014, 2016). This limitation is being overcome by using *V. parviflora* as a bridge species (O'Brien and Drew, 2009).

19.2.7.3 Transgenic Approach

Successful PRSV-resistant transgenic varieties, red-fleshed homozygous line "SunUp" (transgenic Sunset) and yellow-fleshed F_1 hybrid Rainbow (SunUp x non-transgenic Kapoho), have been developed and cultivated successfully in Hawaii since 1998 (Ferreira et al., 1992; Gonsalves, 1998; Gonsalves et al., 2004a,b, 2006). There are unconfirmed reports of transgenic papaya being developed and cultivated in China. There are reports of development of transgenic papaya from other labs, but none has been released as a variety for commercial cultivation. The major limitation of the transgenic approach is its dependence on homology between resistance and the prevailing PRSV-P strains. Moreover, strong global environmental activism is creating an adverse political environment against the use of transgenic plants (Herring, 2008).

19.3 CONCLUSION

Control of PRSV in papaya is a challenge because of the lack of a direct method of control by chemical applications. Therefore integrated management of PRSV has to be utilized involving various strategies.

Minimization of losses by PRSV by following the integrated management strategy is the only option left with papaya growers until an effective solution of PRSV is worked out (Figure 19.6):

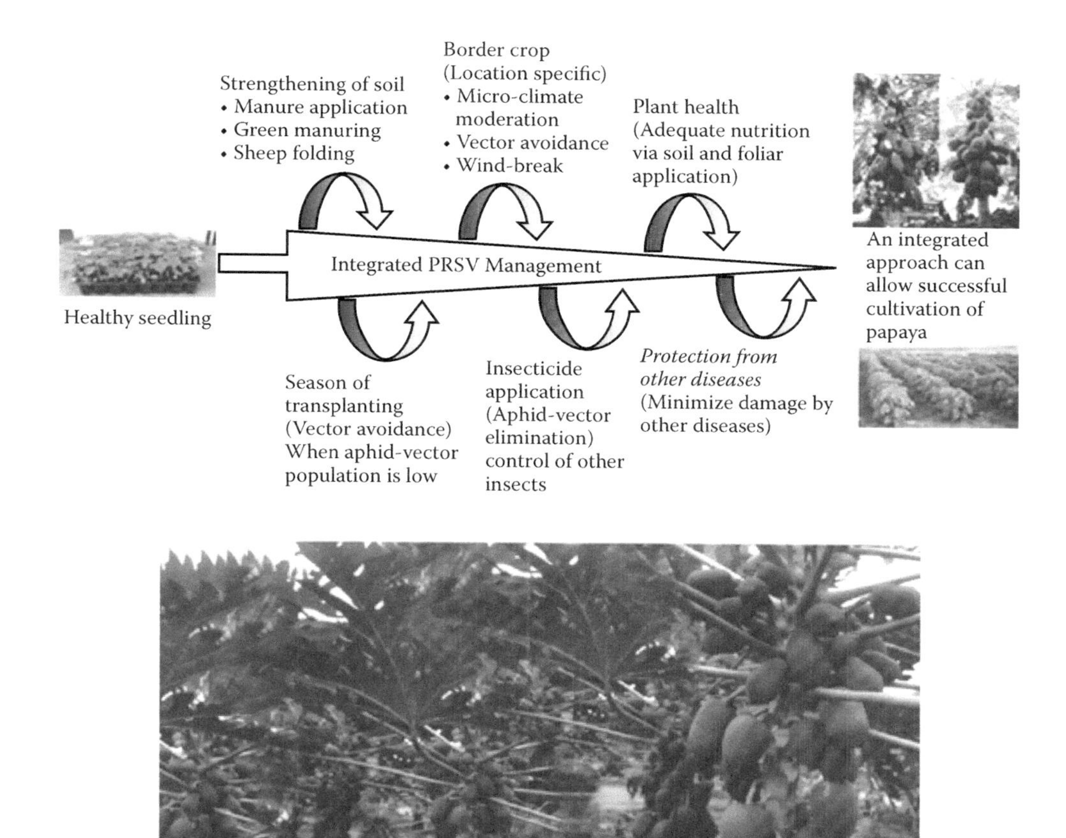

FIGURE 19.6 Schematic representation of integrated PRSV, top, management and its implications in farmers' fields, bottom.

- Selection of suitable cultivars
- Selection of appropriate season of transplantation
- Use of healthy seedlings
- Planting border crops around papaya plantation
- Regular use of insecticides
- Following good agricultural practices by providing adequate nutrition and protection from other pests

REFERENCES

Bhargava KS and Khurana SM Paul (1969). Papaya mosaic control by oil sprays. *Phytopath Z.* 64:338–343.

Datar VV, Singh SJ, Chavan VM, Zote KK, Tomar SPS, Verma R, Sharma SK and Tripathi S (2014). Pune Selection-3 (IC0599272; INGR14016), a papaya (*Carica papaya* L.) germplasm with *Papaya ringspot virus* tolerance and higher yield (30-40 kg of fruits). *Indian Journal of Plant Genetic Resources* 27:192–193.

FAOSTAT (2014). Food and Agricultural Organization of the United Nations Database. http://www.apps.fao.org (accessed January 2017).

Ferreira SA, Mau RFL, Manshardt R, Pitz KY and Gonsalves D (1992). Field evaluation of papaya ringspot virus cross protection. In *Proceedings of the 28th Annual Hawaii Papaya Industry Association Conference*, pp. 14–19. Honolulu.

Gonsalves D (1998). Control of papaya ringspot virus in papaya: A case study. *Annual Review of Phytopathology* 36:415–437.

Gonsalves D and Garsney SM (1989). Cross protection techniques for control of plant virus diseases in the tropics. *Plant Disease* 73:592–597.

Gonsalves D and Ishii M (1980). Purification and serology of papaya ringspot virus. *Phytopathology* 70:1028–1032.

Gonsalves C, Lee DR and Gonsalves D (2004a). Transgenic virus-resistant papaya: The Hawaiian "Rainbow" was rapidly adopted by farmers and is of major importance in Hawaii today. apsnet.org/publications/apsnetfeatures/Pages/PapayaHawaiianRainbow.aspx (accessed April 2017).

Gonsalves D, Gonsalves C, Ferreira S, Pitz K, Fitch M, Manshardt R and Slightom J (2004b). Transgenic virus resistant papaya: From hope to reality for controlling papaya ringspot virus in Hawaii. apsnet.org/publications/apsnetfeatures/Pages/PapayaRingspot.aspx (accessed April 2017).

Gonsalves D, Vegas A, Prasartsee V, Drew R, Suzuki JY and Tripathi S (2006). Developing papaya to control Papaya ringspot virus by transgenic resistance, intergeneric hybridization, and tolerance breeding. *Plant Breeding Reviews* 26:35–73.

Gonsalves D, Tripathi S, Carr JB and Suzuki JY (2010). *Papaya ringspot virus*. The Plant Health instructor APS apsnet.org/edcenter/intropp/lessons/viruses/Pages/PapayaRingspotvirus.aspx (accessed April 2017).

Hernandez-Castro E, Riestra Díaz ED, Garcia Perez E, Ortega Arenas LD and Mosqueda Vazquez R (2000). Response of the papaya ringspot virus (PRSV) in three systems of management. *Manejo Integrado de Plagas* 58:20–27.

Hernandez-Castro E, Riestra-Díaz ED, Villanueva-Jiménez JA and Mosqueda-Vázquez R (2003). Epidemiological analysis of the *Papaya ringspot virus* on cv. MaradolRoja under different densities, application of neem (*Azadirachtaindica* A. Juss.) seed extract sprays, and elimination of diseased plants. *Serie Horticultura* 9:55–68.

Herring RJ (2008). Opposition to transgenic technologies: Ideology, interests and collective action frames. *Nature Reviews Genetics* 9:458–463.

Jain RK, Sharma J, Sivakumar AS, Sharma PK, Byadgi AS, Verma AK and Varma A (2004). Variability in the coat protein gene of *Papaya ringspot virus* isolates from multiple locations in India. *Archives of Virology* 149:2435–2442.

Jensen DD (1949). Papaya ringspot virus and its insect vectors relationships. *Phytopathology* 39:212–220.

Khurana SM Paul (1968). A study on virus diseases of papaya (*Carica papaya* L) in Gorakhpur. PhD thesis approved by GPU, Gorakhpur, India.

Kudada N and Prasad SM (1999). Effect of planting time on the incidence of PRSV disease and yield of fruits. *Indian Phytopathology* 52:224–227.

Kudada N and Prasad SM (2000). Screening of commercial papaya (*Carica papaya*) varieties for tolerance against PRSV. *Journal of Research. Birsa Agricultural University* 12:51–56.

O'Brien CM and Drew RA (2009). Potential for using *Vasconcelle aparviflora* as a bridging species in intergeneric hybridisation between *V. pubescens* and *Carica papaya*. *Australian Journal of Botany* 57:592–601.

Prasad SM and Kudada N (2005). Effect of barrier crops on natural incidence of *Papaya ringspot virus disease* and fruit yield of papaya. *Indian Journal of Virology* 16:24–26.

Purcifull DF, Edwardson J, Hiebert E and Gonsalves D (1984). Papaya ring spot virus. *CMI/AAB Descriptions of Plant Viruses* No. 292 (No. 84 revised).

Ram RD and Singh SJ (1999). Management of papaya ring spot virus through cross protection strategy – A preliminary observation. Abstract. XI Western Zonal Meeting of IPS, Vasant Dada Sugar Institute, Manjari (Pune), 7–8 January, 1999.

Ram RD, Verma R, Tomer SPS and Prakash S (2006). Management of papaya ring spot virus through cross-protection strategy. *Journal of Maharashtra Agricultural University* 31:92–95.

Ray PK, Yadav JP and Kumar A (1999). Effect of transplanting dates and mineral nutrition on yield and susceptibility of papaya to PRSV. *Horticulture Journal* 12:15–26.

Sharma SK and Tripathi S (2014). Overcoming limitations of resistance breeding in *Carica papaya* L. In: Gaur RK, Hohn T, Sharma P (eds). *Plant Virus-Host Interaction: Molecular Approaches and Viral Evolution*. Elsevier, Amsterdam, pp. 177–191.

Sharma SK and Tripathi S (2016). Resistance against *Papaya ringspot virus* in *Vasconcellea* species: Present and potential uses. In: Gaur RK et al. (eds). *Plant Viruses: Evolution and Management*, Springer, Singapore, pp. 215–230.

Sharma SK, Chavan VM, Tomer SPS and Dadade RS (2005).Yield in papaya plants by papaya ring spot virus: Effect of age/stage of plant at the time of infection. Second Global Conference: Plant Health Global Wealth, 25–29 November 2005, MPUA&t, Udaipur, 25–26.

Sharma SK, Zote KK, Kadam UM, Tomar SPS, Dhale MG and Sonawane AU (2010). Integrated management of *Papaya Ringspot Virus*. *Acta Hortic.* 851:473–480.

Sharma SK, Tripathi S, Verma R, Chandrashekar K, Sonawane AU, Singh SJ, Chavan VM, Zote KK, Datar VV and Tomar SPS (2016). Pune Selection-1 (PS-1) (IC0611690; INGR15030), a papaya (*Carica papaya* L.) germplasm with field tolerance to *Papaya ringspot virus* tolerance and yellow pulp colour. *Indian Journal of Plant Genetic Resources* 29:216–217.

Shukla DD, Ward CW and Brunt AA (1994). *The Potyviridae*. Wallingford (UK), CAB International.

Tripathi S, Suzuki JY, Ferreira SA and Gonsalves D (2008). *Papaya ringspot virus*-P: Characteristics, pathogenicity, sequence variability and control. *Molecular Plant Pathology* 9:269–280.

Verghese A, Kumar HRA and Jayanthi PDK (2001). Status and possible management of papaya ringspot virus with special reference to insect vectors. *Pest Management in Horticultural Ecosystems* 7:99–112.

Wolfenbarger DO (1996). Incidence distance and incidence time relationship of papaya virus diseases. *Plant Disease Reporter* 50:908–909.

Yeh SD and Gonsalves D (1984). Evaluation of induced mutants of papaya ring spot virus for control by cross protection. *Phytopathology* 74:1086–1091.

Yeh SD and Gonsalves D (1985). Translation of Papaya ringspot virus RNA *in vitro*: Detection of a possible polyprotein that is processed for capsid protein, cylindrical-inclusion protein, and amorphous-inclusion protein. *Virology* 143:260–271.

Yeh SD, Jan FJ, Chiang CH, Doong PJ, Chen MC, Chung PH and Bau HJ (1992). Complete nucleotide sequence and genetic organization of papaya ringspot virus RNA. *Journal of General Virology* 73:2531–2541.

Yeh SD, Gonsalves D, Wang HW, Namba R and Chiu RJ (1998). Control of papaya ringspot virus by cross protection. *Plant Disease* 72:375–380.

20 Diversity, Host–Pathogen Interactions, and Management of Grapevine Leafroll Disease

Sandeep Kumar, Richa Rai, and Virendra Kumar Baranwal

CONTENTS

20.1 INTRODUCTION

In the contemporary agrihorticultural system, grapevine has acquired the status of the most important fruit crop of the globe. The crop is subjected to attacks by many pests and diseases. Among different pests and pathogens, infectious intracellular obligate parasites such as virus and virus-like agents pose a major threat to the production and quality of the grape fruits (Martelli and Boudon-Padieu, 2006; Martelli, 2014). Currently, 70 virus and virus-like agents have been recorded from different grapevine species worldwide (Martelli, 2014). The vegetative propagation nature of the crop has helped these agents in getting disseminated to the different grape-growing regions of the world by means of the distribution of vegetative cuttings (Naidu et al., 2015). Based on the diseases they cause, grapevine viruses have been categorized into four major groups: viruses associated with degeneration/decline complex, viruses involved in fleck disease complex, viruses associated with rugose wood complex, and viruses involved in leafroll disease complex (Naidu et al., 2014). However, because of its impact and wider occurrence, grapevine leafroll disease (GLD) is considered the most economically important viral disease among the virus and virus-like diseases of grapevine (Naidu et al., 2015). The first mention of GLD can be found in the mid-nineteenth century, and thereafter it was given different names in different languages; however leafroll is the most commonly used word for this disease. Impacts on fruit quality by the disease include reduction in fruit yields, delay in fruit ripening, reduced soluble solids, delayed crop maturity, reduced berry anthocyanin, decreased berry weight, and increased titratable acidity. It also induces the loss of phloem vessels and decrease in photosynthetic potential by affecting various aspects of photosynthesis (Atallah et al., 2012; Bertamini et al., 2004; Freeborough and Burger, 2008; Kumar, 2013; Rayapati et al., 2008). GLD may cause yield loss of 30–68%. It also reduces the life span of the vines from 30 years to

15 years (Golino et al., 2008). In the absence of management practices, and depending upon incidence level of the disease, reduction in yield and impact on fruit quality by GLD can cause a loss of $25,000–$40,000 (Atallah et al., 2012).

Despite being recognized as a major threat for viticulture and related industries, our knowledge about GLD is still limited because of a range of challenges associated with its complexity (Naidu et al., 2014). Because of several intriguing features, GLD stands apart from other plant virus pathosystems and thus it is an exceptionally complex disease of grapevine. It exhibits a variety of unique features with regard to symptoms, causal agents, genome organization and gene expression of the causal agents, and host–vector–vector interactions (Naidu et al., 2015). In this chapter we discuss the diversity of disease at various levels for a comprehensive understanding of the unique complexity of the disease and the associated viruses in the context of current knowledge of the GLD pathosystem.

20.2 SYMPTOMS

Symptomatology is the first level of complexity of GLD. Expression of symptoms varies from cultivar to cultivar and from season to season. GLD exhibits a symptomatic phase alternating with an asymptomatic phase. The symptoms are generally more expressive in red- or dark-fruited cultivars. In the case of dark-fruited cultivars, the interveinal areas of leaf lamina become reddish or reddish-purple. Gradually, with the advancement of the disease the leaf blades become thicker and brittle, and the margins of the infected leaves roll downward. In severe cases, the leaf surface turns deep purple (Figure 20.1). Two anthocynins, *Cyanidin-3-glucoside* and *malvidin-3-glucoside*, synthesized during infection, are responsible for the reddish-purple color of virus-infected leaves of dark-fruited grapevine (Gutha et al., 2010). Symptoms are similar in light- or white-fruited cultivars except for the color of leaf surface, which becomes cholorotic to yellowish, unlike dark-fruited cultivars (Figure 20.1). Symptom development begins with mature leaves at the bottom shoots and gradually extends to the upper leaves (Rayapati et al., 2008). Downward rolling of margins of the leaves of both types of cultivar during advanced stages of infection gives the disease its common name, "leafroll" (Martelli and Boudon-Padieu, 2006; Rayapati et al., 2008). In some cases, symptomless infections; i.e., latent infections, have also been reported. Mixed infections of GLD-causing viruses together with other viruses and viroids further amplifies the symptomatology of the disease. It is yet not understood if the diversity in symptomatology is because of differences in host–pathogen interactions in light-fruited and dark-fruited cultivars or if some other factors are also responsible.

20.3 VIRUSES ASSOCIATED WITH GLD AND THEIR TAXONOMY

Involvement of different viruses and their strains renders the second level of complexity to GLD. Viruses associated with the leafroll disease of grapevine are collectively known as grapevine leafroll-associated viruses (GLRaVs). At one time, 11 filamentous viruses belonging to the family *Closteroviridae* have been reported to be involved with the diseases. These viruses were named GLRaV-1 (*Grapevine leafroll-associated virus* 1), GLRaV-2, GLRaV-3, GLRaV-4, GLRaV-5, GLRaV-6, GLRaV-7, GLRaV-9, GLRaV-Pr (sequence originally deposited in GenBank under the name of GLRaV-10), GLRaV-De (sequence originally deposited in GenBank under the name of GLRaV-11), and GLRaV-Car (Martelli et al., 2012). However, recent studies suggested considering GLRaV-5, GLRaV-6, GLRaV-9, GLRaV-Pr, GLRaV-De, and GLRaV-Car as strains of GLRaV-4, and thus they are written as GLRaV-4 strain 5, GLRaV-4 strain 6, GLRaV-4 strain Pr, GLRaV-4 strain De, and GLRaV-4 strain Car, respectively. Together, these viruses have been suggested to be known as GLRaV-4-like viruses, i.e., GLRaV-4 LV (Martelli et al., 2012; Naidu et al., 2015). Accordingly, the recent taxonomy has grouped GLD causing viruses into five species: GLRaV-1, GLRaV-2, GLRaV-3, GLRaV-4 (and its strains), and GLRaV-7 (Anonymous, 2017). Very recently, a new virus has been added in the list of GLRaVs. This new virus, reported in grapevines showing typical symptoms of GLD from Japan, has been tentatively named GLRaV-13 (Ito and Nakaune, 2016). It exhibits closeness to

FIGURE 20.1 Symptoms of grapevine leafroll disease (GLD) observed during the survey conducted for the study. (a) Vines of cultivar Cabernet Sauvignon in a vineyard of Nashik. (b) Vine of a cultivar Pinot Noir at experimental farm of ICAR-National Research Centre for Grapes (ICAR-NRCG), Pune. (c and d) Close-up views of the leaves of two different vines of cultivar Pinot Noir (from ICAR-NRCG, Pune) found to be positive for GLRaV-1 and GLRaV-3, respectively. (e) Close-up view of leaves of a vine of cultivar Shiraj from Nashik found to be positive for both GLRaV-1 and GLRaV-3. (f) Close-up view of leaves of a vine of light-fruited cultivar Thompson Seedless (from ICAR-NRCG, Pune) found to be positive for GLRaV-3.

GLRaV-1 rather than GLRaV-4 LV. However, the pathogenicity of tentative GLRaV-13 remains unclear (Ito and Nakaune, 2016). More often these viruses infect the host together, which in turn started a debate about whether a single GLRaV infection is responsible for causing the disease or if it needs an association of multiple GLRaVs containing many genomic RNAs and their byproducts. GLRaV-4 has been reported to produce comparatively mild symptoms, while some strains of GLRaV-2 and GLRaV-7 cause no symptoms (Martelli et al., 2012; Naidu et al., 2015). Mixed infections among GLRaVs and with other viruses and viroids not only aggravate the complexity of symptoms of the disease but also amplify the intricacies of other aspects of the disease such as host-virus interactions, vector transmission, etc.

GLD-associated viruses belong to the family *Closteroviridae*, which consists of four genera: *Closterovirus*, *Ampelovirus*, *Crinivirus*, and *Velarivirus* and five unassigned viruses. Recent taxonomy has assigned various GLRaVs into three genera of the family. GLRaV-1, -3, and -4 (GLRaV-4 LV) belong to the genus *Ampelovirus* (derived from the Greek *ampelos*, meaning grapevine, the host for the type species), with GLRaV-3 as the type species. GLRaV-2 has been assigned to the genus *Closterovirus* (derived from the Greek *kloster*, meaning thread) where *Beet yellow virus* (BYV) serves as the type species, whereas GLRaV-7 typified the newly created genus *Velarivirus* (derived from the Latin *velari*, meaning cryptic) (Figure 20.2). Thus it can be seen that except for two (i.e. GLRaV-2 and GLRaV-7), the GLD-causing viruses belong to the genus *Ampelovirus*. The genus *Ampelovirus* has a number of viruses other than those causing GLD, and they are categorized into two subgroups. Subgroup I constitutes viruses with large (in excess of 17,000 nt) and complex (9–12 open reading frams [ORFs]) genome *viz.* GLRaV-3, GLRaV-1, PMWaV-2 (*Pineapple mealybug wilt-associated virus* 2), LChV-2 (*Little cherry virus* 2) and BVBaV (*Blackberry vein banding-associated virus*) (King et al., 2012; Martelli et al., 2012; Naidu et al., 2015), whereas subgroup II includes smaller (approximately 13,000–14,000 nts) and simpler (6 ORFs, 7 genes) genome viral species *viz.* GLRaV-4, PMWaV-1 (*Pineapple mealybug wilt-associated virus* 1), PMWaV-3 (*Pineapple mealybug wilt-associated virus* 3) and PBNSPaV (*Plum bark necrosis stem pitting-associated virus*) (Figure 20.2). An important feature of subgroup II is that they lack the minor coat protein (CPm) (Figure 20.3).

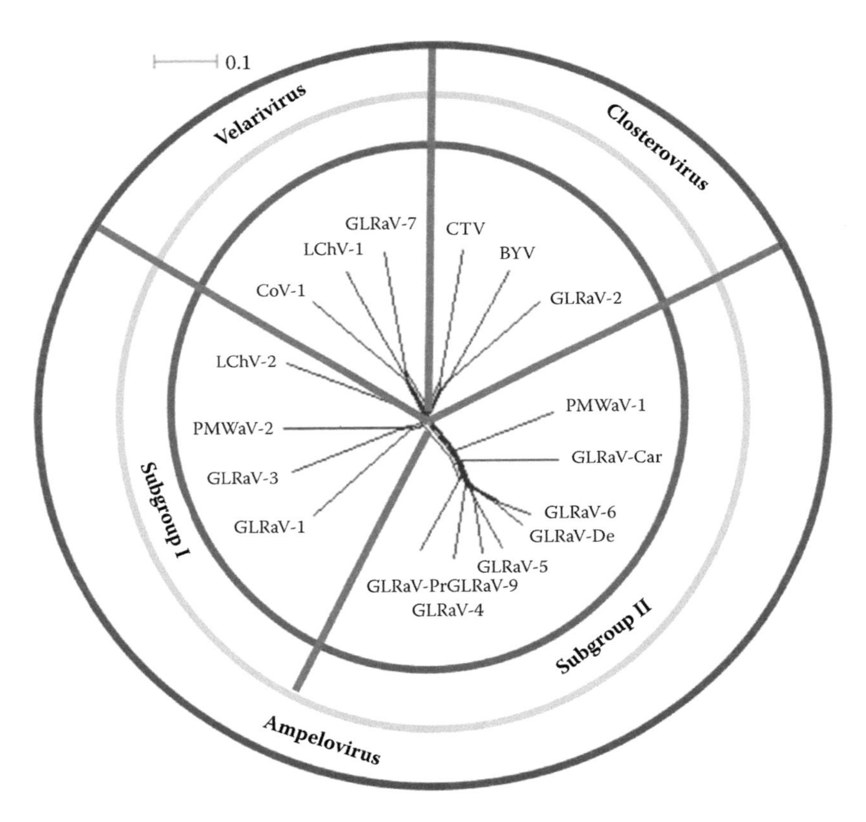

FIGURE 20.2 Neighbor network reconstruction of the complete HSP70h genes of grapevine leafroll disease-associated viruses. Nucleotide sequences were taken from GenBank and the network was constructed using SplitsTreev4 (Huson and Bryant, 2006). Sequences used for constructing the network are: GLRaV-1 (*Grapevine leafroll-associated virus* 1, AF195822), GLRaV-2 (*Grapevine leafroll-associated virus* 2, AF039204), GLRaV-3 (*Grapevine leafroll-associated virus* 3, NC_004667), GLRaV-4 (*Grapevine leafroll-associated virus* 4, FJ467503), GLRaV-5 (*Grapevine leafroll-associated virus* 4 strain 5, NC_016081), GLRaV-6 (*Grapevine leafroll-associated virus* 4 strain 6, FJ467504), GLRaV-9 (*Grapevine leafroll-associated virus* 4 strain 9, AY297819), GLRaV-De (*Grapevine leafroll-associated virus* 4 strain De, AM494935), GLRaV-Car (*Grapevine leafroll-associated virus* 4 strain Car, FJ907331), GLRaV-Pr (*Grapevine leafroll-associated virus* 4 strain Pr, AM182328), GLRaV-7 (*Grapevine leafroll-associated virus* 7, HE588185), PMWaV-1 (*Pineapple mealybug wilt-associated virus* 1), PMWaV-2 (*Pineapple mealybug wilt-associated virus* 2), LChV-1 (*Little cherry virus* 1, NC_001836), LChV-2 (*Little cherry virus* 2, AF416335), CoV-1 (*Cordyline virus* 1, HM588723), CTV (*Citrus tristeza virus*, NC_001661) and BYV (*Beet yellows virus*).

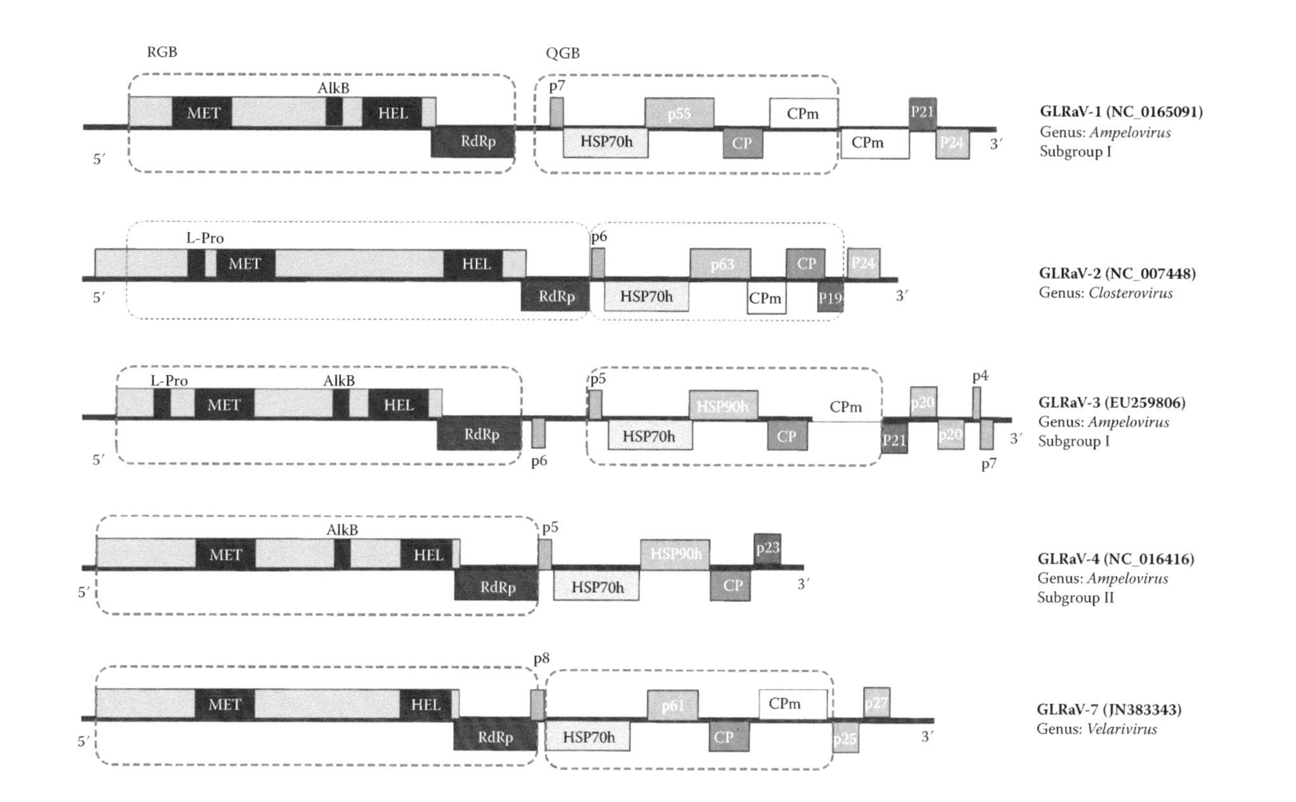

FIGURE 20.3 Schematic representation of the genome organizations of grapevine leafroll disease-associated viruses. GLRaV-1 (*Grapevine leafroll-associated virus* 1, NC_0165091), GLRaV-2 (*Grapevine leafroll-associated virus* 2, NC_007448), GLRaV-3 (*Grapevine leafroll-associated virus* 3, EU259806) GLRaV-4 (*Grapevine leafroll-associated virus* 4, NC_016416), GLRaV-7 (*Grapevine leafroll-associated virus* 7, JN383343). Corresponding genera, subgroups, and accession numbers are indicated to the right side of the genome maps. The open reading frames (ORFs) are shown as boxes with designated protein domains such as L-Pro, papain-like leader protease; AlkB, AlkB domain, MET, methyltransferase; HEL, RNA helicase; and POL, RNA-dependent RNA polymerase domains of the replicase. Conserved ORFs form the replication gene block (RGB) and quintuple gene block (QGB), denoted by dotted line boxes. Abbreviations indicating ORFs are CP, coat protein; CPm, minor coat protein; RdRp, RNA-dependent RNA polymerase. The other ORFs are designated with approximate molecular weight and a common "p" designator. Figures drawn are not to scale.

20.4 GENOME ORGANIZATION OF GLRaVs

The available full-length genome sequences of GLRaVs reveal that they are diverse in genome size, gene content, and genome organization. Thus genomic level diversity of the causal agents is third level of complexity of GLD. Sequence data available in public database indicates that GLRaV-4 strain with genome size 13, 626 nt encoding six ORFs is the smallest, whereas GLRaV-3 with 18, 671 nt encoding twelve ORFs is the largest and the most complex (Naidu et al., 2014, 2015). ORFs located at 5′ end of genomes of GLRaVs are responsible for encoding a characteristic core of replication-associated genes and thus collectively are referred as replication gene block (RGB). RGB comprises of ORF 1a and ORF 1b encoding replication-associated proteins having important domains such as methyltransferase (MET), RNA helicase (HEL) and RNA-dependent RNA polymerase (RdRp). It is interesting to note that ORF 1a of GLD-associated ampeloviruses harbors an AlKB domain having role in reversal of alkylation damage through RNA demethylation, while GLD associated non-ampeloviruses, i.e., GLRaV-2 (genus *Closterovirus*) and GLRaV-7 (genus *Velarivirus*) do not contain this particular domain. ORFs downstream to RGB encode for structural and other accessory proteins. Compared to the ORFs of RGB, these ORFs are more conserved. Five ORFs downstream to RGB together form a block known as quintuple gene block (QGB), which in fact is the signature for the family *Closteroviridae*. First ORF of QGB corresponds to a small transmembrane protein having a role in cell-to-cell movement, while second ORF of QGB encodes for proteins homologous to cellular heat shock protein 70 (HSP70h). Third ORF in the QGB encodes ~60 kDa protein, sometimes denoted as HSP90h (in GLRaV-3 and GLRaV-4 LV). In addition to cooperating in virion head assembly, proteins corresponding to second and third ORFs of QGB also contribute in cell-to-cell movement. Two remaining ORFs of QGB, i.e., CP (coat protein) and CPm encode for CPs and provide the characteristic morphology to the respective virions. In other closterviruses, CPm is known to form the main component of the virion head. CPm gene remains absent in GLRaV-4 LV, while in GLRaV-1 it is duplicated. CP gene is followed by CPm in GLRaV-1, GLRaV-3, and GLRaV-7 while in the case of GLRaV-2, CPm gene is followed by CP. GLRaV-2 follows the other members of the genus *Closterovirus* as far as arrangement of CP and CPm genes is considered. GLRaV-4 LVs do not have a defined QGB as CPm is conspicuous by its absence. At 3′ end of GLRaV-4, ORF corresponding to p23 protein is present. In addition to RGB and QGB, GLRaVs have some other ORFs at their 3′ ends such as duplicated CPm (ORF 7), p21 (ORF 8) and p24 (ORF 9) in GLRaV-1; p24 (ORF 8) in GLRaV-2; p25 (ORF 7) and p27 (ORF 8) in GLRaV-7. Presence of five smaller ORFs, namely p21 (ORF 8), p20A (ORF 9), p20B (ORF 10), p4 (ORF 11), and p7 (ORF 12) at 3′ end and ORF encoding 6 k-Da (ORF 2) protein and a GC-rich intergenic region between ORF 2 and ORF 3 make GLRaV-3 unique among the GLRaVs. Functions of comparatively versatile ORFs at 3′ end are not known, but based on analogies it has been said that they may be coding for proteins having roles in suppression of the host RNA silencing and long-distance transport of the virus (Martelli et al., 2012; Naidu et al., 2014, 2015).

20.5 REPLICATION AND GENE EXPRESSION

The phloem-residing GLRaVs have been known to replicate in phloem companion and parenchyma cells (Kurth et al., 2012). Like all positive-strand RNA viruses in general and members of the family *Closteroviridae* in particular, RGB proteins are expressed from the capped RNA of the virion analogously to BYV, whereas other proteins are expressed from a nested set of the 3′-coterminal subgenomic RNAs (Maree et al., 2013; Naidu et al., 2015). RGB proteins are translated directly into polyproteins, which in turn are processed to give L-Pro and replicase components. ORF 1b is expressed to RdRp via a +1 frameshift translation. Products resulting from the translation of ORF 1a and ORF 1a + 1b are probably processed by a papain-like L-Pro. Replicase mediates the genome replication by initially synthesizing a complementary negative strand RNA. This complementary strand serves as a template for further synthesis of genomic RNA (positive-strand) as well as subgenomic RNAs (sgRNAs). Viral replicase recognizes the internal sgRNA propmoters present at genomic RNA and thus transcribes to form sgRNAs. Each sgRNA allows for monocistronic translation of the 5′-proximal ORF of each sgRNA. sgRNAs are synthesized in a very coordinated and phase-wise manner. During the initial phases of the infection cycle, RNA silencing suppressor-associated sgRNAs are translated, which help the virion countering the host defense systems. Differential expression of ORFs in GLRaV-3 i.e., enhanced expression of ORF 6 (CP), ORF 8 (p21), ORF 9 (p19.6 or p20A), and ORF 10 (p19.7 or p20B) during infection indicate the possible variation in the regulation of sgRNA transcription (Naidu et al., 2014, 2015; Jarugula et al., 2010a; Dolja et al., 2006). The case of GLRaV-3 suggests that there could be differences between the transcription regulation of members of the genus *Ampelovirus* and *Closterovirus*. Variation in QGB further indicates the differences in replication among GLRaVs. The role of lack of CPm in GLRaV-4 (and its strains) and its duplication in GLRaV-1 in virus replication needs to be deciphered. Detailed understanding of replication of GLRaVs is not clear; however, based on similar studies in related viruses some understanding has been come to, and briefly explained above. Variable genome sizes and diverse gene arrangements make GLRaVs a challenging but attractive experimental system to unravel the replication and gene expression of these complex viruses (Naidu et al., 2015). Thus it can be seen that replication and gene expression provide another level of complexity to GLRaVs.

20.6 TRANSMISSION AND HOST RANGE

The first means of transmission of GLRaVs is by propagating materials themselves, as these viruses are found to be graft transmissible. Vegetative cuttings serve the principal means for spreading these viruses from one location to another, even at the intercontinental level, and because of this GLD has been spread to almost every grape-growing region of the world. There has not been any report of mechanical transmission of these viruses. However, with some difficulty, GLRaV-2 can be mechanically transmitted to *Nicotiana benthamiana* (Goszczynski et al., 1996). For experimental purposes, some GLRaVs such as GLRaV-3 and GLRaV-7 can be transmitted by various species of parasitic dodder (Maree et al., 2013; Naidu et al., 2015). Several species of mealybugs and scale insects have been found to disseminate GLRaVs from one vine to another. It is interesting to note that to date, only those GLRaVs belonging to the genus *Ampelovirus* have been shown to be transmitted by insect vectors. So far no insect vectors have been identified for GLRaV-2 and GLRaV-7 belonging to the genera *Closterovirus* or *Velarivirus*. Different species of mealybugs and scale insects transmit these viruses in a semipersistent manner. Works of Naidu et al. (2014) suggest that six species of mealybug (*Helicoccus bohemicus*, *Phenacoccus aceris*, *Pseudococcus viburni*, *Pseudococcus calceolariae*, *Pseudococcus maritimus*, and *Pseudococcus comstocki*) and three species of scale insects (*Pulvinaria vitis*, *Parthenolecanium corni*, and *Neopulvinaria innumerabilis*) transmit GLRaV-1. In the case of GLRaV-3, nine species of mealy bug (*Helicoccus bohemicus*, *Planococcus ficus*, *Planococcus citri*, *Pseudococcus longispinus*, *Pseudococcus calceolariae*, *Pseudococcus maritimus*, *Pseudococcus viburni*, *Pseudococcus comstocki*, and *Phenacoccus aceris*) and seven species of scale insects (*Pulvinaria vitis*, *Neopulvinaria innumerabilis*, *Parthenolecanium corni*, *Coccus hesperidium*, *Coccus longulus*, *Parasaissetia nigra*, and *Ceroplastes rusci*) have been found to transmit the virus. One species of mealy bug, *Planococcus ficus*, and one species of scale insect, *Ceroplastes rusci*, can transmit GLRaV-4 and its strains (Naidu et al., 2014). Studies have indicated that there is no virus-vector specificity in mealybug transmission of GLRaVs. In the case of scale insect transmission biology of GLRaVs, very limited information is available, as it is very challenging to handle these highly sedentary insects (Naidu et al., 2015).

To date, molecular determinants of vector transmission have not been elucidated. However, based on a study of whitefly-transmitted *Lettuce infectious yellows virus* (LIYV) where CPm has been found to be a major determinant of transmission, CPm was thought to play a major role in vector transmission, but because of lack of CPm in GLRaV-4 LV, further investigation is needed to find the exact virus-encoded determinants for each GLRaV. Thus further study is needed to unravel the fifth level of complexity provided by transmission event of GLRaVs. In nature, these viruses are restricted to *Vitis* species only. As mentioned above, *N. benthamiana* can serve as an experimental host exclusively for GLRaV-2. However, in a recent investigation, pomegranate (*Punica granatum* L.) served as an alternate host for GLRaV-1. Discovery of an alternate host for GLRaV-1 may further aggravate the complexity of the biology of GLRaVs (Naidu et al., 2015).

20.7 GENETIC DIVERSITY AND POPULATION STRUCTURE OF ASSOCIATED VIRUSES

Vegetative propagation and the perennial nature of grapevines lead to persistent infections which in turn contribute to accumulation of the complex populations of genetically diverse GLRaVs by grapevine (Naidu et al., 2015). GLRaV-3, the most common and economically the most important virus, is found to be more diverse and more complex, and that is why most of the diversity studies of GLRaVs are focused on GLRaV-3. As per a recent study with up-to-date data, variants of GLRaV-3 clustered in eight distinct clades (Maree et al., 2015). Different gene-based phylogenies have also reported the phenomenon of incongruence for some variants such as Revella-4/12, Revella-4/14, KS-B-7, and Nashik isolates from India (Kumar et al., 2012; Kumar, 2013). Based on sequences of CP and HSP70h genes, the isolates of GLRaV-1 grouped in eight and seven clades, respectively (Fan et al., 2015). Similarly, based on entire CP and partial HSP70h genes, GLRaV-2 isolates clustered in six groups (Fonseca et al., 2016; Jarugula et al., 2010b). Recent studies show the presence of distinct grouping patterns within GLRaV-4 and its strains (Rai et al., 2017; Rubio et al., 2013). Based on p61 gene-based phylogeny, GLRaV-7 isolates clustered in three genetic groups (Lyu et al., 2014). The phenomenon of recombination has also been observed in GLRaV-3, -1, -2, -4 (and its strains) (Farooq et al., 2013; Fonseca et al., 2016; Kumar, 2013; Rai et al., 2017). The error-prone nature of RdRp in GLRaVs, vegetative propagation of grapevine, and absence of natural resistance in *Vitis* spp. generate variability in GLRaVs populations. These genetically variant populations of GLRaVs coexist. In addition, multiple infections of different GLRaVs and their distinct strains over a period of time provide an opportunity to develop the recombinants and lead to the development of complex population structures of GLRaVs. The evolutionary dynamics of their population is countered by bottleneck events during colonization of host and transmission (Gutiérrez et al., 2012). Because of vector transmission, GLRaV-1, -3, and -4 and its strains encounter additional bottleneck events as compared to GLRaV-2 and -7, having no report of vector transmission (Naidu et al., 2015). In addition, there are reports that pre-existing viruses check the secondary infection with the same or closely related viruses, a phenomenon known as superinfection exclusion (SIE) (Foliomonova, 2012). SIE depending upon the virus-coded proteins has recently been discovered in *Citrus tristeza virus* (CTV) and therefore it has been speculated that SIE between coinfecting GLRaVs and their strains occurring at the cellular and whole organism levels could be influencing the distribution and evolutionary processes deciding the virus population in grapevine (Bergua et al., 2014; Naidu et al., 2015).

20.8 HOST–PATHOGEN INTERACTIONS

The perennial-woody nature of grapevine and incomplete functional genomics of GLRaVs limit the knowledge on the vine–virus interaction of GLD. However, generation of infectious cDNA clones in some GLRaVs, namely in GLRaV-2 and GLRaV-3, will help in deciphering the functions of gene products of associated viruses and their corresponding roles in host–pathogen interactions leading to disease development (Jarugula et al., 2012; Kurth et al., 2012). Based on analogies, it has been speculated that in BYV-type GLRaVs, i.e. GLRaV-2, all genes are indispensable, whereas in CTV-type GLRaVs, certain genes located at the 3′ portion may be dispensable without affecting systemic infection in some grapevine species, but in other species the same genes may be essential for infection (Naidu et al., 2015). Thus host–pathogen interactions of GLD are another part of the complexity of the disease. Cultivar level differences in response of hosts (contrastingly different symptoms in light-fruited and dark-fruited cultivars), tissue-specific expression patterns inside a host, seasonal variations, many causal agents, and usually mixed infections of GLRaVs as well as viroids indicate that the host-pathogen interactions in grapevine could be much more complex. Because of the challenges discussed above, very limited studies have been carried out on host–pathogen interactions of GLD, and as they have been carried out in the open field it would not be wise to generalize those experimental results across vineyards (Naidu et al., 2015). The more expressive nature of dark-fruited cultivars in particular has attracted researches to conduct host–pathogen experiments focusing mainly on the response of such cultivars. Therefore some general trends of host–pathogen interactions emerging from the experiments carried out in dark-fruited cultivars can be briefly discussed here.

In dark-fruited cultivars, the phenomena of net photosynthesis and fluorescence of chlorophyll a (Chl a) were indistinguishable between healthy and asymptomatic phases of GLRaV-3 infected vines, but both of these phenomena were found to be significantly reduced during the symptomatic phase of same GLRaV-3-infected vines (Gutha et al., 2012). Bertamini et al. (2004) studied the response of vine post-GLRaV-3 infection in detail and found that electron transport chain of photosystem (PS) II at the donor side was disrupted and thus it affected the electron transport activity of PSII, while there was minimal effect on the same activity of PSI. They further concluded that disruption of the electron transport chain was due to the functional loss of five polypeptides specifically associated with PSII (Bertamini et al., 2004). Comparing the studies in GLD with similar studies in other pathosystems, Naidu et al. (2015) concluded that the response of the host on photosynthesis physiology of GLRaV-3-infected vine is more visible during the symptomatic phase despite the presence of virus within detectable limits during both symptomatic and asymptomatic phases. Downregulation of many chloroplastic genes associated with photosynthesis and biosynthesis of photosynthetic pigments were observed in the leaves of dark-fruited cultivars naturally infected with GLRaV-3. In additon, changes in the expression of genes involved in a wide spectrum of biological functions such as modulation of transcription factors, translation, protein targeting, transport, secondary metabolism, and cell defense have been found (Espinoza et al., 2007a; Naidu et al., 2015). During the infection of GLRaV-3 genes related to host defense and stress have also been found to be induced, and enhanced expression of these genes occurs (Espinoza et al., 2007b). The exact mechanism of induction of defense response of host post-GLRaV-3 infection is not clear. However, based on similar studies in other pathosystems, it has been opined that disruption of photosynthetic electron transport triggers the generation of reactive oxygen species (ROS), leading to oxidative stress in symptomatic leaves and anthocyanins built up in symptomatic leaves protect them from ROS because of their free radical scavenging and antioxidant activities (Naidu et al., 2015).

Host–pathogen interactions discussed in the previous section were concerned with interactions affecting the physiology and genomics of vine leaves. Compatible vine–virus interactions also affect the genomics and metabolomics of berries, leading to impact on fruit chemicals, berry ripening, and finally fruit yield. A transcriptomic study carried out in a dark-fruited cultivar—namely, Cabernet Sauvignon—and GLRaV-3 infection found the downregulation of those genes that are functionally related to signaling, metabolism, cell wall modifications, anthocyanin biosynthesis, and sugar transport. It was further observed that downregulation of these genes eventually led to incomplete berry maturation, delay in berry ripening, reduction in sugar level, and anthocyanin biosynthesis. After onset of the ripening process, the changes became more pronounced. It was thought that delay in fruit ripening could be largely because of altered source/sink balance and disturbed unloading of sugars from phloem tissues, which in turn could be due to alterations in the expression of sugar transporter genes and transcript associated with sugar metabolism. At ripening, the decrease in the anthocyanin content of GLRaV-3-infected berries amounted to 60% (40% lower than healthy berries). It has been suggested that reduced levels of sugars in the berries of infected vines contributes to downregulating the biosynthetic pathway genes (Vega et al., 2011). Proteomic studies carried out in the same year by another group of scientists proved that virus infection in dark-fruited cultivars significantly affects the expression levels of proteins involved in oxidative stress response of berry skin cells and cell structure metabolism in the pulp (Giribaldi et al., 2016; Naidu et al., 2015). Similar types of studies in light-fruited cultivars and with other GLRaVs will further elucidate the interaction mechanisms leading to transcriptome and metabolomic changes post infections, which in turn will provide a better understanding of the host–pathogen interactions in this exceptionally complex disease of grapevine. Significantly less knowledge is available for host-pathogen interactions involving RNA-silencing machinery and its defense mechanisms against virus infections in grapevine. Decoding the functional diversity of viral suppressors of RNA silencing (VSRs) by different GLRaVs and their symptom-specific expression profile will shed more light on RNA silencing-based virus–vine interactions of GLD (Naidu et al., 2015).

20.9 MANAGEMENT

Providing healthy planting materials and checking secondary infection is the best way to manage any disease in general, and viral disease in particular. Since there is no natural source of resistance against GLRaVs in grapevine, strategies for disease management involve a combination of preventive, cultural, sanitary, and vector management (Naidu et al., 2014). Preventive measures consisting of planting new vineyards and replacing the infected vines in established vineyards should be done with planting materials derived from virus-tested mother vines; cultural practices involve roguing, where the infected vines in an established vineyard are selectively removed. Sanitary measures are simply cleanliness of vineyards and nearby areas, i.e., removal of remnant roots during fallow periods and removal of infected fallen leaves. Various methods of mealybug and scale insect management can be employed to check the spread of GLRaVs from infected vines to nearby healthy vines (Almeida et al., 2013; Naidu et al., 2014).

Using healthy planting materials is the first line of defense. Various robust diagnostic techniques are available for screening virus-free planting materials. Among these techniques, serology with ELISA has been the method of choice, and it has been widely used throughout the globe to produce clonally selected and sanitized propagation material. Tissue culture-based techniques have also been used to eliminate viruses from the planting materials. Despite several efforts, no transgenic has been released to date for cultivation purposes, mainly because of the woody-perennial nature of the vines (Kumar, 2013). It has been found that use of certified virus-free planting materials affords more than $50 million per year of economic benefits in the northern coastal region of California (Fuller et al., 2013). Roguing the infected vines and replanting with virus-free planting materials in established vineyards is also one of the best strategies to manage the disease. Roguing and replanting at formative years of the vineyards is more beneficial (Rayapati et al., 2008). According to one estimate, roguing can provide additional benefits of $17,000–$22,000/ha (Atallah et al., 2012). To check the secondary infection by means of insect vectors, it is important to use the management strategies targeting those vectors. Vector management is especially important where there is sustained immigration of mealybugs (Charles et al., 2006). The proper stage to control mealybugs is at their crawling stage. Chemigation with imidacloprid, thiamethoxam, and dinotefuran has been found effective in controlling mealybugs. For dormant applications, foliar sprays of chloropyriphos can also be used (Rayapati et al., 2008). Tsai et al. (2008) recommended using contact and systemic insecticides in combination for vector management in grapevines. Horticultural oils along with acetamiprid and spirotetramat insecticides have also been found to be effective against grape mealybug (*Pseudococcus maritimus*), a primary vector for GLRaVs in North America (Wallingord et al., 2015). Finally, following hygienic practices by the workers and use of sanitized equipment can also check spread of the disease (Naidu et al., 2014; Pieterson et al., 2013).

20.10 CONCLUSION

Expression of symptoms, involvement of many viruses and their strains, genome organization, gene expression and replication of associated viruses, genetic diversity, population structure, and host-7pathogen interactions operating at different levels of the infection process clearly indicate that GLD is an exceptionally complex disease. We have presented up-to-date information in a comprehensive manner as far as possible; however, from the discussion it can be seen that many aspects of the disease are not yet clearly understood. There is a need to carry out functional genomics studies of each GLRaV in both light-fruited and dark-fruited cultivars. Some studies on host-pathogen interaction of GLRaV-3 in dark-fruited cultivars have been conducted, but more comprehensive studies of interactions between GLRaV-3 and other GLRaVs in dark fruited cultivars are needed, and those studies should be extended to light-fruited cultivars as well. Generation of full-length infectious clones in GLRaV-2 and GLRaV-3 would help in deciphering various complex processes such as decoding the functions of virus gene products in disease etiology, various levels of host–pathogen interactions, symptom expression, and the mechanism of RNA silencing and its counter-reactions by hosts during and after infection. The woody-perennial nature of grapevine is a big challenge. However, recent successes in advancement of knowledge in science would definitely help in unraveling the complex processes of the diseases and the response of host. The use of diagnostics, including next-generation sequencing for virus-free certified planting materials, needs to be emphasized in import and export of elite grapevine planting material. The use of certified virus-free planting materials, roguing and replanting, sanitation, and vector control by means of chemicals and botanicals or a combination of both (chemicals and botanicals) can lead to sustainable management of GLD.

ACKNOWLEDGMENT

The authors would like to thank all the researchers working on various aspects of grapevine leafroll disease.

REFERENCES

Almeida R P P, Daane K M, Bell V A, Blaisdell G K, Cooper M L, Herrbach E and Pietersen G (2013). Ecology and management of grapevine leafroll disease. *Front Microbiol.* 4:94.

Anonymous (2017). ictvonline.org/virusTaxonomy.asp (accessed on 25 January 2017).

Atallah S S, Gómez M I, Fuchs M F and Martinson T E (2012). Economic impact of grapevine leafroll disease on *Vitis vinifera* cv. Cabernet franc in Finger Lakes Vineyards of New York. *Am J Enol Vitic.* 63:73–79.

Bergua M, Zwart M P, El-Mohtar C, Shilts T, Elena S F and Folimonova S Y (2014). A viral protein mediates superinfection exclusion at the whole-organism level but is not required for exclusion at the cellular level. *J Virol.* 88:11327–11338.

Bertamini M, Muthuchelian K and Nedunchezhian N (2004). Effect of grapevine leafroll on the photosynthesis of field grown grapevine plants (*Vitis vinifera* L. cv. Lagrein). *J Phytopathol.* 152:145–152.

Charles J G, Cohen D, Walker J T S, Forgie S A, Bell V A and Breen K C (2006). A review of the ecology of *Grapevine leafroll-associated virus* 3 (GLRaV-3). *N.Z. Plant Prot.* 59:330–337.

Dolja V V, Kreuze J F and Valkonen J P T (2006). Comparative and functional genomics of closteroviruses. *Virus Res.* 117:38–51.

Espinoza C, Medina C, Somerville S and Arce-Johnson P (2007b). Senescence-associated genes induced during compatible viral interactions with grapevine and *Arabidopsis*. *J Exp Bot.* 58:3197–3112.

Espinoza C, Vega A, Medina C, Schlauch K, Cramer G and Arce-Johnson P (2007a). Gene expression associated with compatible viral diseases in grapevine cultivars. *Funct Integr Genom.* 7:95–110.

Fan X, Hong N, Dong Y, Ma Y, Zhang Z P, Ren F, Hu G, Zhou J and Wang G (2015). Genetic diversity and recombination analysis of *Grapevine leafroll-associated virus* 1 from China. *Arch Virol.* 160:1669–1678.

Farooq A B, Ma Y X, Wang Z, Zhuo N, Wenxing X, Wang G P and Hong N (2013). Genetic diversity analyses reveal novel recombination events in Grapevine leafroll-associated virus 3 in China. *Virus Res.* 171: 15–21.

Folimonova SY (2012). Superinfection exclusion is an active virus-controlled function that requires a specific viral protein. *J Virol.* 86:5554–5561.

Fonseca F, Esteves F, Teixeira Santos M, Brazão J and Eiras-Dias J E (2016). Genetic variants of *Grapevine leafroll-associated virus 2* infecting Portuguese grapevine cultivars. *Phytopathol Mediterr.* 55:73–88.

Fuller K B, Alston J M and Golino D A (2013). The benefits from certified virus-free nursery stock: A case study of grapevine leafroll-3 in the North Coast region of California. Robert Mondavi Institute-Center for Wine Economics Working Paper number 1306, UC-Davis. p. 35.

Freeborough M J and Burger J T (2008). *Leafroll: Economic implications. Wynboer–A technical guide for wine producers*. December. wynboer.co.za/recentarticles/200812-leafroll.php3.

Giribaldi M, Purrotti M, Pacifico D, Santini D, Mannini F, Caciagli P, Rolle L, Cavallarin L, Giuffrida M G, Marzachì C (2016). A multidisciplinary study on the effects of phloem-limited viruses on the agronomical performance and berry quality of *Vitis vinifera* cv. Nebbiolo. *J Proteomics*. 75:306–315.

Golino D A, Weber E, Sim S and Rowhani A (2008). Leafroll disease is spreading rapidly in a Napa Valley vineyard. *Calif Agric*. 62:156–160.

Goszczynski D E, Kasdorf G G F, Pietersen G and van Tonder H (1996). *Grapevine leafroll-associated virus 2* (GLRaV-2) – mechanical transmission, purification, production and properties of antisera, detection by ELISA. *S Afr J Eno. Vitic*. 17:15–26.

Gutha L R, Alabi O J and Naidu R A (2012). Effects of grapevine leafroll disease on photosynthesis in a redfruited wine grape cultivar. Proc Congr Int Counc *Study Virus Virus-Like Dis. Grapevine (ICVG), 17th, Davis, Calif.*, Oct. 7–14, pp. 168–169. Davis, CA: Found Plant Serv.

Gutha L R, Casassa L F, Harbertson J F and Naidu R A (2010). Modulation of flavonoid biosynthetic pathway genes and anthocyanins due to virus infection in grapevine (*Vitis vinifera* L.) leaves. *BMC Plant Biol.* 10:187.

Gutiérrez S, Michalakis Y and Blanc S (2012). Virus population bottlenecks during within-host progression and host-to-host transmission. *Curr Opin Virol.* 2:546–555.

Huson D H and Bryant D (2006). Application of phylogenetic networks in evolutionary studies. *Mol Biol Evol.* 23:254–267.

Ito T and Nakaune R 2016. Molecular characterization of a novel putative ampelovirus tentatively named grapevine leafroll-associated virus 13. *Arch. Virol.* 161:2555–2559.

Jarugula S, Gowda S, Dawson W O and Naidu R A (2010a). 3′-coterminal subgenomic RNAs and putative cis-acting elements of *Grapevine leafroll-associated virus 3* reveals "unique" features of gene expression strategy in the genus *Ampelovirus. Virol J.* 7:180.

Jarugula S, Alabi O J, Martin R R and Naidu R A (2010b). Genetic variability of natural populations of *Grapevine leafroll-associated virus* 2 in Pacific Northwest vineyards. *Phytopathology*. 100:698–707.

Jarugula S, Gowda S, Dawson W O and Naidu R A (2012). Development of full-length infectious cDNA clone of *Grapevine leafroll-associated virus* 3. Proc Congr Int Counc *Study Virus Virus-Like Dis Grapevine (ICVG), 17th, Davis, Calif.*, Oct. 7–14, pp. 70–71. Davis, CA: Found Plant Serv.

King A M Q, Adams M J, Carstens E B and Lefkowitz E J (2012). Virus taxonomy: Classification and nomenclature of viruses. *Ninth Report of International Committee on Taxonomy of Viruses*. Elsevier Academic Press, San Diego, CA.

Kumar S (2013). Studies on virus(es) associated with grapevine leafroll disease in India. PhD Thesis. Indian Agricultural Research Institute, New Delhi.

Kumar S, Baranwal, V K, Singh P, Jain R K, Sawant S D and Singh S K (2012). Characterization of a *Grapevine leafroll-associated virus 3* from India showing incongruence in its phylogeny. *Virus Genes.* 45:195–200.

Kurth E G, Peremyslov V V, Prokhnevsky A I, Kasschau K D, Miller M, Carrington J C and Dolja V V (2012). Virus-derived gene expression and RNA interference vector for grapevine. *J Virol.* 86:6002–6009.

Lyu M D, Li X M, Guo R, Li M J, Liu X M, Wang Q and Cheng Y Q (2014). Prevalence and distribution of *Grapevine leafroll-associated virus* 7 in China detected by an improved reverse transcription polymerase chain reaction assay. *J Phytopathol.* 63:1168–1176.

Maree H J, Almeida R P P, Bester R, Chooi K M, Cohen D, Dolja V V, Fuchs M F, Golino D A, Jooste A E C, Martelli G P, Naidu R A, Rowhani A, Saldarelli P and Burger J T (2013). *Grapevine leafroll-associated virus* 3. *Front Microbiol.* 4:82.

Maree H J, Pirie M D, Bester R, Oosthuizen K and Burger J T (2015). Phylogenomic analysis reveals deep divergence and recombination in an economically important grapevine virus. *PLOS ONE* 10:e0126819.

Martelli G P (2014). Directory of virus and virus-like diseases of the grapevine and their agents. *J Plant Pathol.* 96(Suppl. 1):1–136.

Martelli G P and Boudon-Padieu E (2006). Directory of infectious diseases of grapevines viruses and virus-like diseases of the grapevine: Bibliographic report 1998–2004. *Options Méditerranéennes*. B55:11–201.

Martelli G P, Abou Ghanem-Sabanadzovic N, Agranovsky A A, Al Rwahnih M Dolja V V, Dovas C I, Fuchs M, Gugerli P, Hu J S, Jelkmann W, Katis N I, Maliogka V I, Melzer M J, Menzel W, Minafra A, Rott M E, Rowhani A, Sabanadzovic S and Saldarelli P (2012). Taxonomic revision of the family Closteroviridae with special reference to the grapevine leafroll-associated members of the genus Ampleovirus and the putative species unassigned to the family. *J Plant Pathol.* 94:7–19.

Naidu R A, Maree H J and Burger J T (2015). Grapevine leafroll disease and associated viruses: A unique pathosystem. *Annu Rev Phytopathol.* 53:613–634.

Naidu R A, Rowhani A, Fuchs M, Golino D and Martelli G P (2014). Grapevine leafroll: A complex viral disease affecting a high-value fruit crop. *Plant Dis.* 98:1172–1185.

Pietersen G, Spreeth N, Oosthuizen T, Van Rensburg A, Van Rensburg M, Lottering D, Rossouw N and Tooth D (2013). Control of grapevine leafroll disease spread at a commercial wine estate in South Africa: A case study. *Am J Enol Vitic.* 64:296–305.

Rai R, Paul Khurana S M, Kumar S, Sharma S K, Watpade S and Baranwal V K (2017). Characterization of *Grapevine leafroll-associated virus* 4 from Indian vineyards. *J Plant Pathol.* 99:255–259.

Rayapati A N, O'Neil S and Walsh D (2008). Grapevine leafroll disease. Washington State University Extension Bulletin 2008. EB2027E. http://cru.cahe.wsu.edu/CEPublications/eb2027e/eb2027e.pdf.

Rubio L, Guerri J, Moreno P (2013). Genetic variability and evolutionary dynamics of viruses of the family *Closteroviridae*. *Front Microbiol.* 4:151.

Tsai C-W, Chau J, Fernandez L, Bosco D, Daane K M and Almeida, R R R (2008). Transmission of *Grapevine leafroll-associated virus* 3 by the vine mealybug (*Planococcus ficus*). *Phytopathology*. 98:1093–1098.

Vega A, Gutierrez R A, Pena-Neira A, Cramer G R and Arce-Johnson P (2011). Compatible GLRaV-3 viral infection affect berry ripening decreasing sugar accumulation and anthocyanin biosynthesis in *Vitis vinifera*. *Plant Mol Biol.* 77:261–274.

Wallingford A K, Fuchs M F, Martinson T, Hesler S and Loeb G M (2015). Slowing the spread of grapevine leafroll-associated viruses in commercial vineyards with insecticide control of the vector, *Pseudococcus maritimus* (Hemiptera: Pseudococcidae). *J Insect Sci.* 15:112.

21 Virus-Induced Gene Silencing

Applying Knowledge of Plant–Virus Interactions for High-Throughput Plant Functional Genomics

Ravi Kant and Indranil Dasgupta

CONTENTS

21.1 INTRODUCTION

Viruses of plants have long attracted the attention of plant scientists, mainly because of their pathogenic potential and the loss of crop yields due to virus infections. Two important properties of viruses, different from other pathogenic microorganisms, have played an important role in their emergence as tools for basic molecular biology—their obligate intracellular nature and the small size of their genetic material. These properties not only make them relatively easy to understand from the point of molecular biology, but also make them a convenient material to start modifying their genomes to obtain useful tools for further scientific pursuits. One such use which has been realized from plant viruses is virus-induced gene silencing (VIGS). The concept of VIGS has emerged from the deeper understanding of the manner in which viral genomes replicate, spread, and accumulate in the host plants, and new methods for introducing cloned viral genomes in plants. VIGS leads to rapid silencing of selected genes of the plant, leading to a change in the phenotype and thus revealing the functions of the genes silenced. The phenomenon of VIGS is based on a natural defense reaction of the host plant, which it mounts against the invading virus. It illustrates a beautiful example of a technology emerging from basic studies on plant–virus interactions. In this chapter, the origins of VIGS, its advantages, the viruses that have been used, the genes that have been silenced, and its future potential as a versatile gene-silencing tool for plant functional genomics are discussed.

21.2 VIGS

21.2.1 Origins of VIGS

VIGS has been one of the potential reverse genetics tools for the functional genomics in plants for the past two decades. It utilizes recombinant viral vectors carrying the host target gene fragment to induce the host RNA interference (RNAi) pathway, a defense pathway for the functional analysis of target genes in plants. RNA-mediated gene silencing is one of the innate pathways, used in wide variety of organisms, termed RNAi in animals, post-transcriptional gene silencing (PTGS) in plants, and quelling in fungi. Several gene-silencing tools have been used so far to knock down the expression of target genes; e.g., T-DNA insertion, transposon tagging, and RNAi knockdown transgenics, to study the function of plant genes. In *Arabidopsis*, two insertional mutagenesis approaches, T-DNA (Krysan et al., 1999) and transposon tagging (Parinov et al., 1999; Speulman et al., 1999) had been applied to disrupt the gene function in order to generate novel mutants. However, due to inability to decipher the function of duplicated genes, inability to reach genome saturation, and multiple insertion sites, these approaches have some limitations.

Viruses have always been a major threat to crops because of their high multiplicity and virulent nature to elicit various disease symptoms resulting in yield losses. Protecting crop plants against viruses was the prime objective for plant biologists in the early days of developing disease resistance. Plants that were exposed to infection by viruses became resistant to subsequent infection by the same or closely related viruses; this phenomenon was termed "cross-protection" (McKinney, 1929). Based on this observation, introduction of viral genetic material into the plant genome to enable them to developresistance was described as pathogen-derived resistance (PDR). Some plants engineered for PDR exhibited a recovery phenotype; they initially developed disease symptoms upon virus infection, but these symptoms were absent in newly emerging tissue (Lindbo et al., 1993). The molecular pathway responsible for cross-protection, PDR, and recovery was reported as PTGS (Ratcliff et al., 1997). The molecule that triggers the initiation of PTGS was found to be double-stranded RNA (dsRNA) (Fire et al., 1998; Waterhouse et al., 1998; Klahre et al., 2002). Later, Hamilton and Baulcombe (1999) demonstrated that a class of small RNA of about 25 nucleotides works as the key to trigger the signal for gene silencing event, which is produced by cleavage of the dsRNA (Hamilton and Baulcombe, 1999; Martinez et al., 2002). Various assumptions were made for PTGS to be an antiviral mechanism in plants. It was soon observed that viruses by themselves can trigger PTGS in certain plants (Ratcliff et al., 1997, 1999; Al-Kaff et al., 1998; Marathe et al., 2000). It was also demonstrated that PTGS-compromised *Arabidopsis* mutants were highly susceptible to virus infection (Dalmay et al., 2000; Mourrain et al., 2000).

The term VIGS was coined to describe the resistance against virus infection (van Kammen, 1997), utilizing host-induced PTGS machinery. The VIGS system was developed for the first time using an RNA virus, TMV (*Tobacco mosaic virus*) to study the inhibition of gene expression based on homology between viral vector with host target gene fragment and endogenously expressed gene (Lindbo et al., 1993; Kumagai et al., 1995). On the basis of involvement of cytoplasmic RNA viruses, it was inferred that the mechanism of VIGS involves destabilization of the target mRNAs in a homology-dependent manner in the cytoplasm (Smith et al., 1994; Baulcombe, 1999). From the above-mentioned reports of versatility of VIGS, several viruses were utilized for the development of VIGS vectors for functional determination of many important genes responsible for plant growth and development, biotic and abiotic stress resistance, fruit ripening, flowering time, and so forth.

21.2.2 The Need for VIGS Technology

Various methods for functional characterization of genes have been employed by plant biologists. Some of the widely used techniques such as chemical mutagenesis and use of transposons or *Agrobacterium* T-DNA insertion were the most established techniques for loss-of-function studies.

However, due to certain limitations, these methods were not put forward. One of the major limitations for the development of mutant plants was generation of large populations to screen for mutations in a gene of interest (Parinov and Sundaresan, 2000; Bouche and Bouchez, 2001). Identification of knockout mutation was not easy, even with new technologies such as Targeting Induced Local Lesions In Genomes (TILLING; McCallum et al., 2000). Another limitation, due to which many point mutations and insertions did not generate obvious phenotypes, was the presence of large gene families and gene duplications in plant genomes (Bouche and Bouchez, 2001). Determination of appropriate mutants became difficult at times due to one or multiple insertion sites in plants, which led to complications in the interpretation of a knockout phenotype (Henikoff and Comai, 2003). VIGS avoids most of these limitations, and therefore has been adopted widely by scientific communities as an excellent alternative approach. The peculiar and durable properties such as cost-effectiveness and its rapid and robust nature enable VIGS technology as a revolutionary breakthrough in the field of plant science for the easy, cost-effective, rapid, and high-throughput functional characterization of plant genes.

VIGS has also been utilized successfully for functional genomics studies, especially in plants that are recalcitrant to transformation. It provides an outstanding platform for large-scale screening, targeting multiple genes simultaneously, targeting gene family and ability to silence embryo lethal genes. Genes associated with embryonic development or essential housekeeping genes in plants can also be silenced using VIGS technology (Tao and Zhou, 2004; Ding et al., 2006). Thus VIGS provides a great avenue for functional analysis of plant genes devoted to high-yield, growth and development and stress resistance in crop and model plants (Pflieger et al., 2008; Unver, 2008).

21.2.3 Mechanism of VIGS

Plants are infected by multiple pathogens during their life cycle; to sustain themselves against pathogens, they have developed several defense mechanisms. RNAi, which is a sequence-dependent RNA-mediated phenomenon leading to the degradation or translation arrest of transcripts or change in the methylation pattern of DNA, is one of most important defense pathways induced by plants in response to pathogen infection. Viruses (DNA and RNA viruses) are the potential initiator of RNAi in plants; upon infection, the transcribed viral RNA (in the case of DNA virus) is converted into dsRNA structure by host-derived RNA-dependent RNA polymerase (RdRp) enzyme. These dsRNAs are recognized and cleaved by another host-encoded RNase III such as the enzymes known as Dicer-like proteins (DCLs), into 21–24 nucleotides double-stranded siRNAs. Each ds-siRNA generated unwound to single-stranded siRNAs passenger strand and guide strand, respectively. The passenger strand is degraded and the guide strand binds to a specific domain of multiprotein complex, RNA-induced silencing complex (RISC), where it targets and binds to the viral cognate mRNA in a sequence-specific manner, thus degrading viral mRNA (Baulcombe, 2004; Vaucheret, 2006). As discussed earlier, VIGS exploits the host RNAi pathway for functional analysis of genes in plants. When the VIGS vector carrying the target gene fragment is inoculated in plants, recombinant transcripts of viral genome as well as host candidate gene fragments are processed to generate siRNAs. The guide strand siRNAs specific to the host candidate gene are incorporated into RISC complex and pair with complementary host endogenous transcripts, thus downregulating candidate gene expression, which leads to transient loss of function of the host target gene (Figure 21.1).

The first step for the VIGS-mediated gene silencing approach is to develop a self-replicating VIGS vector by cloning minimum essential viral genome between the left and right borders of the binary vector. The developed VIGS vector must contain multiple cloning sites for the easy cloning of the host target genes and should be devoid of causing disease symptoms in plants. Selection of target gene region and insert length also play vital roles as far as silencing efficiency is concerned. To avoid off-target silencing, a unique region of the corresponding gene should be selected to generate a recombinant VIGS vector.

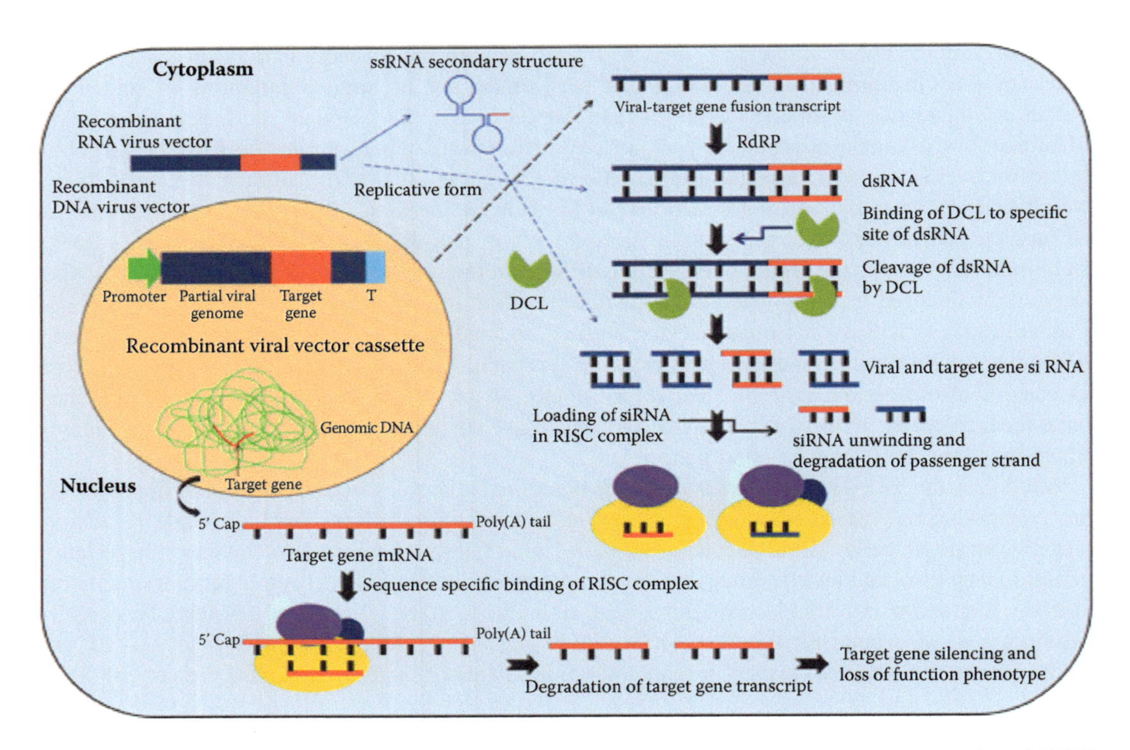

FIGURE 21.1 Diagrammatic representation of mechanism of VIGS. Modified viral genome-based VIGS vectors containing fragments (DNA virus in nucleus and RNA virus in cytoplasm) of host target gene undergoing ssRNA secondary structure formation and replication (RNA virus vector transcripts). The replication intermediate/dsRNA structures containing the host target gene transcripts are recognized and cleaved into siRNAs by DCL. The guide strands of siRNAs are loaded into RISC targeting viral as well as target gene transcripts in a sequence-specific manner, leading to downregulation of target gene expression.

The VIGS system was developed nearly two decades ago for the functional analysis of plant genes responsible for various processes. It was initially used to reveal the gene functions in *Solanaceous* species such as tobacco, tomato, and in the model plant, *Arabidopsis*. However, broad adoptability by plant biologists and extensive applications of VIGS for gene function analysis in several plant species are usual today. Some important DNA and RNA viruses developed into VIGS vector are discussed here briefly.

TMV, an RNA virus, was modified for the first time to develop as a VIGS vector for silencing PDS gene through mechanical inoculation of *in vitro* transcript in *Nicotiana benthamiana* and *N. tabacum* (Kumagai et al., 1995). Another RNA virus, *Tobacco rattle virus* (TRV) was also modified in a VIGS vector for gene silencing in *N. benthamiana* and in tomato (Liu et al., 2002). Retinoblastoma-related (RBR) gene, a plant root development gene and a gene in meristem (ML1) have been functionally characterized by TRV-VIGS (Valentine et al., 2004). The DNA viruses used as VIGS vectors for the first time were *Tomato golden mosaic virus* (TGMV) and *Cabbage leaf curl virus* (CaLCuV), both bipartite DNA-containing geminiviruses (Peele et al., 2001; Turnage et al., 2002). There are other viruses modified for silencing of dicotyledonous plants such as *African cassava mosaic virus* (ACMV) in cassava, *Pea early browning virus* (PEBV) in pea, and *Bean pod mottle virus* (PBMV) in soybean. Biolistic inoculation of cassava plants with the ACMV-based vector carrying the gene encoding a subunit for the chlorophyll synthesis enzyme magnesium chelatase (*su*) resulted in the development of yellow-white spots in the leaves characteristic of *su* silencing (Fofana et al., 2004). Using PEBV-VIGS, several genes such as *phytoene desatuase* (*pds*), *uni* (involved in regulating compound leaf architecture) and *korrigan1* (encoding endo-1, 4-β-D-glucanase, involved in maintaining cell wall architecture) were efficiently silenced (Constantin et al., 2004). PBMV was used for VIGS in soybean and has been tested by silencing *pds* (Zhang and Ghabrial, 2006).

An RNA virus, *Cucumber mosaic virus* (CMV)-based VIGS vector has been developed for silencing of chalcone synthase (*chs*) and *sf3h1* genes in soybean and *pds*, *IspH*, *ATG3*, *and ATG8a* genes in *Zea mays* (Nagamatsu et al., 2007; Kim et al., 2011; Lee et al., 2015; Wang et al., 2016). Another RNA virus associated with Citrus sp., *Citrus leaf blotch virus* (CLBV) was developed as a gene silencing vector; *pds*, *actin*, and *su* genes were efficiently silenced using this vector (Aguero et al., 2012). A tree-infecting RNA virus, *Apple latent spherical virus* (ALSV) has also been developed into a VIGS vector to silence *pds* and *su* among a broad range of plants including tobacco, tomato, Arabidopsis, cucurbits, and legumes (Yaegashi et al., 2007; Igarashi et al., 2009) as well as *CPN60a*, *EF-1a*, and *TFL1* in tree plants apple, pear, and Japanese pear (Sasaki et al., 2011). VIGS vector developed from *Cotton leaf crumple virus* (CLCrV) has also revealed its versatility to silence *pds* and *chlI* genes in cotton by particle bombardment (Tuttle et al., 2008). Moreover, two DNA viruses, *Beet curly top virus* (BCTV) and *Abutilon mosaic virus* (AbMV), have been successfully developed into silencing vector for the functional analysis of *rbcS*, *TK*, and *chlI* genes in *Spinacea oleracea* and *S. lycopersicum* (Golenberg et al., 2009). *Pepper huasteco yellow veins virus* (PHYVV)-derived vector was generated for silencing three genes: Comt (encoding a caffeic acid O-methyltransferase), pAmt (a putative aminotransferase), and Kas (a β-keto-acyl-[acyl carrier-protein] synthase), which are involved in the biosynthesis of capsaicinoids and responsible for the pungent taste of chilli pepper fruits of *Capsicum* species (Abraham-Juarez et al., 2008).

The scientific achievements of VIGS technology are not limited to dicot plants; some viruses infecting monocots have also been developed into VIGS vectors. *Barley stripe mosaic virus* (BSMV)-VIGS was used for efficient silencing of *pds* gene in barley (Holzberg et al., 2002) and wheat genes (Scofield et al., 2005). Another monocot infecting virus, *Brome mosaic virus* (BMV) was modified as a vector, which efficiently silenced *pds*, *actin1*, and *rubisco activase* genes in barley, rice, and maize (Ding et al., 2006). Another monocot DNA virus, *Rice tungro bacilliform virus* (RTBV)-based VIGS vector was developed for silencing a marker gene, *pds*, in rice (Purkayastha et al., 2010; Kant et al., 2015). Recently, *Foxtail mosaic virus* (FoMV)-based VIGS vector was developed and employed for functional characterization of *pds*, *chlH*, *CLA1*, and *IspH* genes in rice, maize, and wheat (Mei et al., 2016). A dicot infecting virus, CMV-based VIGS vector was also recently developed, which displayed efficient silencing of *pds*, *IspH*, *ATG3*, and *ATG8a* genes in *Zea mays* (Lee et al., 2015; Wang et al., 2016). Many such viruses have been developed and used as VIGS vectors for gene function analysis in monocots and dicots to date. Table 21.1 provides a detailed list of important plant virus-based VIGS vectors, indicating their genomic structure, silencing host, and genes silenced.

In the case of gene families, the conserved regions can be targeted by VIGS to avoid inefficient silencing due to functional redundancy (Igarashi et al., 2009; Ramegowda et al., 2014). According to previous reports VIGS-based downregulation of gene expression was found to be efficient using a gene fragment length of 150–800 bp. Recent studies suggest that for optimum VIGS, insert lengths should be in the range of 200–350 bp (Zhang et al., 2010; Rodrigo et al., 2011). However, a gene fragment of 1500 bp can also induce gene silencing (Padmanabhan et al., 2009; Kurth et al., 2012). Similarly, TRV-VIGS-mediated gene silencing of *pds* in *N. benthamiana* was reported using insert length of 192–1304 bp (Liu and Page, 2008). Another factor that plays an important role in VIGS is orientation of target gene insert. Inverted repeats (VIGS vector carrying host gene fragment in hairpin orientation) have shown to be more efficient than in antisense orientation, which in turn is more efficient than sense orientation (Lacomme et al., 2003; Hein et al., 2005). The efficiency of VIGS also depends on environmental conditions such as temperature, light intensity, humidity, and so forth. Silencing of PDS gene was found to be highly temperature-sensitive, and silencing was enhanced under low ambient temperature (less than 24°C), low light intensity (250 mmol m^{-2} s^{-1}), and high humidity (85–90%) using TRV-VIGS (Nethra et al., 2006).

TABLE 21.1
Characteristics of Some Important Plant Viruses Developed as VIGS Vectors

Virus	Genus, Nature of Genome	Inoculation Method	Silencing Hosts	Genes Silenced	References
Tobacco mosaic virus	Tobamovirus, positive-strand RNA, monopartite	Mechanical, in vitro transcripts	*N. benthamiana*, *N. tabacum*	*pds, psy*	Kumagai et al., 1995
Tobacco rattle virus	Tobravirus, positive-strand RNA, bipartite	Agroinoculation	*N. benthamiana, A. thaliana, S. Lycopersicon, D. stramonium*	*pds, rbcS*, NFL, CTR1, CTR2, PMT	Liu et al., 2002; Ratcliff et al., 2001; Eftekhariyan et al., 2014; Liu and Page, 2008
Apple latent spherical virus	Cheravirus, positive-strand RNA, bipartite	Mechanical, in vitro transcripts, particle bombardment	*N. tabacum, N. occidentalis, N. benthamiana, N. glutinosa, Solanum lycopersicon, A. thaliana Apple, Cucurbit* sp., peer, few legume species	*pds, su, pcna, CPN60a, EF-1a, MdTFL1*	Igarashi et al., 2009; Sasaki et al., 2011
Barley stripe mosaic virus	Hordeivirus, positive-strand RNA, tripartite	Agroinfiltration, mechanical inoculation, sap inoculation, spray inoculation	*Hordeum vulgare, N. benthamiana*, wheat, barley, *B. distachyon*	*pds, chlH, TK, PMR5, RAR1, SGT1, HSP90*	Holzberg et al., 2002; Yuan et al., 2011; Scofield et al., 2005; Lee et al., 2012
Brome mosaic virus	Bromovirus, Positive strand RNA, tripartite	Vascular puncture inoculation, sap inoculation	*Hordeum vulgare, Oryza sativa, wheat, Zea mays*	*pds, actin1, rubisco activase, tps6/11*	Ding et al., 2006; van der Linde et al., 2010
African cassava mosaic virus	Begomovirus, ssDNA, bipartite	Agroinoculation, particle bombardment	*N. benthamiana, M. esculenta*	*pds, su, cyp79d2*	Fofana et al., 2004
Bean pod mottlevirus	Comovirus, positive-strand RNA, bipartite	Rub inoculation of in vitro transcript, particle bombardment	*G. max, P. vulgaris*	*pds, gfp, Nod22, SACPD, SHMT*	Zhang et al., 2010; Kandoth et al., 2013
Cabbage leaf curl virus	Begomovirus, ssDNA, bipartite	Agroinoculation	*A. thaliana, N. benthamiana*	*gfp, CH42, pds, Su, CLA1, SGT1*	Turnage et al., 2002; Tang et al., 2010

(Continued)

TABLE 21.1 (CONTINUED)
Characteristics of Some Important Plant Viruses Developed as VIGS Vectors

Virus	Genus, Nature of Genome	Inoculation Method	Silencing Hosts	Genes Silenced	References
Pea early browning virus	Tobravirus, positive-strand RNA, bipartite	Agroinoculation	*P. sativum*	*pds, uni, kor, sym 19, PT4, Nin*	Constantin et al., 2004; Grolund et al., 2010
Cucumber mosaic virus	Cucomovirus, positive-strand RNA, tripartite	Rub inoculation, sap inoculation, agroinfiltration followed by vascular puncture in maize seeds	*G. max, A. majus, Zea mays*	*chs, sf3h1, Am-ANT, pds, IspH, ATG3, ATG8a*	Nagamatsu et al., 2007; Kim et al., 2011; Lee et al., 2015; Wang et al., 2016
Tobacco curly shoot virus DNA1 component	DNA satellite-like virus, DNA, satellite	agroinoculation	*N. tabacum, Solanum lycopersicon, Petunia hybrida, N. benthamiana*	*EDS1, gfp, su, chs, pcna*	Huang et al., 2009
Tomato golden mosaic virus	Begomovirus, ssDNA, bipartite	Particle bombardment	*N. benthamiana*	*su, luc*	Peele et al., 2001
Turnip yellow mosaic virus	Tymovirus, positive-strand RNA, monopartite	Rub inoculation, in vitro transcript	*A. thaliana*	*pds, lfy*	Pflieger et al., 2008
Poplar mosaic virus	Carlavirus, positive-strand RNA, monopartite	Agroinoculation	*N. benthamiana*	*gfp*	Naylor et al., 2005
Potato virus X	Potexvirus, positive-strand RNA, monopartite	Agroinoculation	*N. benthamiana, A. thaliana*	*gus, pds, DWARF, SSU, NFL, LFY*	Ruiz et al., 1998
Rice tungro bacilliform virus	Tungrovirus, double-stranded DNA	Agronioculation	*O. sativa*	*pds*	Purkayastha et al., 2010; Kant et al., 2015
Pepper huasteco yellow veins virus	Begomovirus, single-stranded RNA, bipartite	Particle bombardment	*Capsicum* spp.	*Comt, Kas, Amt*	del Rosario et al., 2008
Grapevine virus A	*Vitivirus*, single-stranded RNA	Agroinoculation	*N. benthamiana, V. vinifera*	*pds*	Muruganantham et al., 2009
Cotton leaf crumple virus	Begomovirus, ssDNA, bipartite	Particle bombardment, agroinoculation	*G. hirsutum, N. benthamiana*	*pds, chlI, gfp*	Tuttle et al., 2008

(Continued)

TABLE 21.1 (CONTINUED)
Characteristics of Some Important Plant Viruses Developed as VIGS Vectors

Virus	Genus, Nature of Genome	Inoculation Method	Silencing Hosts	Genes Silenced	References
Tomato yellow leaf curl China virus-associated βDNA satellite	Begomovirus, ssDNA, satellite	Agroinoculation	*N. benthamiana, S. lycopersicon, N. glutinosa, N. tabacum*	*Pcna, pds, su, gfp*	Tao et al., 2004
Beet curly top virus	Curtovirus, ssDNA, monopartite	Particle bombardment, agroinoculation	*Spinacea oleracea, S. lycopersicum*	*rbcS, TK, chlI*	Golenberg et al., 2009
Abutilon mosaic virus	Begomovirus, ssDNA, bipartite	Agroinoculation	*N. benthamiana*	*pds*	Krenz et al., 2010
Plum pox virus	Potyvirus, positive-sense ssRNA, monopartite	Agroinoculation	*N. benthamiana*	*gfp*	Vaistij and Jones, 2009
Cotton leaf curl Multan betasatellite	Begomovirus, ssDNA, monopartite	Agroinoculation	*N. benthamiana, N. tabacum, S. lycopersicon, A. thaliana, G. hirsutum*	*Su, uidA*	Kumar et al., 2014
Citrus leaf blotch virus	Citrivirus, positive-sense ssRNA	Agroinfiltration, stem slashing, graft inoculation	*N. benthamiana*, citrus	*pds, actin, Su*	Aguero et al., 2012
Foxtail mosaic virus	Potexvirus, positive-sense RNA, monopartite	Mechanical inoculation/ sap inoculation	*Oryza sativa, Triticum aestivum, Zea mays, Setaria viridis, Sorghum bicolor*	*pds, chlH, CLA1, IspH, lesion mimic22, brown midrib3, iojap*	Mei et al., 2016; Liu et al., 2016

21.2.4 Methods of Inoculation

A number of inoculation methods are practiced in different plant species in order to achieve efficient and high-throughput VIGS-mediated gene silencing (Table 21.1). The common methods used for inoculation are DNA virus-based agro-inoculation (Purkayashta et al., 2010), biolistic particle bombardment (Fofana et al., 2004), and RNA virus-based rub-inoculation of RNA transcripts (Lee et al., 2012; Kandoth et al., 2013). BSMV-based VIGS vector is highly efficient to infect monocots such as rice, maize, wheat, *B. distachyon*, etc. as well as dicots such as *N. benthamiana*, spinach, etc. (Yuan et al., 2011). This property of BSMV is used for determining the gene function in non-host plants by inoculating the recombinant vector in *N. benthamiana*, then using its sap as a secondary inoculum to infect the other non-host plants. Therefore sap inoculation can also be used as a potential tool in plants for which VIGS vectors have not been developed. Agrodrench, a method of watering the plant roots with agrobacterium-containing recombinant VIGS vector, has also reported to show high silencing efficiency (Ryu et al., 2004). Liu et al. (2002) successfully induced the TRV vector into tomato by spraying a TRV agroculture using an airbrush. Efficient gene silencing could be obtained by vacuum agro-infiltration in plants that are hard to inoculate by conventional methods. Loss-of-function phenotype has also been observed by gene silencing in some fruits by direct injection with an agroculture (Fu et al., 2005). In addition, efficient silencing could be induced by injecting plucked tomato and strawberry with an agroculture-containing VIGS vector (Chai et al., 2011; Romero et al., 2011). A novel method of inoculation was observed to silence *pds* gene in maize using BMV-based VIGS vector by vascular puncture method (Benavente et al., 2012).

21.3 THE FUTURE OF VIGS TECHNOLOGY

21.3.1 Improvements in VIGS

Several novel VIGS vectors have been developed using many important plant viruses in a wide range of crop species and have been extensively in practice for nearly two decades. However, several improvements have been taken into account in order to ameliorate VIGS-mediated functional analysis of plant genes. Some of the important improvements are described here. A green fluorescent protein (*GFP*) gene has been tagged to the 3′ terminus of coat protein gene of TRV2 vector for easy identification of silenced tissue to trace only green fluorescent tissues that have the recombinant virus (Ramegowda et al., 2014). Recently, to improve silencing efficiency, the RNA1 component of the bipartite TRV-vector was modified to serve as a VIGS vector which can infect plants systemically in the absence of RNA2 (which contain silencing suppressor protein) (Deng et al., 2013). Some VIGS vectors have been derived to induce transcriptional gene silencing (TGS) instead of PTGS by cloning endogenous target gene promoter (Kanazawa et al., 2011). VIGS protocol has also been modified to perform silencing in different tissues including detached plant parts such as petals (Dai et al., 2012), leaves, and fruits (Romero et al., 2011; Ramegowda et al., 2013).

The need for mutants or stable RNAi lines can be avoided by using the VIGS system, as it has been reported that VIGS can persist for several years, and possibly until the death of the plant in appropriate conditions (Senthil-Kumar and Mysore, 2011).

Similarly, BSMV-mediated VIGS has been shown to be transmitted (Bruun Rasmussen et al., 2007) for up to six generations in barley (*Hordeum vulgare*) and wheat (Bennypaul et al., 2012). Therefore, seed transmissibility of VIGS to progenies can be considered as an alternative to stable gene silencing methods. Moreover, developing VIGS vector for gene silencing in many different plant species could be of paramount importance for plant biologists. For example, ASLV-VIGS vector has the potential to silence genes in 15 plant species representing *Brassicaceae*, *Leguminosae*, *Cucurbitaceae*, and *Solanaceae* (Igarashi et al., 2009). Such improvements in VIGS will provide a great avenue for development of sustainable, stress resistant, and high-yielding food crops and food security.

21.3.2 Limitations of VIGS

Despite several advantages of VIGS, a few limitations linked to it compromise the gene silencing and functional genomics approaches. The following points must be considered while using the VIGS vector for gene silencing in plants.

- The VIGS vector should not develop disease symptoms in inoculated plants during infection; therefore disease symptom-determinant genes must be removed to study the efficient silencing effects.
- A few VIGS vectors have been found to be unstable during multiplication and spread (Bruun-Rasmussen et al., 2007). In addition, the size of the viral vector and target gene fragment affect loss of viral genes and insert; therefore, the minimum possible size of viral vector should be constructed.
- Off-target silencing due to functional redundancy of conserved sequences between gene family members may lead to inappropriate results, which can be overcome by selecting the target sequence from unique non-redundant sites (Xu et al., 2006).
- Performance of VIGS is affected by host species; hence specific standardization of a VIGS protocol is required for each genotype in some plant species.
- Silencing can be affected by changes in environmental conditions, temperature, relative humidity, and light (Fu et al., 2005; Kotakis et al., 2010). To overcome these problems, optimum environmental conditions should be standardized for the viral multiplication and maintenance the VIGS.

21.4 FUTURE PROSPECTS

VIGS is a powerful technique for the functional characterization of genes in a wide range of plant species, and this could be extended to delineate molecular mechanisms of several defense, regulatory, and metabolic pathways of crop plants. The efficiency and reliability of most of VIGS vectors have some limitations with respect to wide host range. Therefore the prime objective of the plant biologist should be to develop VIGS vectors such as TRV-, ASLV-, and BSMV-based vectors having wide host ranges and significant gene silencing efficiency. Recent availability of developed full genome of tomato will open various avenues for high-throughput screening of various genes in solanaceous plants (The Tomato Genome Consortium, 2012). While some of the geminiviruses have been deployed for VIGS, their use should be extended, as their genome structure is conserved and they infect a wide range of crop plants including soybean, cotton, and vegetable crops. Recently, both PTGS- and TGS-based VIGS were shown to be transmissible to progeny seedlings and persistent for a long duration. However, heritable and long-duration VIGS needs to be studied for silencing effects in many important genes.

BSMV-based VIGS has been applied for characterization of wheat genes, which has produced successful results for analysis of meiosis and seed-specific genes (Bennypaul et al., 2012). Hence to enhance the agronomic value of crop plants, functions of various flowering related genes, flowering time-related genes, and seed-specific genes must be studied. The advancement of molecular techniques has greatly assisted in high-throughput functional genomics; therefore microarray and next-generation sequencing technologies should be linked with VIGS to accelerate the functional relevance of genes in a high-throughput manner. The most important application of VIGS to improve genetic transformation of crop plants that are recalcitrant to transform can be enhanced by deciphering the genes involved in negative regulation of *Agrobacterium*-mediated plant transformation (Anand et al., 2007). Recently, microRNA-mediated VIGS has also been incorporated for revealing the various regulatory and developmental pathways important in playing various roles in stress tolerance and plant growth. MIR-VIGS is achieved by virus vector-mediated artificial microRNA expression in plants (Tang et al., 2010). This approach may be a useful tool for silencing genes in plants where VIGS vectors are yet to be developed. Overall, along with mutant, RNAi, and overexpression approaches, VIGS can contribute to deciphering the development of biotic and abiotic stress-tolerant varieties with improved yield.

ACKNOWLEDGMENTS

RK gratefully acknowledges the research fellowship from the Indian Council of Medical Research, New Delhi. The funding available from the Department of Biotechnology, Government of India; DU-DST PURSE Grant, and the R&D Grant of University of Delhi to ID are also acknowledged.

REFERENCES

Abraham-Juarez MR, Rocha-Granados MC, Lopez MG, Rivera-Bustamante RF, Ochoa-Alejo N (2008). Virus-induced silencing of Comt, pAmt and Kas genes results in a reduction of capsaicinoid accumulation in chili pepper fruits. *Planta* 227, 681–695.

Agüero J, Ruiz-Ruiz S, Del Carmen Vives et al. (2012). Development of viral vectors based on Citrus leaf blotch virus to express foreign proteins or analyze gene function in citrus plants. *Mol Plant Microbe Interact* 25, 1326–1337.

Al-Kaff NS, Covey SN, Kreike MM, Page AM, Pinder R, Dale PJ (1998). Transcriptional and post-transcriptional plant gene silencing in response to a pathogen. *Science* 279, 2113–2115.

Anand A, Vaghchhipawala Z, Ryu CM, Kang L, Wang K, Del-Pozo O et al. (2007). Identification and characterization of plant genes involved in Agrobacterium-mediated plant transformation by virus-induced gene silencing. *Mol Plant Microbe Interact* 20, 41–52.

Baulcombe DC (1999). Fast forward genetics based on virus-induced gene silencing. *Curr Opin Plant Biol* 2, 109–113.

Baulcombe D (2004). RNA silencing in plants. *Nature* 431, 356–363.

Benavente LM, Ding XS, Redinbaugh MG, Nelson RS, Balint-Kurti PJ (2012). Virus-induced gene silencing in diverse maize lines using the Brome mosaic virus-based silencing vector. *Maydica* 57, 205–213.

Bennypaul HS, Mutti JS, Rustgi S et al. (2012). Virus-induced gene silencing (VIGS) of genes expressed in root, leaf, and meiotic tissues of wheat. *Funct Integr Genomics* 12, 143–156.

Bouche N, Bouchez D (2001). Arabidopsis gene knockout: Phenotypes wanted. *Curr Opin Plant Biol* 4, 111–117.

Bruun-Rasmussen M, Madsen CT, Jessing S Albrechtsen M (2007). Stability of Barley stripe mosaic virus-induced gene silencing in barley. *Mol Plant Microbe Interact* 20, 1323–1331.

Chai YM, Jia HF, Li CL, Dong QH, Shen YY (2011). FaPYR1 is involved in strawberry fruit ripening. *J Exp Bot* 62, 5079–5089.

Constantin GD, Krath BN, MacFarlane SA, Nicolaisen M, Johansen IE, Lund OS (2004). Virus-induced gene silencing as a tool for functional genomics in a legume species. *Plant J* 40, 622–631.

Dai F, Zhang C, Jiang X, Kang M, Yin X, Lü P et al. (2012). RhNAC2 and RhEXPA4 are involved in the regulation of dehydration tolerance during the expansion of rose petals. *Plant Physiol* 160, 2064–2082.

Dalmay T, Hamilton A, Rudd S, Angell S, Baulcombe DC (2000). An RNA-dependent RNA polymerase gene in Arabidopsis is required for posttranscriptional gene silencing mediated by a transgene but not by a virus. *Cell* 101, 543–553.

del Rosario Abraham-Juarez M, del Carmen Rocha-Granados M, Lopez MG et al. (2008). Virus-induced silencing of Comt, pAmt and Kas genes results in a reduction of capsaicinoid accumulation in chili pepper fruits. *Planta* 227, 681–695.

Deng X, Kelloniemi J, Haikonen T, Vuorinen AL, Elomaa P, Teeri TH et al. (2013). Modification of Tobacco rattle virus RNA1 to serve as a VIGS vector reveals that the 29K movement protein is an RNA silencing suppressor of the virus. *Mol Plant Microbe Interact* 26, 503–514.

Ding XS, Schneider WL, Chaluvadi SR, Rouf Mian RM, Nelson RS (2006). Characterization of a Brome mosaic virus strain and its use as a vector for gene silencing in monocotyledonous hosts. *Mol Plant Microbe Interact* 19, 1229–1239.

Eftekhariyan GMR, Karimi F, Mousavi GSL, Hosseini TSA, Salami SA (2014). Assessing the tobacco-rattle-virus-based vectors system as an efficient gene silencing technique in Datura stramonium (Solanaceae). *Virus Genes* 49, 512–516.

Fire A, Xu S, Montgomery MK, Kostas SA, Driver SE, Mello CC (1998). Potent and specific genetic interference by double-stranded RNA in Caenorhabditis elegans. *Nature* 391, 806–811.

Fofana IB, Sangare A, Collier R, Taylor C, Fauquet CM (2004). A geminivirus-induced gene silencing system for gene function validation in cassava. *Plant Mol Biol* 56, 613–624.

Fu DQ, Zhu BZ, Zhu HL, Jiang WB, Luo YB (2005). Virus-induced gene silencing in tomato fruit. *Plant J* 43, 299–308.

Golenberg EM, Sather DN, Hancock LC et al. (2009). Development of a gene silencing DNA vector derived from a broad host range geminivirus. *Plant Methods* 5, 9.

Hamilton AJ, Baulcombe DC (1999). A species of small antisense RNA in post-transcriptional gene silencing in plants. *Science* 286, 950–952.

Hein I, Barciszewska-Pacak M, Hrubikova K et al. (2005). Virus-induced gene silencing-based functional characterization of genes associated with powdery mildew resistance in barley. *Plant Physiol* 138, 2155–2164.

Henikoff S, Comai L (2003). Single-nucleotide mutations for plant functional genomics. *Annu Rev Plant Physiol Plant Mol Biol* 54, 375–401.

Holzberg S, Brosio P, Gross C, Pogue GP (2002). Barley stripe mosaic virus-induced gene silencing in a monocot plant. *Plant J* 30(3), 315–327.

Huang C, Xie Y, Zhou X (2009). Efficient virus-induced gene silencing in plants using a modified geminivirus DNA1 component. *Plant Biotechnol J* 7, 254–265.

Igarashi A, Yamagata K, Sugai T et al. (2009). Apple latent spherical virus vectors for reliable and effective virus-induced gene silencing among a broad range of plants including tobacco, tomato, Arabidopsis thaliana, cucurbits, and legumes. *Virology* 386, 407–416.

Kanazawa A, Inaba J, Kasai M, Shimura H, Masuta C (2011). RNA-mediated epigenetic modifications of an endogenous gene targeted by a viral vector: A potent gene silencing system to produce a plant that does not carry a transgene but has altered traits. *Plant Signal Behav* 6, 1090–1093.

Kandoth PK, Heinz R, Yeckel G et al. (2013). A virus-induced gene silencing method to study soybean cyst nematode parasitism in Glycine max. *BMC Res Notes* 6, 255.

Kant R, Sharma S, Dasgupta I (2015). Virus-induced gene silencing (VIGS) for functional genomics in rice using rice tungro bacilliform virus (RTBV) as a vector. In: K Mysore, M Senthil-Kumar (Eds.) *Plant Gene Silencing. Methods in Molecular Biology*. New York: Humana Press, 1287, 201–217.

Kim BM, Inaba J, Masuta C (2011). Virus induced gene silencing in Antirrhinum majus using the Cucumber mosaic virus vector: Functional Analysis of the AINTEGUMENTA (Am-ANT) gene of A. Majus. *Hort Environ Biotechnol* 52, 176–182.

Klahre U, Crete P, Leuenberger SA, Iglesias VA, Meins Jr F (2002). High molecular weight RNAs and small interfering RNAs induce systemic posttranscriptional gene silencing in plants. *Proc Natl Acad Sci* 99, 11981–11986.

Kotakis C, Vrettos N, Kotsis D, Tsagris M, Kotzabasis K, Kalantidis K (2010). Light intensity affects RNA silencing of a transgene in Nicotiana benthamiana plants. *BMC Plant Biol* 10, 220.

Krenz B, Windeisen V, Wege C, Jeske H, Kleinow T (2010). A plastid-targeted heat shock cognate 70 kDa protein interacts with the Abutilon mosaic virus movement protein. *Virology* 401, 6–17.

Krysan PJ, Young JC, Sussman MR (1999). T-DNA as an insertional mutagen in Arabidopsis. *Plant Cell* 11, 2283–2290.

Kumagai MH, Donson J, Della-Cioppa G, Harvey D, Hanley K, Grill LK (1995). Cytoplasmic inhibition of carotenoid biosynthesis with virus-derived RNA. *PNAS* 92, 1679–1683.

Kumar J, Gunapati S, Kumar J, Kumari A, Kumar A, Tuli R, Singh SP (2014). Virus-induced gene silencing using a modified betasatellite: A potential candidate for functional genomics of crops. *Arch Virol* 159, 2109–2113.

Kurth EG, Peremyslov VV, Prokhnevsky AI, Kasschau KD, Miller M, Carrington JC et al. (2012). Virus-derived gene expression and RNA interference vector for grapevine. *J Virol* 86, 6002–6009.

Lacomme C, Hrubikova K, Hein I (2003). Enhancement of virus-induced gene silencing through viral-based production of inverted-repeats. *Plant J* 34, 543–553.

Lee WS, Hammond-Kosack KE, Kanyuka K (2012). Barley stripe mosaic virus-mediated tools for investigating gene function in cereal plants and their pathogens: VIGS, HIGS and VOX. *Plant Physiol* 160, 582–590.

Lee WS, Hammond-Kosack KE, Kanyuka K (2015). In planta transient expression systems for monocots. In: K Azhakanandam, A Silverstone, H Daniell, MR Davey (eds). *Recent Advancements in Gene Expression and Enabling Technologies in Crop Plants*. Springer, New York, pp. 391–422.

Lindbo JA, Silva-Rosales F, Proebsting WM, Dougherty WG (1993). Induction of a highly specific antiviral state in transgenic plants: Implications for regulation of gene expression and virus resistance. *Plant Cell* 5, 1749–1759.
Liu E, Page J (2008). Optimized cDNA libraries for virus-induced gene silencing (VIGS) using Tobacco rattle virus. *Plant Methods* 4, 5.
Liu N, Xie K, Jia Q et al. (2016). Foxtail Mosaic virus-induced gene silencing in monocot plants. *Plant Physiol.* 171(3), 1801–1807.
Liu Y, Schiff M, Dinesh-Kumar SP (2002). Virus-induced gene silencing in tomato. *Plant J* 31, 777–786.
Marathe R, Anandalakshmi R, Smith T, Pruss G, Vance, V (2000). RNA viruses as inducers, suppressors and targets of posttranscriptional gene silencing. *Plant Mol Biol* 43, 295–306.
Martinez J, Patkaniowska A, Urlaub H, Lührmann R, Tuschl T (2002). Single-stranded antisense siRNAs guide target RNA cleavage in RNAi. *Cell* 110, 563–574.
McCallum MC, Comai L, Greene EA, Henikoff S (2000). Targeting Induced Local Lesions IN Genomes (TILLING) for plant functional genomics. *Plant Physiol* 123, 439–442.
McKinney HH (1929). Mosaic diseases in the Canary Islands, West Africa, and Gibraltar. *J Agric Res* 39, 557–578.
Mei Y, Zhang C, Kernodle BM et al. (2016). A Foxtail mosaic virus vector for virus-induced gene silencing in maize. *Plant Physiol* 171, 760–772.
Mourrain P, Beclin C, Elmayan T, Feuerbach F, Godon C, Morel JB, Jouette D, Lacombe AM, Nikic S, Picaul N (2000). Arabidopsis SGS2 and SGS3 genes are required for posttranscriptional gene silencing and natural virus resistance. *Cell* 101, 533–542.
Muruganantham M, Moskovitz Y, Haviv S, Horesh T, Fenigstein A, Preez J, Stephan D, Burger JT, Mawassi M (2009). Grapevine virus A-mediated gene silencing in Nicotiana benthamiana and Vitis vinifera. *J Virol Methods* 155, 167–174.
Nagamatsu A, Masuta C, Senda M, Kasai HMA, Hong J, Kitamura K, Abe J, Kanazawa A (2007). Functional analysis of soybean genes involved in flavonoid biosynthesis by virus-induced gene silencing. *Plant Biotech J* 5, 778–790.
Naylor M, Reeves J, Cooper JI, Edwards ML, Wang H (2005). Construction and properties of a gene-silencing vector based on Poplar mosaic virus (genus Carlavirus). *J Virol Methods* 124, 27–36.
Nethra P, Nataraja KN, Rama N, Udaya Kumar M (2006). Standardization of environmental conditions for induction and retention of post-transcriptional gene silencing using Tobacco rattle virus vector. *Curr Sci* 90, 431–435.
Padmanabhan M, Dinesh-Kumar SP (2009). Virus-induced gene silencing as a tool for delivery of dsRNA into plants. *Cold Spring Harb Protoc* 4, 2.
Parinov S, Sundaresan V (2000). Functional genomics in Arabidopsis: Large-scale insertional mutagenesis complements the genome sequencing project. *Curr Opin Biotechnol* 11, 157–161.
Parinov S, Sevugan M, De Y, Yang WC, Kumaran M, Sundaresan V (1999). Analysis of flanking sequences from dissociation insertion lines: A database for reverse genetics in Arabidopsis. *Plant Cell* 11, 2263–2270.
Peele C, Jordan CV, Muangsan N, Turnage M, Egelkrout E, Eagle P, Hanley-Bowdoin L, Robertson D (2001). Silencing of a meristematic gene using geminivirus-derived vectors. *Plant J* 27, 357–366.
Pflieger S, Blanchet S, Camborde L, Drugeon G, Rousseau A, Noizet M, Planchais S, Jupin I (2008). Efficient virus-induced gene silencing in Arabidopsis using a "one-step" TYMV-derived vector. *Plant J* 56, 678–690.
Purkayastha A, Mathur S, Verma V, Sharma S, Dasgupta I (2010). Virus-induced gene silencing in rice using a vector derived from a DNA virus. *Planta* 232, 1531–1540.
Ramegowda V, Mysore KS, Senthil-Kumar M (2014). Virus-induced gene silencing is a versatile tool for unraveling the functional relevance of multiple abiotic-stress-responsive genes in crop plants. *Front Plant Sci* 5, 323.
Ratcliff F, Harrison BD, Baulcombe DC (1997). A similarity between viral defense and gene silencing in plants. *Science* 276, 1558–1560.
Ratcliff F, Martin-Hernandez AM, Baulcombe DC (2001). Tobacco rattle virus as a vector for analysis of gene function by silencing. *Plant J* 25, 237–245.
Rodrigo G, Carrera J, Jaramillo A, Elena SF (2011). Optimal viral strategies for bypassing RNA silencing. *JR Soc Interface* 8, 257–268.
Romero I, Tikunov Y, Bovy A (2011). Virus-induced gene silencing in detached tomatoes and biochemical effects of phytoene desaturase gene silencing. *J Plant Physiol* 168, 1129–1135.

Ruiz MT, Voinnet O, Baulcombe DC (1998). Initiation and maintenance of virus-induced gene silencing. *Plant Cell* 10, 937–946.

Ryu CM, Anand A, Kang L, Mysore KS (2004). Agrodrench: A novel and effective agroinoculation method for virus-induced gene silencing in roots and diverse Solanaceous species. *Plant J* 40, 322–331.

Sasaki S, Yamagishi N, Yoshikawa N (2011). Efficient virus-induced gene silencing in apple, pear and Japanese pear using Apple latent spherical virus vectors. *Plant Methods* 7, 15.

Scofield SR, Huang L, Brandt AS, Gill BS (2005). Development of a virus-induced gene-silencing system for hexaploid wheat and its use in functional analysis of the Lr21-mediated leaf rust resistance pathway. *Plant Physiol* 138, 2165–2173.

Senthil-Kumar M, Mysore KS (2011). Virus-induced gene silencing can persist for more than 2 years and also be transmitted to progeny seedlings in Nicotiana benthamiana and tomato. *Plant Biotechnol J* 9, 797–806.

Smith HA, Swaney SL, Parks TD, Wernsman EA, Dougherty WG (1994). Transgenic plant virus resistance mediated by untranslatable sense RNAs: Expression, regulation, and fate of nonessential RNAs. *Plant Cell* 6, 1441–1453.

Speulman E, Metz PLJ, Van Arkel G, Hekkert BTL, Stiekema WJ, Pereira A (1999). A two-component enhancer-inhibitor transposon mutagenesis system for the functional analysis of the Arabidopsis genome. *Plant Cell* 11, 1853–1866.

Tang Y, Wang F, Zhao J, Xie K, Hong Y, Liu Y (2010). Virus-based microRNA expression for gene functional analysis in plants. *Plant Physiol* 153, 632–641.

Tao X, Zhou X (2004). A modified viral satellite DNA that suppresses gene expression in plants. *Plant J* 38, 850–860.

The Tomato Genome Consortium (2012). The tomato genome sequence provides insights into fleshy fruit tomato. *Nature* 485, 635–641.

Turnage MA, Muangsan N, Peele CG, Robertson D (2002). Geminivirus-based vectors for gene silencing in Arabidopsis. *Plant J* 30, 107–117.

Tuttle JR, Idris AM, Brown JK, Haigler CH, Robertson D (2008). Geminivirus-mediated gene silencing from Cotton leaf crumple virus is enhanced by low temperature in cotton. *Plant Physiol* 148, 41–50.

Unver T (2008). Detection and characterization of plant genes involved ın various biotic and abiotic stress conditions using DDRTPCR and isolation of interacting proteins. PhD thesis, Middle East Technical University, Institute of Natural and Applied Sciences, Ankara, Turkey.

Vaistij FE, Jones L (2009). Compromised virus-induced gene silencing in RDR6-deficient plants. *Plant Physiol* 149, 1399–1407.

Valentine T, Shaw J, Blok VC, Phillips MS, Oparka KJ, Lacomme C (2004). Efficient virus induced gene silencing in roots using a modified Tobacco rattle virus vector. *Plant Physiol* 136, 3999–4009.

van der Linde K et al. (2010). Systemic virus-induced gene silencing allows functional characterization of maize genes during biotrophic interaction with Ustilago maydis. *New Phytol* 189, 471–483.

van Kammen A (1997). Virus-induced gene silencing in infected and transgenic plants. *Trends Plant Science* 2, 409–411.

Vaucheret H (2006). Post-transcriptional small RNA pathways in plants: Mechanisms and regulations. *Genes Devel* 20, 759–771.

Wang R, Yang X, Wang N et al. (2016). An efficient virus-induced gene silencing vector or maize functional genomics research. *Plant J* 86(1), 102–115.

Waterhouse PM, Graham MW, Wang MB (1998). Virus resistance and gene silencing in plants can be induced by simultaneous expression of sense and antisense RNA. *Proc Natl Acad Sci* 95, 13959–13964.

Xu P, Zhang Y, Kang L, Roossinck MJ, Mysore KS (2006). Computational estimation and experimental verification of off-target silencing during posttranscriptional gene silencing in plants. *Plant Physiol* 142, 429–440.

Yaegashi H, Yamatsuta T, Takahashi T, Li C, Isogai M, Kobori T, Ohki S, Yoshikawa N (2007). Characterization of virus-induced gene silencing in tobacco plants infected with Apple latent spherical virus. *Arch Virol* 152, 1839–1849.

Yuan C, Li C, Yan L, Jackson AO, Liu Z, Han C, Yu J, Li, D (2011). A high throughput Barley stripe mosaic virus vector for virus induced gene silencing in monocots and dicots. *PLoS ONE* 6, e26468.

Zhang C, Bradshaw JD, Whitham SA et al. (2010). The development of an efficient multipurpose bean pod mottle virus viral vector set for foreign gene expression and RNA silencing. *Plant Physiol* 153, 52–65.

Zhang C, Ghabrial SA (2006). Development of Bean pod mottle virus-based vectors for stable protein expression and sequence-specific virus-induced gene silencing in soybean. *Virology* 344, 401–411.

22 Exclusion of Plant Viruses for Ensuring Biosecurity

A Critical Appraisal

V. Celia Chalam and R.K. Khetarpal

CONTENTS

22.1 INTRODUCTION

The principles of management of plant diseases include chemical, cultural, and biological control, development of host resistance, and exclusion or avoidance of the disease. In the management of plant viruses, it is the development of host resistance through conventional breeding or development of transgenics and the exclusion or avoidance of disease through quarantine and certification procedures that assumes greatest importance. The nature and extent of the severe economic losses with special reference to plant viruses have been well documented (Waterworth and Hadidi 1998). The National Plant Protection Organizations (NPPOs) assumes responsibility for protecting their countries from the unwanted entry of new viruses and for coordinating programs to eradicate those that have recently arrived and are still sufficiently confined for their elimination to be realistic. The exclusion can be achieved by a combination of regulatory and technical approaches that can ensure biosecurity for a region.

Biosecurity is a strategic and integrated approach that encompasses the policy and regulatory frameworks (including instruments and activities) that analyze and manage risks in the sectors of food safety, animal life and health, and plant life and health, including associated environmental risk (FAO, www.fao.org/Biosecurity). The quarantine measures are critical for ensuring biosecurity by excluding transboundary movement of viruses. Plant quarantine is defined as all activities designed to prevent the introduction and/or spread of quarantine pests or to ensure their official control. A quarantine pest is a pest of potential economic importance to the area endangered thereby and not yet present there, or present but not widely distributed and being officially controlled (FAO 2016). Plant biosafety also has to be ensured, which entails the management of risks arising from genetically modified organisms using recombinant DNA technology (Khetarpal and Gupta 2007).

Recent years have witnessed a significant growth in national and global trade of agri-horticultural crops. The national movement of agri-horticultural produce can spread diseases to regions not known to be previously infected, and the global movement has the potential for introducing new pests, including viruses which may pose potential risks to the agriculture of the importing country. The devastating effects resulting from viruses introduced along with international movement of seed and other planting material are well documented. The worldwide distribution of many economically important viruses such as *Bean common mosaic virus*, *Soybean mosaic virus*, *Pea seed-borne mosaic virus*, *Wheat streak mosaic virus*, *Peanut mottle virus*, etc. is attributed to the unrestricted exchange of seed lots.

A number of exotic plant viruses have been introduced into India along with imported planting material. These include the devastating *Banana bunchy top virus* (BBTV), *Banana streak mosaic virus, Peanut stripe virus*, etc. BBTV was probably introduced into India from Sri Lanka in 1940 (Magee 1953) and has since spread widely in the country. The international spread of BBTV is primarily through infected planting material (Wardlaw 1961). In India, an annual loss of Rs.400 million due to BBTV has been estimated in Kerala alone.

22.2 FRAMEWORK FOR ENSURING BIOSECURITY

22.2.1 International Scenario for Excluding Movement of Viruses

The recent trade-related developments in international activities and the thrust of the World Trade Organization (WTO) Agreements imply that countries need to update their quarantine/plant health services to facilitate pest-free import/export. The establishment of the WTO in 1995 has provided unlimited opportunities for international trade of agricultural products. Also, legal standards have come up in the form of sanitary and phytosanitary (SPS) measures for regulating international trade. SPS measures concern the application of food safety and human, animal, and plant health regulations.

SPS measures are defined as any measure applied within the territory of the Member State: (a) to protect animal or plant life or health from risks arising from the entry, establishment or spread of pests, diseases, disease- carrying/causing organisms; (b) to protect human or animal life or health from risks arising from additives, contaminants, toxins or disease-causing organisms in food, beverages, or foodstuffs; (c) to protect human life or health from risks arising from diseases carried by animals, plants or their products, or from the entry, establishment/spread of pests; or (d) to prevent or limit other damage from the entry, establishment, or spread of pests.

The International Plant Protection Convention (IPPC) of Food and Agriculture Organization (FAO) of the United Nations develops the International Standards for Phytosanitary Measures (ISPMs), which provide guidelines on pest prevention, detection, and eradication. To date, 37 ISPMs (International Standards for Phytosanitary Measures, n.d.) have been developed (see Appendix I). Prior to the establishment of the WTO, governments could adopt international standards, guidelines, recommendations, and other advisory texts on a voluntary basis. Although these norms shall remain voluntary, a new status has been conferred upon them by the SPS agreement. A WTO member adopting such norms is presumed to be in full compliance with the SPS agreement.

22.2.2 National Scenario for Excluding Movement of Viruses

22.2.2.1 Import Quarantine

As early as 1914, the government of India passed the Destructive Insects and Pests (DIP) Act, to regulate or prohibit the import of any article into India likely to carry any pest that may be destructive to any crop, or from one state to another. The DIP Act has since undergone several amendments. The Plant Quarantine (Regulation of Import into India) Order 2003 (hereafter referred to as PQ Order), came into force on January 1, 2004 to comply with the SPS Agreement of WTO (Khetarpal et al. 2006a). A number of amendments of the PQ Order were notified and the revised list under Schedules VI and VII include 693 and 294 crops, respectively. Schedule IV includes 14 crops, Schedule V includes 17 crops, and Schedule VIII includes 31 quarantine weed species. The PQ Order ensures the incorporation of "Additional/Special Declarations" for import commodities free from quarantine pests on the basis of pest risk analysis following international norms, particularly for seed/planting material (Plant Quarantine Information System, n.d.).

Schedule IV includes 14 crops and countries from where import is prohibited along, with the name(s) of pest(s). Out of 14 crops, five crops are due to viruses; three to phytoplasma, two to viruses and phytoplasma, one to viroids, and one to phytoplasma and viroids. Schedule V includes 17 crops with restricted import permissible only with the recommendation of authorized institutions with additional declarations and special conditions. Schedule VI includes 693 crops permitted to be imported with additional declarations required to be incorporated into Phytosanitary Certificate and special conditions. As per the PQ Order, 285 viruses, 18 phytoplasma, and five viroids are regulated pests which are of quarantine significance for India. Schedule IV includes 32 viruses, five phytoplasma, and one viroid affecting 14 crops. Schedule V includes 43 viruses, seven phytoplasma, and two viroids affecting 17 crops. Schedule VI includes 242 viruses, 11 phytoplasma, and five viroids affecting 693 crops.

The Directorate of Plant Protection, Quarantine and Storage (DPPQS) under the Ministry of Agriculture and Farmers Welfare (MoA&FW) is responsible for enforcing quarantine regulations and for quarantine inspection and disinfestation of agricultural commodities. The quarantine processing of bulk consignments of grain/pulses, etc. for consumption and seed/planting material for sowing are undertaken by the 57 plant quarantine stations located in different parts of the country, and many pests have been intercepted in imported consignments. Import of bulk material for sowing/planting purposes are authorized only through five plant quarantine stations located in New Delhi, Mumbai, Chennai, Kolkata, and Amritsar. There are 41 inspection authorities who inspect the consignment being grown in isolation in different parts of the country for presence of exotic pests including plant viruses. In addition, DPPQS has developed 22 national standards on various phytosanitary issues such as pest risk analysis, pest-free areas for fruit flies and stone weevils, certification of facilities for treatment of wood packaging material, methyl bromide fumigation, etc. Also, six Standard Operating Procedures have been developed and notified including export inspection and phytosanitary certification of plants/plant products and other regulated articles, post-entry quarantine inspection, etc. Some important viruses were detected and intercepted in planting material imported into India for commercial purposes (Sushil 2016; plantquarantineindia.nic.in/PQISMain/Default.aspx).

The Indian Council of Agriculture Research-National Bureau of Plant Genetic Resources (ICAR-NBPGR), the nodal institution for exchange of plant genetic resources (PGR) has been empowered under the Plant Quarantine (PQ) Order to handle quarantine processing of germplasm including transgenic planting material imported for research purposes into the country by both public and private sectors. ICAR-NBPGR has developed well-equipped laboratories and a post-entry quarantine greenhouse complex. Keeping in view the biosafety requirements, a national containment facility level-4 (CL-4) has also been established at ICAR-NBPGR to ensure that no viable biological material/pollen/pathogen enters or leaves the facility during quarantine processing of transgenics. Adopting a workable strategy such as post-entry quarantine (PEQ) growing in PEQ greenhouses/ containment facilities and inspection, PEQ inspection at indenter's site, electron microscopy, enzyme-linked immunosorbent assay (ELISA) and reverse transcription-polymerase chain reaction (RT-PCR), 41 viruses of great economic and quarantine importance have been intercepted in exotic germplasm including transgenics. Examples of interceptions include 17 viruses not yet reported from India—*Barley stripe mosaic virus* (BSMV), *Bean mild mosaic virus* (BMMV), *Bean pod mottle virus* (BPMV), *Broad bean mottle virus* (BBMV), *Broad bean stain virus* (BBSV), *Broad bean true mosaic virus* (BBTMV), *Cherry leaf roll virus* (CLRV), *Cowpea mottle virus* (CPMoV), *Cowpea severe mosaic virus* (CPSMV), High plains virus (HPV), *Maize chlorotic mottle virus* (MCMV),

Pea enation mosaic virus (PEMV), *Peanut stunt virus* (PSV), *Pepino mosaic virus* (PepMV), *Raspberry ringspot virus* (RpRSV), *Tomato ringspot virus* (ToRSV), and *Wheat streak mosaic virus* (WSMV). In addition, 19 viruses not known to occur on particular host(s) in India have been intercepted and these are also of quarantine significance for India. Twenty viruses have been intercepted in germplasm imported from Consultative Group on International Agricultural Research (CGIAR) centers (Anonymous 2016; Chalam 2014, 2016; Chalam et al. 2005a, 2008b, 2009, 2014; Chalam and Khetarpal 2008; Khetarpal et al. 2001; Prasada Rao et al. 2012; Singh et al. 2003).

Seventeen intercepted viruses are not known to occur in India, and their potential vectors exist and so also the congenial conditions for them to multiply, disseminate, and spread the destructive exotic viruses/strains. The risk of introduction of 41 viruses or their strains into India was thus eliminated. All the plants infected by the viruses were uprooted and incinerated and the virus-free material was only used for further distribution and conservation. If not intercepted, some of the above quarantine viruses could have been introduced into India's agricultural fields and caused havoc to production. Thus, in addition to eliminating the introduction of exotic viruses from our crop improvement programs, the harvest obtained from virus-free plants ensured conservation of virus-free exotic germplasm in the National Genebank.

22.2.2.2 Export Quarantine

The DPPQS under the MoA&FW is responsible for enforcing quarantine regulations and for quarantine inspection and disinfestation of agri-horticultural commodities. All material meant for export should be accompanied by a Phytosanitary Certificate giving the details of the material and treatment, as in the model certificate prescribed under the IPPC of the FAO. Export certification is carried out in accordance with the provision of Article V of IPPC. The export inspection is conducted as per the "Standard Operating Procedures for Export Inspection and Phytosanitary Certification" developed and notified by MoA&FW and also in line with the relevant international standards, such as ISPM 7-Export Certification System; ISPM 12-Guidelines for Phytosanitary Certificates, and ISPM 23-Guidelines for Inspection. The export inspections are conducted at exporters' premises also to facilitate exports for agricultural commodities meant for consumption. The MoA&FW, Government of India has notified 161 officers to grant Phytosanitary Certificates for export of plants and plant materials. The ICAR-NBPGR is vested with the authority to issue Phytosanitary Certificates for seed material and plant propagules of germplasm meant for export for research purposes (Chalam and Mandal 2013; Jain and Chalam 2013).

22.2.2.3 Domestic Quarantine

Domestic quarantine or internal quarantine is aimed at preventing the spread of introduced exotic species or an indigenous key pest to clean (pest-free) areas within the country. The legislative measures to prevent the introduction and spread of destructive pests of crops are operative through the DIP Act, 1914. So far, about 30 pest species have been introduced into India while notifications have been issued against the spread of nine introduced pests only; namely, fluted scale, San Jose scale, codling moth, coffee berry borer, potato wart disease, potato cyst nematode, apple, BBTV, and *Banana mosaic virus* (Khetarpal et al. 2006a). According to notifications issued under the DIP Act, an introduced pest, for example, BBTV, has been declared a pest in states of Assam, Kerala, Orissa, Tamil Nadu (TN), and West Bengal (WB), and bananas that come out of these states have to be accompanied by a certificate of health from the state pathologist or other competent authorities that the plants are free from it. However, due to absence of domestic quarantine, BBTV has spread to most banana growing areas in the country. The limitations and constraints of domestic quarantine include lack of basic information on the occurrence and distribution of major key pests in the country; in other words, pest distribution maps are lacking for most of the key pests; there is an absence of concerted action and enforcement of internal quarantine regulations by the state governments; lack of interstate border quarantine checkposts at rail and road lines greatly adds to the free movement of planting material across the states; lack of close cooperation and effective coordination between state governments and centers for timely notification of introduced pests, organizing pest detection surveys for delineating the affected areas and immediate launching of eradication campaigns in affected areas; lack of public awareness; lack of rapid diagnostic tools/kits for quick detection/identification of exotic pests at the field level; and lack of rigorous seed/stock certification or nursery inspection programs to make available the pest-free seed/planting material for farmers (Bhalla et al. 2014).

There is a need to review the status of existing domestic quarantine for establishment of interstate quarantine checkposts for monitoring movement of viruses of significance. Also, the list of viruses to be regulated under domestic quarantine needs to be reviewed and updated. For example, BBTV and Banana mosaic virus (*Cucumber mosaic virus*) need to be deleted as regulated pests under domestic quarantine, as they are widespread in different parts of India.

22.2.2.4 The Agricultural Biosecurity Bill, 2013

The Agricultural Biosecurity Bill, 2013 was introduced in the Lok Sabha of India on March 11, 2013. The Bill seeks to set up an autonomous authority encompassing the four sectors of agricultural biosecurity: plant health, animal health, living aquatic resources (fisheries, etc.), and agriculturally important micro-organisms. It provides for modernizing the legal framework to regulate safe movement of plants and animals within the country and in international trade, and harmonize the legal requirements of the various sectors of agricultural biosecurity. The proposed legislation will ensure agricultural biosecurity of the country for common benefit and for safeguarding the agricultural economy.

The DIP Act, 1914 and the Livestock Importation Act, 1898 are subsidiary to the Customs Act, 1962, which does not give direct power to quarantine officers to deport, destroy, or confiscate the consignment or lodge complaints under the Indian Penal Code. The Bill repeals the DIP Act, 1914 and the Livestock Importation Act, 1898.

The Bill establishes the Agricultural Biosecurity Authority of India (Authority), whose head office shall be located at Faridabad, Haryana, India (Agricultural Biosecurity Bill 2013). As per the Bill, the functions of the Authority shall include: (i) regulating the import and export of plants, animals and related products; (ii) preventing the introduction of quarantine pests from outside India; and (iii) implementing post-entry quarantine measures. The control of existing Plant Quarantine Stations, Central Integrated Pest Management Centers, and other laboratories under the DPPQS shall be transferred to and vested in the Authority.

a. *Salient Features of the Bill*

- integration of plant and animal quarantine services
- establishment of an Authority for prevention, control, eradication and management of pests and diseases of plants and animals and unwanted organisms for ensuring agricultural biosecurity
- to meet international obligations of India for facilitating imports and exports of plants, plant products, animals, animal products, aquatic organisms and regulation of agriculturally important micro-organisms
- prevention and control of pest infestation or infection, including declaration of an area as "controlled area" for this purpose and measures for control of such infestation or infection
- provision for inspection, taking samples, entry and search of premises, checking of conveyances to ensure compliance of phytosanitary and sanitary measures and also seizure, treatment and disposal of plants, animals and their products to prevent spread of pests by designated officers
- declaration of biosecurity emergency in case of outbreak of organisms threatening biosecurity and actions and procedures to deal with it
- removal of plant, animals, their products and other objects imported in violation of the provisions of the proposed legislation

b. *Functions of the Authority*

- prevent the introduction of quarantine pests in India from outside the country by regulating the import of plants, animals and plant products or animal products and other objects
- regulate the export of plants, animals, plant products or animal products and other objects, to meet the importing country's requirements in accordance with international agreements, and to discharge such obligations under those international agreements
- declare, by notification, any place to be a controlled area under clause
- regulate the spread of pests and diseases of plants and animals from one State to another
- regulate the introduction of new or beneficial organisms into the country
- implement such post entry quarantine measures wherever necessary, either by itself or through research institutes, or jointly with such research institutes, as may be provided by regulations
- undertake pest risk analysis
- undertake regular review and revision with a view to update and harmonise sanitary or phytosanitary measures
- undertake surveys and surveillance of pests and diseases of plants and animals in India
- interact with international, regional or national plant protection organisations
- interact with research institutes and State Governments on matters relating to plant and animal protection and quarantine
- provide such technical guidance and assistance as it considers necessary to agriculture, horticulture, animal husbandry and fisheries departments of State Governments and other statutory bodies
- arrange training programs and hold workshops, seminars and conferences periodically to review status of pests and pathogens, and to spread awareness on plant and animal quarantine through mass media
- frame guidelines for the import and export of plants, animals, plant products or animal products and other objects, whether for trade or research
- regulate the import of transgenic materials with respect to sanitary and phytosanitary matters
- establish plant and animal quarantine stations, pest management centres or other units at such places as may be deemed necessary
- promote integrated pest management
- watch and control locusts in such areas as the Central Government may, by notification, specify
- take steps to ensure availability of safer and effective pesticides and their quality control
- contribute towards development of human resource in plant and animal protection technology
- advise and assist the Central Government on all matters including international obligations related to plant and animal protection
- establish and maintain diagnostic laboratories related to pests and diseases of plants and animals
- charge such fees for the services provided under the proposed legislation, as may be specified by regulations
- recommend to the Central Government to issue directions to the State Governments for the purpose of enforcing obligations under international agreements
- undertake such other activities as may be prescribed

22.2.2.5 Certification of Planting Material

a. Seed Certification

Seed certification for a crop comprises legal norms to be qualified for ensuring genetic identity, physical purity, germinability, and freedom from seed-transmitted pathogens and weeds. International Seed Testing Association (ISTA), the Association of Official Seed Certifying Agencies (AOSCA), and the Central Seed Certification Board, Government of India, among others have introduced minimum seed certification standards.

i. Methodology for Quality Control of Seeds

In seed testing stations, many seed lots need to be tested and in case of viruses even very low rates of infection have to be detected in large samples. Biological assays require time for standardization, are laborious, and time- and space-consuming for working with bulk samples. The testing of seeds in groups thus becomes imperative. Maury and Khetarpal (1997) discussed in depth the use of ELISA for detecting viruses in single embryo, determination of seed transmission by coupling it with group analysis, mode of eliminating the interference of non-embryonic tissues (which do not play a role in transmission of virus through seeds) in routine assessment of seed transmission rate and its role in seed certification programs.

ii. Group Testing of Seeds for Quality Control of Seed-Transmitted Viruses

A large number of seeds of a bulk seed lot is divided into a number of groups of equal size for group testing. Different groups are tested in ELISA as individual composite samples. The decision on the acceptance or rejection of a seed lot can be made either on the basis of assessment of seed transmission or by positioning the seed lot in relation to a level of tolerance. These two alternative approaches are (a) quality control by assessing the seed transmission rate, and (b) quality control by positioning the seed lot in relation to a level of tolerance.

iii. Seed Health Certification in India

In India, the Seeds Act, 1966 (including the Seeds Bill, 2004) does not require a mandatory seed certification against any pathogens including viruses, which are the most dangerous pathogens as they cannot be controlled by ordinary physicochemical methods and require sophisticated techniques for proper detection and identification. In addition, out of 110 crops for which seed certification standards are prescribed, seed health standards for seed-borne diseases are available for only 43 crops (59 fungal diseases, 17 bacterial diseases, 14 viral diseases, and one phytoplasma disease) by seed crop inspection at field stage, for two crops (two fungal diseases and two bacterial diseases) by seed sample analysis at seed stage, and for seven crops by both field inspection and seed analysis. Thus, only in nine crops, including potato and sweet potato, post-harvest pathology is related to seed certification. These crops cover 16 fungal diseases, four bacterial diseases, one nematode disease, and one bacteria + nematode complex only (Khetarpal et al. 2006b). The National Seed Research and Training Centre (NSRTC) located at Varanasi, Uttar Pradesh (UP) imparts training on seed health testing to the officials working in seed certification agencies (20), seed testing laboratories (101), seed law enforcement agencies (35 states), agricultural universities, and other institutes dealing with seeds. The NSRTC has a full-fledged seed testing laboratory which functions as a Central Seed Testing Laboratory under clause 4(1) of the Seeds Act, 1966. The Central Seed Testing Laboratory acts as a referral lab as and when disputes arise in the court of law with regard to quality of seed. The laboratory at NSRTC is expected to test 30,000 samples per year and performs on par with ISTA with regard to seed testing.

About 130 plant viruses are known to be seed-transmitted, of which one-third have great economic importance (Power and Flecker 2003), but there are no seed health standards prescribed for viral diseases at seed stage (Khetarpal et al. 2006b). Also, seed certification for pathogen infection during storage is not mandatory with regard to certified packed seed in store/under storage. Seed analysis is carried out essentially by dry seed examination, though many advanced detection techniques are available.

iv. A Case Study of Developing Certification Norms for Seed-Transmitted Viruses of Grain Legumes

Certification is an important means of managing seed-transmitted viral diseases, which are not easy to control. Keeping in view the high economic significance of seed-transmitted viruses and the complete absence of seed certification standards for them, initiatives were taken in the year 2000 at ICAR-NBPGR, New Delhi to develop a model system for seed certification for viruses in collaboration with Gujarat Agricultural University, Anand and University of Mysore, on important seed-transmitted viruses of grain legumes such as *Bean common mosaic virus* (BCMV) and urdbean leaf crinkle disease (ULCD) of black gram and green gram, *Black-eye cowpea mosaic virus* (BlCMV, now a strain of BCMV) and *Cowpea aphid-borne mosaic virus* (CABMV) of cowpea, *Soybean mosaic virus* (SMV) of soybean, and *Pea seed-borne mosaic virus* (PSbMV) of pea for generating information on epidemiological parameters to be used in a quality control program and to develop a model system of seed certification for some of them. Based on extensive surveys carried out for three years in nine major legume-growing states in India complemented by testing 972 seed samples collected from diverse agencies from 21 states, a national map on prevalence of seed-transmitted viruses of grain legumes was prepared, and studies revealed that the disease incidence varied with the location and the crop variety. The detection and identification of viruses both in leaves and seeds were done by deploying a combination of growing-on test, infectivity assay, electron microscopy, ELISA, and RT-PCR. The results gave preliminary indication on a number of sites in different states that were found to be free from certain viral diseases. A correlation in viral disease incidence with aphid vector population, and appreciable losses in seed yield were observed. Based on virus spread using a known level of initial seed/seedling infection, the seed standards for certification against viruses of cowpea and soybean were proposed to be 0.5%, and for pea 2%. ELISA-based diagnostic kits against BlCMV and SMV were prepared to be efficiently utilized for quality control of seeds. For testing seed samples in bulk, further studies on group testing of seeds coupled with ELISA is needed on case-by-case basis. It is expected that the results would contribute in developing a national program for seed certification of grain legumes (Chalam et al. 2004, 2008a, 2016; Chalam and Khetarpal 2007).

b. National Certification System for Tissue Culture-Raised Plants

A National Certification System for Tissue Culture-Raised Plants (NCS-TCP) has been developed for the first time in the world by the Department of Biotechnology (DBT), Government of India, where currently no such organized structure exists, for certification of tissue culture material. The DBT has been authorized as the Certification Agency by MoA&FW through the Gazette of India Notification dated 10 March 2006 under Section 8 of the Seeds Act, 1966. Accordingly, DBT has established NCS-TCP to facilitate certification of the tissue culture plants raised in the laboratory. NCS-TCP is a unique quality management system for tissue culture industry. This is a very comprehensive system closely associated with all the stakeholders, namely the Tissue Culture Certification Agency DBT, Accreditation Unit and Project Management Unit at Biotech Consortium India Limited, referral centers including ICAR-Indian Agricultural Research Institute, New Delhi (for virus indexing) and the National Research Centre on Plant Biotechnology, New Delhi (for genetic fidelity testing), accredited test laboratories (ATLs), recognized tissue culture production facilities, and state agriculture/horticulture departments to ensure production and distribution of quality tissue culture plants.

Five test laboratories—Central Potato Research Institute, Shimla; National Research Centre for Banana, Tiruchirapalli; University of Agricultural Sciences, GKVK, Bangalore; Vasantdada Sugarcane Institute, Pune, and Indian Institute of Sugarcane Research, Lucknow—have been accredited as ATLs by DBT for testing and certification of tissue culture–raised plants. As of now, 94 tissue culture production facilities in Assam, AP, Bihar, Chattisgarh, Gujarat, Haryana, HP, Karnataka, MH, Madhya Pradesh, Orissa, Punjab, Rajasthan, TN, Telangana, UP, and WB have been recognized based on infrastructure, technical competency, and package and practice of production of tissue culture plants. ATLs will test and certify the tissue culture plants produced by recognized tissue culture production facilities (Chalam et al. 2016).

22.2.3 Technical Challenges in Ensuring Biosecurity

Testing of seeds and other planting material for viruses is demanding in both cost and labor, compared to, for example, testing for fungi. The issues related to quarantine methodology and the challenges in plant virus disease diagnosis in quarantine were analyzed by Khetarpal (2004) and Chalam and Khetarpal (2008).

22.2.3.1 Challenges Prior to Import

Pest risk analysis (PRA) is now mandatory for import of new commodities into India. The import permit will not be issued for commodities not covered under Schedule-V, VI, and VII under the PQ Order. Hence for import of new commodities in bulk for sowing/planting, the importer should apply to the Plant Protection Adviser (PPA) to the Government of India for conducting PRA. In the case of germplasm, import permit shall be issued by the director, ICAR-NBPGR, after conducting PRA based on international standards.

The IPPC has published 37 ISPMs, and ISPM-2, ISPM-11, and ISPM-21 deal with the guidelines for PRA, PRA for quarantine pests including analysis of environmental risks and living modified organisms, and PRA for regulated non-quarantine pests, respectively (Crop Protection Compendium, n.d.). The process requires detailed information on pest scenarios at national and international levels. ICAR-NBPGR has published compilations on pests including viruses of quarantine significance in cereals (Dev et al. 2005; Chalam et al. 2005b), grain legumes (Chalam et al. 2012c, 2012d), and edible oilseeds (Gupta et al. 2013; Chalam et al. 2013) for India. *The Crop Protection Compendium* from CAB International, UK, is a useful asset to scan for global pest data.

22.2.3.2 Challenges on Import

a. Applicability of Appropriate Virus Detection Techniques

A combination of techniques including biological tests (growing-on test, infectivity test), electron microscopy, ELISA, and PCR-based protocols are being used to detect viruses in plants and planting material. The judicious use and application of various techniques in the context of import are presented below.

***Biological tests*:** The earliest methods used for virus indexing were examination of seedlings/plants raised in isolation/post-entry quarantine greenhouses and infectivity test. These tests are time-consuming and labor intensive.

***Electron microscopy*:** The transmission electron microscope is very expensive equipment and is often not available. Moreover, electron microscopy is not suited for routine virus indexing, whereas the highly sensitive immunosorbent electron microscopy (ISEM) developed by Derrick (1973) is occasionally used to detect viruses in seeds or to verify results of other detection methods.

***Serological assays*:** ELISA is by far the most common immunodiagnostic technique. It has been consistently used for virus detection since the 1970s (Clark and Adams 1977), long before DNA-based techniques were available. There are over 800 different antisera available for plant viruses through the American Type Culture Collection (Schaad et al. 2003). Polyclonal and monoclonal antisera for many viruses are available commercially (Agdia, USA; A.C. Diagnostics Inc., USA; Bio-Reba, Switzerland; Loewe Diagnostics, Germany, and Neogen Europe Ltd., UK) or in individual labs. These antibodies are used in numerous protocols, including ELISA, immunostrips, and ISEM, to identify viruses. For rapid identification, immunostrips/lateral flow strips are very useful and are user-friendly.

The specificity of all immunoassays can be improved by using monoclonal antibodies. However, increased specificity means that some target strains may be missed (false negative). Many viruses exist at low or variable titer levels that are difficult to consistently detect using ELISA. In addition, the time involved in good antibody production, the possibilities for false positives or false negatives, and the inability to differentiate between closely related viruses also reduce the effectiveness of ELISA detection protocols. However, ELISA remains the consistent protocol of choice for viruses in diagnostic labs due to its high throughput capability, simplicity, relative cheapness, suitability for large-scale testing, and partial automation (Khetarpal et al. 2003; Maury and Khetarpal 1989; Chalam and Khetarpal 2008).

Nucleic acid-based methods: The nucleic acid-based methods, especially PCR and RT-PCR, with their high specificity and extremely high sensitivity, are increasingly used in plant virus detection, including seed health testing for viruses. As we face challenges to crops from intentional or unintentional introductions of plant viruses, speed and accuracy of detection become paramount. PCR-based technologies will undoubtedly figure strongly in any preventive detection plans. Many researchers have recently taken advantage of the speed, sensitivity, and quantitative nature of real-time PCR, by which detection of the PCR products takes place during amplification (Siljo et al. 2014; Chalam et al. 2004). The variant of PCR known as immunocapture RT-PCR has also been used for detection of viruses in seeds (Phan et al. 1998).

Multiplex PCR, where several viruses are diagnosed in a single reaction, has been used to detect a number of plant viruses (Bhat and Siju 2007; Chalam et al. 2012a). Array technology, which is typically used to study the expression of multiple genes simultaneously, has also been used for viral strain diagnosis. In the last decade, various methods such as microarray (Boonham et al. 2007), loop-mediated isothermal amplification (LAMP; Arif et al. 2012; Bhat et al. 2013; Siljo and Bhat 2014), helicase dependent amplification (HDA; Chalam et al. 2012b), and next-generation sequencing (NGS; Baranwal et al. 2015; Visser et al. 2016) were developed which have specific advantages of either sensitivity or discovery of novel viruses. Despite the obvious advantages, nucleic acid-based methods still lack simplicity of use and suitability for large-scale testing in plant quarantine compared with ELISA. However, intense efforts in many places are underway to improve and simplify molecular detection techniques.

b. Sample Size

The size of consignment received is very critical in quarantine from the processing point of view. Bulk seed samples of seed lots need to be tested by drawing workable samples as per norms. The prescribed sampling procedures need to be followed strictly, and there is a need to develop or adapt protocols for batch testing, instead of individual seed analysis (Maury et al. 1985; Maury and Khetarpal 1989). The germplasm samples are usually received as a few seeds/sample, and thus extreme precaution is needed to ensure that whatever result is obtained in the tested part should as far as possible not denote a false positive or a false negative sample. Importers need to be encouraged to get as much sample as possible to allow effective processing. More attention needs to be given to non-destructive techniques wherever possible, as in case of groundnut, where the whole seed is not used for ELISA testing for viruses and only the cotyledonary part of the seed is analyzed (Chalam and Khetarpal 2008).

c. Detecting an Unknown/Exotic Virus

It is essential to have a database of viruses present in the country and in other parts of the world in order to prevent the entry of exotic viruses into the country. It is also important to have the information on different strains/isolates present in the country, so as to prevent the entry of virulent strains. In quarantine the detection of exotic viruses is the prime concern, and for specialized detection and identification of exotic viruses it is mandatory to have the corresponding antisera for carrying out serological assays. Similarly, for detection by PCR the sequence of the exotic virus or a part of its genome is necessary in order to design and synthesize the corresponding primers for their detection. To develop nucleic acid-based protocols, the availability of reference material for exotic viruses is a limiting factor, and very few kits are available commercially. This requires an antisera bank of exotic viruses and a database on sequences of viruses/primers, as well as a repository of seeds of indicator hosts for easy access for quarantine officials, as they often have to work with time constraints.

For poorly characterized or unknown viruses, detection methods either are not available or are not cost-effective. NGS has proven to be a valuable tool for virus detection, discovery, or diversity (Baranwal et al. 2015; Visser et al. 2016), and Boonham et al. (2007) reviewed the application of microarrays for rapid identification of plant viruses. However, use of NGS and microarrays in quarantine are not cost-effective. Nucleic acid-based methods still lack simplicity of use and suitability for large-scale testing, as mentioned earlier. Therefore, detecting an unknown virus in bulk consignments needs proper sampling and development/adaption of protocols for batch testing.

d. Urgency of Clearance of the Sample

Removal of exotic viruses from germplasm by growing in PEQ greenhouses inevitably causes a delay in the release of seeds. It takes a crop season to release the harvest only from the indexed virus-free plants from PEQ greenhouses/nurseries. Samples received after the stipulated sowing time would require the indenter to wait for another season. Non-destructive testing of the seeds could shorten this time.

e. Maintaining Genebanks Free from Exotic Viruses

Of the germplasm handled by the 15 CGIAR centers, about 95% is conserved and exchanged in the form of true seeds. Germplasm collections usually contain seeds from regions of the world that are not only the centers of origin of the crop, but may also be centers for genetic diversity of crop-specific pathogens. Seeds from these regions may therefore be contaminated by pathogenic strains or pathotypes exotic to the recipient location (Diekmann 1997; Maury et al. 1998). Similar efforts for establishing virus-free accessions of germplasm seed have been made in the United States for cowpea (Gillaspie et al. 1995) and in Brazil for groundnut (Pio-Ribeiro et al. 2000). Unfortunately, plantings with virus-free seeds, in regions with widespread incidence of viruses that are both seed-borne and insect-borne, may also quickly succumb to virus infection and serious crop losses (Albrechsten 2006).

Similarly, the Australian post-entry quarantine found several legume germplasm lines imported from large germplasm seed banks to be infected (Jones 1987). In this case, among lines introduced in 1978–1981, 29 of 302 (9.6%) of *Vigna* spp. and 54 of 309 lines (17.5%) of *Glycine* max and *Phaseolus* spp. were infected with seed-transmitted viruses.

Testing facilities are particularly inadequate in developing and least developed countries, as virus testing is cost intensive, and this results in perpetuation of seed-transmitted viruses during germplasm exchange, loss of valuable PGR, and unrecognized international distribution of viruses. Therefore efforts should be made to conserve virus-free germplasm in Genebanks of different countries and CGIAR centers. At ICAR-NBPGR, if seeds would not have been harvested from virus-free plants during post-entry quarantine, the Indian National Genebank could have gotten exotic germplasm accessions infected by exotic viruses/strain.

22.3 THE WAY FORWARD

Plant viruses have great potential to spread locally and globally due to the liberalized trade and exchange of research material and germplasm if stringent quarantine measures are not followed as per SPS/WTO norms. In India, the introduced viruses could get established, as virus vectors are present in the country. There is also a need to strengthen the domestic quarantine system to prevent the spread of viruses with limited distribution within the country. The way forward to ensure biosecurity is thus highlighted below.

- Strengthen plant quarantine stations dealing with bulk samples in terms of manpower, infrastructure (well-equipped laboratories, treatment facilities, and greenhouses), and expertise, with special emphasis on advanced techniques for detection of viruses and their strains in bulk samples through regular training.
- Initiate pre-import inspection and strengthen the PEQ, growing, and inspection of imported material. The inspection authorities should be given adequate support in terms of manpower and funds for undertaking this work. Also, impart training to inspection authorities on plant quarantine issues, with special emphasis on post-entry quarantine requirements and methodology as per the notified Standard Operating Procedures on the same.
- Undertake regular survey and surveillance program to get a realistic picture of the status of viral diseases in the country and for authentic mapping of endemic viruses present in localized pockets; this in turn will help in identification of virus-free areas, and include these viruses

under domestic quarantine. There is a need for development of virus eradication strategies for recently introduced viruses and also for viruses with limited distribution. It would give a boost to our exports when the importing country is assured by a certification agency that the produce is from virus-free areas.

- Detection and diagnosis of viruses are crucial for application of mitigation strategies, trade, and for exchange of germplasm. There is a need to accredit diagnostic laboratories at the central and state levels for quick and accurate identification of viruses, and also for the National Certification Programme for Seed Health in line with NCS-TCP, and a need to review seed certification standards proposed in the Seeds Bill, 2004. A national repository of diagnostics for viral diseases including antisera bank, database of primers, seeds of indicator hosts, virus reference collections (lyophilized positive controls), user-friendly diagnostics (such as lateral flow strips/dip sticks which can detect multiple viruses, multiplex RT-PCR protocols, LAMP and HDA protocols for detection of viruses in the field and at ports of entry, microarrays, DNA barcoding and ultimately, a cost-effective national biosecurity chip for diagnosis of all current threats to crop plants) would be the backbone for strengthening the program on biosecurity for plant viruses. There is an urgent need to develop a national plant pests diagnostic and certification network linking research laboratories with seed/vegetative planting material testing laboratories and quarantine stations, which would be the backbone for strengthening the program on biosecurity from plant pests, including viruses.
- Review the national regulatory framework and develop a mechanism for distribution or sale of virus-free seeds/plants/planting material within the country, be it seed distribution for multilocation testing under All India Coordinated Research Projects, inland supply of germplasm by ICAR-NBPGR, or seed distribution by the national and state seed corporations and private organizations. Also there is a need to develop a national mechanism to monitor the movement of vegetatively propagated material and tissue culture-raised plants across the states. State certification mechanism needs to be strengthened to ensure the supply of virus-free nursery material.
- A database on all viral diseases, including information on host range, geographical distribution, strains, etc. should be made available to scientists, extension workers, and quarantine personnel.
- Proper authenticity for reports of new viruses and deposition of reference cultures in the national repositories should be established and made mandatory. Exports may suffer due to incorect identification of viruses and their reporting as their new record.
- Use simulation models for developing an early warning system to predict outbreaks of viral diseases. Remote sensing may also be used for this.

APPENDIX I

1. ISPM 1: Principles of plant quarantine as related to international trade
2. ISPM 2: Guidelines for pest risk analysis
3. ISPM 3: Code of conduct for the import and release of exotic biological control agents
4. ISPM 4: Requirements for the establishment of pest free areas
5. ISPM 5: Glossary of phytosanitary terms
6. ISPM 6: Guidelines for surveillance
7. ISPM 7: Export certification system
8. ISPM 8: Determination of pest status in an area
9. ISPM 9: Guidelines for pest eradication programmes
10. ISPM 10: Requirements for the establishment of pest free places of production and pest-free production site

11. ISPM 11: Pest risk analysis for quarantine pests including analysis of environmental risks and living modified organisms
12. ISPM 12: Guidelines for phytosanitary certificates
13. ISPM 13: Guidelines for the notification of noncompliance and emergency action
14. ISPM 14: The use of integrated measure in a systems approach for pest risk management
15. ISPM 15: Guidelines for regulating wood packaging material in international trade
16. ISPM 16: Regulated non-quarantine pests: concept and application
17. ISPM 17: Pest reporting
18. ISPM 18: Guidelines for the use of irradiation as a phytosanitary measure
19. ISPM 19: Guidelines on list of regulated pests
20. ISPM 20: Guidelines for phytosanitary import regulatory system
21. ISPM 21: Pest risk analysis for regulated non-quarantine pests
22. ISPM 22: Requirements for the establishment of areas of low pest prevalence
23. ISPM 23: Guidelines for inspection
24. ISPM 24: Guidelines for the determination and recognition of equivalence of phytosanitary measures
25. ISPM 25: Consignments in transit
26. ISPM 26: Establishment of pest-free areas for fruit flies (*Tephritidae*)
27. ISPM 27: Diagnostic protocols for regulated pests
28. ISPM 28: Phytosanitary treatments for regulated pests
29. ISPM 29: Recognition of pest-free areas and areas of low pest prevalence
30. ISPM 30: Establishment of areas of low pest prevalence for fruit flies (*Tephritidae*)
31. ISPM 31: Methodologies for sampling of consignments
32. ISPM-32: Categorization of commodities according to their pest risk
33. ISPM 33: Pest free potato (*Solanum* spp.) micropropagative material and minitubers for international trade
34. ISPM 34: Design and operation of post-entry quarantine stations for plants
35. ISPM 35: Systems approach for pest risk management of fruit flies (*Tephritidae*)
36. ISPM 36: Integrated measures for plants for planting
37. ISPM 37: Determination of host status of fruit to fruit flies (*Tephritidae*)

REFERENCES

Agricultural Biosecurity Bill (2013). Ministry of Agriculture and Farmers Welfare Government of India. www.indiaenvironmentportalorgin/files/file/AgriculturalBiosecurityBill.pdf (accessed on March 28, 2017).

Albrechsten SE (2006). *Testing Methods for Seed-Transmitted Viruses: Principles and Protocols*. Wallingford: CAB International.

Anonymous (2016). Annual report of the ICAR-National Bureau of Plant Genetic Resources 2015–16. ICAR-NBPGR Pusa Campus, New Delhi, India.

Arif MJ, Daniels VC, Chalam J, Fletcher FM, Ochoa Corona (2012). Detection of *High plains virus* with loop-mediated isothermal amplification. In: 2012 American Phytopathological Society Meeting Abstracts of Presentations, August 4–8, Rhode Island, USA. *Phytopathology* 102(4):7.

Baranwal VKP, Jain RK, Saritha RK, Jain NK, Gautam (2015). Detection and partial characterization of *Cowpea mild mottle virus* in mungbean and urdbean by deep sequencing and RT-PCR. *Crop Prot* 75:77–79.

Bhalla S, Chalam VC, Tyagi V, Lal A, Agarwal PC, Bisht IS (2014). *Teaching Manual on Germplasm Exchange and Plant Quarantine*. National Bureau of Plant Genetic Resources, New Delhi, India.

Bhat AI, Siju S (2007). Development of a single tube multiplex RT-PCR for the simultaneous detection of Cucumber mosaic virus and Piper yellow mottle virus associated with stunt disease of black pepper. *Current Sci* 93:973–976.

Bhat AI, Siljo A, Deeshma KP (2013). Rapid detection of Piper yellow mottle virus and Cucumber mosaic virus infecting black pepper (Piper nigrum) by loop-mediated isothermal amplification (LAMP). *J Virol Methods* 193:190–196.

Boonham N, Tomlinson J, Mumford R (2007). Microarrays for rapid identification of plant viruses. *Ann Rev Phytopathol* 45:307–328.

Chalam VC (2014). Biotechnological interventions for biosecuring Indian agriculture from transboundary plant viruses. In: *Current Trends and Future Challenges in Biotechnology and Biomedicine for Human Welfare and Sustainable Development. Part II.* SK Mishra (ed), pp. 16–24. Rewa MP, India: Department of Botany & Biotechnology Govt New Science College.

Chalam VC (2016). Challenges for plant virus diagnosis in certification and transboundary movement of planting material. In: *Souvenir and Abstracts of Zonal Annual Meeting of Delhi Zone Indian Phytopathological Society (IPS) and National Symposium on Biosecurity in Food Value Chain.* February 20, pp. 5–17. VC Chalam. J Akhtar, SC Dubey (eds). New Delhi, India: Delhi Zone IPS and ICAR-National Bureau of Plant Genetic Resources.

Chalam VC, Khetarpal RK (2007). Quality control of seeds for viruses: A case study of legume viruses. In: *Seed Health Testing and Certification: Need for Marching Ahead.* November 17, pp. 23–28. New Delhi, India: Indian Phytopathological Society and National Bureau of Plant Genetic Resources.

Chalam VC, Khetarpal RK (2008). A critical appraisal of challenges in exclusion of plant viruses during transboundary movement of seeds. *Indian J Virol* 19:139–149.

Chalam VC, Mandal B (2013). Phytosanitary issues for export of ornamental plants. In: *Export Oriented Horticulture.* T Janakiram, KV Prasad, KP Singh, K Swaroop, DVS Raju, Ritu Jain and Namita (eds), pp. 108–116. New Delhi, India: Indian Agricultural Research Institute.

Chalam VC, Pappu HR, Druffel KL, Khetarpal RK (2004). Detection of potyviruses in legume seeds by RT-PCR and real-time RT-PCR. *Proceedings of National Symposium on Molecular Diagnostics for the Management of Viral Diseases.* October 14–16 (2004). New Delhi, India: Indian Agricultural Research Institute and Indian Virological Society.

Chalam VC, Khetarpal RK, Parakh DB, Maurya AK, Jain A, Singh S (2005a). Interception of seed-transmitted viruses in French bean germplasm imported during 2002-03. *Indian J Plant Prot* 33:134–138.

Chalam VC, Parakh DB, Singh S, Khetarpal RK (2005b). Viruses of quarantine significance in cereals. In: *Potential Quarantine Pests for India: Cereals.* Dev U, Khetarpal RK, Agarwal PC, Lal A, Kapur ML, Gupta K, Parakh DB (eds), pp. 82–92. New Delhi, India: National Bureau of Plant Genetic Resources.

Chalam VC, Khetarpal RK, Prakash HS, Mishra A (2008a). Quality control of seeds for management of seed-transmitted viral diseases of grain legumes in India. In: *Food Legumes for Nutritional Security and Sustainable Agriculture. Vol 2. Proceedings of the Fourth International Food Legumes Research Conference (IFLRC-IV).* Kharakwal MC (ed), pp. 468–474. New Delhi, India: Indian Society of Genetics and Plant Breeding.

Chalam VC, Parakh DB, Khetarpal RK, Maurya AK, Jain A, Singh S (2008b). Interception of seed-transmitted viruses in cowpea and mungbean germplasm imported during 2003. *Indian J Virol* 19:12–16.

Chalam VC, Parakh DB, Khetarpal RK, Maurya AK, Pal D (2009). Interception of seed-transmitted viruses in broad bean (*Vicia faba* L) germplasm imported into India during 1996-2006. *Indian J Virol* 20:83–87.

Chalam VC, Arif M, Caasi DR, Fletcher J, Ochoa-Corona FM (2012a). Discrimination among CLRV GFLV and ToRSV using multiplex RT-PCR. In: American Phytopathological Society Meeting Abstracts of Presentations, August 4–8. Rhode Island, USA. *Phytopathology* 102(4):7.

Chalam VC, Arif M, Fletcher J, Ochoa-Corona FM (2012b). Detection of *Bean pod mottle virus* using RT-PCR RT-qPCR and isothermal amplification. In: American Phytopathological Society Meeting Abstracts of Presentations, August 4–8. Rhode Island, USA *Phytopathology* 102(4):7.

Chalam VC, Bhalla S, Singh B, Rajan (eds) (2012c). *Potential Quarantine Pests for India in Grain Legumes.* National Bureau of Plant Genetic Resources New Delhi, India.

Chalam VC, Parakh DB et al. (2012d). Viruses of quarantine significance in grain legumes. In: *Potential Quarantine Pests for India in Grain Legumes.* Chalam VC, S Bhalla S, Singh B, Rajan (eds), pp. 174–211. New Delhi, India: National Bureau of Plant Genetic Resources.

Chalam VC, Parakh DB, Kumar A, Maurya AK (2013). Viruses viroids and phytoplasma of quarantine significancein edible oilseeds. In: *Potential Quarantine Pests for India in Edible Oilseeds.* Gupta K, Singh B, Akhtar J, Singh MC (eds), pp. 175–212. New Delhi, India: National Bureau of Plant Genetic Resources.

Chalam VC, Parakh DB, Maurya AK, Singh S, Khetarpal RK (2014). Biosecuring India from seed-transmitted viruses: The case of quarantine monitoring of legume germplasm imported during (2001–2010). *Indian J Plant Prot* 42(3):270–279.

Chalam VC, Khetarpal RK, Agarwal PC (2016). Addressing plant health management using diagnostics and certification issues. In: *Plant Health Management for Food Security Issues and Approaches.* Katti G, Kodaru A, Somasekhar N, Laha GS, Sarath Babu B, Varaprasad KS (eds), pp. 39–59. New Delhi, India: Astral International Pvt Ltd.

Clark MF, Adams AN (1977). Characteristics of the microplate method of enzyme-linked immunosorbent assay for the detection of plant viruses. *J Gen Virol* 334:475–483.

Crop Protection Compendium (n.d.). Wallingford, UK: CAB International. www.cabiorg/cpc/.

Derrick KS (1973). Quantitaive assay for plant viruses using serologically specific electron microscopy. *Virology* 56:652–653.

Dev U, Khetarpal RK, Agarwal PC et al. (eds) (2005). *Pests of Quarantine Significance in Cereals*. New Delhi, India: National Bureau of Plant Genetic Resources.

Diekmann M (1997). Activities at the international agricultural research centers to control pathogens in germplasm. In: *Plant Pathogens and the Worldwide Movement of Seeds*. McGee DC (ed). St. Paul, MN: APS Press.

FAO (Food and Agriculture Organizaton of the United Nations) (2016). ISPM 5 Glossary of Phytosanitary Terms. ippcint/static/media/files/publication/en/(2016)/06/ISPM_05_(2016)_En_(2016)-06-03_c6w6Iq3pdf.

FAO (Food and Agriculture Organizaton of the United Nations). www.faoorg/Biosecurity (accessed on March 28, 2017).

Gillaspie AG, Hopkins MS Jr, Pinnow DL, Hampton RO (1995). Seed-borne viruses in preintroduction cowpea seed lots and establishment of virus-free accessions. *Plant Dis* 79:388–391.

Gupta K, Singh B, Akhtar J, Singh MC (eds) (2013). *Potential Quarantine Pests for India in Edible Oilseeds*. New Delhi, India: National Bureau of Plant Genetic Resources.

International Standards for Phytosanitary Measures (n.d.). International Plant Protection Convention Food and Agriculture Organization of the United Nations Rome. www.ippcint/en/core-activities/standards-setting /ispms (accessed on March 28, 2017).

Jain RK, Chalam VC (2013). Pests: Threat to seed exports. In: *Souvenir of Indian Seed Congress*, February 8–9, pp. 3–8. Gurgaon, India: National Seed Association of India.

Jones DR (1987). Seed borne diseases and the international transfer of plant genetic resources: An Australian perspective. *Seed Sci Technol* 15:765–776.

Khetarpal RK (2004). A critical appraisal of seed health certification and transboundary movement of seeds under WTO regime. *Indian Phytopathol* 57:408–421.

Khetarpal RK, Gupta K (2007). Plant biosecurity in India: Status and strategy. *Asian Biotech Dev Rev* 9(2): 39–63.

Khetarpal RK, Singh S, Parakh DB, Maurya AK, Chalam VC (2001). Viruses intercepted in exotic germplasm during 1991-2000 in quarantine. *Indian J Plant Genet Resour* 14:127–129.

Khetarpal RK, Parakh DB, Chalam VC (2003). Virus indexing of plant germplasm. In: *Conservation Biotechnology of Plant Germplasm*. Mandal BB, Chaudhury R, Engelmann F, Bhag Mal Tao KL, Dhillon BS (eds), pp. 99–104. New Delhi, India: National Bureau of Plant Genetic Resources IPGRI Rome Italy/ FAO Rome Italy.

Khetarpal RK, Lal A et al. (2006a). Quarantine for safe exchange of plant genetic resources. In: *Hundred Years of Plant Genetic Resources Management in India*. Singh AK, Srinivasan K, Saxena S, Dhillon BS (eds), pp. 83–108. New Delhi, India: National Bureau of Plant Genetic Resources.

Khetarpal RK, Sankaran V, Chalam VC, Gupta K (2006b). Seed health testing for certification and SPS/WTO requirements. In: *Seed: A Global Perspective*. Kalloo G, Jain SK, Alice Vani K, U Srivastava U (eds), pp. 239–258. New Delhi, India: Indian Society of Seed Technology.

Magee CJP (1953). Some aspects of the bunchy top disease of banana and other *Musa* spp. *J Proc Royal Soc NSW* 87:3–18.

Maury Y, Khetarpal RK (1989). Testing seeds for viruses using ELISA. In: *Perspectives in phytopathology*. Agnihotri VP, Singh N, Chaube HU, Singh US, Dwivedi TS (eds), pp. 31–49. New Delhi, India: Today and Tomorrows Printers and Publishers.

Maury Y, Khetarpal RK (1997). Quality control of seed for viruses: Present status and future prospects. In: *Seed Health Testing: Progress Towards the 21st Century*. Hutchins JD, Reeves JE (eds), pp. 243–252. Wallingford, UK: CAB International.

Maury Y, Duby C, Bossenec JM, Boudazin G (1985). Group analysis using ELISA: Determination of the level of transmission of soybean mosaic virus in soybean seed. *Agronomie* 5:405–415.

Maury Y, Duby C, Khetarpal RK (1998). Seed certification for viruses In: *Plant Virus Disease Control*. Hadidi A, Khetarpal RK, Koganezawa H (eds), pp. 237–248. St Paul, MN: American Phytopathological Society Press.

Phan TTH, Khetarpal RK, Le TAH, Maury Y (1998). Comparison of immunocapture-PCR and ELISA in quality control of pea seed for pea seed borne mosaic potyvirus. In: *Seed Health Testing: Progress Towards the 21st Century*. Hutchins JD, Reeves JE (eds), pp. 193–199. Wallingford, UK: CAB International.

Pio-Riberio G, Andrade GP et al. (2000). Virus eradication from peanut germplasm based on serological indexing of seeds and analyses of seed multiplication fields. *Fitopatol Bras* 25:42–48.

Plant Quarantine Information System (n.d.). Directorate of Plant Protection Quarantine and Storage Ministry of Agriculture and Farmers Welfare Government of India. www.plantquarantineindianicin/PQISMain /Defaultaspx (accessed on March 28, 2017).

Power AG, Flecker AS (2003). Virus specificity in disease systems: are species redundant? In: *The Importance Of Species: Perspectives on Expendability and Triage*. Kareiva P, Levin SA (eds), pp. 330–346. Princeton, NJ: Princeton University Press.

Prasada Rao RDVJ, Anipetha K, Chakrabarty SK et al. (2012). Quarantine pathogen interceptions on crop germplasm in India during 1986-2010 and their possible economic impact. *Indian J Agric Sci* 82(5):436–441.

Schaad N, Frederick WRD, Shaw J, Schneider WL, Hickson R, Petrillo DM, Luster DG (2003). Advances in molecular-based diagnostics in meeting crop biosecurity and phytosanitary issues. *Annu Rev Phytopathol* 41:305–324.

Siljo A, Bhat AI (2014). Reverse transcription loop-mediated isothermal amplification assay for rapid and sensitive detection of *Banana bract mosaic virus* in cardamom (*Elettaria cardamomum*). *Eur J Plant Pathol* 138:209–214.

Siljo A, Bhat AI, Biju CN (2014). Detection of *Cardamom mosaic virus* and *Banana bract mosaic virus* in cardamom using SYBR Green based reverse transcription-quantitative PCR. *Virus Dis* 25:137–141.

Singh BR, Bhalla S, Chalam VC, Pandey BM, Singh SK, Kumar N, Khetarpal RK (2003). Quarantine processing of imported transgenic planting material. *Indian J Agric Sci* 73(2):97–100.

Sushil SN (2016). Agricultural biosecurity system in India In: *Souvenir and Abstracts of Zonal Annual Meeting of Delhi Zone Indian Phytopathological Society (IPS) and National Symposium on Biosecurity in Food Value Chain*, February 20. Chalam VC, Akhtar J, and Dubey SC (eds), pp. 19–25. New Delhi, India: Delhi Zone IPS and ICAR-National Bureau of Plant Genetic Resources.

Visser M, Bester R, Burger JT, Maree Hans J (2016). Next-generation sequencing for virus detection: Covering all the bases. *Virology J* 13:85.

Wardlaw CW (1961). *Banana Diseases Including Plantain and Abaca*. London: Longmans, Green and Co Ltd.

Waterworth HE, Hadidi A (1998). Economic losses due to plant viruses. In *Plant Virus Disease Control*. Hadidi A, Khetarpal RK, Koganezawa H (eds), pp. 1–13. St Paul, MN: American Phytopathological Society Press.

Index

A

B

C

D

E

M

N

O

P

Q

R

S

T

U

V

W

Y

Z